Electric Machines

Electric Machines

Steady State and Performance with MATLAB®

Second Edition

Ion Boldea and Lucian N. Tutelea

CRC Press
Taylor & Francis Group
Boca Raton London New York

CRC Press is an imprint of the
Taylor & Francis Group, an **informa** business

Second edition published 2022

by CRC Press
6000 Broken Sound Parkway NW, Suite 300, Boca Raton, FL 33487-2742

and by CRC Press
2 Park Square, Milton Park, Abingdon, Oxon, OX14 4RN

First edition published by Willan 2009

CRC Press is an imprint of Taylor & Francis Group, LLC

Library of Congress Cataloging-in-Publication Data
Names: Boldea, I., author. | Tutelea, Lucian, author.
Title: Electric machines : steady state and performance with MATLAB / Ion Boldea, Lucian N. Tutelea.
Description: Second edition. | Boca Raton : CRC Press, 2022. | Includes bibliographical references and index.
Identifiers: LCCN 2021035704 (print) | LCCN 2021035705 (ebook) | ISBN 9780367374716 (hardback) | ISBN 9781032102689 (paperback) | ISBN 9781003214519 (ebook)
Subjects: LCSH: Electric machinery--Design and construction--Data processing. | MATLAB.
Classification: LCC TK2331 .B58 2022 (print) | LCC TK2331 (ebook) | DDC 621.31/042--dc23
LC record available at https://lccn.loc.gov/2021035704
LC ebook record available at https://lccn.loc.gov/2021035705

ISBN: 978-0-367-37471-6 (hbk)
ISBN: 978-1-032-10268-9 (pbk)
ISBN: 978-1-003-21451-9 (ebk)

DOI: 10.1201/9781003214519

Typeset in Times
by SPi Technologies India Pvt Ltd (Straive)

Contents

Preface xi
Authors xiv

Chapter 1 Introduction 1

1.1 Electric Energy and Electric Machines 1
1.2 Basic Types of Transformers and Electric Machines 3
1.3 Losses and Efficiency 11
1.4 Physical Limitations and Ratings 14
1.5 Nameplate Ratings 17
1.6 Methods of Analysis 19
1.7 State of the Art and Perspective 20
1.8 Summary 22
1.9 Proposed Problems 23
References 24

Chapter 2 Electric Transformers 27

2.1 AC Coil with Magnetic Core and Transformer Principles 28
2.2 Magnetic Materials in EMs and Their Losses 34
2.2.1 Magnetization Curve and Hysteresis Cycle 34
2.2.2 Permanent Magnets 37
2.2.3 Losses in Soft Magnetic Materials 38
2.3 Electric Conductors and Their Skin Effects 40
2.4 Components of Single- and Three-Phase Transformers 46
2.4.1 Cores 47
2.4.2 Windings 50
2.5 Flux Linkages and Inductances of Single-Phase Transformers 53
2.5.1 Leakage Inductances of Cylindrical Windings 55
2.5.2 Leakage Inductances of Alternate Windings 57
2.6 Circuit Equations of Single-Phase Transformers with Core Losses 58
2.7 Steady State and Equivalent Circuit 60
2.8 No-Load Steady State ($I_2 = 0$)/Lab 2.1 61
2.8.1 Magnetic Saturation under No Load 64
2.9 Steady-State Short-Circuit Mode/Lab 2.2 65
2.10 Single-Phase Transformers: Steady-State Operation on Load/Lab 2.3 68
2.11 Three-Phase Transformers: Phase Connections 72

2.12 Particulars of Three-Phase Transformers on No Load 75
2.12.1 No-Load Current Asymmetry 75
2.12.2 Y Primary Connection for the Three-Limb Core 77
2.13 General Equations of Three-Phase Transformers 78
2.13.1 Inductance Measurement/Lab 2.4 79
2.14 Unbalanced Load Steady State in 3-Phase Transformers/Lab 2.5 80
2.15 Paralleling 3-Phase Transformers 84
2.16 Transients in Transformers 86
2.16.1 Electromagnetic (R,L) Transients 87
2.16.2 Inrush Current Transients/Lab 2.6 88
2.16.3 Sudden Short Circuit from No Load $(V_2' = 0)$/ Lab 2.7 89
2.16.4 Forces at Peak Short-Circuit Current 90
2.16.5 Electrostatic (C,R) Ultrafast Transients 92
2.16.6 Protection Measures Against Overvoltage Electrostatic Transients 95
2.17 Instrument Transformers 95
2.18 Autotransformers 96
2.19 Transformers and Inductances for Power Electronics 98
2.20 Preliminary Transformer Design (Sizing) by Example 100
2.20.1 Specifications 101
2.20.2 Deliverables 101
2.20.3 Magnetic Circuit Sizing 101
2.20.4 Windings Sizing 102
2.20.5 Losses and Efficiency 104
2.20.6 No-Load Current 105
2.20.7 Active Material Weight 105
2.20.8 Equivalent Circuit 106
2.21 Summary 106
2.22 Proposed Problems 108
References 110

Chapter 3 Energy Conversion and Types of Electric Machines 111

3.1 Energy Conversion in Electric Machines 111
3.2 Electromagnetic Torque 112
3.3 Passive Rotor Electric Machines 114
3.4 Active Rotor Electric Machines 117
3.4.1 DC Rotor and AC Stator Currents 117
3.4.2 AC Currents in the Rotor and the Stator 118
3.4.3 DC (PM) Stator and AC Rotor 119
3.5 Fix Magnetic Field (Brush–Commutator) Electric Machines 120

3.6 Traveling Field Electric Machines.. 122
3.7 Types of Linear Electric Machines....................................... 123
3.8 Flux – Modulation Electric Machines: A New Breed 129
3.9 Summary.. 135
3.10 Proposed Problems 136
References 137

Chapter 4 Brush–Commutator Machines: Steady State.................................. 139

4.1 Introduction 139
4.1.1 Stator and Rotor Construction Elements.................. 139
4.2 Brush–Commutator Armature Windings 141
4.2.1 Simple Lap Windings by Example: $N_s = 16$, $2p_1 = 4$.. 144
4.2.2 Simple Wave Windings by Example: $N_s = 9$, $2p_1 = 2$... 146
4.3 The Brush–Commutator 148
4.4 Airgap Flux Density of Stator Excitation MMF.................... 149
4.5 No-Load Magnetization Curve by Example.......................... 150
4.6 PM Airgap Flux Density and Armature Reaction by Example 156
4.7 The Commutation Process.. 160
4.7.1 The Coil Commutation Inductance 162
4.8 EMF 163
4.9 Equivalent Circuit and Excitation Connections..................... 165
4.10 D.C. Brush Motor/Generator with Separate (or PM) Excitation/Lab 4.1 166
4.11 D.C. Brush PM Motor Steady-State and Speed Control Methods/Lab 4.2... 169
4.11.1 Speed Control Methods.. 171
4.12 D.C. Brush Series Motor/Lab 4.3 177
4.12.1 Starting and Speed Control 178
4.13 A.C. Brush Series Universal Motor...................................... 179
4.14 Testing Brush–Commutator Machines/Lab 4.4..................... 182
4.14.1 D.C. Brush PM Motor Losses, Efficiency, and Cogging Torque 183
4.15 Preliminary Design of a D.C. Brush PM Automotive Small Motor by Example.. 186
4.15.1 PM Stator Geometry... 186
4.15.2 Rotor Slot and Winding Design.............................. 188
4.16 Summary... 190
4.17 Proposed Problems 192
References 194

Chapter 5 Induction Machines: Steady State 195

5.1 Introduction: Applications and Topologies 195
5.2 Construction Elements 195
5.3 AC Distributed Windings 197
5.3.1 Traveling MMF of AC Distributed Windings 198
5.3.2 Primitive Single-Layer Distributed Windings ($q \geq 1$, Integer) 201
5.3.3 Primitive Two-Layer Three-Phase Distributed Windings (q = Integer) 202
5.3.4 MMF Space Harmonics for Integer q (Slots/Pole/Phase) 203
5.3.5 Practical One-Layer AC Three-Phase Distributed Windings 207
5.3.6 Pole Count Changing AC Three-Phase Distributed Windings 211
5.3.7 Two-Phase AC Windings 211
5.3.8 Cage Rotor Windings 212
5.3.9 EMF of AC Windings 215
5.4 Induction Machine Inductances 216
5.4.1 Main Inductance 216
5.4.2 Leakage Inductance 218
5.5 Rotor Cage Reduction to the Stator 221
5.6 Wound Rotor Reduction to the Stator 221
5.7 Three-Phase Induction Machine Circuit Equations 222
5.8 Symmetric Steady State of Three-Phase IMs 224
5.9 Ideal No-Load Operation/Lab 5.1 226
5.10 Zero Speed Operation ($S = 1$)/Lab 5.2 228
5.11 No-Load Motor Operation (Free Shaft)/Lab 5.3 232
5.12 Motor Operation on Load ($1 > S > 0$)/Lab 5.4 234
5.13 Generating at Power Grid ($n > f1/p1, S < 0$)/Lab 5.5 235
5.14 Autonomous Generator Mode ($S < 0$)/Lab 5.6 236
5.15 Electromagnetic Torque and Motor Characteristics 237
5.16 Deep-Bar and Dual-Cage Rotors 243
5.17 Parasitic (Space Harmonics) Torques 244
5.18 Starting Methods 247
5.18.1 Direct Starting (Cage Rotor) 248
5.18.2 Reduced Stator Voltage 249
5.18.3 Additional Rotor Resistance Starting 250
5.19 Speed Control Methods 251
5.19.1 Wound Rotor IM Speed Control 251
5.20 Unbalanced Supply Voltages 255
5.21 One Stator Phase Open by Example/ Lab 5.7 257
5.22 One Rotor Phase Open 260
5.23 Capacitor Split-Phase Induction Motors/ Lab 5.8 262

5.24 Linear Induction Motors 267
5.24.1 End and Edge Effects in LIMs 267
5.25 Regenerative and Virtual Load Testing of IMs/Lab 5.7 271
5.26 Preliminary Electromagnetic IM Design by Example 273
5.26.1 Magnetic Circuit 274
5.26.2 Electric Circuit 278
5.26.3 Parameters 279
5.26.3.1 Leakage Reactances 281
5.26.4 Starting Current and Torque 282
5.26.5 Breakdown Slip and Torque 283
5.26.6 Magnetization Reactance, X_m, and Core Losses, p_{iron} 283
5.26.7 No-Load and Rated Currents, I_0 and I_n 285
5.26.8 Efficiency and Power Factor 286
5.26.9 Final Remarks 286
5.27 Dual Stator Windings Induction Generators (DWIG) 287
5.28 Summary 290
5.29 Proposed Problems 293
References 294

Chapter 6 Synchronous Machines: Steady State 297

6.1 Introduction: Applications and Topologies 297
6.2 Stator (Armature) Windings for SMs 300
6.2.1 Nonoverlapping (Concentrated) Coil SM Armature Windings 304
6.3 SM Rotors: Airgap Flux Density Distribution and EMF 310
6.3.1 PM Rotor Airgap Flux Density 314
6.4 Two-Reaction Principle via Generator Mode 315
6.5 Armature Reaction and Magnetization Reactances, X_{dm} and X_{qm} 318
6.6 Symmetric Steady-State Equations and Phasor Diagram 320
6.7 Autonomous Synchronous Generators 322
6.7.1 No-Load Saturation Curve/Lab 6.1 322
6.7.2 Short-Circuit Curve: $(I_{sc}(I_F))$/Lab 6.2 322
6.7.3 Load Curve: $V_s(I_s)$/Lab 6.3 325
6.8 Synchronous Generators at Power Grid/Lab 6.4 328
6.8.1 Active Power/Angle Curves: $P_e(\delta_V)$ 329
6.8.2 V-Shaped Curves 330
6.8.3 Reactive Power Capability Curves 331
6.9 Basic Static- and Dynamic-Stability Concepts 332
6.10 Unbalanced Load Steady State of SGs/Lab 6.5 336
6.10.1 Measuring X_d, X_q, Z_-, and X_0/Lab 6.6 337
6.11 Large Synchronous Motors 341
6.11.1 Power Balance 341

6.12 PM Synchronous Motors: Steady State 342
6.13 Load Torque Pulsations Handling by Synchronous Motors/Generators 346
6.14 Asynchronous Starting of SMs and Their Self-Synchronization to Power Grid 348
6.15 Single-Phase and Split-Phase Capacitor PM Synchronous Motors 350
6.15.1 Steady State of Single-Phase Cageless-Rotor PMSMs 350
6.16 Preliminary Design Methodology of a Three-Phase Small Automotive PMSM by Example 354
6.17 Single Phase PM Autonomous A.C. Generator with Step–Capacitor Voltage Control: A Case Study 361
6.17.1 Introduction 361
6.17.2 The Proposed Configuration Characterization 363
6.17.3 Sample Step Capacitor Results 365
6.17.4 Experimental Effort 365
6.17.5 Experimental Effort 370
6.18 Summary 370
6.19 Proposed Problems 376
References 378

Index 381

Preface

Electric energy is paramount to prosperity while inspiring social responsibility of industry.

Intelligent use of electric energy implies better electric generators to "produce" it in modern distributed power systems, with day by day more renewable energy penetration.

It also implies better electric motors used more and more in variable speed drives via power electronics and digital control to save energy and increase productivity in all industries where the output varies in time.

Electric machines are reversible in the sense of switching easily from motoring to generating. New industries such as e-transport, robotics, e-building appliances, renewable (wind) energy have extended quickly recently with similar trends in the near and distant future.

The present textbook is intended as for the first semester treating electric transformers, rotary and linear machines steady-state modeling and performance computation, preliminary dimensioning and testing standardized and innovative techniques. Numerical examples/problems are scattered all over the book to bring a feeling of magnitudes. Proposed problems are available at the end of all 6 chapters; also 9 MATLAB/SIMULINK® user-friendly computer programs that illustrate main machine characteristics are available online. The textbook may directly be used also by R&D engineers in industry as all machine parameters and characteristics are calculated by ready to use in industrial design mathematical expressions.

Educating electrical engineering students and R&D engineers in industry in electric machinery and drives faces new challenges as the recent technical progress in the field has been staggering.

CONTENTS

The textbook is structured in 6 chapters which are presented briefly in what follows.

Chapter 1 Introduction

Introduction deals with electric energy "production" and consumption up to date trends and energy conversion in basic types of electric transformers and rotary and linear motion electric machines with their principles, power ranges, applications in modern power systems (with renewable energy large penetration) and in various industries, from e-transport to industrial automation (robotics), home appliances and info—gadgets.

Chapter 2 Electric transformers

Though motion-less, electric transformers, which modify a.c. voltage level (up and down), being based also on Faraday's law (as rotary and linear machines) and being used together with electric machines and power electronics in all applications, are treated here comprehensively for single-phase and three-phase power configurations.

Three- and single-phase transformer main industrial topologies (and materials) steady-state modeling, performance, testing (for laboratory sessions), electromagnetic and electrostatic transients and a preliminary electromagnetic design case study, are all treated in this Chapter.

Chapter 3 Energy Conversion and Types of Electric Machines

This chapter deals with energy conversion as the fundamental concept in electric machines and deciphers their topological variants: with d.c. (or a.c.) or passive rotors, with a.c. or d.c. currents, single phase and multiphase, with fix or moving magnetic fields. New trends such as "flux modulation" machines with 1(2) rotors are included. This way the Chapter offers a panoramic view of electric machine up-to-date status.

Chapter 4, 5, 6 treat rather extensively:

* Brush-commutator machines
* Induction machines
* Synchronous machines,

In terms of principles circuit modeling, characteristics and performance characteristics under steady state, testing techniques and preliminary electromagnetic-thermic dimensioning with lots of solved numerical examples and special cases to illustrate new such electric machines with strong industrialization potential.

MATLAB/SIMULINK Computer programs (online) refer to:

EXAMPLE 1 – The magnetic circuit

EXAMPLE 2 – Transformer steady state

EXAMPLE 3 – Transformer unbalanced load currents

EXAMPLE 4 – Transformer unbalanced impedances

EXAMPLE 5 – Separately-excited d.c. – brush motor

EXAMPLE 6 – Series-excited d.c. – brush motor

EXAMPLE 7 – Induction motor characteristics

EXAMPLE 8 – Synchronous motor characteristics

This second edition is characterized by a thorough verification of the text, numerical examples, adding numerous new References that refer to recent Subjects in all Chapters. Also new Paragraphs are added in Chapters 3, 5, 6 on:

- Flux modulation electric machines (Chapter 3)
- Dual stator winding induction generators (Chapter 5)
- Premium efficiency line-start PM and reluctance induction machines (Chapter 6)

The above-mentioned 8 MATLAB/SIMULINK user-friendly Computer Programs that illustrate performance characteristics in Chapters 4-6, are available online and may be used as home-work in order to facilitate a deep grasping of fundamental issues.

The Paragraphs marked with "Lab." describe testing technics (most standardized) that may be used directly as text for Lab. Sessions.

Ion Boldea

Lucian Tutelea
Timişoara, Romania

Authors

Ion Boldea is a full professor of Electrical Engineering at the University Politechnica of Timişoara, Romania. He has spent approximately 5 years as visiting professor of Electrical Engineering in both Kentucky and Oregon, USA since 1973, when he was a senior Fullbright Scholar for 10 months. He was also a visiting professor in the UK at UMIST and Glasgow University. He is a full member of the Romanian Academy of Technical Sciences, a full member of the European Academy of Sciences and Arts of Salzburg, Austria, and a full member of the Romanian Academy. He has delivered IEEE-IAS Distinguished Lectures since 2008. He has given keynote speeches, tutorial courses, intensive courses, technical consulting in the USA, South America, E.U, South Korea, and China based on his numerous books and IEEE Trans. and Conference papers over the last 45 years in the field of rotating and linear electric machines and drives for renewable energy, vehicular, industrial, and residential applications. Professor Boldea is a life fellow of IEEE. He won the IEEE 2015 Nikola Tesla Award for "contributions to the design and control of rotating and linear electric machines for industry applications."

Lucian N. Tutelea (M'07) received the B.S. and Ph.D. degrees in electrical engineering from the Politehnica University Timişoara, Timişoara, Romania, in 1989 and 1997, respectively. He was a visiting researcher with the Institute of Energy Technology, Aalborg University, Denmark, in 1997, 1999, 2000, and 2006, as well as the Department of Electrical Engineering, Hanyang University, South Korea, in 2004. He is currently a professor with the Department of Electric Engineering, Politehnica University Timişoara. His main research interests include design, modeling, and control of electric machines and drives. Professor Tutelea published more than 80 papers indexed IEEE Xplore or in Web of Science with more than 400 citations.

1 Introduction

1.1 ELECTRIC ENERGY AND ELECTRIC MACHINES

Electric energy represents a key element in modern society. Fossil fuels such as coal, natural gas, or nuclear fuel are burned in a combustor to produce heat, which is then transformed into mechanical energy in a turbine (prime mover). Alternatively, wind and hydroturbines transform wind and hydro energies into mechanical energy. An electric generator is driven directly or through transmission by the turbine to produce electric energy.

Electric energy is measured in joule or kWh:

$$1\text{kWh} = 3.6 \times 10^6 \text{ J} \tag{1.1}$$

The current global energy consumption is about 20×10^{12} kWh per year with a projected increase of 2%–3% per year.

Electric power is measured either in W, kW (1 kW = 10^3 W) and MW (1 MW = 10^6 W), or in GW (1 GW = 10^9 W).

The total installed power in power plants all over the world today is around 4000 GW (800 GW in the United States), out of which 400 GW is installed in wind turbine generators. Installed power tends to increase faster (above 4% per year) than the consumption of electric energy (in kWh) because of the limited availability of various fuel power plants and the daily (or monthly) peak power requirements (Figure 1.1).

With the exception of solar and fuel cells, which contribute negligibly, practically all electric energy is "produced" or rather converted from mechanical energy through electric generators: constant-speed-regulated a.c. (synchronous) generators are mostly used, but in recent times variable speed-regulated a.c. (synchronous and induction) or d.c. output (switched reluctance) generators are being used for small hydro and wind energies.

Electric energy is converted into controlled mechanical work again in electric motors (about 60%), or into controlled heat (for lighting, cooling, heating, etc.). Electric machines are reversible and can function either as generators or as motors (Figure 1.2). They either convert mechanical energy into electric energy (generator mode) or the process is reversed (motor mode).

In both modes, energy conversion ratios and electric machine costs are paramount, because, ultimately, a more costly electric machine means more active materials, which in turn means more energy to produce them, leading to more thermal and chemical pollution.

Variable speed control of electric motors through power electronics is currently the key solution to increased productivity, consuming less energy in both residential

DOI: 10.1201/9781003214519-1

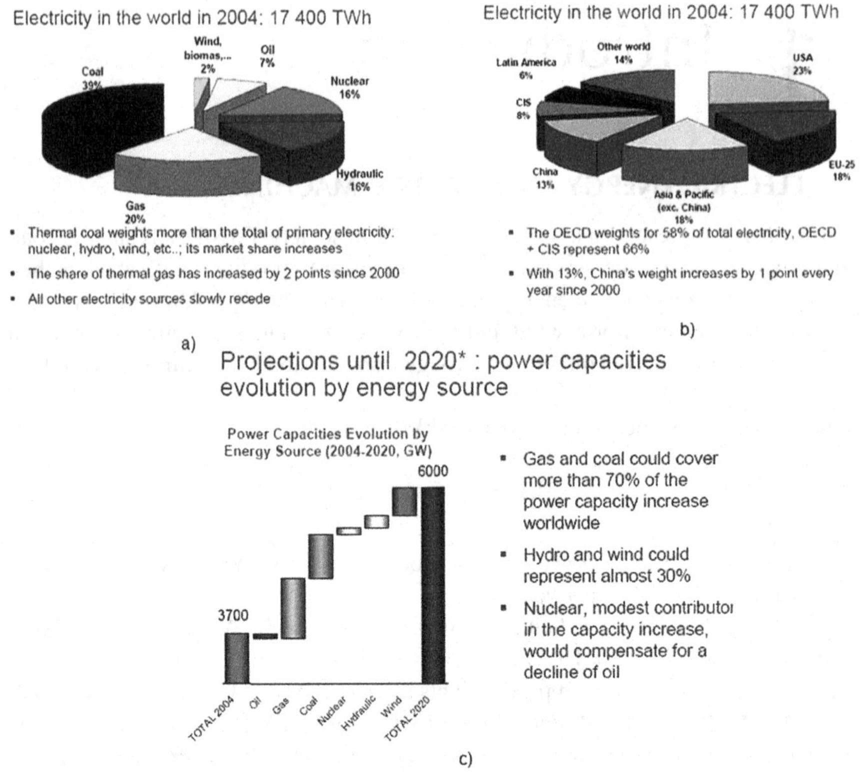
Electricity in the world in 2004: 17 400 TWh
Wind, biomas,... 2%
Oil 7%
Coal 39%
Nuclear 16%
Hydraulic 16%
Gas 20%
Thermal coal weights more than the total of primary electricity: nuclear, hydro, wind, etc..; its market share increases
The share of thermal gas has increased by 2 points since 2000
All other electricity sources slowly recede
a)
Electricity in the world in 2004: 17 400 TWh
Other world 14%
Latin America 6%
USA 23%
CIS 8%
China 13%
EU-25 18%
Asia & Pacific (exc. China) 18%
The OECD weights for 58% of total electricity, OECD + CIS represent 66%
With 13%, China's weight increases by 1 point every year since 2000
b)
Projections until 2020* : power capacities evolution by energy source
Power Capacities Evolution by Energy Source (2004-2020, GW)
6000
3700
TOTAL 2004
Oil
Gas
Coal
Nuclear
Hydraulic
Wind
TOTAL 2020
Gas and coal could cover more than 70% of the power capacity increase worldwide
Hydro and wind could represent almost 30%
Nuclear, modest contributor in the capacity increase, would compensate for a decline of oil
c)

FIGURE 1.1 Electric energy in the world.

* *Note*: Forecasts extracted from the EnerFuture Forecasts Service.

and industrial applications: from info gadgets and home appliances to the new electric or hybrid automobiles, public transportation, pumps, compressors, and industrial drives (Figure 1.3). Refs. [1–34] deal with the core topics of this book in more detail.

Electric machines do convert mechanical energy into electric energy, but not directly. They need to store energy in a magnetic form. Electrostatic machines are seldomly used and thus they are not treated here.

Electric machines are systems of coupled electric and magnetic circuits that convert energy, based on the electromagnetic induction (Faraday's) law for bodies in relative motion.

They have, in general, a fixed part called the stator and a movable part, called the rotor, with an airgap layer of 0.2 mm (for smallest, sub-Watt units) to about 20 mm (for largest power turbine generators of 1700 MW/unit).

As electric generators produce electric power at low and medium voltages (below 28 kV in general), and, as electric energy transmission at hundreds of kilometers requires high voltage (to reduce copper losses and weight and costs of transmission

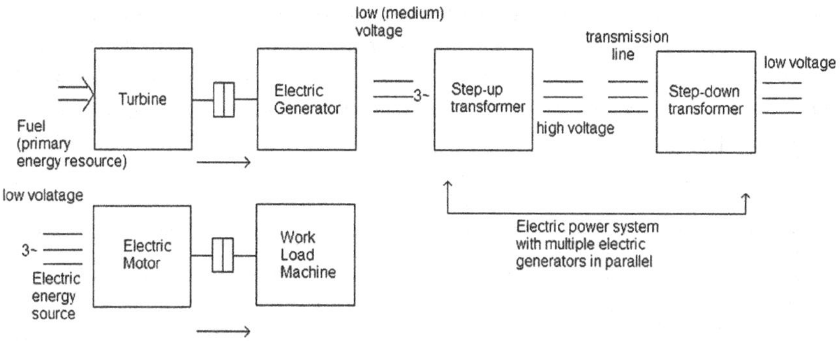

FIGURE 1.2 Generator/motor operation mode.

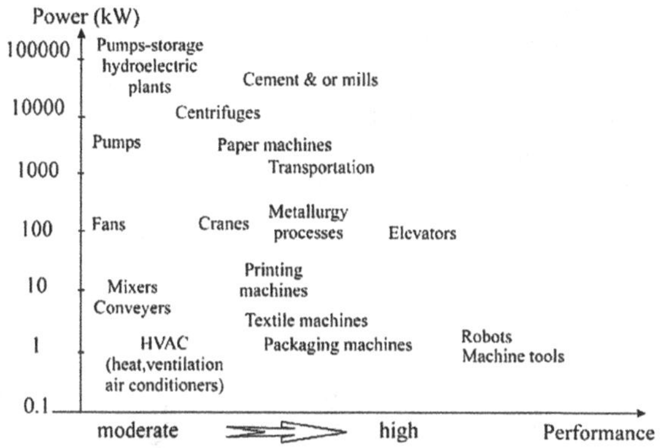

FIGURE 1.3 Electric motor drive applications.

lines), while consumers require low-voltage electric energy (for lower costs and human safety), voltage step-up and step-down are required. Voltage step-up and step-down are also required to match a local source of electric energy to motor voltage requirements, such as those of electric automobiles or other passive loads with (or without) power electronics, for example, furnaces in metallurgy, etc. Also, electric separation for equipment safety is required in many applications.

The electric transformer, which is a system of coupled electric and magnetic circuits, performs voltage (or current) step-up or step-down based on Faraday's law, but for relatively fixed bodies (with no mechanical to electric energy conversion).

Electric power transformers, based on Faraday's law, are primarily used with electric machines, and, therefore, are dealt with in detail. Power electronics transformers are dealt with in a separate paragraph, as they deserve special attention.

1.2 BASIC TYPES OF TRANSFORMERS AND ELECTRIC MACHINES

This book introduces the basic types of transformers and electric machines; representative illustrations of the main types of transformers (by applicability) and electric machines (by principle) are also provided.

It is almost needless to say that transformers and electric machines represent a worldwide business regulated technically by national and international standards. The IEC (International Electrotechnical Commission) and the IEEE (Institute of Electrical and Electronics Engineers) issue some of the most approved international standards in the field.

The main types of electric transformers are as follows:

- *Power transformers* (Figure 1.4)—three-phase and one-phase—at 50 (60) Hz for power transportation and distribution to consumers (residential or industrial) at the required voltage level [1]
- *Power transformers for special applications*: autotransformers, phase-shifting transformers, HVDC (high voltage direct current) transmission line transformers, industrial power electronics motor drive transformers, traction transformers, reactors, earthing transformers, welding transformers, locomotive transformers, furnace transformers, etc. [1]
- Voltage and current measurement transformers (to measure high a.c. voltages and currents with 100 V, 5 A instruments)
- Power electronics power and control transformers and reactors (at high switching frequency—from kHz to MHz) [2]

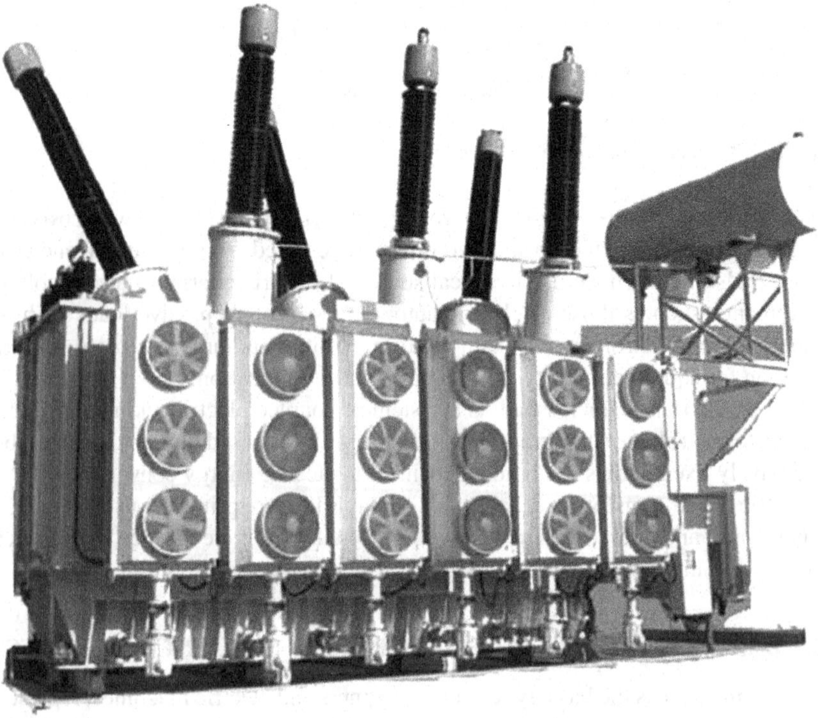

FIGURE 1.4 1100 MVA, 19/345 kV, three-phase generator step-up transformer.

Electric machines may be classified based on principle into two main categories:

- Fixed magnetic fields of the stator and the rotor in the airgap (brush–commutator electric machines)
- Moving magnetic fields in the airgap

For all electric machines, the objective is to produce an electromagnetic torque (or electric power) that does not show ripple during steady-state operation. For pure traveling and fixed magnetic field machines, this is feasible. But when the moving magnetic field speed is not constant in time, this is not possible (such as in single-phase alternating current machines).

Brush–commutator cylindrical electric machine: (a) d.c.-excited (two poles: $2p_1 = 2$), (b) PM (d.c.)-excited (two poles: $2p_1 = 2$), (c) a.c.-series-excited (universal motor) (two poles: $2p_1 = 2$), and (d) electric and physical position of brushes.

- *Fixed magnetic field machines* (Figure 1.5) are all provided on the rotor shaft with a cylindrical (or disk-shaped) mechanical commutator realized with copper sectors that are insulated from each other and that connect all rotor coils (placed in a slotted laminated cylindrical or disk-shaped core) in series and on which act mechanically pressured electric brushes that collect (for generator mode) or "introduce" direct current (d.c.) in the rotor to (from) a d.c. power source. The stator is made of a thin laminated soft-iron core with salient poles ($2p_1$ poles)—semi-periods—that hold d.c.-fed coils (or permanent magnets [PMs]) that produce a fixed heteropolar magnetic flux density distribution in the airgap, aligned with stator poles. On the other hand, the mechanical

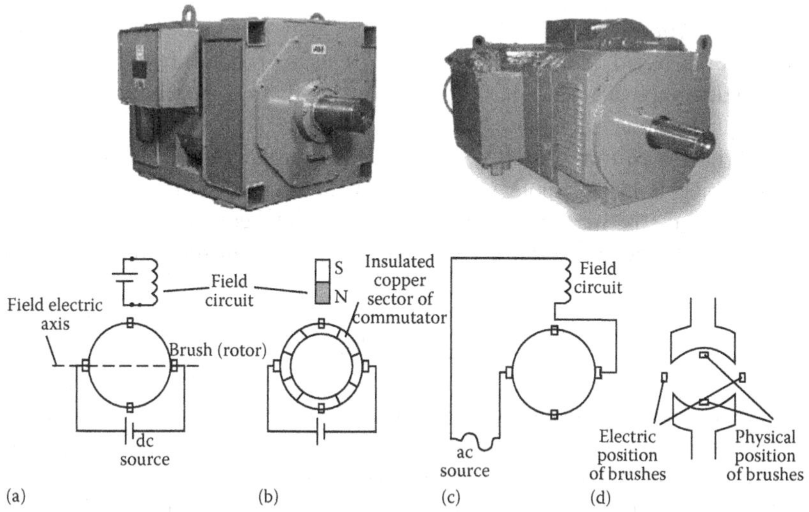

FIGURE 1.5 Brush–commutator cylindrical electric machine: (a) d.c.-excited (two poles: $2p_1 = 2$), (b) PM (d.c.)-excited (two poles: $2p_1 = 2$), (c) a.c.-series-excited (universal motor) (two poles: $2p_1 = 2$), and (d) electric and physical position of brushes.

brush–commutator changes the brush d.c. currents into a.c. currents in the rotor ($f = n \cdot p_1$, where f is the frequency in Hz and n is the speed in rps), whose magnetic flux density in the airgap is fixed, physically 90 electrical degrees shifted to brushes (if the coils are symmetric).

To produce maximum torque (based on the $\overline{f} = \overline{j} \times \overline{B}$ principle), the two field axes (maximums) have to be shifted by 90 electric degrees ($\alpha_e = p_1 \cdot \alpha_m$, where α_m is the mechanical angle and α_e is the electric angle), that is, for symmetric rotor coils, the brush physical axis coincides with the stator pole axis (Figure 1.5d). The actual position of brushes is, in fact, along the rotor current field axis.

Brush–commutator machines are connected to d.c. sources, but can also be connected to a.c. (one-phase) sources, with the field excitation circuit connected in series to brushes (universal motor). Also, d.c. series excitation brush–commutator machines are still used by urban public transportation and in electric or diesel locomotives (where, as multiple d.c.-excited generators, still exist in some parts of the world, generating powers up to 6–8 MW/unit).

The speed and power limitations are rather severe as dictated by the safe (spark-free) mechanical commutation of rotor currents from d.c. to a.c. in the mechanical commutator. The most popular contemporary brush–commutator machines are small-power machines with PM excitation to drive printers, small ventilators on info gadgets, and auxiliaries (shield wipers, fuel pumps, door openers) on electric vehicles or in small robots. Though the tendency is to replace them with brushless (moving field) electric motors, they may last long, up to 1 kW and 30,000 rpm in motors used as ac fed series brush (universal) motors in construction tools (vibrators, etc.) and for various home appliances (dryers, vacuum cleaners, some washing machines, kitchen mixers, etc.).

- *Moving (traveling) field electric machines* may be classified into two main categories based on the way the rotor currents are produced:
  a. Induction machines
  b. Synchronous machines
  Both have a uniformly slotted, laminated, soft-iron cylindrical core in the stator which holds $2p_1$-pole (semiperiods) three-phase winding coils, which, when an alternative (sinusoidal) voltage is fed at frequency f_1, produce in the airgap a magnetic flux density wave field at a constant speed (called synchronous speed) n_1:

$$n_1 = f_1/p_1 \tag{1.2}$$

The rotor of the induction machine (IM) holds, in the uniformly slotted laminated rotor core, aluminum (copper) bars short-circuited by end rings (cage rotor) or a wound rotor with three-phase windings (as in the stator), connected to insulated copper rings and then to stator brushes, which are not commutators (but only a mechanical power switch) as they do not change the frequency of current in the rotor (Figure 1.6).

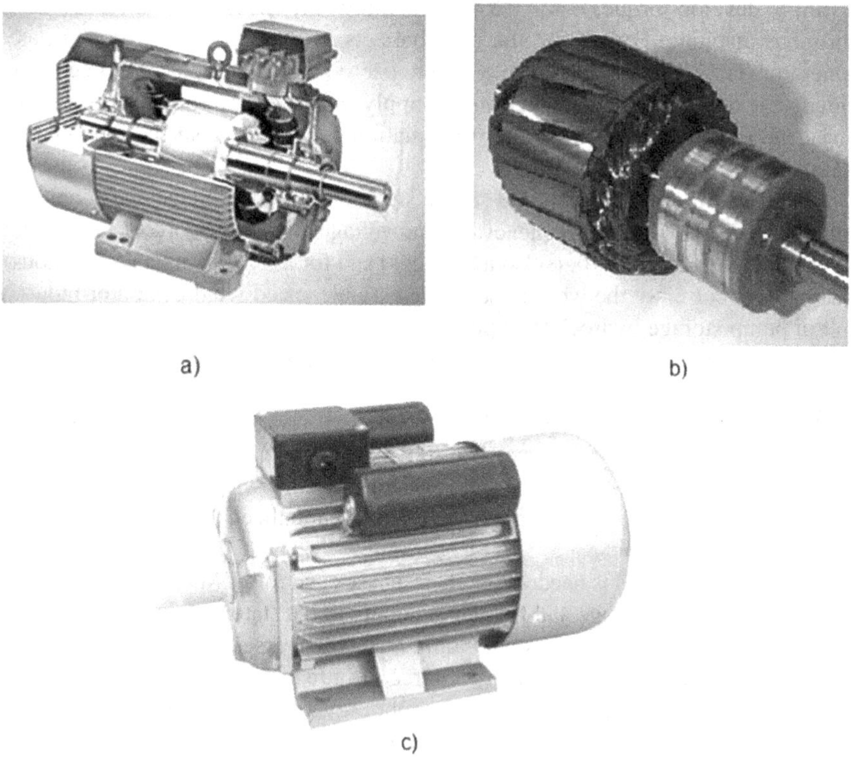

FIGURE 1.6 Induction machines (IM): (a) three-phase with cage rotor, (b) three-phase with wound rotor, and (c) one-phase supply capacitor IM.

The wound rotor may be connected to a variable impedance or, through a frequency changer, to the same a.c. power source as the stator.

The traveling field (at n_1, rps) of stator currents produces in the rotor, which rotates at speed n (rps), "electromagnetic forces" (emfs) at frequency f_2:

$$f_2 = f_1 - n \cdot p_1 = S \cdot f_1; \quad S = 1 - \frac{n \cdot p_1}{f_1} \tag{1.3}$$

where S is called slip.

Now, the corresponding rotor currents will have the same frequency f_2 but their traveling field speed with respect to the stator will be the same, $n_1 = f_1/p_1$, as that of the stator currents, under steady state.

So the magnetic fields produced by the stator and the rotor in the airgap will be at standstill with each other at any speed. Consequently, the machine develops a constant steady-state torque at any speed if the rotor currents are nonzero. The rotor current is zero for the cage rotor at the rotor speed, $n_0 = n_1 = f_1/p_1$, called the standard ideal no-load (or synchronous) speed.

So, to develop torque, $n \neq n_1$ ($S \neq 0$), and thus the IM is also called the asynchronous machine. For $n < n_1$, the machine acts as a motor and, for $n > n_1$, as a generator. The wound rotor IM may reach a zero rotor current at any speed, provided that the rotor power electronics supply can produce (or absorb) electric power at a frequency f_2, according to Equation 1.3, called also the "theorem of frequencies."

The wound rotor IM doubly fed (in the stator at constant voltage and frequency, and in the rotor at variable frequency (f_2) and voltage (V_2)) may work as a motor and a generator both for $n < f_1/p_1$ (subsynchronous) and for $n > f_1/p_1$ (supersynchronous). The latter is currently the workhorse of the variable speed wind generator industry and in pump storage hydroelectric power plants (which generate electricity at peak consumption hours and pump water back into the upper reservoir during off-peak hours) generating up to 400 MW/unit.

At low powers (up to 2–3 kW) in hand tools, small-power compressors, and washing machines, the one-phase power grid (50 (60) Hz)–supplied IM, with a main and an auxiliary winding in the stator and cage rotor, is still widely used, owing to its ruggedness and overall low cost (of motor and capitalized losses), and constant speed.

When variable speed is needed, the three-phase cage rotor IM is used, with power electronics variable voltage and frequency supply to the stator.

The cage rotor IM is, thus, not only the workhorse but has also recently become (at variable V_1 and f_1) the racehorse of industry. A synchronous machine (Figure 1.7) has about the same stator morphology as IM but the rotor completely resembles the stator of the brush–commutator machine. That is, the rotor has a laminated core with the d.c.-fed electric coils placed around salient poles or in slots or has permanent magnets to produce a heteropolar magnetic flux density in the airgap, with the same number of poles, $2p_1$, as the a.c. stator winding currents.

The rotor frequency is $f_2 = 0$, and, thus, mandatorily, the rotor speed n is

$$n = n_1 = f_1 / p_1 \tag{1.4}$$

The slip is $S = 0$ during steady state and this is why the machine is called synchronous (SM). The trouble is that the rotor field is fixed to the rotor and thus the machine can work only at synchronism, $n = n_1 = f_1/p_1$. To change the speed, the stator frequency has to be changed accordingly, from zero, for synchronous starting.

So the SM fed from the standard 50 (60) or 400 Hz sources (the latter on aircraft) cannot be started as such. It may start as an induction motor first, up to a speed $n < n_1$, with a cage placed on rotor poles and the excitation winding connected to a 10/1 resistance. Then, the self-synchronization starts by switching the field winding to the d.c. supply. DC excitation may be produced by slip rings and brushes but the synchronous motor may be also brushless.

For the cage rotor PMSM, the self-starting at the power grid as an induction motor and self-synchronization take place in one step.

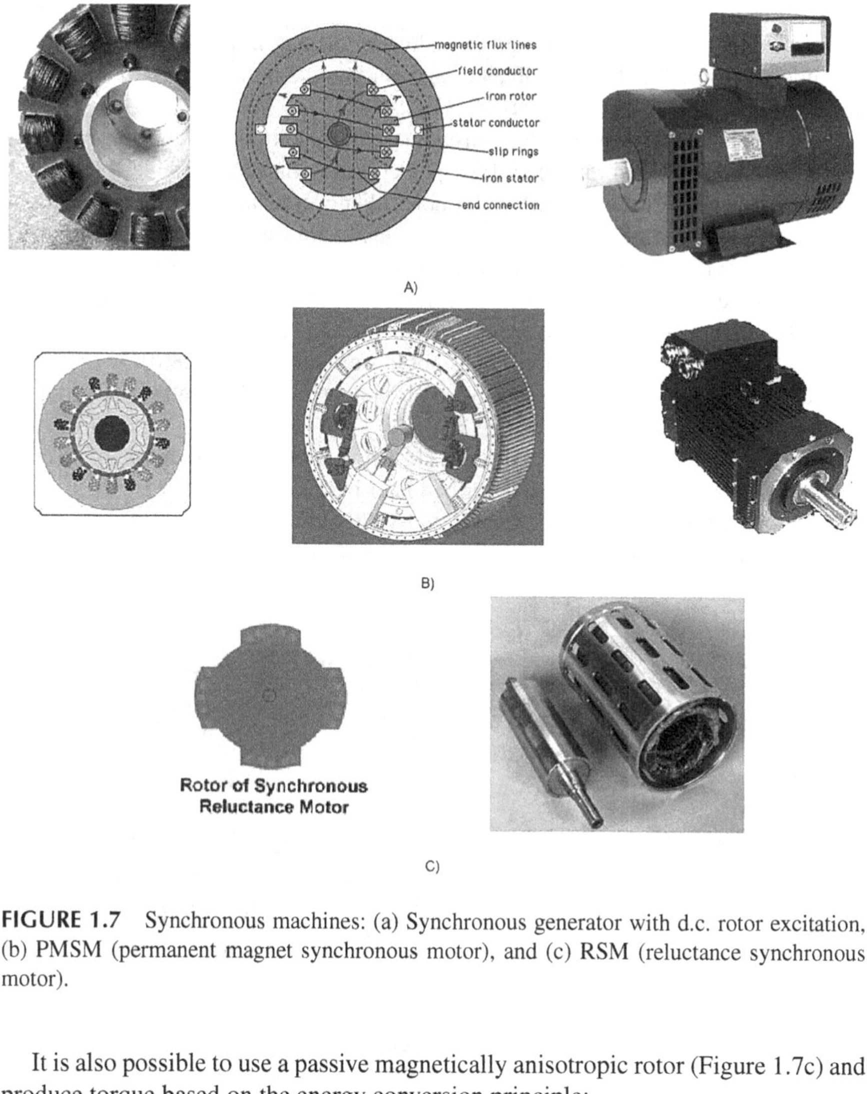

FIGURE 1.7 Synchronous machines: (a) Synchronous generator with d.c. rotor excitation, (b) PMSM (permanent magnet synchronous motor), and (c) RSM (reluctance synchronous motor).

It is also possible to use a passive magnetically anisotropic rotor (Figure 1.7c) and produce torque based on the energy conversion principle:

$$T_e = -\frac{\partial W_m}{\partial \theta_r} \tag{1.5}$$

When W_m—magnetic energy stored in the machine—varies with the rotor position due to rotor anisotropy, electromagnetic (reluctance) torque occurs.

This is the so-called reluctance synchronous machine (RSM), which, due to large magnetic saliency in the rotor, is now considered competitive, based on good performance at lower cost, for various power grid and variable speed applications at powers up to 500 kW and down to 200 W or less.

With the same PM or anisotropic rotor, with or without a cage, the SM may have only a main and an auxiliary capacitor winding in the stator, when only a one-phase a.c. supply is available. More appliances and automotive light-starting torque applications are the target of such motors.

Now, if IMs and SMs qualify for constant speed (traveling) field machines, which may be supplied either from standard a.c. power grid or through variable voltage and frequency power electronics, there are double saliency machines with passive anisotropic rotors without a cage, which have jumping stator field and are totally dependent on power electronics. They need stator-position-triggered current pulsed control. They are called switched reluctance machines (Figure 1.8a) -SRM- or stepper motors (when the current pulse sequence is independent of the rotor position, but frequency ramping is limited so that the motor does not loose steps (Figure 1.8b)).

In thermally or chemically aggressive environments, the SRM and stepper motors have found good markets.

This section has attempted a qualitative classification (characterization) of electric machines, with their applications, described later in this book.

For more info on this subject see [3] and visit www.abb.com/motors, www.siemens.com, www.ge.com. Please also note that for every rotary-motion electric

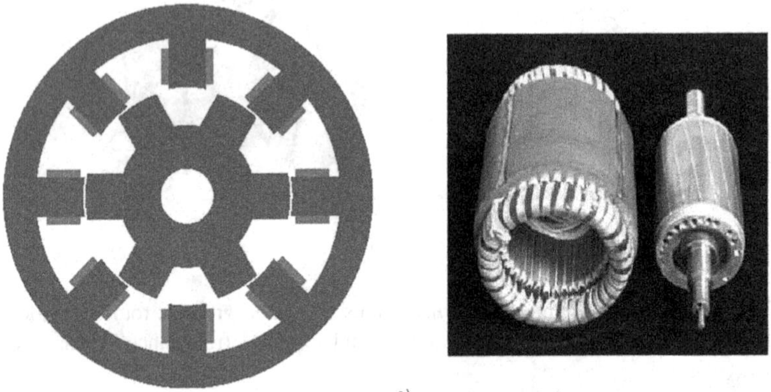

a)

b)

FIGURE 1.8 (a) Switched reluctance motors and (b) stepper motors.

machine there is a linear-motion counterpart. They will be treated in the following chapters. For more on linear electric machines, see [4]. The torque production of various electric machines will be revisited in a more rigorous mathematical manner in Chapter 3.

In the last decade reluctance electric machines using a flux-modulation second rotor part have been investigated worldwide, with applications at small speeds, to produce higher torque density at good efficiency and moderate initial cost [35].

1.3 LOSSES AND EFFICIENCY

Energy losses in electric machines produce heat, harm the environment, and cost money, as more input energy is required. Losses in transformers are electric in nature and occur as copper (winding), P_{copper}, and core (magnetic), P_{core}, losses; electric machines add mechanical losses, p_{mec}.

The power efficiency of an electric machine, η_e, is defined as the output/input power:

$$\eta_e = \frac{\text{output power}}{\text{input power}} = \frac{\text{output power}}{\text{output power} + \sum \text{losses}} = \frac{P_2}{P_2 + \sum p} \quad (1.6)$$

Also $\eta_e = \frac{P_2}{P_1}$.

$$\sum p = p_{copper} + p_{core} + p_{mec} \quad (1.7)$$

For an electric machine with frequent acceleration–deceleration sequences, energy efficiency per duty cycle, EE, is more relevant

$$\text{EE} = \frac{\text{energy output}}{\text{energy output} + \text{energy losses}} = \frac{W_2}{W_2 + \sum W} \quad (1.8)$$

with

$$W_2 = \int P_2 \, dt; \quad \sum W = \sum \int p \, dt \quad (1.9)$$

P_2 in efficiency formula (1.6) is active power, or time average in a.c., per cycle period T, in a.c. machines. For three-phase machines as generators, P_{2e} is the electric power:

$$P_{2\,\text{electric}} = \frac{1}{T} \sum_{n}^{1} \int_0^T v_i(t) \cdot i_i(t) \, dt \quad (1.10)$$

For sinusoidal stator voltages and currents,

$$v_i(t) = V_1\sqrt{2}\cos\left(\omega_1 t - (i-1)\frac{2\pi}{3}\right); \quad i = 1,2,3 \quad (1.11)$$

$$i_i(t) = I_1\sqrt{2}\cos\left(\omega_1 t - (i-1)\frac{2\pi}{3} - \varphi_1\right); \quad i = 1,2,3 \tag{1.12}$$

$$P_{2e} = 3V_1 I_1 \cos\varphi_1 \tag{1.13}$$

where

V_1 and I_1 are the rms values of sinusoidal voltages and currents, respectively, and
φ_1 is the time lag angle between phase voltage and current phasors.

For d.c. current (d.c. brush–commutator) machines as generators, the output power P_{2e} is

$$P_{2e} = V_{dc} I_{dc} \tag{1.14}$$

For an electric motor operation, P_{2e} is the mechanical (shaft) power P_{2m}:

$$P_{2m} = T_{shaft}\, 2\pi n \tag{1.15}$$

where

T_{shaft} is the shaft torque [Nm] and
n is the speed (rps).

The electromagnetic torque, T_e, is

$$T_e = T_{shaft} \pm \frac{p_{mec}}{2\pi n} \tag{1.16}$$

The "+" sign is valid for motor operations and the "–" sign for generator operations.

For a.c. motors and for motors supplied from power electronics, more energy conversion performance indexes may be added.

The first one is EEF, the ratio between the active output power, P_2, and the apparent RMS input power, S_{RMS}, or peak apparent power, S_{peak}:

$$EEF_{RMS} = \frac{P_2}{S_{RMS}}; \quad EEF_{peak} = \frac{P_2}{S_{peak}} \tag{1.17}$$

$$S_{RMS} = 3V_{RMS} I_{RMS} \tag{1.18}$$

$$S_{peak} = 3V_{peak} I_{peak} \tag{1.19}$$

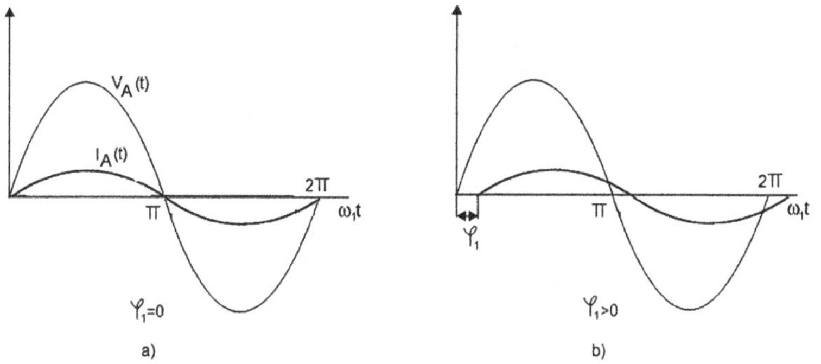

FIGURE 1.9 (a) Unity power factor (synchronous motors) and (b) lagging power factor in a.c. machines (induction motors).

For a sinusoidal operation,

$$S_{RMS} = 3V_1I_1; \quad EEF_{RMS} = \eta \cos \varphi_1 \tag{1.20}$$

Operation at the unity power factor ($\varphi_1 = 0$, Figure 1.9a) for sinusoidal voltages and currents is ideal in a.c. machines because the absorbed stator current is minimum, for a given active power, and thus the copper losses in the a.c. supply grid are minimum. For SMs with a d.c. rotor excitation, this is feasible for all load levels if the d.c. field current can be controlled. Not so is the case in power-grid-connected IMs, which always show a lagging power factor angle (Figure 1.9b).

The peak apparent power, S_{peak}, is very useful in the rating of power switches or of power electronics converters that connect the electric machines to the power grid.

Example 1.1

A directly driven permanent magnet synchronous wind generator (PMSG) of S_n = 4.5 MVA is connected to a V_{nl} = 3.5 kV (line voltage), 50 Hz power grid, and works at a rated efficiency $\eta = 0.96$ and power factor $\cos \varphi = 0.5$.

Calculate

a. Rated stator phase current (star connection): I_n
b. Number of pole pairs p_1, if the rated speed n_n = 15 rpm
c. Total losses and shaft (input) power
d. Rated shaft torque

Solution:

a. From Equation 1.20

(Continued)

Example 1.1: (Continued)

$$S_n = \sqrt{3}V_{nl}I_n = 4.5\times10^6 \text{ VA}$$

$$I_n = \frac{4.5\times10^6}{\sqrt{3}\times3500} = 743.18 \text{ A}$$

b. Based on Equation 1.4

$$p_1 = \frac{f_1}{n_1} = \frac{50}{15/60} = 200 \text{ pole pairs!}$$

So the machine shows 400 poles.

c. The total losses, $\sum p$, from Equation 1.17

$$\sum p = P_{2e}\cdot\left(\frac{1}{\eta_n}-1\right) = S_n\cos\varphi_n\left(\frac{1}{\eta_n}-1\right)$$

$$= 4.5\times10^6\times0.5\left(\frac{1}{0.96}-1\right) = 0.09375\times10^6 \text{ W}$$

The shaft power, P_{1m}, is

$$P_{1m} = \frac{\sum p}{1-\eta_n} = \frac{0.09375\times10^6}{1-0.96} = 2.343\times10^6 \text{ W} = 2.343 \text{ MW}$$

d. The shaft torque, T_{shaft}, from Equation 1.16 is

$$T_{shaft} = \frac{P_{1m}}{2\pi n} = \frac{2.343\times10^6}{2\pi}\times\left(\frac{15}{60}\right)^{-1} = 1.492\times10^6 \text{ Nm!}$$

1.4 PHYSICAL LIMITATIONS AND RATINGS

Physical limitations on electric machines are of electric, mechanical, and thermal origins. The maximum and time-average temperature of electric conductors is limited by the class of electric insulation. PMs irreversibly lose their hard magnetic properties above a certain temperature (except for Ferrite PMs).

The insulation materials have been classified (standardized) into four main classes whose maximum safe temperature limits, given by the IEC 317 Standard, are as follows:

Class A: 105C (less and less used today)

Class B: 130C

Class F: 155C

Class H: 180C

With special insulation materials, even higher safe temperatures are feasible. Electric conductor insulation materials are dealt with in IEC standards 317-20,

51, 13, 26 for the insulation classes mentioned above. Among the mechanical limitations, we mention here the maximum rotor shear stress, f_t (N/cm^2), that guarantees the rotor mechanical integrity, dynamic airgap error, and maximum safe speed (in rps).

The electromagnetic rotor shear stress, f_t, may be calculated as

$$f_t = K_e \cdot A_1 \cdot B_{g1} \cdot \cos\gamma_1 \tag{1.21}$$

$$T_e = f_t \cdot \pi D_r L \cdot \frac{D_r}{2} \tag{1.22}$$

where

- L is the axial lamination stack length (m),
- D_r is the rotor diameter (m),
- A_1 is the fundamental of stator slot ampere-turns/m of rotor periphery (peak value) (from 2×10^3 to $2 \times 10^5 A_{turns}$/m),
- B_{g1} is the peak value of fundamental resultant flux density in the airgap (from 0.2 T to maximum 1.1 T),
- B_{g1} is limited by the magnetic saturation in the machine magnetic cores, which corresponds to a $B_{coremax} \approx (1.2 - 2.3)$ T as the airgap magnetic flux passes from the stator to the rotor core. Values above 2.0 T correspond to special soft magnetic materials such as Hyperco.50,
- K_e is a form factor; for sinusoidal A and B_g, and
- $K_e = 1$ (it is around 1/2 for a one-phase machine)

The phase angle between A_1 and B_{g1} is γ and its optimum value is $\gamma_1 = 0$, which, in a.c. machines, occurs only in synchronous machines.

For practical machines, the rotor shear stress—or tangential specific force—is f_t = 0.1–10 N/cm^2, due to the magnetic saturation, $B_{coremax}$, and winding temperature limitation, $A_{1\,max}$; mechanically, the rotor materials can handle more, with the exception of heavily loaded electric machines with extremely high torque densities (Nm/m^3 or Nm/kg of rotor or of the entire machine). From Equation 1.22

$$T_e = 2 f_t \times (\text{rotor volume}) \tag{1.23}$$

Note that f_t value increases with the rotor diameter.

The stator outer to rotor diameter is, in general,

$$\frac{D_{out}}{D_r} \approx 2 - K_p\,(p_1 - 1) \tag{1.24}$$

For p_1 = 1–4, $K_p \approx$ 0.1–0.2.

So, in fact, the torque per motor volume is approximately

$$\frac{T_e}{\text{motor volume}} \approx \frac{2f_t}{\left(2 - K_p\left(p_1 - 1\right)\right)^2} \tag{1.25}$$

With an average specific weight of $\gamma_{an} = 8 \times 10$ kg/m³, the torque/weight of motor (Nm/kg) ranges from 0.2 Nm/kg, in small- or high-speed motors, to 3 Nm/kg in typical 1000–3600 rpm kW motors and to considerably more in large-torque (diameter) generators/motors.

The temperature limitations are also related to the current density in the windings and to the cooling system, torque, and speed of the electric machine. In transformer design, current densities vary from 2.5 to 4 A/mm², while in forced air-cooled electric motors, it is 5–8 A/m². For forced water (or air)-cooled electric machines, average values of j_{cor} = 8–16 A/mm² are typical.

It is evident that high torque/volume is in contradiction to low losses (and high-efficiency) attempts. So lower active material weight (and cost) is in contradiction to low losses (and temperatures). This is why a global cost function is to be defined and then minimized through design optimization.

$$\begin{aligned}&\text{Global cost in U.S. dollars}\left(\text{euro}\right) = \text{materials and fabrication manpower costs}\\&\quad + \text{capitalized loss energy costs}\left(\sum p_a t_{an} \times \text{energy costs}\right)\\&\quad + \text{capitalized maintenance and repair costs}\end{aligned}$$

For very few small-power applications, the motor initial cost is largest in the global cost. It depends also on the average number of hours/day for the average life of the electric machine (above 10–15 years, in general).

Alternatively, all components in global cost may be converted into joules and, even better, in CO_2 pollution weight as machine materials fabrication, losses, maintenance, and repair means energy (and, consequently, pollution).

Such a joule global cost analysis has demonstrated recently that a 100 L refrigerator is superior to a desktop computer. In complex (say vehicular) applications, not the electric motor but the entire vehicle (system) global cost (in U.S. dollars or euro or joule or CO_2 weight) should be subject to optimization.

Example 1.2

A small three-phase PM synchronous motor, used for active power steering in a modern car, has a rotor diameter, D_r = 0.030 m, and a stack length (stator and rotor average), L = 0.06 m. The sintered NeFeB PMs produce a peak (sinusoidal) airgap flux density, B_g = 0.7 T, and, for rated stator current, the specific tangential force is f_t = 1.2 N/cm².

Calculate

a. Rated electromagnetic torque, T_e
b. With zero mechanical losses, the rated (mechanical) power P_{2m} at 3000 rpm

c. If the rated efficiency is $\eta_n = 0.9$ and power factor $\cos(\varphi_n) = 0.8$, $V_{nl} = 18\sqrt{3}$ V (star connection) – 42 V d.c. bus—determine the rated phase current

d. Rated stator Ampere-turns/m (A_1)

Solution:

a. From Equation 1.23 the electromagnetic torque T_e is

$$T_e = f_t \pi D_r L \frac{D_r}{2} = 1.2 \times 10^4 \times \pi \times 0.03 \times 0.06 \times \frac{0.03}{2} = 1.01736 \text{ Nm}$$

b. The electromagnetic power, $P_{2e} = P_{2m}$, (zero mechanical losses) is

$$P_{2m} = T_e \ 2\pi n = 1.01736 \ 2\pi \frac{3000}{60} = 319.45 \text{ W}$$

c. The input electric power from Equation 1.7:

$$P_{1e} = \frac{P_{2m}}{\eta_n} = \sqrt{3} V_{nl} I_n \cos\varphi_n; \quad I_n = \frac{319.45}{0.9\sqrt{3} \times 18 \times 0.8} \approx 14.245 \text{ A}$$

d. From Equation 1.22

$$A_1 = \frac{f_t}{K_e B_{gPM}} = \frac{1.2 \times 10^2}{1 \times 0.7} = 1.714 \times 10^4 \text{ A/m}$$

1.5 NAMEPLATE RATINGS

Sample nameplates for a transformer and an IM are given in Figure 1.10.

The manufacturers provide basic information for a transformer (Figure 1.10a) such as

- Voltage rating, V (kV)
- Apparent power, VA (kVA or MVA)
- Current rating, A
- Temperature rise, °C
- Short-circuited voltage rating, %
- No-load current rating, %
- Connection diagrams (such as Y_{dx}; $x = 1, 3, 5, 7, 9, 11$ or Y_{yx}; $x = 0, 2, 4, 6, 8, 10, 12$)
- Serial number
- Weight, kg
- Insulation class
- Cooling information

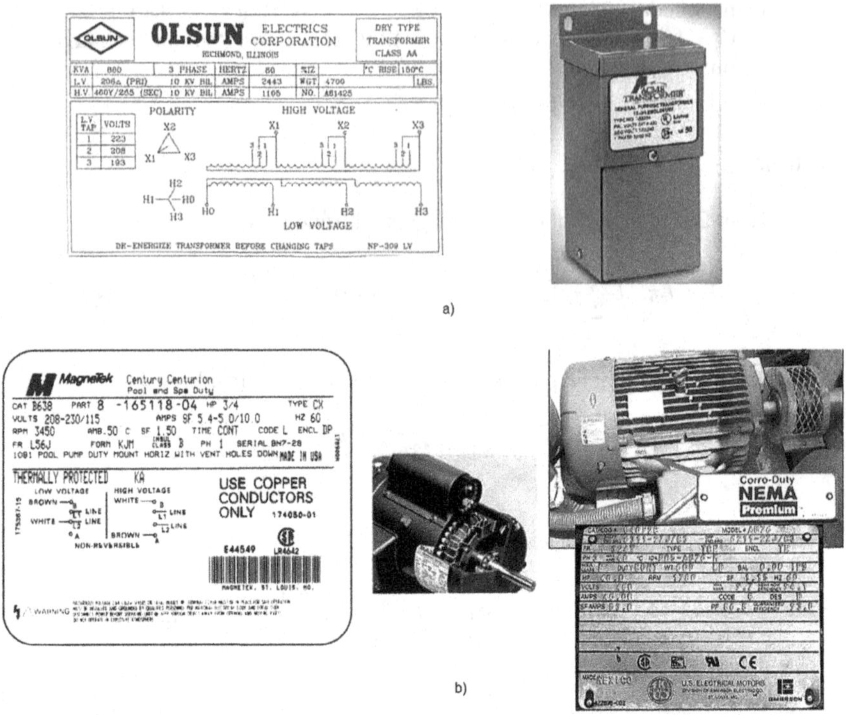

FIGURE 1.10 Sample nameplates: (a) A transformer and (b) an induction machine.

- High- and low-voltage markings (Hi, Xi, ANSI standards, UVW and, respectively, uvw in IE-IEC standards; ABC, abc in some national standards)

For electric machines (Figure 1.10b), the nameplates contain data such as

- Power: (rated, active, mechanical) power for motors and apparent power (in kVA or MVA) for power grid-connected electric machines; for variable speed power electronics fed electric machine the base power, P_b, at base speed, n_b, concept is used and corresponds to the full converter voltage, assigned duty cycle load torque at rated motor winding temperature.
- Information on motor environment and heat transfer marked as OPEN (drip-proof, splash-proof, dust-proof, water-cooled, encapsulated).
- Speed (rated and synchronous) in revolutions per minute for grid-connected (constant speed) motors and base speed, n_b, and maximum speed, n_{max}, for variable speed machines.
- Line-to-line stator (and rotor, if applied) voltage.
- Rated (line stator and rotor (if applied)) currents.
- Efficiency at full load (and at 25% load).

- Volt amperes in a.c. machines.
- Maximum allowable temperature rise in the hot spot.
- Extreme ambient temperature and altitude.
- Service factor indicates how much overrated power the machine can continuously sustain without overheating: this is 1.15 for many motors.
- Supply frequency in Hz (constant or variable, in stator or rotor).
- Torque is given only for variable speed motors at base and maximum speed.
- Rotor inertia (in kg m^2).

1.6 METHODS OF ANALYSIS

Again, electric transformer and electric machines represent systems of coupled electric and magnetic circuits at standstill, and, respectively, in relative motion with each other.

So the magnetic field spatial distribution and time variation in the magnetic circuits and the spatial distribution of windings and their current time variations in the electric circuits have to be solved first.

Accounting for magnetic saturation in magnetic circuits and for skin (field) effect in electric circuits is also necessary as they notably affect the performance. The magnetic field distribution may be approached by analytical and by numerical (finite element) methods.

Then, from calculated magnetic energy or flux in various parts of the machine, the machine's various self-(L_i) and mutual inductances are calculated (W_m is the magnetic energy (Joule) and I_i, current (A))

$$L_i = \frac{2W_{mi}}{I_i^2} = \frac{\Psi_i}{I_i}; \qquad R_i = \rho\,\frac{l_{con}}{A}\,K_{skin} \tag{1.26}$$

Resistances of machine phases can be calculated based on winding geometry, accounting for skin effect by a frequency-dependent coefficient $K_{skin} \geq 1.0$. (ρ is the electric resistivity (Ω m), l_{con} is the conductor length (m), and A is the conductor cross section (m^2)).

Moreover, the electric transformers or electric machines are reduced to electric circuits that are only coupled electrically and that contain resistances, inductances, and "electromagnetic forces" (emfs) produced by relative motion between magnetic field axes and windings physical axes.

Thus, circuit models of transformers and electric machines are used in a few, now widely recognized, forms [5–34]:

- Phase coordinate models (with phase phasor models for sinusoidal steady state in a.c. machines)
- Orthogonal (*dq*) axes models [5]
- Space vector (complex variable) models [12–14]
- Spiral vector models [28]

In this book, we will use the above models as follows:

Part 1—steady state: analytical field model (circuit model) and phasor form for a.c. machines

Part 2—transients: orthogonal (dq) and space vector models

Part 3—finite-element (FE) analysis and analytical and FE field/circuit models for optimal design

1.7 STATE OF THE ART AND PERSPECTIVE

Invented in the nineteenth century (d.c. brush machine by Faraday in 1831), two-phase IM by Ferraris in 1886 and Tesla in 1887, 3 phase IM and transformer development by Dolivo Dobrovolski in 1890, etc.), electric machines have become a mature field by 1930 when strong electric power systems became available all over the world.

Brush–commutator machines have been used as variable speed motors before 1965 as they required only variable d.c. voltage from brush–commutator generators (Ward-Leonard machine group).

The development of power electronics after 1965 (based on thyristors, then GTOs or bipolar transistors) led to variable speed drives with a.c. motors of medium–large power.

But the development of IGBT and MOSFETs after 1980 produced a revolution in sub-kW to MW/unit variable speed drives with a.c. motors in the foreground.

This era coincided with the industrial development of permanent magnet brush–commutator and synchronous motors and the latter have gained widespread acceptance with torques from milli-Nm (for mobile phones) to more than mega-Nm (for directly driven wind generators).

Advanced-vector and direct torque and flux control field oriented (FOC, DTFC) of a.c. motor drives with IGBTs (or IGCTs for very large powers) pulse-width-modulated (PWM) static power convertors (power electronics) have led to fast (millisecond range) and robust torque control response.

A very good proportion of all electric motors is used in variable speed drives for various applications (Figure 1.11). The electric car is one such example (Figure 1.12).

Variable speed generators [34] with power electronics, cage rotor and wound rotor induction or d.c.-excited or PM synchronous (up to 8 MW/unit) are already in use (500 GW total installed wind power) with a steady (above 10%) annual growth, to produce more "green" energy.

Electric machines are tightly standardized—see IEEE and IEC standards—from specifications to installment, maintenance, and repair.

The numerous electric machine manufacturers have their proprietary design methodologies. For FE analysis and design backup with power electronics control, there are distinct, small engineering companies that produce and upgrade their dedicated software annually (such as "Vector Fields", "Ansoft", "ANSYS", "CEDRAT", "FEMM", "SPEED Consortium", motor design for design code, etc.)

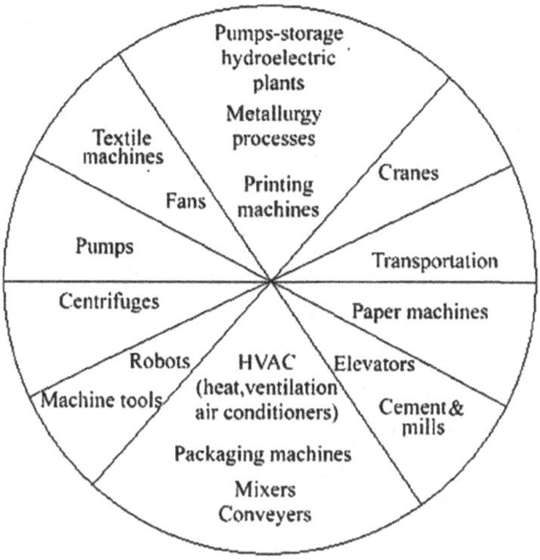

FIGURE 1.11 Variable speed drives applications.

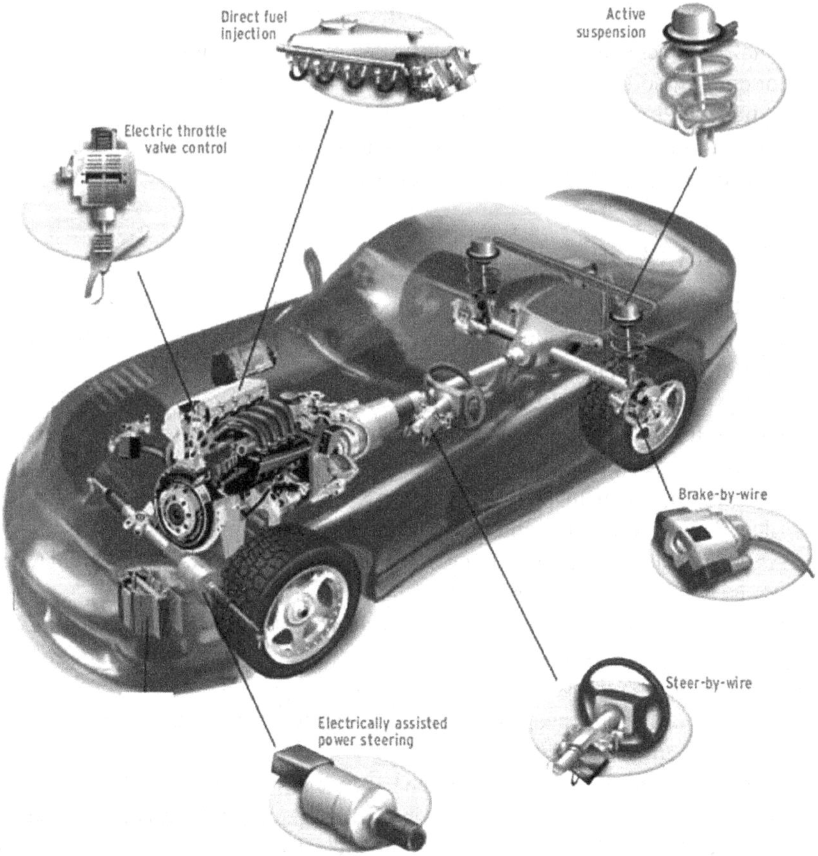

FIGURE 1.12 Automotive electric motor applications.

The following subjects are considered as "hot" in the field of electric machines:

- Less time-consuming FE-coupled circuit models for analysis, to be directly used for modeling in optimal design methodologies.
- Lower-cost high-performance PMs with remnant flux density well above Br = 1.3 T.
- Distributed (weaker) power systems will lead to lower average power/unit generators that will have, in part, to operate at variable speed to be stable and flexible, with operation for lower losses. Design, fabrication, and application of variable speed generator systems, especially for wind and small hydro applications, seems to be a promising field.
- Higher-efficiency, lower-weight (and cost) PM synchronous and switched reluctance motors for home appliance microrobots, and, even more, for the more electric automobiles, aircraft, and vessels.
- Strong developments in synchronous and flux-modulation reluctance machines.
- Magnetic composite soft materials for high-frequency (speed) electric motors with relative permeability above 500–800 and lower specific losses and costs.
- Small displacement linear oscillatory PM brushless motor/generator systems for compressors and hybrid electric vehicle (HEV) actuators.
- Better power grid and high-frequency electric transformers in various applications, from distributed power systems to industrial- and automotive-power-electronics-controlled systems for better power quality.

1.8 SUMMARY

- Electric energy is a key element of the civilization level.
- Electric energy is "produced" in electric power plants where a prime mover (turbine) rotates an electric generator.
- The prime mover is driven by the heat of a combustor that burns fossil or nuclear fuels, or by the kinetic energy of wind or water.
- Electric energy conversion produces heat, while burning fuels, and thus pollutes the environment chemically.
- Electric energy is thus limited, for sustainable development.
- As 60% of electric energy is converted to mechanical work, it is essential to use it wisely in electric motors.
- Electric motors/generators are used in all industries from home appliances, info gadgets, and robotics to transportation, pumps, ventilators, compressors, and industrial processes.
- Electric machines convert mechanical energy to electric energy or vice versa, with magnetic energy storage.
- Electric transformers step up (at generators end) or step down (at consumers end) the voltage (and current level) in alternating current (a.c.) power systems. They do not contain parts in motion, but as electric machines, they work on the principle of Faraday's law of electromagnetic induction and are associated, in

most cases, with electric machines, in applications. This is why they are dealt with here.

- Electric transformers are used in electric power systems, various industries, power electronics, and for a.c. voltage and current measurements.
- Electric machines and transformers are limited by the maximum allowable temperature of insulation materials, magnetic loading (due to magnetic saturation), and the skin effect and temperature of electric conductors, but more importantly by the mechanical tangential and radial stresses on the rotor and stator materials. Thus, limited N/cm^2, Nm/m^3, Nm/kg, and kVA/kg are typical to electric transformers and electric machines [36].
- The initial cost of electric machines tends to be in conflict with the capitalized losses costs and maintenance/repair costs over the operation life of the machine. Thus, global costs optimization is required.
- Global costs may be expressed not only in U.S. dollars (euro, etc.) but also in joule, or even better, in kilograms of CO_2 leaked in the environment during the fabrication of machine materials, the fabrication of machines, human lifestyle, CO_2 by-production, capitalized losses translated back into energy capacity and thus CO_2 exhaust, and, finally, maintenance and repair costs translated into kilograms of CO_2, as per the entire life of the respective electric machine.
- The minimum total energy in joule/person/year and the total CO_2 weight/person/year criteria for a target living standard seems a wiser way to approach technologies of the future, including electric machines.

1.9 PROPOSED PROBLEMS

1.1 A large lossless (three-phase) 1-MW, 6-kV line voltage synchronous motor (star connection) operates at 3600 rpm at 60 Hz. Calculate

   a. Input active power, P_{1e}
   b. Number of pole pairs, p_1
   c. Rated torque
   d. Stator phase current at $\cos\varphi_1 = 1$ and $\cos\varphi_1 = 0.9$

   Hints: Equations 1.13 through 1.15.

1.2 An old electric motor of 50 kW and rated efficiency $\eta_n = 0.91$ is replaced by a new one with 0.94 rated efficiency. For 2500 h per year of operation at full power, calculate

   a. Input energy/year in the two cases
   b. Total energy losses/year in the two cases
   c. Energy cost savings per year in U.S. dollars if 1 kWh costs \$0.1

   Hints: Example 1.1, loss costs = $\sum p(W)$ hours 10^{-3} energy cost (USD/kWh).

1.3 The new motor in Example 1.2 costs \$2000 and should be in operation for 15 years. If the energy costs increase by 1% per year from \$0.1/kWh in the first year and the maintenance and repair costs over the 15 years of life are about \$1000 for the old motor and \$500 for the new one, for 2500 h/year at full power, calculate the overall cost savings with the new motor over the 15 years of the motor's life.

Hints: Add the cost of losses for 15 years (considering the energy cost increase per year) to the maintenance cost for the new and old motor and find the difference; then add the initial cost (for the new motor only).

1.4 An electric three-phase hydro generator of $S_n = 215$ MVA at 15 kV line voltage (star connection) operates at the power grid at $f_1 = 50$ Hz, at $n_n = 75$ rpm. The copper losses in the rotor field winding is $P_{exc} = 0.01$ Sn, the stator copper losses $P_{cos} = 0.0033$ Sn at $\cos\varphi_1 = 1$ and $p_{core} = p_{mec} = 0.002$ Sn. Calculate

   a. Total rated losses at unity power factor ($\cos\varphi_1 = 1$)
   b. Rated efficiency
   c. Rated shaft torque
   d. Electromagnetic torque
   e. Rated phase current
   f. Number of pole pairs

   Hints: Add all losses, use Equation 1.6 for efficiency, Equation 1.15 for shaft torque, Equation 1.16 for electromagnetic torque, Equation 1.13 for rated current, and Equation 1.1 for the number of pole pairs.

1.5 A three-phase PM synchronous electric motor is designed with $A_1 = 2 \times 10^4$ Aturns/m and an airgap PM flux density fundamental $B_{gPM1} = 0.8$ T, for a specific tangential force $f_t = 2$ N/cm^2. For a 100 Nm torque and a rotor diameter D_r to stack length L ratio $D_r/L_1 = 1$, the following is required:

   a. Rotor diameter D_r and stack length
   b. The ratio torque/rotor volume
   c. The ratio power/rotor volume at $n = 6000$ rpm

   Hints: Example 1.2

REFERENCES

1. *ABB-Transformer Handbook*, ABB Power Technologies Management Ltd., Baden, Switzerland (www.abb.acom/transformers).
2. A. Van den Bossche and V.C. Valchev, *Inductors and Transformers for Power Electronics*/CRC Press/Taylor & Francis Group, Boca Raton, FL, 2004.
3. *ABB-Induction Motors Handbook*, ABB Power Technologies Management Ltd., Baden, Switzerland (www.abb.com/motors).
4. I. Boldea and S.A. Nasar, *Linear Electric Actuators and Generators*, Cambridge University Press, Cambridge, U.K., 1997.
5. R.H. Park, Two reaction theory of synchronous machines, *AIEE Transaction* 48, 1929, 716–727.
6. E. Clarke, *Circuit Analysis of A-C Power Systems*, Vol. 1, Wiley, New York, 1943.
7. Ch. Concordia, *Synchronous Machines*, Wiley, New York, 1951.
8. R. Richter, *Electric Machines*, Vols. 1–6, Birkhauser Verlag, Basel, Switzerland, 1951–1958 (in German).
9. C.D. White and H.H. Woodson, *Electromechanical Energy Conversion*, John Wiley & Sons, New York, 1953.
10. C.G. Veinott, *Theory and Design of Small Induction Motors*, McGraw-Hill, New York, 1959.

11. G. Kron, *Equivalent Circuits of Electric Machinery*, Wiley, New York, 1951 (with a new preface published by Dover, New York, 1967, 278 pp).
12. K.P. Kovacs, *Symmetrical Components in AC Machinery*, Birkhauser Verlag, Basel, Switzerland, 1962, in German (in English, by Springer Verlag, New York, 1985, as *Transients of AC Machinery*).
13. V.A. Venicov, *Transient Processes in Electrical Power Systems*, MIR Publishers, Moscow, Russia, 1964 (in Russian).
14. K. Stepina, Fundamental equations of the space vector analysis of electrical machines, *ACTA Technica CSAV, Prague* 13, 184–198, 1968.
15. P.L. Alger, *The Nature of the Induction Machine*, 2nd edition, Gordon and Breach, New York, 1970 (new edition 1995).
16. S. Yamamura, *Theory of Linear Induction Motors*, John Wiley & Sons, New York, 1972.
17. S.A. Nasar and I. Boldea, *Linear Motion Electric Machines*, John Wiley & Sons, New York, 1976.
18. M. Poloujadoff, *The Theory of Linear Induction Machines*, Clarendon Press, Oxford, U.K., 1980.
19. I. Boldea and S.A. Nasar, *Linear Motion Electromagnetic Systems*, John Wiley & Sons, New York, 1985.
20. T. Kenjo and S. Nagamori, *Permanent-Magnet and Brushless DC Motors*, Clarendon Press, Oxford, U.K., 1985.
21. P.C. Krause, *Analysis of Electric Machinery*, McGraw-Hill, New York, 1986.
22. P.L. Cochran, *Polyphase Induction Motors*, Marcel Dekker, New York, 1989.
23. A.E. Fitzgerald, Ch. Kingsley Jr., and S.D. Umans, *Electric Machinery*, McGraw-Hill, New York 1990, 1983, 1971, 1961, 1952.
24. S.A. Nasar and I. Boldea, *Electric Machines Steady-State Operation*, Taylor & Francis, New York, 1990.
25. I. Boldea and S.A. Nasar, *Electric Machines, Dynamics and Control*, CRC Press/Taylor & Francis Group, Boca Raton, FL, 1993 (translated in Spanish).
26. P. Vas, *Electrical Machines and Drives: A Space-Vector Theory Approach*, Clarendon Press, Oxford, U.K., 1992.
27. I. Boldea and S.A. Nasar, *Vector Control of AC Drives*, CRC Press, Boca Raton, FL, 1992.
28. S. Yamamura, *Spiral Vector Theory of AC Circuits and Machines*, Oxford University Press, Oxford, U.K., 1992.
29. T.J.E. Miller, *Switched Reluctance Motors and Their Control*, Magna Physics Publishing and Oxford University Press, London, U.K., 1993.
30. D.W. Novotny and T.A. Lipo, *Vector Control and Dynamics of AC Drives*, Oxford University Press, Oxford, U.K., 1996.
31. I. Boldea and S.A. Nasar, *Induction Machine Handbook*, CRC Press/Taylor & Francis Group, New York, 2001, 2nd edition, 2010.
32. I. Boldea and S.A. Nasar, *Linear Motion Electromagnetic Devices*, Taylor & Francis Group, New York, 2001.
33. J.F. Gieras and M. Wing, *Permanent Magnet Motors Technology*, 2nd edition, Marcel Dekker, New York, 2002.
34. I. Boldea, *Electric Generators Handbook, Vol. 1, Synchronous Generators, Vol. 2, Variable Speed Generators*, CRC Press/Taylor & Francis Group, New York, 2006.
35. I. Boldea, L. Tutelea, *Reluctance Electric Machines Design and Control*, CRC Press/Taylor & Francis Group, New York, 2018.
36. R. Fischer, *Electrical Machines*, 17th edition, Hanser Verlag Munchen, 2017 (in German).

2 Electric Transformers

Electric power transformers are static devices, made of ensembles of electric and magnetic circuits which step-up or step-down the a.c. input voltage (single or multi-phase) to transmit the input power minus losses (core and winding losses), based on Faraday (electromagnetic induction) Law.

To present the principle, the primary a.c. coil with magnetic core is described first in the case of open secondary a.c. coil (on same magnetic core), followed by the presentation of magnetic and electric materials with their characteristics and a.c. losses.

Further on the single transformer circuit model and no-load and full-load characteristics are treated. Three-phase transformer topologies, their connections, general circuit equations, balanced and unbalanced load operation, conditions to parallel them are all investigated in detail and with numerical examples.

Transformer transients such as inrush current, sudden short-circuit, dynamic forces and transformer behavior (model) for ultrafast voltage pulses (of commutation and of atmospheric nature) have been given special quantitative analysis with case studies.

Finally, measurement transformers, transformers with 3+ windings, and transformers (and coils) for power electronics are dealt with, to end the Chapter (with 58 figures and 190 equations) with a preliminary detail design (sizing) methodology on a case study, together with a solid Summary and a few proposed problems with solving hints.

Electric transformers step up or step down the input a.c. voltage, transmitting the input power minus losses, $\sum p$. Transformers have thus a primary and a secondary. A typical one-phase transformer has a laminated silicon–iron core and a primary and a secondary coil (winding) (Figure 2.1). This chapter highlights refer to the following main issues.

- To understand the voltage step-up or step-down ability of the transformer, we consider here the secondary switch open (zero secondary current: $I_2 = 0$). In other words, we have an a.c. (primary) coil with a laminated soft iron (lossless) core for the time being.
- After elucidating the principles, we treat the main characteristics (and loss mechanisms) of magnetic core and electric conductors magnetic fields (or currents).
- Then we discuss the construction of single-phase transformers and their main and leakage inductance expressions.
- The circuit model equations of the single-phase transformer are derived, and then used to describe no-load, short-circuit, and on-load operation modes.
- Three-phase transformer topologies, connections, and general equations are introduced later.

DOI: 10.1201/9781003214519-2

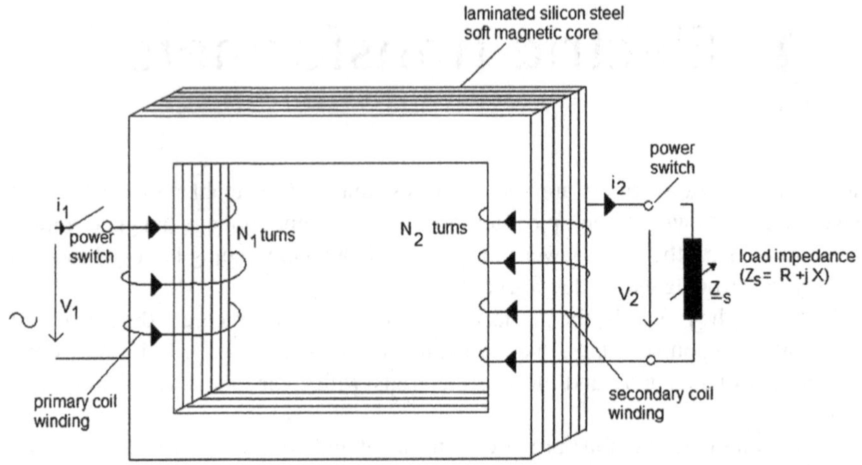

FIGURE 2.1 Single-phase transformer.

- Unbalanced load operation of three-phase transformers and the conditions to connect them in parallel are treated in some detail.
- Electric transformer transients are treated in general, with applications for inrush current, sudden short circuit, electrodynamic forces, and transformer behavior at ultra-fast voltage pulses (of commutation and of atmospherical nature).
- Special transformers, such as autotransformers, three-winding transformers, and transformers for power electronics, are also introduced.
- An example of the preliminary electromagnetic design methodology is given at the end of this chapter.

2.1 AC COIL WITH MAGNETIC CORE AND TRANSFORMER PRINCIPLES

The a.c. coil with ideal (lossless) magnetic core in Figure 2.2, which corresponds to a transformer on no load, contains an airgap, which brings more generality to the subject, as even practical transformers have small (0.1–0.2 mm) airgaps between

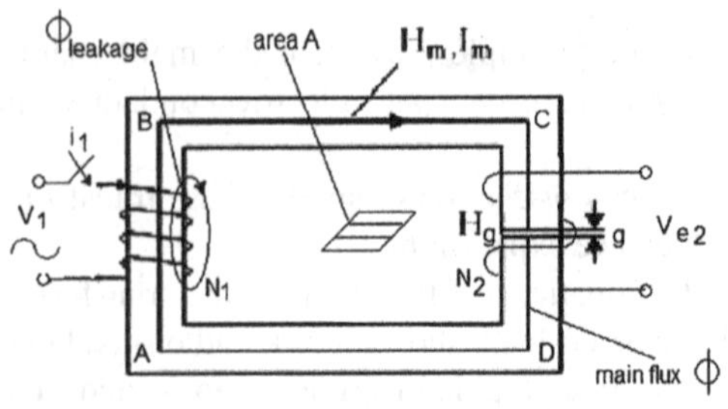

FIGURE 2.2 The magnetic core a.c. coil (transformer on no load).

laminations, while all rotary and linear electric machines have airgaps from 0.2 to 20 mm, in general.

As the frequency is considered to be less than 30 kHz (in most cases much lower: 50(60) Hz in today's electric power systems), the magnetic circuit (Ampere's) law, for the average magnetic field path ABCD in Figure 2.2, is written as

$$\oint \bar{H} \cdot \overline{dl} = \Sigma N_i \cdot I_i \tag{2.1}$$

or approximately

$$H_{\mathrm{m}} \cdot l_{\mathrm{m}} + H_{\mathrm{g}} \cdot g = N_1 \cdot i_{10} \tag{2.2}$$

where

H_{m} and H_{g} are the magnetic fields (in A/m) in the magnetic core and airgap, respectively
l_{m} and g are their path lengths in the core and the airgap, respectively

The magnetic core, B_{m} (H_{m}), relationship is given here as

$$B_{\mathrm{m}} = \mu_{\mathrm{m}}(H_{\mathrm{m}}) \cdot H_m \tag{2.3}$$

where

B_{m} is the flux density (in T)
μ_{m} is the magnetic permeability (in H/m)

For soft magnetic materials, typically found in transformers and electric machines, μ_{m} is large, that is, $\mu_{\mathrm{m}} > 1000\mu_0$, where $\mu_0 = 1.256 \times 10^{-6}$ H/m is the air permeability. The magnetic permeability of soft magnetic materials, μ_{m}, decreases with H_{m} and B_{m}, a phenomenon called magnetic saturation. The flux density is thus limited to B_{max} = 1.6–1.8 T in 50(60) Hz standard transformers, to limit the maximum value of H_{m}, and thus (from Equation 2.2) the no-load current, i_{10}.

Now, if the cross-section core area is A, the flux (Gauss) law at the airgap is

$$\Phi_{\mathrm{A}} = \iint_A B_{\mathrm{m}}\, \mathrm{d}A \tag{2.4}$$

or approximately

$$\Phi_{\mathrm{A}} = B_{\mathrm{m}} \cdot A \approx B_{\mathrm{g}} \cdot A \Rightarrow B_{\mathrm{m}} \approx B_{\mathrm{g}} \tag{2.5}$$

Combining Equations 2.2, 2.3, and 2.5 yields

$$N_1 \cdot i_{10} = \Phi\left[R_{\mathrm{mm}} + R_{\mathrm{mg}}\right]; \quad R_{\mathrm{mm}} = \frac{l_{\mathrm{m}}}{\mu_{\mathrm{m}} \cdot A}, R_{\mathrm{mg}} = \frac{g}{\mu_0 \cdot A} \tag{2.6}$$

where R_{mm} and R_{mg} are the so-called magnetic reluctances of the magnetic core and the airgap zone, respectively. They resemble the electric resistances. Φ is now the "current" and $N_1 i_{10}$ is now the "voltage" in d.c. circuits under steady state (Figure 2.3).

For a.c. current in the coil, R_{mm} (μ_m) depends on the momentary value of the current; but, in general, for practical designs, a single value of flux density in the magnetic core is used (in general, at $B_{max} \times 0.867$).

The a.c. current produces an a.c. magnetic field (H_m and B_m are a.c.), say, sinusoidal as the supply voltage V_1 is sinusoidal:

$$V_1(t) = V\sqrt{2} \cdot \cos(\omega_1 t + \gamma_0) \tag{2.7}$$

The a.c. magnetic field crosses the area of each coil turn, shown in Figure 2.2, once and, as it varies with frequency f_1, it induces an electromagnetic force (emf)-induced voltage, V_e, according to Faraday's law:

$$\oint \bar{E} \cdot \overline{dl} = -N_1 \cdot \frac{d\Phi}{dt} \tag{2.8}$$

or

$$V_{e1} = -N_1 \cdot \frac{d\Phi}{dt} \tag{2.9}$$

Making use of Equation 2.6 in Equation 2.9 yields the emf V_e expression:

$$V_{e1} = \frac{-N_1^2 \cdot \dfrac{di_{10}}{dt}}{R_{mm} + R_{mg}} = -L_{1m} \cdot \frac{di_{10}}{dt} \tag{2.10}$$

L_{1m} is the so-called main inductance:

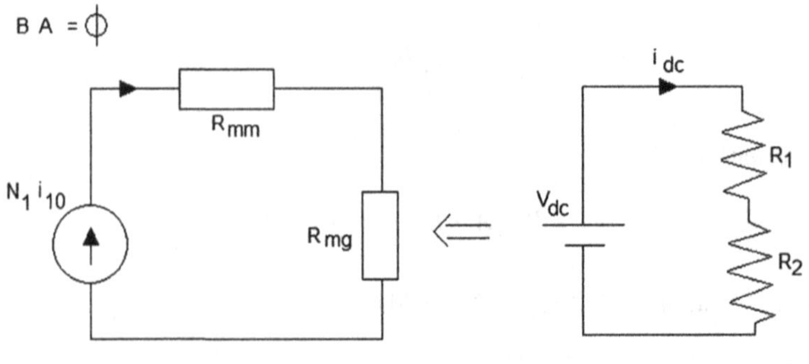

FIGURE 2.3 The equivalent electric circuit of a magnetic core coil with airgap.

$$L_{1m} = \frac{N_1^2}{R_{\text{mm}} + R_{\text{mg}}} \tag{2.11}$$

V_{e1} is called self-induced voltage, but a similar induced voltage, produced by the same flux, Φ, occurs in the secondary coil (Figure 2.2):

$$V_{e2} = -N_2 \cdot \frac{d\Phi}{dt} = -\frac{N_2}{N_1} \cdot L_{1m} \cdot \frac{di_{10}}{dt} \tag{2.12}$$

The ratio of the two emfs is

$$\frac{V_{e2}}{V_{e1}} = \frac{N_2 \cdot \phi}{N_1 \cdot \Phi} = \frac{N_2}{N_1} \tag{2.13}$$

because the same a.c. magnetic flux is involved.

Now coming back to Faraday's law (Equation 2.8), for the coil complete circuit as a sink, we find

$$i_{10} \cdot R_1 - V_1 = V_{e1} - L_{1l} \cdot \frac{di_1}{dt} \tag{2.14}$$

The additional term in Equation 2.14 is related to the leakage (partly in air) flux, Φ_l in Figure 2.2, which does not embrace the secondary coil. In general, $L_{1l} < L_m/500$ but is, still, very important in operation under load.

With an a.c. voltage, V_1 (Equation 2.7), the steady-state solution of Equation 2.14 is obtainable in complex numbers:

$$\underline{V}_1 = V\sqrt{2} \cdot e^{j(\omega_1 t + \gamma_0)}; \quad \underline{I}_{10} = \frac{\underline{V}_1}{\underline{Z}_{10}}; \quad \underline{Z}_{10} = R_1 + j\omega_1 \left(L_{1l} + L_{1m}\right) \tag{2.15}$$

or

$$\underline{I}_{10} = I_{10}\sqrt{2} \cdot \cos\left(\omega_1 t + \gamma_0 - \varphi_0\right); \quad I_{10} = \frac{V}{\left|\underline{Z}_{10}\right|} \tag{2.16}$$

where φ_0 is the voltage/current (power factor) angle and is in general large ($\cos \varphi_0 < 0.05$), as the coil resistance should be small in comparison with the large reactance due to the large magnetic permeability of the soft magnetic core.

A phasor diagram corresponding to the a.c. (sinusoidal current) coil with magnetic core under steady state is shown in Figure 2.4.

Remarks:

- The a.c. coil with soft lossless magnetic core is a high inductance *R*, *L* circuit and may be treated as such.

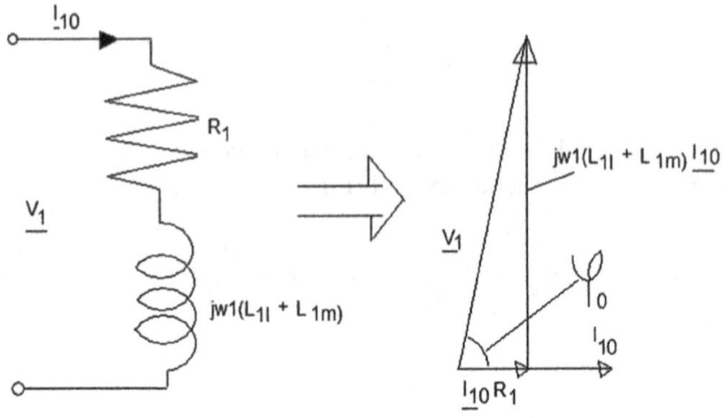

FIGURE 2.4 (a) Equivalent circuit and (b) the phasor diagram of the a.c. coil with soft ideal (lossless) magnetic core.

- The presence of the airgap, g, decreases the inductance, L, and thus increases the current absorbed for a given voltage, frequency, and coil geometric data.
- For an ideal magnetic core (no losses and univoque B_m (H_m) relationship), the a.c. magnetic field, H_m, is in phase with the coil current, I_1, and both vary sinusoidally in time.
- The equivalent magnetic circuit of the coil with ideal magnetic core, though valid also in a.c., resembles the purely resistive electric circuit. With magnetic flux, Φ, replacing the current, the coil magnetomotive force, $N_1 \cdot i_{10}$, replaces the voltage and the role of electric resistance (Figure 2.4) is taken by the magnetic reluctance.
- For sinusoidal voltage, and ideal magnetic core, the a.c. coil draws sinusoidal current under steady state and behaves as an $R_1 + j \cdot X_0$ impedance, which may be illustrated in complex variables (phasors) with the corresponding phasor diagram (Figure 2.4b).
- As the primary a.c. coil "sends" about the same flux, Φ, through the secondary coil, the latter induces an emf in the secondary V_{e2}. Approximately,

$$V_{e2} = V_{e1} \cdot \frac{N_1}{N_2} \approx V_1 \cdot \frac{N_2}{N_1} \tag{2.17}$$

  because the voltage drop on the coil resistance and leakage reactance ($\omega_1 \cdot L_{1l}$) are less than 0.1% of V_1.
- With $N_2 > N_1$ the voltage is stepped up and with $N_2 < N_1$ the voltage is stepped down.
- Now, neglecting the transformer losses and leakage fluxes, for the time being, the input and output power are equal to each other:

$$V_1 \cdot I_1 \approx V_2 \cdot I_2 \tag{2.18}$$

So, stepping up the voltage means stepping down the current and vice versa.

This illustrates the transformer basic principles.

Example 2.1

An ideal transformer, with a magnetic core uniform cross section, $A = 1.5 \cdot 10^{-2}$ m, mean flux path length $l_m = 1$ m, and $N_1 = 50$ turns, has a manufacturing airgap, $g = 2 \cdot 10^{-4}$ m, and operates at no load and a core flux density, $B_m = 1.6$ T (peak value), when supplied from an a.c. voltage source at 60 Hz. Calculate the following:

a. The primary voltage, V_1, equal to emf V_{e1} (RMS)
b. For an average magnetic core permeability, $\mu_m = 2000 \cdot \mu_0$, calculate the primary (magnetization) current, I_{10} (RMS)
c. Neglecting the leakage inductance, calculate the primary coil inductance, $L_0 = L_{1m}$
d. With a design current density, $j_{cor} = 3$ A/mm², for a rated current, $I_n = 100 \cdot I_{10}$, and a mean turn length of $l_{ct} = 0.6$m, determine the primary coil resistance, R_1, and then the circuit power factor on no load
e. The number of turns required in the secondary to halve the primary voltage

Solution:

a. From Equation 2.9, $\underline{V}_{e1} = -N_1 \cdot \frac{d\underline{\Phi}}{dt} = -N_1 \cdot A \cdot \frac{d\underline{B}_m}{dt} = -N_1 \cdot A \cdot \omega_1 \cdot \underline{B}_m$ (because $d/dt \to j\omega_1$ in complex numbers (phasors) for sinusoidal variables)

$$\left(V_{e1}\right)_{peak} = 50 \cdot 1.5 \cdot 10^{-2} \cdot 2\pi \cdot 60 \cdot 1.6 = 452.16 \text{ V}$$

or

$$\left(V_{e1}\right)_{RMS} = \frac{\left(V_{e1}\right)_{peak}}{\sqrt{(2)}} = 320.6 \text{ V}$$

From Equation 2.6, the primary (magnetization) current on no load, i_{10} (RMS), is

$$\left(i_{10}\right)_{RMS} = \frac{\Phi_{max}}{N_1\sqrt{2}}\left(R_{mm} + R_{mg}\right) = \frac{B_{max}}{N_1\sqrt{2}}\left(\frac{l_m}{\mu_m} + \frac{g}{\mu_0}\right)$$
$$= \frac{1.6}{50\sqrt{2}}\left(\frac{1}{2000} + 2 \cdot 10^{-4}\right) \cdot \frac{1}{1.256 \cdot 10^{-6}} = 12.65 \text{ A}$$

b. The main primary coil inductance, L_{1m}, is (Equation 2.11)

$$L_{1m} = \frac{N_1 \cdot \frac{\Phi_{max}}{\sqrt{2}}}{\left(i_{10}\right)_{RMS}} = \frac{N_1 \cdot \frac{B_{max} \cdot A}{\sqrt{2}}}{\left(i_{10}\right)_{RMS}} = \frac{50 \cdot \frac{1.5}{\sqrt{2}} \cdot 10^{-2} \cdot 1.6}{12.65} = 0.0672 \text{ H}$$

c. The primary winding (coil) resistance is known as

$$R_1 = \rho_{Co} \cdot \frac{l_{ct} \cdot N_1}{A_{Co}} = \rho_{Co} \cdot \frac{l_{ct} \cdot N_1}{I_n / j_{cor}} = \frac{2.1 \cdot 10^{-8} \cdot 0.6 \cdot 50}{100 \cdot 12.65 / \left(3 \cdot 10^6\right)} = 1.494 \cdot 10^{-3}\ \Omega$$

The power factor, $\cos\varphi_1$, is (from Figure 2.4b)

$$\cos\varphi_{10} = \frac{R_1 \cdot (i_{10})_{RMS}}{(V_1)_{RMS}} = \frac{1.494 \cdot 10^{-3} \cdot 12.65}{320.6} = 5.9 \cdot 10^{-5}$$

The ideal magnetic core coil—or transformer on no load—is essentially a large reactance with a very small coil resistance. Here, the core loss is neglected, which explains why $\cos\varphi_{10}$ is much smaller than 0.05.

d. The number of turns to halve the voltage in the secondary is, from (2.13),

$$N_2 = N_1 \cdot \frac{V_{e2}}{V_{e1}} = 50 \cdot \frac{1}{2} = 25 \text{ turns}$$

Note: When an a.c. coil with magnetic core and airgap (eventually multiple airgaps) is used as a reactor in power systems or in power electronics (for voltage boosting), the total airgap is notably larger to allow large currents without running in too heavy magnetic saturation, which, as we will show in the next paragraph, means very large core losses, that is, overheating.

2.2 MAGNETIC MATERIALS IN EMS AND THEIR LOSSES

Magnetic core materials in electric machines are defined by a few characteristics out of which the flux density (B_m, in T) and the magnetic field (H_m, in A/m) are paramount (Equation 2.3).

Magnetic permeability, μ_m ($\mu_m = B_m/H_m$), is defined as a scalar in homogeneous materials and as a tensor in nonhomogeneous materials.

2.2.1 Magnetization Curve and Hysteresis Cycle

A magnetic material is characterized by its relative magnetic permeability, μ_{mrel}:

$$\mu_{mrel} = \frac{\mu_m}{\mu_0} \tag{2.19}$$

Some magnetic materials have $\mu_{mrel} > 1$ (they are ferromagnetic or soft) while nonmagnetic materials have $\mu_{mrel} < 1$ (they are paramagnetic with $\mu_{mrel} \approx 1$ or superconducting with $\mu_{mrel} \approx 0$). Soft magnetic materials that make the magnetic cores of transformers and electric machines include alloys of iron, nickel, cobalt, and one rare earth element or they are soft steels with silicon, with $\mu_{mrel} > 2000$ at $B_m = 1.0$ T.

For frequencies above 500 Hz, compressed, injected, soft powder materials that contain iron particles suspended in an epoxy or a plastic matrix are used.

Soft magnetic materials are characterized by

- B_m (H_m) and H_m (B_m) curves
- Saturation flux density, B_{sat}
- Temperature variation of permeability

- Hysteresis cycle
- Electric conductivity
- Curie temperature
- Loss coefficients

As, from power transformers to power electronics coils and transformers, frequency ranges from 50 (60) Hz to 1 MHz, choosing adequate soft magnetic materials is a very challenging task [1]. We will detail more here on soft magnetic materials for power transformers and electric machines, where fundamental frequency is lower than 4 kHz (in 240,000 rpm small electric motors).

Silicon steel laminations and soft magnetic powders are used for the purpose.

A graphical representation of the magnetization curve, B_m (H_m), and of the hysteresis cycle of a standard silicon (3.5) steel M19 are shown in Figure 2.5a and b.

The magnetization process implies the magnetization dipoles in the material being gradually oriented by an external magnetic field (mmf). For a monotonous rising and falling of mmf level, the hysteresis cycle is obtained. So, the magnetization curve represents either an average curve through the middle of the hysteresis cycles or the tips of subsequent hysteresis cycles of lower and lower magnitudes.

The presence of the hysteresis cycle is related to the energy consumed in the material to orient the magnetic microdipoles. Hence, hysteresis losses occur; they are reasonably small in soft magnetic materials but increase with frequency as the number of cycles per second increases and because the hysteresis cycle itself becomes larger as frequency increases (Figure 2.5b).

The presence of magnetic saturation and, implicitly, the B_m (H_m) nonlinearity with the hysteresis cycle suggest three different magnetic permeabilities (Figure 2.6).

- The normal permeability, μ_n, is

$$\mu_n = \frac{B_m}{H_m} = \tan \alpha_n; \quad \mu_{nrel} = \frac{\mu_n}{\mu_0} \tag{2.20}$$

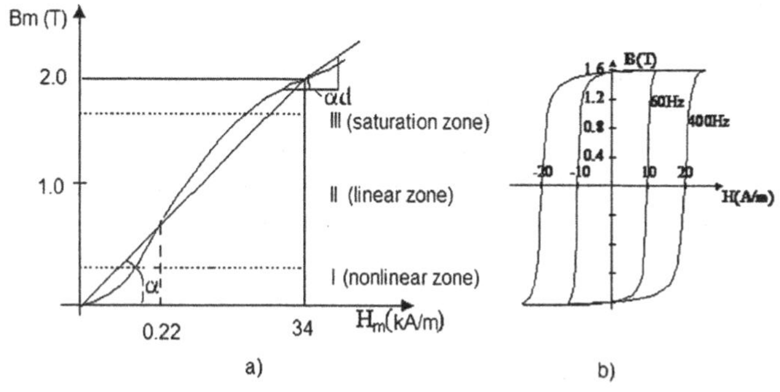

FIGURE 2.5 (a) Magnetization curve and (b) hysteresis cycle of deltamax tape-wound core 0.5 mm strip.

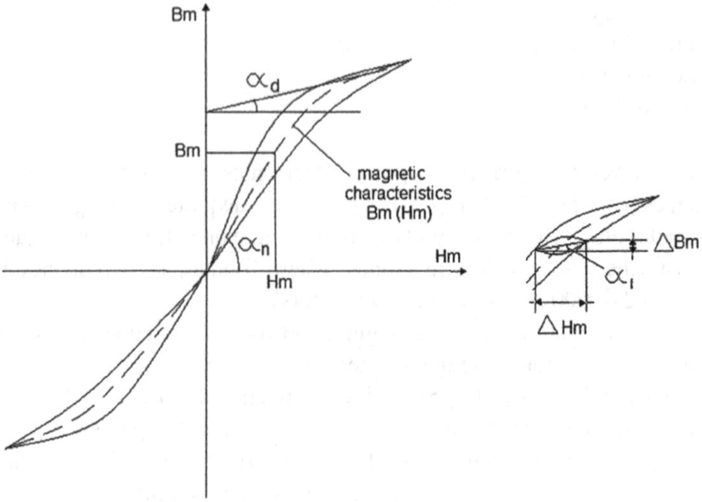

FIGURE 2.6 The three different permeabilities in soft magnetic materials.

- The differential permeability, μ_d, is

$$\mu_d = \frac{dB_m}{dH_m} = \tan\alpha_d; \quad \mu_{drel} = \frac{\mu_d}{\mu_0} \tag{2.21}$$

- The incremental permeability, μ_i, is

$$\mu_i = \frac{\Delta B_i}{\Delta H_i} = \tan\alpha_i; \quad \mu_{irel} = \frac{\mu_i}{\mu_0} \tag{2.22}$$

Close to origin and as the material saturates (Figure 2.5a), the nonlinearity of the magnetization curve makes the three permeabilities different from each other. Only for the linear zone II in Figure 2.5a, they are identical (Figure 2.7).

A few remarks are in order:

1. All magnetic permeabilities decrease notably for silicon iron above 1.2 T or so, but the differential permeability, which shows up in large a.c. transients, and the incremental permeability, which appears in small deviation a.c. transients over d.c. magnetization, are about equal to each other and notably smaller than the normal permeability, corresponding to d.c. magnetization.
2. As, in electric machines (at least), mixed d.c.–a.c. magnetization is common, all three permeabilities are to be used adequately if correct current response is expected.
3. Also, the magnetization curve varies with frequency and, for frequencies notably above 50–60 Hz, pertinent measurements are needed. In essence, the permeability decreases with frequency and in contrast to hysteresis cycle area. Smaller peak flux densities are adopted for frequencies above 200 Hz.

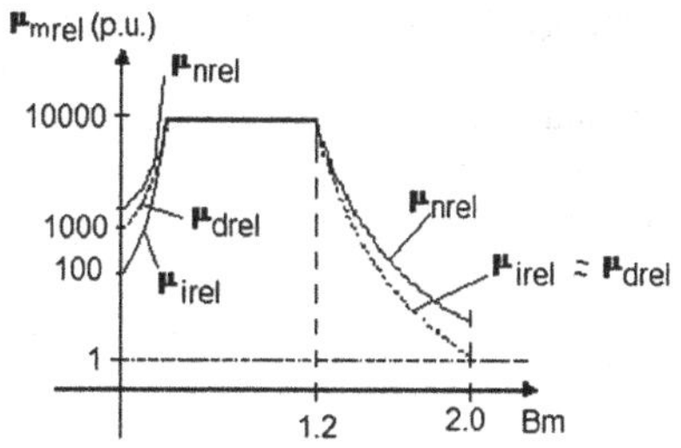

FIGURE 2.7 The three magnetic permeabilities of soft silicon steel core sheets.

2.2.2 Permanent Magnets

PMs are solid, ferro-magnetic materials with an extremely wide hysteresis cycle and a recoil permeability, $\mu_{rec} \approx (1.05\text{–}1.3) \cdot \mu_0$ (Figure 2.8) [2].

Only the second quadrant of the PM hysteresis cycle is given, with the "knee" (demagnetization) point (K_1, K_2, K_4) in the third quadrant. The remnant flux density, B_r ($H_m = 0$), and the coercive field, H_c ($B_m = 0$), add to characterize the PM fully.

Also, the B_m/H_m curve moves downward for sintered and bonded NdFeB and Sm_xCo_y materials and upward (they improve!) for hard ferrites, when the PM temperature rises. Sm_xCo_y may work up to 300°C while NdFeB only to 120°C, before very important demagnetization will occur. Eddy currents from external fields produce losses in PMs. PMs are used to produce d.c.-type magnetic fields to replace d.c. (excitation coils) in brush–commutator and in synchronous machines.

They are magnetized in a special magnetic enclosure capable of producing flux densities in the magnet of $3B_r$, with magnetic fields of $3H_c$ in a few millisecond-long pulses. The low recoil permeability, $\mu_{rec}/\mu_0 = (1.05 - 1.3)$, allows the PMs to hold a lot of stored magnetic energy for years, unless temperature or too large overcurrent mmfs demagnetize them. For more on PMs in electric machines, see [3].

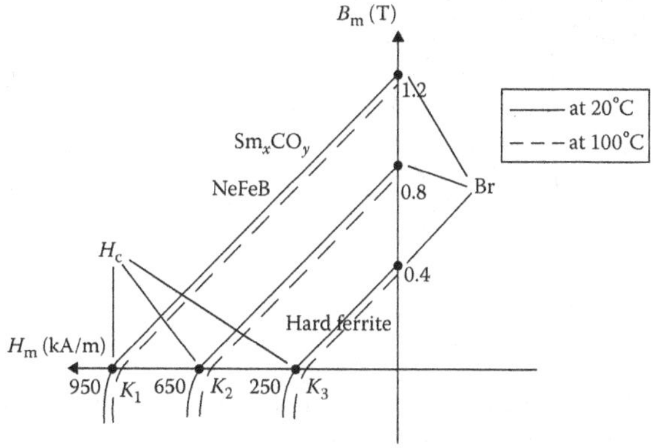

FIGURE 2.8 Permanent magnet characteristics.

2.2.3 Losses in Soft Magnetic Materials

Traditionally, soft magnetic material losses have been divided into hysteresis losses, P_h (in W/kg or W/m³), and eddy current losses, P_{eddy}, (in W/kg or W/m³). Hysteresis losses per hysteresis cycle are proportional to hysteresis area and the frequency of the magnetic field, f, (current) in sinusoidal operation mode:

$$P_h \approx K_h \cdot f \cdot B_m^2 \left[\mathrm{W/kg} \right] \tag{2.23}$$

where

B_m is the maximum a.c. flux density
K_h accounts for hysteresis-involved loop contour (shape) and for frequency (more on hysteresis cycle approximations in [4])

Hysteresis losses depend also on the magnetic field character such as a.c. in transformers and in fix field machines and as a moving field as in induction and synchronous machines. They are 10%–30% larger in traveling fields than in a.c. fields for $B_m < 1.6$ T. However, in traveling fields, core losses have a maximum at around 1.6 T and then decrease to smaller values by 2.0 T in silicon steel sheets.

Eddy current losses in soft magnetic materials are produced by a.c. fields parallel to soft iron sheets that cannot penetrate the sheets completely because the eddy currents are induced by this field in a plane perpendicular to the field direction (Figure 2.9a). The induced current density paths are closed (because div $\bar{J} = 0$), and its amplitude decreases along the sheet depth.

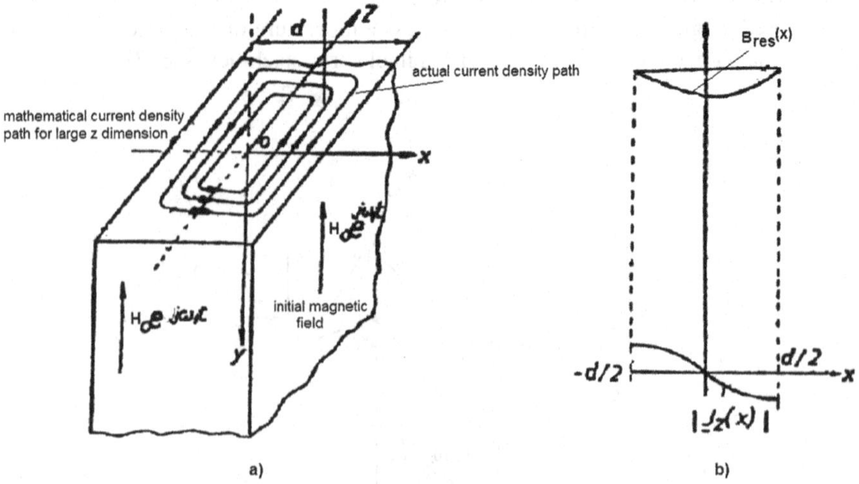

FIGURE 2.9 (a) Eddy current path in a soft iron sheet and (b) resultant flux density, B_{res}, and eddy current density versus sheet depth.

Making use of Ampere's and Faraday's laws, the induced magnetic field, H_y, equations are

$$\text{rot}\,\bar{H} = \bar{J};\quad \text{rot}\left(\sigma_{\text{iron}}^{-1}\bar{J}\right) = -\frac{dB_{\text{res}}}{dt}$$

or, with single-dimensional current density, J_z,

$$\frac{\partial H_y}{\partial x} = J_z;\quad H_{0y} = H_0 \cdot e^{j\omega_1 t} \tag{2.24}$$

$$\frac{1}{\sigma_{\text{iron}}} \cdot \frac{\partial J_z}{\partial x} = j\omega_1 \cdot \mu_{\text{m}} \cdot \left(H_{0y} + H_y\right) = j\omega_1 B_{\text{res}}$$

where

B_{res} is the resultant flux density and
H_{0y} is the initial external magnetic field.

Eliminating H_y from Equation 2.24 we obtain

$$\frac{\partial^2 J_z}{\partial x^2} - j\omega_1 \cdot \mu_{\text{m}} \cdot \sigma_{\text{iron}} \cdot J_z = j\omega_1 \cdot \sigma_{\text{iron}} \cdot B_0 \tag{2.25}$$

The boundary condition (Figure 2.9b) is

$$\left(\partial J_z\right)_{x=\pm d/2} = 0 \tag{2.26}$$

This way, the eddy current density within the sheet is obtained; then, power dissipated per unit weight, with γ_{iron} as iron mass density (in kg/m^3), is

$$P_{\text{eddy}} = \frac{2 \cdot \gamma_{\text{iron}}}{d \cdot \sigma_{\text{iron}}} \cdot \frac{1}{2} \cdot \int_{d/2}^{0} \left(J_z(x)\right)^2 dx$$
$$= \frac{\gamma_{\text{iron}} \cdot d \cdot \omega_1}{\delta \cdot \mu_{\text{m}}} B_0^2 \cdot \left(\frac{\sinh\frac{d}{\delta} - \sin\frac{d}{\delta}}{\cosh\frac{d}{\delta} - \cos\frac{d}{\delta}} \right); \quad \frac{\text{W}}{\text{kg}} \tag{2.27}$$

With δ, the so-called depth of field penetration in the sheet:

$$\delta = \sqrt{\frac{2}{\omega_1 \cdot \mu_{\text{m}} \cdot \sigma_{\text{iron}}}} \tag{2.28}$$

The depth of field penetration means, in fact, an $e(2.781)$ times reduction of current density from the sheet surface.

For well-designed, low losses, sheets, $\delta \gg d/2$. With $\sigma_{iron} = 10^6(\Omega \cdot m)^{-1}$, $\mu_m = 2000 \cdot \mu_0, f_1 = 60$ Hz, and $\delta = 2.055 \cdot 10^{-3}$ m. So 0.35 to 0.5 mm thick silicon iron sheets provide for small skin (Field) effect. For $\delta \gg d/2$,

$$P_{eddy} \approx K_w \cdot \omega_1^2 \cdot B_m^2; \quad \frac{W}{kg}; \quad k_w = \frac{\sigma_{iron} \cdot d_{iron}^2}{24} \tag{2.29}$$

Equation 2.29 is valid for a.c. fields. For traveling fields of same amplitude, the eddy current losses are about two times larger. As traveling and a.c. fields coexist in most electric machines (not in transformers), it is recommended to run tests on the magnetic core in conditions very similar to those in the respective machine.

A more complete formula of total soft material losses is given as [5]

$$\begin{aligned} P_{iron} &\approx K_h \cdot f \cdot B_m^2 \cdot K(B_m) + \frac{\sigma_{iron}}{12} \cdot \frac{d^2 \cdot f}{\gamma_{iron}} \cdot \int_{1/f} \left(\frac{dB}{dt}\right)^2 \cdot dt \\ &+ K_{ex} \cdot f \int_{1/f} \left(\frac{dB}{dt}\right)^{1.5} \cdot dt; \quad \frac{W}{kg} \end{aligned} \tag{2.30}$$

where

$$K(B_m) = 1 + \frac{0.65}{B_m} \cdot \sum_{1}^{n} \Delta B_i$$

B_m is the maximum flux density,
f is the frequency, and
K_{ex} is the excess loss coefficient.
ΔB_i being flux density variation during integration time step

The formula is valid for $\delta > d/2$ conditions and includes an excess loss term while it accepts even nonsinusoidal field variation. It has, however, to be ranked against experiments that, by regression methods, may lead to best K_h, K_{ex}, K choices for the assigned frequency range.

Original data for typical M19, 0.5 mm thick sheet magnetization curve and losses (for a.c. fields) are given in Tables 2.1 and 2.2.

Note: As properties of magnetic materials improve by the year, the reader is urged to visit the sites of leading magnetic-materials producers such as Hitachi, Vacuum Schmelze, mag-inc.com, Magnet sales manufacturing Inc., Hoganas AB, etc.

2.3 ELECTRIC CONDUCTORS AND THEIR SKIN EFFECTS

Electric currents in transformers and in electric machines flow in electric conductors characterized by high electrical conductivity and by the magnetic permeability of the air.

TABLE 2.1
***B–H* Curve for Silicon (3.5%) Steel (0.5 mm thick) at 50 Hz**

B (T)	0.05	0.1	0.15	0.2	0.25	0.3	0.35	0.4	0.45	0.5
H (A/m)	22.8	35	45	49	57	65	70	76	83	90
B (T)	0.55	0.6	0.65	0.7	0.75	0.8	0.85	0.9	0.95	1
H (A/m)	98	106	115	124	135	148	162	177	198	220
B (T)	1.05	1.1	1.15	1.2	1.25	1.3	1.35	1.4	1.45	1.5
H (A/m)	237	273	310	356	417	482	585	760	1050	1340
B (T)	1.55	1.6	1.65	1.7	1.75	1.8	1.85	1.9	1.95	2.0
H (A/m)	1760	2460	3460	4800	6160	8270	11170	15220	22000	34000

Pure electrolytic copper (seldom aluminum) is used to make electric conductors. The electric resistivity of copper conductor, ρ_{Co}, is

$$\rho_{Co} \approx 1.8 \cdot 10^{-8} \cdot \left(1 + \frac{1}{273} \cdot \left(T - 20^{\circ}\right)\right), \qquad \left[\Omega \cdot m\right] \tag{2.31}$$

For d.c. currents, the current spreads uniformly over the electric conductor cross section. Round wire conductors are produced with up to 3 mm uninsulated diameter. Above 6 mm^2 cross section, rectangular cross-section conductors are used. Round copper (magnetic) bare diameters are standardized from 0.3 up to 3 mm. Values from 0.3 to 1.5 mm are as follows: 0.3, 0.32, 0.33, 0.35, 0.38, 0.40, 0.42, 0.45, 0.48, 0.5, 0.53, 0.55, 0.58, 0.6, 0.63, 0.65, 0.67, 0.7, 0.71, 0.75, 0.8, 0.85, 0.9, 0.95, 1.0, 1.05, 1.1, 1.12, 1.15, 1.18, 1.2, 1.25, 1.3, 1.32, 1.35, 1.40, 1.45, 1.5.

For d.c. current conductors, the d.c. resistance only produces the known copper losses, $(P_{Co})_{d.c.}$:

$$\left(P_{Co}\right)_{dc} = R_{dc} \cdot I_{dc}^2 \tag{2.32}$$

There is d.c. current in the stator of brush–commutator motor and in the synchronous machine rotor. The vicinity of the magnetic circuit (slot walls) does not alter the uniform distribution of the d.c. (excitation) current in the conductors in slots. In contrast, in transformers and all electric machines, either on rotor or on stator, a.c. currents flow into copper-conductor coils connected in windings. Even if the conductor is surrounded by air, as the frequency increases and once the conductor radius approaches the field penetration depth, formula (2.28), for copper, gives

$$\delta_{Co} = \sqrt{\frac{2}{\mu_0 \cdot \omega_1 \cdot \sigma_{copper}}} \tag{2.33}$$

Skin effect occurs in the sense that the current density decreases with the radius, inside the conductor.

TABLE 2.2
Typical Core Loss—W/lb) of As-sheared 29 Gage M19 Fully Processed CRNO at Various Frequencies

Induction (KG)	50 Hz	60 Hz	100 Hz	150 Hz	200 Hz	300 Hz	400 Hz	600 Hz	1000 Hz	1500 Hz	2000 Hz
1.0	0.008	0.009	0.017	0.029	0.042	0.074	0.112	0.205	0.465	0.900	1.451
2.0	0.031	0.039	0.072	0.119	0.173	0.300	0.451	0.812	1.786	3.370	5.318
4.0	0.109	0.134	0.252	0.424	0.621	1.085	1.635	2.960	6.340	11.834	18.523
7.0	0.273	0.340	0.647	1.106	1.640	2.920	4.450	8.180	17.753	33.720	53.971
10.0	0.404	0.617	1.182	2.040	3.060	5.530	8.590	16.180	36.303	71.529	116.702
12.0	0.687	0.858	1.648	2.860	4.290	7.830	12.203	23.500	54.258	108.995	179.321
13.0	0.812	1.014	1.942	3.360	5.060	9.230	14.409	27.810	65.100	131.918	
14.0	0.969	1.209	2.310	4.000	6.000	10.920	17.000				
15.0	1.161	1.447	2.770	4.760	7.150	13.000	20.144				
15.5	1.256	1.559	2.990	5.150	7.710	13.942	21.619				
16.0	1.342	1.667	3.179	5.466	8.189						
16.5	1.420	1.763	3.375	5.788	8.674						
17.0	1.492	1.852	3.540	6.089	9.129						

For $\sigma_{copper} = 5.55 \cdot 10^7$ S, $f_1 = 60$ Hz, the penetration depth, $\delta = 6.16 \cdot 10^{-3}$ m. So, if the conductor diameter, $d_{con} \ll \delta$, and, for maximum 3 mm at 60 Hz, all standardized round magnetic wires qualify for low skin effect. The conductors in air such as in the end connections of coils in electric machines are typical for the case in point. In the vicinity of laminated magnetic cores, in the transformer's windows or in machine slots, Figure 2.10a and b, the situation differs.

It seems obvious that for both situations in Figure 2.10, the conductor(s) traveled by a.c. current of frequency, f, are placed in some kind of open slot (with three soft iron core walls). In reality, for electric machinery, the slots may be semi-closed, or even closed, but these cases will be treated in the respective chapters in the book. For the single conductor (bar) in the rotor slot, applying the same Ampere's and Faraday's laws at point A (Figure 2.10b), we get

$$\frac{\partial H_y}{\partial x} = J_z; \quad \frac{1}{\sigma_{Al}} \cdot \frac{\partial J_z}{\partial x} = \mu_0 \cdot \frac{\partial H_y}{\partial t} \tag{2.34}$$

with the boundary condition:

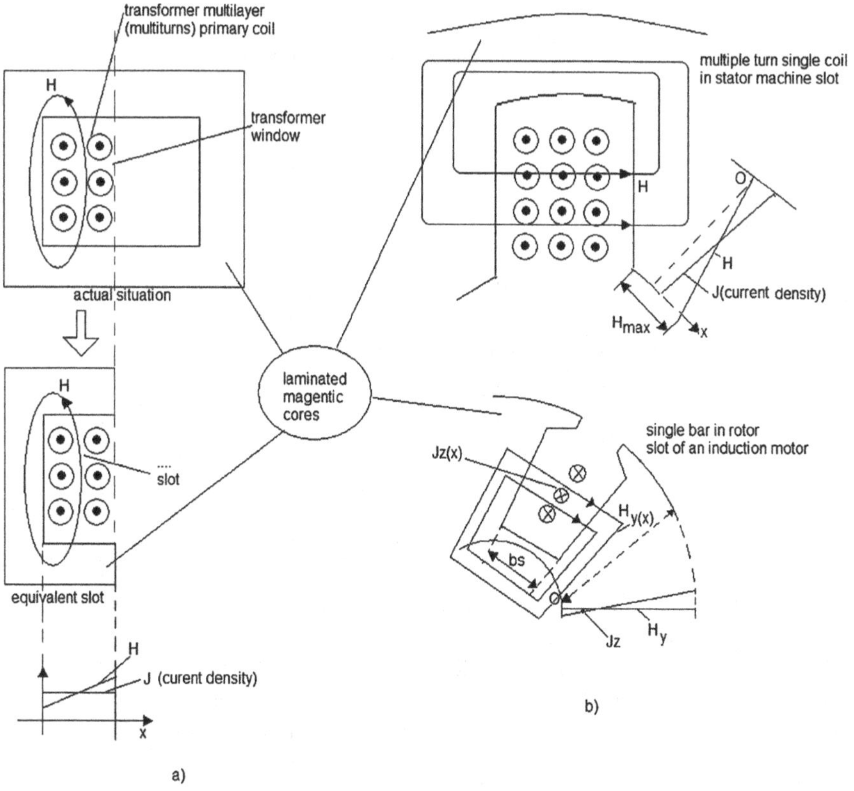

FIGURE 2.10 Placement of a.c. coils in the vicinity of laminated iron core walls (a) in transformers and (b) in electric machines.

$$\left(H_y\right)_{x=0} = 0 \quad \text{and} \quad b_s \cdot \int_0^h J_z(x)\mathrm{d}x = I\sqrt{2} \tag{2.35}$$

The a.c. current in the conductor bar in Figure 2.10b may be expressed in complex number terms:

$$\underline{I} = I\sqrt{2} \cdot e^{j\omega_1 t} \tag{2.36}$$

As the bar current is sinusoidal, the magnetic field $\underline{H}_y$ varies the same way in time. So,

$$\underline{H}_y = H_y(x) \cdot e^{j\omega_1 t} \tag{2.37}$$

Consequently, from Equation 2.34, after eliminating J_z, we obtain an equation similar to Equation 2.26:

$$\frac{\mathrm{d}^2\underline{H}_y(x)}{\mathrm{d}x^2} = \underline{\gamma}^2 \cdot \underline{H}_y(x); \quad \underline{\gamma} = \pm\beta \cdot (1+j); \quad \beta = \sqrt{\frac{\omega \cdot \mu_0 \cdot \sigma_{\mathrm{Al}}}{2}} \tag{2.38}$$

Again, the depth of its own field penetration in the conductor bar, δ_{Al}, is

$$\delta_{\mathrm{Al}} = \frac{1}{\beta} = \sqrt{\frac{2}{\omega \cdot \mu_0 \cdot \sigma_{\mathrm{Al}}}} \tag{2.39}$$

The solution of Equation 2.38 is

$$\underline{H}_y(x) = \underline{A}_1 \cdot \sinh\beta(1+j)x + \underline{A}_2 \cdot \cosh\beta \cdot (1+j)x \tag{2.40}$$

Finally, with the boundary conditions (Equation 2.35), we get

$$\begin{aligned} \underline{H}_y(x) &= \frac{I\sqrt{2}}{b_s} \cdot \frac{\sinh\beta \cdot (1+j) \cdot x}{\sinh\beta \cdot (1+j) \cdot h}; \\ \underline{J}_z(x) &= \frac{I\sqrt{2}}{b_s} \cdot \beta \cdot (1+j) \cdot \frac{\cosh\beta \cdot (1+j) \cdot x}{\cosh\beta \cdot (1+j) \cdot h} \end{aligned} \tag{2.41}$$

The gradual amplitude reduction of current density, J_z, and its magnetic field, H_y, with slot depth is visible also in Figure 2.10b.

When frequency goes down, the field penetration depth δ_{Al} increases, and the current density becomes more uniform over the entire cross section of the conductor.

To calculate the active and reactive powers that penetrate the slot, the Poyting vector S definition is used:

$$\underline{S} = \frac{1}{2} \cdot \int_{A_{\mathrm{upper}}} \left(\underline{E}_z \cdot \underline{H}_y\right)\mathrm{d}A = P + j \cdot Q = P_{\mathrm{dc}} \cdot \varphi(\xi) + j \cdot Q_{\mathrm{dc}} \cdot \Psi(\xi) \tag{2.42}$$

with

$$\sigma_{\text{Al}} \cdot \underline{E}_z = \underline{J}_z \tag{2.43}$$

From Equations 2.40 through 2.42 we obtain

$$\begin{aligned} &\xi = \beta \cdot h; \quad \varphi(\xi) = \xi \frac{\sinh(2\xi) + \sin(2\xi)}{\cosh(2\xi) - \cos(2\xi)}; \\ &\Psi(\xi) = \frac{3}{2\xi} \cdot \frac{\sinh(2\xi) - \sin(2\xi)}{\cosh(2\xi) - \cos(2\xi)}; \quad \xi = \frac{h}{\delta_{Al}} \end{aligned} \tag{2.44}$$

$$P_{\text{dc}} = \frac{L_{\text{stack}} \cdot I^2}{\sigma_{\text{Al}} \cdot h \cdot b_s} = R_{\text{dc}} \cdot I^2; \quad Q_{\text{dc}} = \frac{\omega \cdot \mu_0 \cdot L_{\text{stack}} \cdot h_s}{3b_s} \cdot I^2 = \frac{1}{2} \cdot X_{\text{sldc}} \cdot I^2 \tag{2.45}$$

Equating Equation 2.41 with Equation 2.44, we may identify the a.c. resistance, R, and slot leakage reactance, X_{sl}:

$$R_{\text{mono}} = R_{\text{dc}} \cdot \varphi(\xi); \quad X_{\text{slmono}} = X_{\text{sldc}} \cdot \Psi(\xi) \tag{2.46}$$

A graphical representation of the a.c. resistance and slot leakage reactance skin effect coefficients $\varphi(\xi)$ and $\Psi(\xi)$ dependence on ξ, the ratio between conductor height, h, and its own a.c. field penetration depth, δ, is shown in Figure 2.11.

While the single conductor case is typical to cage rotors of induction motors, in the stator of electric machines and in transformers (Figure 2.10a), there are multiple electric conductors in series in coils, arranged into, say, m layers.

For the situation in Figure 2.12, the a.c. resistance factor, $K_{\text{R}p}$, for the conductors in layer p, on same rationale, is

$$K_{\text{R}p} = \varphi(\xi) + \frac{I_u \cdot \left(I_u + I_p \cdot \cos(\gamma_{up})\right)}{I_p^2} \cdot \Psi'(\xi) \tag{2.47}$$

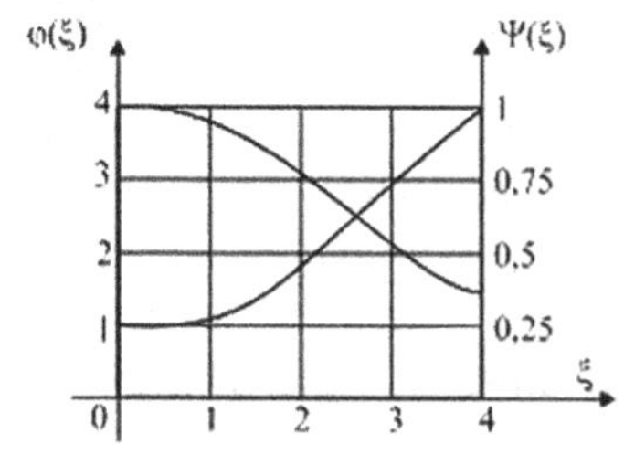

FIGURE 2.11 Skin effect a.c. resistance and reactance coefficients, $\varphi(\xi)$ and $\Psi(\xi)$, of a single conductor in slot.

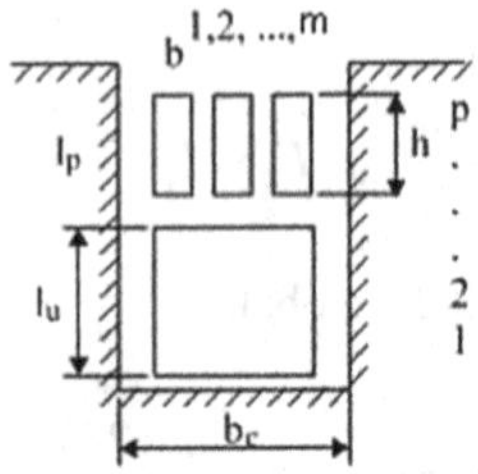

FIGURE 2.12 Multiple-layer conductors in slots.

with

$$\Psi'(\xi) = \frac{\sinh\xi + \sin(\xi)}{\xi \cdot (\cosh(\xi) + \cos(\xi))} \tag{2.48}$$

And γ_{up} is the time lag angle between the currents in the p layer and in the layers beneath it.

If all the conductors are connected in series in the slot, ($I_u = (p-1) \cdot I_p$),

$$K_{Rp} = \varphi(\xi) + p \cdot (p-1) \cdot \Psi'(\xi) \tag{2.49}$$

The average value of K_{Rp} for all m layers, K_{Rm}, is

$$K_{Rm} = \varphi(\xi) + \left(\frac{m^2 - 1}{3}\right) \cdot \Psi'(\xi); \tag{2.50}$$

Note: In most electric machine stators, there are two multiple-turn typical coils in a slot, which do not necessarily belong to the same phase; this influences the a.c. resistance coefficient, K_{Rm}.

For a given slot as in Figure 2.12 (total slot height: $h_{slot} = m \cdot h$), the number, m, of conductor layers increases, so the allowable conductor height, h, decreases; consequently, there is an optimal conductor height, when K_{Rm} is minimum, called critical conductor height, $h_{critical}$.

For large transformers and electric machines at power grid, (50(60) Hz), or for small power transformers at higher frequencies or small power high speed (frequency) motors, a single turn (bar) is made of quite a few conductors in parallel. To attenuate the proximity effect (mutually induced eddy currents), the elementary conductors are all placed in turn in all positions within the slot. This way, the Roebel bar was born (Figure 2.13). For Roebel bars, $K_{Rm} \approx \varphi(\xi)$, which is a great advantage as by design K_{Rm} is kept as $1 < K_{Rm} \leq 1.1$, to limit a.c. skin effect losses.

2.4 COMPONENTS OF SINGLE- AND THREE-PHASE TRANSFORMERS

Again, the transformer is a static apparatus for electric a.c. energy (power) transfer, with step-up or step-down voltage, by magnetically and (eventually) electrically coupled electric circuits, at a given frequency.

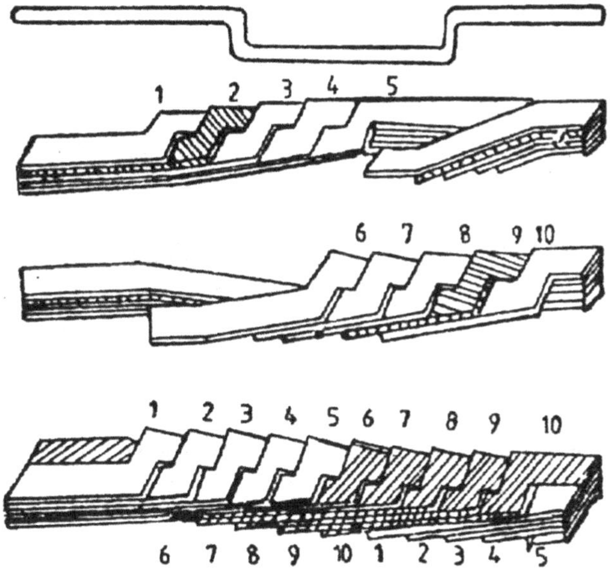

FIGURE 2.13 Transposed elementary conductors (Roebel bar).

The single-phase power transformer (at 50(60) Hz frequency) contains quite a few components such as [6]

- The closed laminated soft iron core
- The low- and one (or more) high-voltage windings
- Oil tank (if any) with conservator
- Terminals, low- and high-voltage oil to air bushings
- Cable connections
- Coolers
- Radiators
- Fans
- Forced oil (forced air) heat exchangers
- Oil pumps
- Off-circuit and on-load tap changers
- Accessories

We will treat mainly the magnetic core and the windings in some detail.

2.4.1 Cores

There are two practical laminated core configurations for single-phase power transformers: core type and shell type (Figure 2.14a and b).

It is evident that the shell configuration allows for a lower height and wider topology, which may be advantageous for transportation permits.

The three-phase power transformers are built with three limbs (Figure 2.15a) or with five limbs (Figure 2.15b). The limbs are surrounded by the low- (high-) voltage

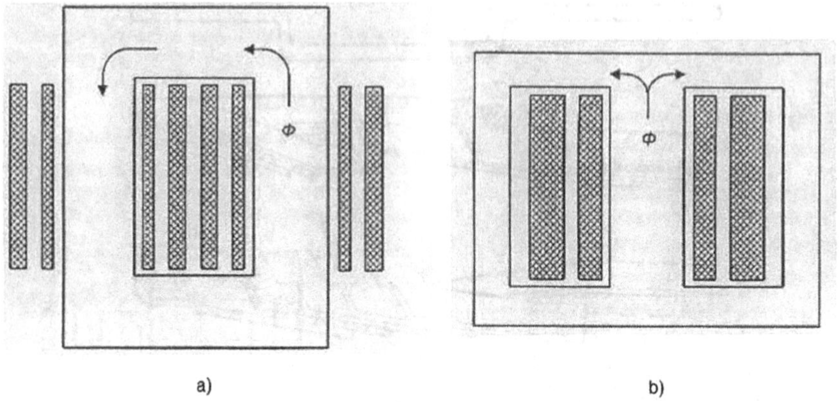

FIGURE 2.14 Single-phase power transformer magnetic cores (a) core type and (b) shell type.

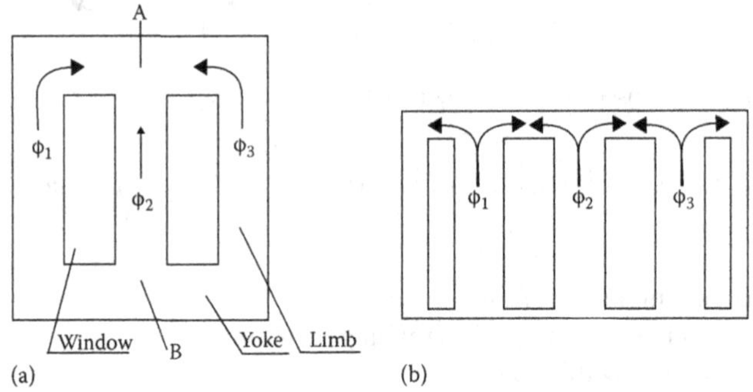

FIGURE 2.15 Three-phase power transformer cores (a) with three limbs and (b) with five limbs.

windings of one phase. In the five-limb configuration, the height of the yoke is halved and thus, again, the transformer is less tall, but wider. In normal, symmetric, operations, the phase voltages and currents are symmetric (same amplitude, 120° phase shift between phases). So, the magnetic flux in all limbs is sinusoidal, and their summation in points A and B (Figure 2.15a) is zero, according to Gauss's flux law.

The outer limbs and the yokes of the five-limb three-phase transformer "see" half of the central limb's flux. The same is true for the single-phase shell core (Figure 2.14a).

The thin (0.5 mm thick for 50(60) Hz) oriented-grain silicon steel sheets that make the magnetic cores are stacked to achieve low-core losses and low-magnetization current.

The connection between limbs (columns) and yokes is arranged in general by 45° joints, to achieve a large cross-over section and low-flux density in the non-preferred magnetization direction.

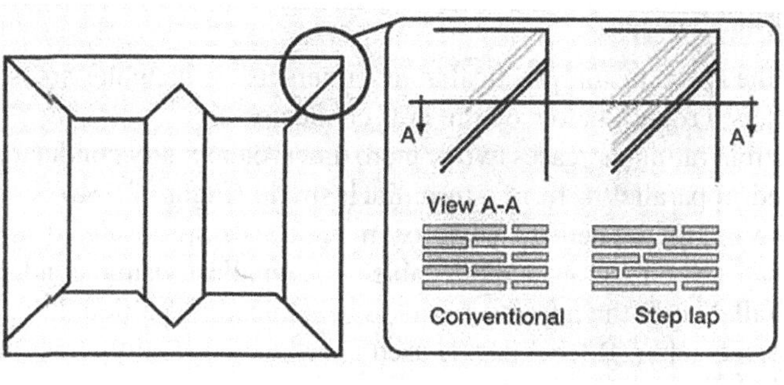

FIGURE 2.16 Stacking pattern with conventional or step-up joints.

They are laid in packets of two or four where each packet (layer) has the joint displaced relative to the adjacent one (Figure 2.16).

The core is earthed, in general, in one point. The magnetic core made of thin insulated lamination layers is glued together for small power and wrapped with steel straps around the limbs or epoxy-cured stacked for large powers. Holes through the lamination core are to be avoided to reduce additional losses. Clamps with curved tie bolts keep the yoke laminations tight.

The cross section of limbs is quadratic or polygonal but the yoke's interior sides are straight to allow the winding location in the transformer window. While the stacked cores are typical for large power, single-phase distribution transformer use wound (less expensive) cores (Figure 2.17).

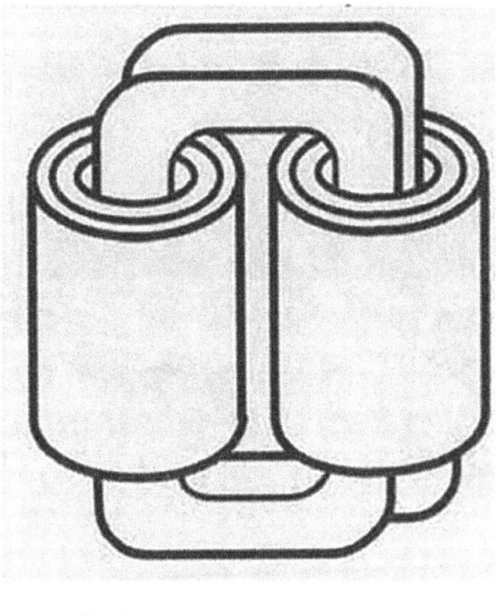

FIGURE 2.17 Single-phase distribution transformer with wound core.

2.4.2 Windings

To increase the space factor, practical transformers use rectangular cross section (or the least flattened round) below 6 mm^2 area conductors.

As the current rating increases two or more times, elementary conductors (strands) are connected in parallel to form a turn. Each strand (Figure 2.18a) is insulated by paper lapping or by an enamel. If two or more strands are insulated in a common paper covering, they are considered a cable—the smallest visible conductor. A few cables in parallel carry the phase currents. As explained in Section 2.3, the continuously transposed cable (Roebel bar) is used to reduce proximity effect in large current rating transformers (Figure 2.18b).

By transposing at a 10 cm pitch the insulated strands (up to a hundred!), all of them, experience the same emf (produced by their currents) and thus circulating currents (proximity effects) are avoided.

Windings are divided into four main types:

- Layer (cylindrical) windings
- Helical windings
- Disc windings
- Foil windings

The level of current and the number of turns per phase determine the winding type.

Layer windings (Figure 2.19a) have turns that are arranged axially, close to each other in single or multiple layers and are used mainly for small and medium power/unit, or, in large transformers, as regulating windings (Figure 2.19a).

The helical winding (Figure 2.19b) is similar to layer windings but with spacers between each turn or thread and is suitable for large currents that share several parallel strands.

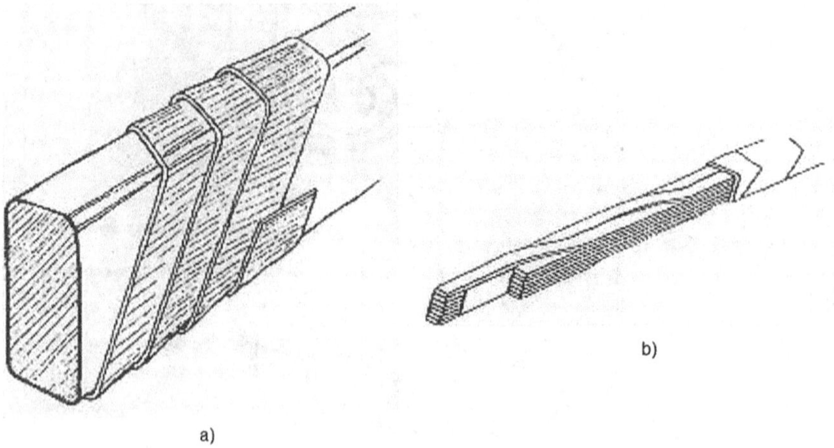

FIGURE 2.18 (a) Conductor strand with insulation paper lapping and (b) continuously transposed cable.

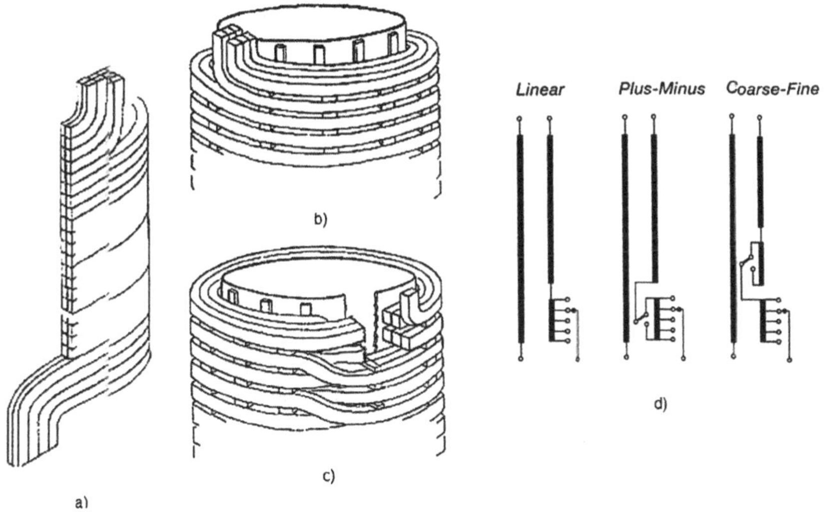

FIGURE 2.19 Transformer windings: (a) Regulating winding in layer type design, (b) double-threaded helical winding, (c) conventional and interleaved disc winding, and (d) three basic ways to arrange voltage regulation.

All cables (made of one or more strands in a common paper cover) in a disc belong to the same turn and are in parallel.

To avoid circulating currents between strands, each of them changes position along the winding transposition zone, to experience to same total a.c. magnetic field. The helical winding has a high space factor and is mechanically robust and easy to manufacture from a continuously transposed cable. Disc windings (Figure 2.19c) are used for a large number of turns (and lower currents). It is built with a number of discs connected in series. The turns in a disc are wound radially like a spiral.

Helical windings have one turn per disc while disc windings have two or more turns per disc. The capacitance between segments of conventional disc windings are lower than that between them and earth and thus the distribution of fast front (atmospherically or commutation) voltage pulse will be nonuniform. To counteract this demerit, interleaved disc windings are used (Figure 2.19c).

Foil windings are made of thin aluminum or copper sheets from 0.1 to 1.2 mm in thickness, with the main merit of small electrodynamic axial forces during transformer sudden short circuit, due to the beneficial influence of eddy currents induced in the neighboring foils by the current in one foil, at the cost of additional losses. Low voltage, distribution transformers have foil windings, due to ease in manufacturing and large space factor and in large transformers that experience frequent strong overcurrents.

Tapping for turn-ratio limited range regulation of the voltage is feasible for not-so-large currents (Figure 2.19d). In this case, a small part of the winding will be left without current, but the uncompensated large axial force during large overcurrents has to be considered in the mechanical design. For large regulating range (as in the locomotive transformers, for example), the layer and helical regulating turns are arranged in a separate winding shell whose height is almost equal to the main

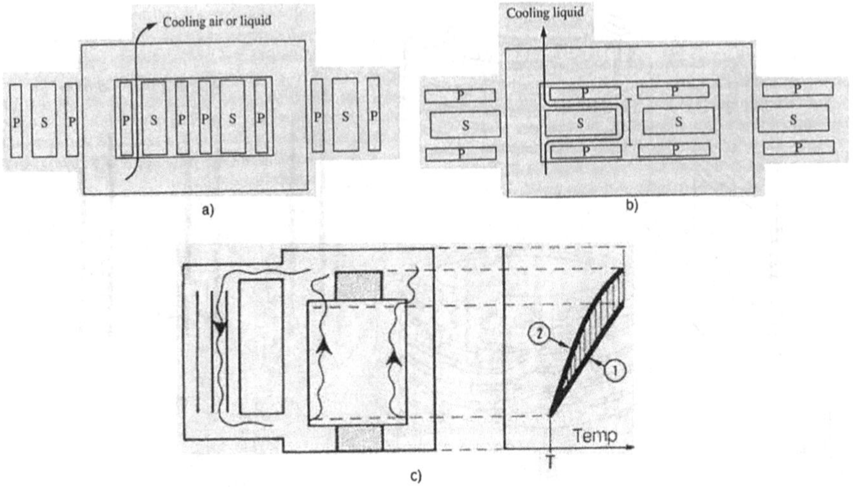

FIGURE 2.20 (a) Cylindrical windings, (b) alternated windings, and (c) cooling oil circulation.

winding's axial height, to avoid uncompensated large axial forces. They are located by the winding neutral star point where the electric potential between phases is small. There are no-load and on-load tap changers. The latter evolved dramatically, from mechanical to electronic (thyristor) commutation under load.

Windings may also be classified as

- Concentric and biconcentric cylindrical (Figure 2.20a)
- Alternate windings (Figure 2.20b)

Alternate windings are used in large transformers to reduce leakage reactance as the rather opposing currents in subsequent primary/secondary coils reduce the leakage (air field) magnetic field and energy. This way, the voltage regulation (voltage reduction with load) is decreased, as needed in large voltage a.c. transmission power lines.

The windings and the cores produce heat by winding and core losses and oil, with high heat capacity (1.8 kW s/kg K), is needed to transport this heat to the heat exchangers. For a 20 inlet–outlet oil temperature differential and 180 kW of losses (30–50 MVA transformer) an $180/(1.8 \times 20) = 5$ kg/s oil flow rate is required. Ten times more is needed for a 400 MVA transformer.

The oil circulation through heat exchangers is done naturally through thermo siphon effect or it is pumped in a controllable way. Unfortunately, the loss of auxiliary power source that supplies the electric motor pumps leads to immediate transformer tripping. The transformer tank and its oil conservator are shown in Figure 2.21a and b.

The tank is designed to cope with temperature and oil expansion during operation, and, in most cases, a separate expansion vessel, called conservator, is added (Figure 2.21b).

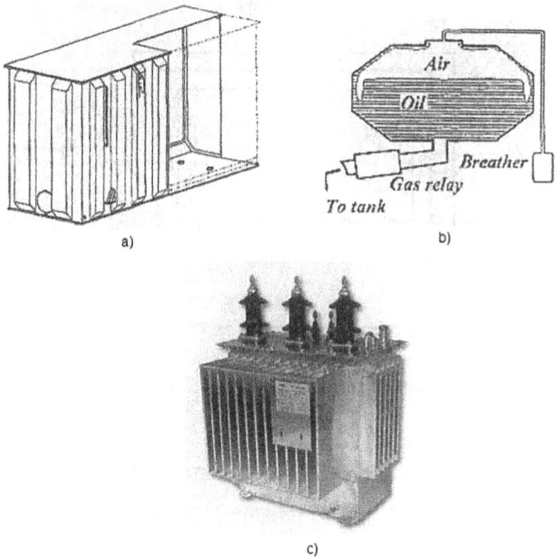

FIGURE 2.21 (a) Transformer tank, (b) oil conservator, and (c) corrugated tank.

For small and medium powers, a corrugated tank is used. Figure 2.21c shows the cover, corrugated walls (which act as enhanced-area heat transmitters), and the bottom box. They are hermetically sealed, in general. The lack of oil contact with air humidity is a definite advantage of corrugated tanks. Oil to air bushings are clearly visible in Figure 2.21c and their construction and geometry depend on the current and voltage levels.

High-current terminals with low voltage and high current (30 kA) for furnace and electrolysis transformers are in the form of flat bar palms or cylindrical studs mounted in panels of plastic laminate.

2.5 FLUX LINKAGES AND INDUCTANCES OF SINGLE-PHASE TRANSFORMERS

As already inferred in Section 2.1, the transformer operates based on Faraday's, Ampere's, and Gauss's laws applied to coupled magnetic/electric circuits at standstill.

Single-phase transformer main, Φ_m, and leakage fluxes, Φ_{1l} and Φ_{2l}.

According to Figure 2.22, under load, there is a current, I_2, in the transformer, that is its secondary. The primary and secondary mmfs together produce the main magnetic flux (coupling) through the core, $\Phi_{\mathrm{m}} = B_{\mathrm{m}} \cdot A$, where A is the average laminated core cross-section area and B_{m} is the average flux density.

So the emfs induced in the transformer primary and secondary, V_{e1} and V_{e2}, are proportional to the number of turns, N_1 and N_2:

$$V_{e1} = -N_1 \cdot \frac{\mathrm{d}\Phi_{\mathrm{m}}}{\mathrm{d}t}; \quad V_{e2} = -N_2 \cdot \frac{\mathrm{d}\Phi_{\mathrm{m}}}{\mathrm{d}t} \tag{2.51}$$

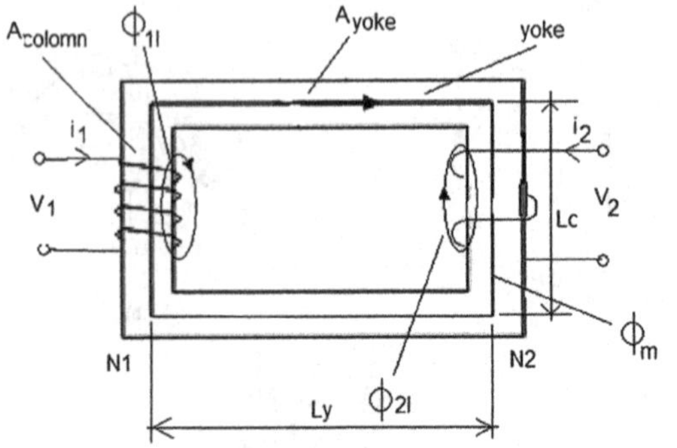

FIGURE 2.22 Single-phase transformer main, Φ_m, and leakage fluxes, Φ_{1l} and Φ_{2l}.

We may thus define main self L_{11m}, L_{22m} and mutual L_{12m} inductances for transformers in relation to main flux linkages, Ψ_{11m}, Ψ_{12m}, Ψ_{22m}:

$$\begin{aligned}
\Psi_{11m} &= L_{11m} \cdot i_1 = \left(B_m \cdot A\right)_{i_2=0} \cdot N_1; \quad L_{11m} = \frac{N_1^2}{R_m} \\
\Psi_{12m} &= L_{12m} \cdot i_1 = \left(B_m \cdot A\right)_{i_2=0} \cdot N_2; \quad L_{12m} = \frac{N_1 \cdot N_2}{R_m} \\
\Psi_{22m} &= L_{22m} \cdot i_2 = \left(B_m \cdot A\right)_{i_1=0} \cdot N_2; \quad L_{22m} = \frac{N_2^2}{R_m}
\end{aligned} \tag{2.52}$$

where R_m is the resultant magnetic reluctance for the flux lines that flow in the laminated core and embrace both windings:

$$R_m \approx \frac{2 \cdot L_{column}}{\mu_{rc} \cdot \mu_0 \cdot A_{column}} + \frac{2 \cdot L_y}{\mu_{ry} \cdot \mu_0 \cdot A_{yoke}} + \frac{g_{cy}}{\mu_0 \cdot A_{yoke}} \tag{2.53}$$

where

μ_{rc} and μ_{ry} are relative (P.U.) values of magnetic permeability in the limb (column) and yoke, respectively

g_{cy} is the equivalent airgap due to yoke/column stacking of laminations,

A_{column} and A_{yoke} are the cross-section cores of coil limbs and yokes

It is evident that μ_{rc} and μ_{ry} depend on the average flux density, B_{mc} and B_{my}, in the limb and yoke. According to Ampere's law (Figure 2.21),

$$N_1 \cdot i_1 + N_2 \cdot i_2 = R_m \cdot \Phi_m = N_1 \cdot i_{01} \tag{2.54}$$

The current, i_{01}, is called the magnetizing current and is, in general, less than 2% of the rated primary current, I_{1n} ($I_{01}/I_{1n} < 0.02$). The rated current, I_{1n}, is defined as the current in the primary under full load that, by design, allows for safe operation under a given overtemperature of a transformer over its entire life (above 10–15 years in general), for the average duty cycle (load power versus time).

Now if, at first approximation, i_{01} is neglected in Equation 2.53,

$$N_1 \cdot i_1 + N_2 \cdot i_2 = 0; \quad i_2 = -\frac{N_1}{N_2} \cdot i_1 \tag{2.55}$$

Consequently, and evidently, from Equations 2.51 and 2.55,

$$V_{e1} \cdot i_1 = -V_{e2} \cdot i_2 \tag{2.56}$$

Now, as V_{e1} and V_{e2} are in phase (see Equation 2.51), it follows from Equation 2.56 that the currents, i_1 and i_2, are shifted in time at ideally 180° (in reality a little less or more), and that their instantaneous mmfs are almost equal in amplitude.

The sign ⊖ comes from the fact that we adopted the sink/source association of power signs for primary/secondary (Figure 2.22).

It is also evident from Figure 2.22 that a part of the magnetic flux lines of both primary and secondary mmfs flow partly through air without embracing the other winding. These are called flux leakage lines and they produce corresponding fluxes, Φ_{1l} and Φ_{2l}, and magnetic energy (in the air within each winding and the space between the two), W_{m1l} and W_{m2l}, for which we may define leakage inductances, L_{1l} and L_{2l}:

$$L_{1l} = \frac{N_1 \cdot \Phi_{1l}}{i_1} = \frac{\Psi_{1l}}{i_1} = \frac{2W_{m1l}}{i_1^2}; \quad L_{2l} = \frac{N_2 \cdot \Phi_{2l}}{i_2} = \frac{\Psi_{2l}}{i_2} = \frac{2W_{m2l}}{i_2^2} \tag{2.57}$$

These leakage inductances are calculated in Sections 2.5.1 and 2.5.2 for cylindrical (layer) and alternate windings, respectively.

2.5.1 Leakage Inductances of Cylindrical Windings

Because the magnetic core is rather flat and the coils are circular, the leakage flux lines show true three-dimensional paths (Figure 2.23a).

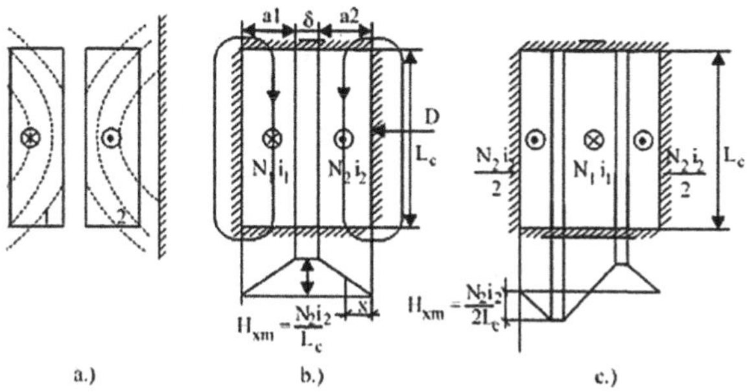

FIGURE 2.23 Leakage field of cylindrical windings: (a) Actual flux path, (b) computational flux path (concentric winding), and (c) the case of biconcentric winding.

To a first approximation, inside and outside the transformer window, the magnetic field path may be considered linear (vertical), as shown in Figure 2.23b for concentric windings and in Figure 2.23c for biconcentric windings.

With this gross simplification (to be corrected finally by Rogowski's coefficient for the leakage inductance), the magnetic field, H_x, varies only with x (horizontal variable):

$$H_x \cdot L_c = N_2 \cdot i_2 \cdot \frac{x}{a_2} \tag{2.58}$$

where a_1 and a_2 are the radial thickness of the two windings, respectively. In between the windings space (δ):

$$H_{xm} = \frac{N_2 \cdot i_2}{L_c} \approx -\frac{N_1 \cdot i_1}{L_c}, \quad a_2 < x < a_2 + \delta \tag{2.59}$$

The magnetic field paths of each winding fill their own volume plus half of the interval between them, and thus the magnetic energy of each of them may be calculated separately:

$$\begin{aligned} L_{2l} &= K_{\text{Rog2}} \cdot \frac{2W_{m2l}}{i_2^2} = \frac{2K_{\text{Rog2}}}{i_2^2} \cdot \frac{1}{2} \cdot \mu_0 \cdot \int^{a_{20}+\frac{\delta}{2}} H_x^2 \cdot \pi \cdot (D + 2x) \cdot L_c \cdot dx \\ &= \frac{\mu_0 \cdot N_2^2}{L_c} \cdot \pi \cdot D_{2\text{av}} \cdot a_{r2} \cdot K_{\text{Rog2}}; \\ a_{r2} &= \frac{a_2}{3} + \frac{\delta}{2} \end{aligned} \tag{2.60}$$

In a similar way, for the primary,

$$L_{1l} = \mu_0 \cdot \frac{N_1^2}{L_c} \cdot \pi \cdot D_{1\text{av}} \cdot a_{r1} \cdot K_{\text{Rog1}}; \quad a_{r1} = \frac{a_1}{3} + \frac{\delta}{2} \tag{2.61}$$

The average diameters of turns, $D_{2\text{av}}$ and $D_{1\text{av}}$, are

$$D_{2\text{av}} \approx D + 3\frac{a_2}{2} \tag{2.62}$$

$$D_{1am} \approx D + 2 \cdot (a_2 + \delta) + 3 \cdot \frac{a_1}{2} \tag{2.63}$$

where

$D_{2\text{av}} < D_{1\text{av}}$ corresponds to the lower voltage winding, which is placed closer to the core,

D is the outside diameter of the core insulation cylinder, and
K_{Rog1} and K_{Rog2} are the Rogowski's coefficients, which are larger than 1.

(Today, finite element 3D analysis allows us to calculate with more precision the leakage inductances at the cost of considerable computation time).

A few remarks are in order:

- Full use of almost 180° phase shift of i_1 and i_2 was made to calculate L_{1l} and L_{2l}.
- The leakage inductance is inversely proportional to the column (limbs) height, L_c, and is proportional to turn average diameter and radial thickness of the windings, a_1 and a_2, and the insulation layer, δ, between them.
- To reduce the leakage inductances, the transformer core should be slim and tall.
- To increase leakage inductances (and reduce short-circuit current), the transformer with cylindrical windings should be less tall and wide.
- The biconcentric winding—where the two halves of the low-voltage winding alternate radially around the high-voltage winding—reduces two times the maximum leakage field between the windings and thus reduces more than two times the leakage inductance.
- Reducing the leakage magnetic energy by large L_c, the electrodynamic forces between windings will also be reduced.

2.5.2 Leakage Inductances of Alternate Windings

The alternate winding with its actual and computational leakage field paths and leakage field variation along the winding are all shown in Figure 2.24, for the division of both windings in q separately insulated coils in series (2 q half coils).

The leakage magnetic field, H_x, alternates along vertical dimension, and is

$$H_x = \frac{N_2 \cdot i_2}{2 \cdot q \cdot L_y} \cdot \frac{2x}{a_2}; \quad 0 \le x \le \frac{a_2}{2} \tag{2.64}$$

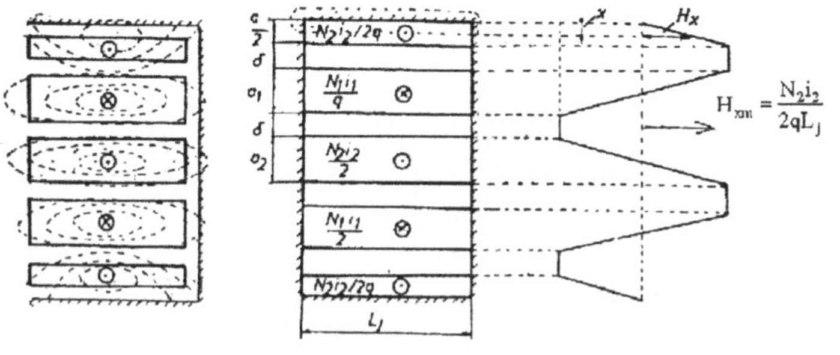

FIGURE 2.24 Alternate windings leakage field: (a) actual field paths, (b) computational field paths, and (c) field vertical distribution.

$$H_x = \frac{N_2 \cdot i_2}{2 \cdot q \cdot L_y}; \quad \frac{a_2}{2} < x < \frac{a_2}{2} + \delta \tag{2.65}$$

Proceeding as for the cylindrical windings, we get the two leakage inductances:

$$L_{2l} = \frac{\mu_0 \cdot N_2^2 \cdot}{2 \cdot q \cdot L_y} \cdot \pi \cdot D_{\text{av}} \cdot a_{r1} \cdot K_{\text{Rog}}; \quad a_{r2} = \frac{a_2}{6} + \frac{\delta}{2} \tag{2.66}$$

$$L_{1l} = \frac{\mu_0 \cdot N_1^2 \cdot}{2 \cdot q \cdot L_y} \cdot \pi \cdot D_{\text{av}} \cdot a_{r2} \cdot K_{\text{Rog}}; \quad a_{r1} = \frac{a_1}{6} + \frac{\delta}{2} \tag{2.67}$$

The average turn diameter, D_{av}, is now the same, and so is the Rogowski's coefficient, K_{Rog}.

Note: It is already evident that the leakage inductances are much smaller for the alternate windings, more than $2q$ times smaller, for $L_y = L_c$. For power transmission transformer and other applications, when low-voltage regulation (voltage reduction with load) is necessary, alternate windings are advantageous, at the costs of higher short-circuit current (in relative values: p.u.). The inductance expressions in relation to transformer geometry are to be essential even in the preliminary electromagnetic design, described at the end of this chapter.

While the above equations are valid for both transients and steady state, we now continue with steady state.

2.6 CIRCUIT EQUATIONS OF SINGLE-PHASE TRANSFORMERS WITH CORE LOSSES

The circuit equations of a single-phase transformer stem from the circuit form of Faraday's law applied to primary and secondary circuits of a transformer (Figure 2.23) whose main self and mutual main inductances, L_{11m}, L_{22m}, and L_{12m}, and leakage inductances, L_{1l} and L_{2l}, have been defined and calculated in the previous section.

$$i_1 \cdot R_1 - V_1 = -\frac{d\Psi_1}{dt}; \quad \text{sink} \tag{2.68}$$

$$i_2 \cdot R_2 + V_2 = -\frac{d\Psi_2}{dt}; \quad \text{source} \tag{2.69}$$

The total flux linkages, in the absence of the core loss effect on them, are

$$\Psi_1 = \Psi_{1m0} + L_{1l} \cdot i_1; \quad \Psi_2 = \Psi_{2m0} + L_{2l} \cdot i_2 \tag{2.70}$$

where Ψ_{1m0} and Ψ_{2m0} are the main flux linkages.

$$\Psi_{1m0} = L_{11m} \cdot i_1 + L_{12m} \cdot i_2; \quad \Psi_{2m0} = L_{22m} \cdot i_2 + L_{12m} \cdot i_1 \tag{2.71}$$

Now, the iron losses may be considered as produced in a purely resistive (R_{iron}) short-circuited fictitious winding that embraces the yoke. Then, we may write

$$-\frac{d\Psi_{1m0}}{dt} = R_{iron} \cdot i_{iron} \tag{2.72}$$

Equation 2.72 shows that the fictitious core loss winding current produces the core loss in the resistance, R_{iron}, reduced to the primary (same number of turns: N_1). We may now admit that the core loss eddy current, i_{iron}, produces a reaction field through the main inductance, L_{11m}. Consequently, the resultant main flux, Ψ_{1m}, is

$$\Psi_{1m} = \Psi_{1m0} + L_{11m} \cdot i_{iron} \tag{2.73}$$

Eliminating i_{iron} from Equations 2.72 and 2.73 yields

$$\Psi_{1m} = \Psi_{1m0} - \frac{L_{11m}}{R_{iron}} \cdot \frac{d\Psi_{1m0}}{dt} \tag{2.74}$$

The Ψ_{1m} and Ψ_{2m} will replace Ψ_{1m0} and Ψ_{2m0} in Equation 2.70 to yield from Equations 2.68 and 2.69:

$$i_1 \cdot R_1 + L_{1l}\frac{di_1}{dt} - V_1 = V_{e1} = -\frac{d\Psi_{1m}}{dt} \tag{2.75}$$

$$i_2 \cdot R_2 + L_{2l}\frac{di_2}{dt} + V_2 = V_{e2} = -\frac{d\Psi_{2m}}{dt} = \frac{N_2}{N_1} \cdot V_{e1} \tag{2.76}$$

and

$$V_{e1} = -\frac{d\Psi_{1m}}{dt} = -L_{11m} \cdot \frac{di_{01}}{dt} + \frac{L_{11m}^2}{R_{iron}} \cdot \frac{d^2 i_{01}}{dt}; \tag{2.77}$$

$$i_{01} = i_1 + i_2'; \quad i_2' = \frac{N_2}{N_1} \cdot i_2 \tag{2.78}$$

Multiplying Equation 2.76 by N_1/N_2 we obtain

$$i_2^2 \cdot R_2' + L_{2l}' \cdot \frac{di_2'}{dt} + V_2' = V_{e1} \tag{2.79}$$

$$R_2' = R_2 \cdot \frac{N_1^2}{N_2^2}; \quad L_{2l}' = L_{2l} \cdot \frac{N_1^2}{N_2^2}; \quad V_2' = V_2 \cdot \frac{N_1}{N_2} \tag{2.80}$$

Now, as both the primary and the secondary windings show the same emf, V_{e1}, it means that the new Equations 2.75 and 2.79 refer to the transformer with the same number of turns (N_1) or with the secondary reduced to the primary. It is evident that the actual and reduced secondaries are equivalent in terms of losses.

2.7 STEADY STATE AND EQUIVALENT CIRCUIT

Under steady state, the load is unchanged and the input voltage, $V_1(t)$, is sinusoidal in time. If we neglect magnetic saturation (and hysteresis cycle) nonlinearities, the output voltage, V_2, and the input and output currents, i_1 and i_2, are also sinusoidal in time:

$$V_1(t) = V_1\sqrt{2}\cdot\cos(\omega_1 t + \gamma_0) \tag{2.81}$$

Consequently, complex variables may be used: $V_1(t) \to \underline{V}_1$, with $d/dt = j\omega_1$. So, in complex variables (phasors), Equations 2.74, 2.75, and 2.79 become

$$\begin{gathered}\underline{I}_1\underline{Z}_1 - \underline{V}_1 = \underline{V}_{e1};\quad \underline{Z}_1 = R_1 + jX_{1l};\quad X_{1l} = \omega_1 L_{1l};\quad \underline{I}_2'\underline{Z}_2' + \underline{V}_2' = \underline{V}_{e1}\\ \underline{Z}_2' = R_2' + jX_{2l}';\quad X_{2l}' = \omega_1 L_{2l}';\quad \underline{V}_{e1} = -\underline{Z}_{1m}\underline{I}_{01}\\ \underline{Z}_{1m} = R_{1m} + jX_{1m};\quad X_{1m} = \omega_1 L_{1m}\end{gathered} \tag{2.82}$$

with $R_{1m} = \omega_1^2 \dfrac{L_{11m}^2}{R_{\text{iron}}}$.

The load equation is

$$\underline{V}_2' = \underline{Z}_s'\underline{I}_2';\quad \underline{Z}_s' = \underline{Z}_s\frac{N_1^2}{N_1^2} \tag{2.83}$$

with $\underline{Z}_s$ being the load impedance characterized by amplitude, Z_s, and phase angle, φ_2.

As in the first two equations of (Equation 2.82) $\underline{V}_{e1}$ shows twice, it means that the primary and reduced secondary magnetically coupled electric circuits may be electrically connected, to form the equivalent circuit of the transformer under steady state with R_{1m}, a series resistance which relates to core losses (Figure 2.25).

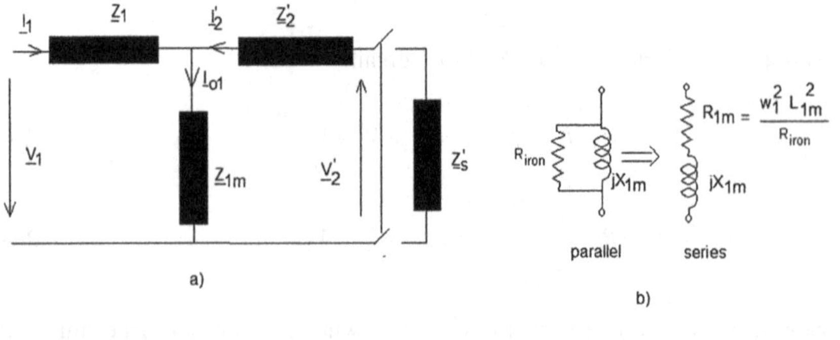

FIGURE 2.25 Transformer steady-state (a) equivalent circuit and (b) core loss parallel resistance, R_{iron}, and series resistance, R_{1m}.

R_{1m} is the equivalent core loss series resistance (Figure 2.25b) that is convenient to use with the equivalent circuit, as it is calculated from measured (or precalculated in the design stage) core losses:

$$p_{iron} = I_{01}^2 R_{1m} = \frac{(X_{1m} I_{01})^2}{R_{iron}} \tag{2.84}$$

A few remarks are in order:

- In regular transformers, L_{1l} and L'_{2l} are not far away from each other in value $(X_{1l} \approx X'_{2l})$ and are at least 100 times smaller than main inductance, L_{11m}; so $X_{1l}(X'_{2l}) \ll X_{1m}$.
- The core loss series resistance, $R_{1m} \ll R_{iron}$ but much smaller than X_{1m} in value and a few tens of times larger than the primary winding resistance, R_1.
- Now R_1 and are not far away from each other in numbers $(R_1 \approx R'_2)$ as the secondary is reduced to primary and the design current density is not much different for the two; so the primary and secondary winding losses are not far away from each other.

For steady-state thorough analysis, we first discuss no-load and short-circuit operation modes and tests.

2.8 NO-LOAD STEADY STATE ($I_2 = 0$)/LAB 2.1

Under no load, the secondary current is zero or the secondary terminals are disconnected from any load. In fact, the transformer degenerates into the case of an a.c. coil with magnetic (laminated) core.

From Equation 2.82, with $\underline{I}'_2 = 0 R'_2$ we obtain

$$\underline{I}_{10}\underline{Z}_1 - \underline{V}_1 = \underline{V}_{e10}; \quad \underline{V}_{e10} = -\underline{Z}_{1m}\underline{I}_{10}; \quad \underline{V}'_{20} = \underline{V}_{e1} \tag{2.85}$$

Finally,

$$\underline{I}_{10} = \frac{\underline{V}_1}{\underline{Z}_0}; \quad \underline{Z}_0 = \underline{Z}_1 + \underline{Z}_{1m} = R_0 + j\omega_1 L_0 \tag{2.86}$$

with

$$R_0 = R_1 + R_{1m}\ (R_1 \ll R_{1m}) \text{ and } L_0 = L_{1l} + L_{1m} \tag{2.87}$$

The equivalent circuit under no load (from Equation 2.25) is shown in Figure 2.26a. R_0 and $L_0(X_0)$ are the no-load resistance and inductance (reactance), and, for standard power transformers (at 50(60) Hz), $X_0/R_0 > 20$, so the power factor at no load is very small.

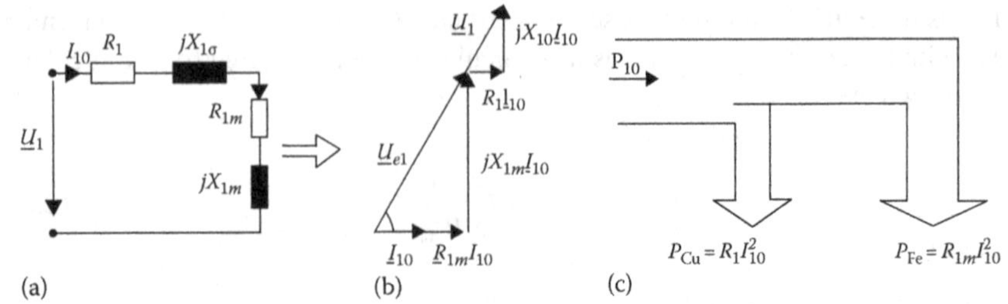

FIGURE 2.26 No-load case: (a) Equivalent circuit, (b) phasor diagram, and (c) power breakdown.

The no-load current at rated voltage, V_{1n}, is almost equal to magnetizing current, I_{01}, under rated load in most power transformers because the voltage drop on $\underline{Z}_1$ is small (less than 1%–2%).

Consequently, the magnetic flux density in the core is about the same under no load and load. And so are the core losses.

So, it is sufficient to measure the core losses under no load and then use them as constant under various loads, for same (rated) frequency and voltage, V_1, level.

Now, the iron losses are much larger than copper losses because $I_{10} < 0.02I_{1n}$, though $R_{1m} > R_1$. So the measured no-load power, P_0, is

$$P_0 = R_1 I_{10}^2 + R_{1m} I_{10}^2 \approx R_{1m} I_{10}^2 = p_{\text{iron}} \tag{2.88}$$

The no-load test at full voltage serves to measure, in fact, core losses ($P_0 = P_{\text{iron}}$), valid also for load conditions.

The no-load characteristics to be drawn, after measurements, refer to P_0 and I_{10} measured directly for growing values of input voltage, V_1, from $0.1V_{1n}$ to $1.05V_{1n}$ in 0.02–0.05 pu steps by using a variable output voltage additional transformer source (Figure 2.27).

From the measured I_{10}, P_0, and voltage, V_1, we may calculate

$$R_1 + R_{1m} = f_1(V_1) \approx R_{1m} \quad \text{and} \quad X_{1m} + X_{1l} = f_2(V_1) \approx X_{1m} \tag{2.89}$$

$$R_1 + R_{1m} \approx R_{1m} = \frac{P_0}{I_{10}^2}; \quad X_{1m} + X_{1l} \approx X_{1m} = \sqrt{\frac{P_0^2}{I_{10}^2} - \left(R_{1m} + R_{1m}\right)^2} \tag{2.90}$$

The function, $X_{1m}(V_1)$, may be considered the no-load magnetization curve of the transformer and may be used to verify the design calculations of main inductance variation with load (magnetization current)

$$L_{11m}(I_{10}) = \frac{X_{1m}(I_{10})}{\omega_1} = \frac{N_1^2}{R_m(I_{10})} \tag{2.91}$$

where, again, R_m is the magnetic core reluctance, which is dependent on the level of flux density, B_m, and on the stacking airgap, g_e, during core assembly from laminations.

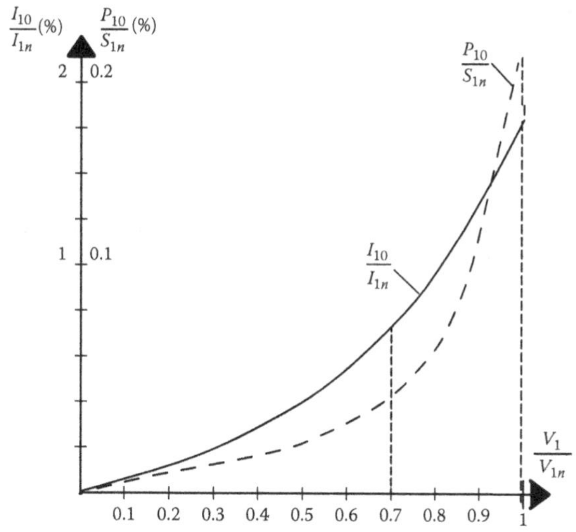

FIGURE 2.27 No-load current, I_{10}, and core losses vs. input voltage.

Example 2.2: No-Load Transformer

A one-phase transformer is rated $S_n = 10$ MVA with $V_{1n}/V_{2n} = 35/6$ kV, and has the design primary resistance, $R_1 = 0.612\ \Omega$ and leakage primary reactance, $X_{1l} = 25\ \Omega$; under no load and at rated voltage, the transformer absorbs the power, $P_e = 25$ kW, for the no-load current, $I_{10} = 0.02I_{1n}$ (I_{1n}: rated current). The magnetization (core loss) series and parallel resistances, R_{1m} and R_{iron}, magnetization reactance, X_{1m}, and power factor, cos φ_0, are all required (Figure 2.28).

Solution:

The definition of rated apparent power is

$$S_n = V_{1n}I_{1n} \approx V_{2n}I_{2n}$$

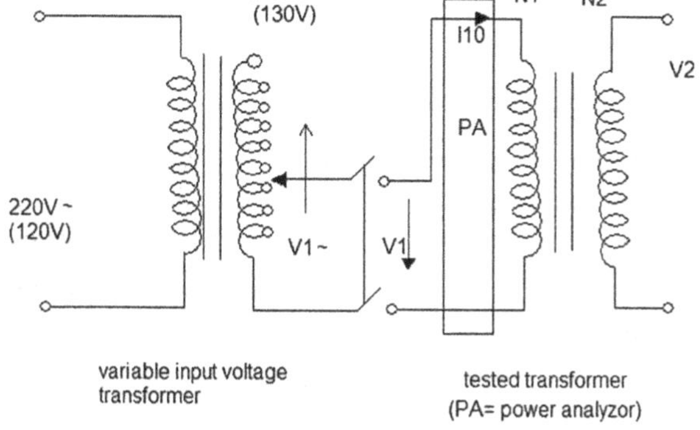

FIGURE 2.28 Lab arrangement for no-load transformer testing.

(Continued)

Example 2.2: (Continued)

Consequently, the rated current, I_{1n}, is found first:

$$I_{1n} = \frac{S_n}{V_{1n}} = \frac{10 \cdot 10^6 \text{ VA}}{35 \cdot 10^3 \text{ V}} = 285.7 \text{ A}$$

So, the rated no-load current, $I_{10} = 0.02 I_{1n} = 0.02 \cdot 285.7 = 5.714$ A.
From Equation 2.89,

$$R_{1m} = \frac{P_0}{I_{10n}^2} - R_1 = \frac{22 \cdot 10^3}{5.714^2} - 0.612 = 673.8 - 0.612 = 673.2\ \Omega$$

So,

$$R_{1m} \gg R_1$$

Also

$$X_{1m} = \sqrt{\frac{V_{1n}^2}{I_{10n}^2} - (R_1 + R_{1m})^2} - X_{1l} = \sqrt{\frac{(35 \cdot 10^3)^2}{5.714^2} - 673.8^2} - 2.5$$
$$= 6088 - 2.5 = 6085\ \Omega$$

Notice that $X_{1m} \approx 10R_{1m}$ in our case.
So, the power factor, $\cos \varphi_0$, at no load is simply

$$\cos \varphi_0 = \frac{P_0}{V_{1n} I_{10}} = \frac{22{,}000}{35{,}000 \cdot 5.714} = 0.11$$

As expected, the power factor at no load is low because the core losses are moderate and the machine core is properly saturated. Also,

$$\frac{X_{1l}}{X_{1m}} = \frac{L_{1l}}{L_{11m}} = \frac{25}{6063} = \frac{1}{242}!.$$

2.8.1 Magnetic Saturation under No Load

The presence of magnetic saturation, let alone the hysteresis cycle (delay), leads not only to a reduction in the magnetization reactance (inductance), $X_{1m}(L_{11m})$, and an increase in the no-load current at rated voltage, but the nonlinear dependence of main flux Ψ_{1m} of i_{10} (Figure 2.26): $L_{11m} = \dfrac{\Psi_{1m}}{\Psi_{10}}$, results in a nonsinusoidal current waveform for sinusoidal voltage input:

$$V_1(t) = V_1\sqrt{2} \cdot \cos(\omega_1 t) = \frac{d\Psi_{10}}{dt} \approx \frac{d\Psi_{1m}}{dt} \tag{2.92}$$

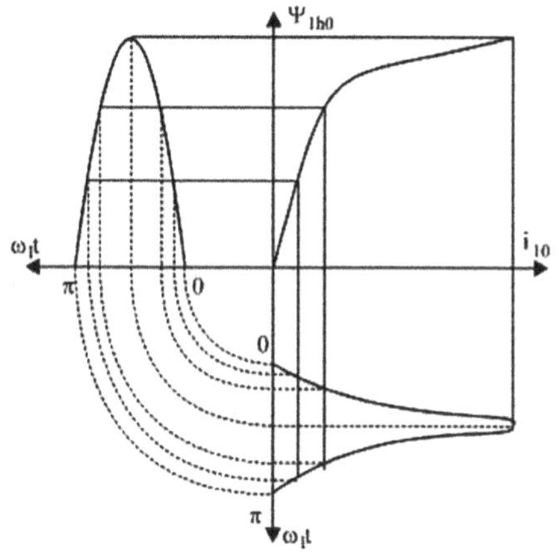

FIGURE 2.29 No-load current actual waveform with magnetic saturation considered.

So

$$\Psi_{1m}(t) \approx \frac{V\sqrt{2}}{\omega_1}\sin\omega_1 t \tag{2.93}$$

When we look for each instantaneous value of flux, $\Psi_{1m}(t)$, in the nonlinear magnetization curve, $\Psi_{1m}(I_{10})$, for the corresponding instantaneous current value, a non-sinusoidal current waveform is obtained (Figure 2.29).

It is evident that 3rd, 5th, 7th (odd numbers in general) harmonics are present and, using a digital oscilloscope or a wide frequency band current sensor with galvanic separation and a computer interface, the $I_{10}(t)$ wave can be captured and analyzed as part of the "no-load transformer test." Now, if the current instrument, in the previous current and power measurements, measures the RMS value of current, then, still, the R_{1m} and X_{1m} calculated values are acceptable even for industrial tests, which are standardized (see IEEE, NEMA, and IEC standards on transformer testing).

The influence of hysteresis cycle produces also flux/current delays and additional harmonics in the current wave form. But, once the transformer is loaded, as $I_{10n} < 0.02I_{1n}$, the influence of the no-load (magnetization) current becomes too small to really count.

2.9 STEADY-STATE SHORT-CIRCUIT MODE/LAB 2.2

The steady-state short circuit refers to the already short-circuited secondary situation $V_2' = 0$ of transformer under sinusoidal input voltage.

From Equation 2.82 we obtain

$$\underline{I}_{1sc}\underline{Z}_1 - \underline{V}_1 = \underline{V}_{e1};\quad \underline{V}_{e1} = -\underline{Z}_{1m}\underline{I}_{01sc} = \underline{I}_{2sc}'\underline{Z}_2';\quad \underline{I}_{01sc} = \underline{I}_{1sc} + \underline{I}_{2sc}' \tag{2.94}$$

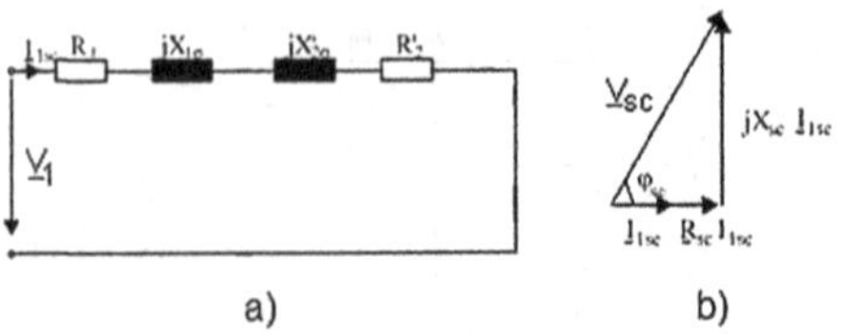

FIGURE 2.30 Steady state of short-circuited transformer. (a) Equivalent circuit and (b) phasor diagram.

Now, if we consider $\underline{I}_{01sc} \approx 0$, because I_{1sc} and $I'_{2sc} \gg I_{10}$, from Equation 2.94 we obtain

$$\underline{I}_{1sc} \approx -\underline{I}'_{2sc}; \quad \underline{I}_{1sc} = \frac{\underline{V}_1}{\underline{Z}_{sc}}; \quad \underline{Z}_{sc} = \underline{Z}_1 + \underline{Z}'_2 \tag{2.95}$$

So the transformer (with reduced secondary) degenerates into its short-circuit impedance, Z_{sc} (Figure 2.30).

As the short-circuit impedance is small, the steady-state short-circuit current is 10–30 times larger than the rated current.

So we should not supply the short-circuited transformer with the rated voltage, but at a much lower voltage, such that not to surpass rated current, I_{1n}.

In fact, the rated short-circuit voltage, V_{1scn}, is defined as the low voltage for which, under short circuit, the transformer absorbs rated current, $I_{1n} \approx I'_{2n}$.

So (from Equation 2.94),

$$V_{1nsc} = \left|Z_{sc}\right| I_{1n} \approx \left(0.03 - 0.12\right) V_{1n} \tag{2.96}$$

Under short-circuit tests we again use the arrangement in Figure 2.28, below 12% rated voltage (but with the secondary shortcircuited) and up to rated primary current I_{1n}!

Again, we measure power, P_{sc}, current, I_{1sc}, and voltage, V_{1sc}, to calculate the short-circuit impedance (from Equation 2.95), $|Z_{sc}|$, resistance, R_{sc}, and reactance, X_{sc}:

$$R_{sc} = \frac{P_{sc}}{I_{1sc}^2}; \quad X_{sc} = \sqrt{\frac{V_{1sc}^2}{I_{1sc}^2} - R_{sc}^2} \tag{2.97}$$

And power factor, $\cos \varphi_{sc}$, is

$$0.3 > \cos \varphi_{sc} = \frac{R_{sc}}{\left|\underline{Z}_{sc}\right|} > \cos \varphi_0 \tag{2.98}$$

The power factor at short circuit is essential, when paralleling transformers, to avoid circulating currents between them.

Example 2.3: Transformer Under Short Circuit

For the transformer in Example 2.2, the rated short-circuit voltage, $V_{1nsc} = 0.04\,V_{1n}$, while the active power measured at rated current, I_{1n} is $P_{scn} = 100$ kW. Calculate Z_{sc}, R_{sc}, X_{sc}, $\cos\varphi_{sc}$, R_2', X_2'.

Solution:

From Equation 2.96, at rated current, the short-circuit impedance, $|Z_{sc}|$ is

$$|\underline{Z}_{sc}| = \frac{V_{1scn}}{I_{1n}} = \frac{0.04 \cdot 35{,}000\ \text{V}}{285.7\ \text{A}} = \frac{1400}{285.7} = 4.9\ \Omega$$

The short-circuit resistance, R_{sc} (Equation 2.97) is

$$R_{sc} = \frac{P_{scn}}{I_{1n}^2} = \frac{100 \cdot 10^3}{285.7^2} = 1.225\ \Omega$$

As $R_1 = 0.612\ \Omega$ (from Example 2.2), $R_2' = R_{sc} - R_1 = 1.225 - 0.612 = 0.613\ \Omega \approx R_1$, as expected.

Now, the power factor at short circuit is (Equation 2.98)

$$\cos_{\varphi_{sc}} = \frac{R_{sc}}{Z_{sc}} = \frac{1.225}{4.9} = 0.25$$

The short-circuit reactance, X_{sc}, is

$$X_{sc} = Z_{sc}\sin_{\varphi_{sc}} = 4.9 \cdot 0.968 = 4.744\ \Omega$$

With $X_{1l} = 2.5\ \Omega$, $X_{2l}' = X_{sc} - X_{1l} = 4.744 - 2.5 = 2.244\ \Omega$, not far away from X_{1l}.

In general, in measurements, we may not separate R_1 from R_2' and X_{1l} from X_{2l}'; so $R_1 \approx R_2'$ and $X_{1l} \approx X_{2l}'$ in measured R_{sc} and X_{sc}. Now, the tests may by repeated at voltages up to V_{1nsc} such that the current is $(0.25, 0.5, 0.75, 1.0) \cdot I_{1n}$, and for each current value, R_{sc} and X_{sc} are calculated. Finally, the average values of R_{sc} and X_{sc} are taken as the final results. The a.c. winding losses at rated current (and short-circuit rated voltage tests) are the same as those at rated load. So, at given load,

$$p_{copper} = P_{scn}\frac{I_1^2}{I_{1n}^2} \tag{2.99}$$

The short-circuit test, up to rated current is useful to determine R_{sc} and copper winding a.c. losses P_{copper} for given load (current: I_1).

The no-load and short-circuit tests serve thus to separate iron and copper losses, to be used in assessing efficiency under load.

2.10 SINGLE-PHASE TRANSFORMERS: STEADY-STATE OPERATION ON LOAD/LAB 2.3

On (under) load the transformer, supplied at given (rated) voltage, delivers power to an a.c. impedance load in the secondary. The energy transfer is done electromagnetically, with iron and copper losses.

Active and reactive power break down under load is shown in Figure 2.31.

The operation under load is characterized by efficiency η versus load factor, $K_s = I_1/I_{1n}$ and load voltage regulation (drop) for given power factor, $\cos\varphi_2$ (1, 0.8, 0.6).

The efficiency η is defined as

$$\eta = \frac{\text{output active power}}{\text{output active power}} = \frac{P_2}{P_2 + p_{\text{copper}} + p_{\text{iron}}} \tag{2.100}$$

Based on no-load and short-circuit mode considerations, Equation 2.99 becomes

$$\eta = \frac{V_2' I_2' \cos\varphi_2}{V_2' I_2' \cos\varphi_2 + p_{0n} + p_{sc}\dfrac{I_1^2}{I_{1n}^2}} \tag{2.101}$$

Now, if $I_2' \approx I_1$ (which means $I_1/I_{1n} > 0.2$) and, with load factor, $K_s = I_1/I_{1n}$, Equation 2.101 becomes

$$\eta = \frac{S_{2n} K_s \cos\varphi_2}{S_{2n} K_s \cos\varphi_2 + p_{0n} + p_{scn} K_s^2} \tag{2.102}$$

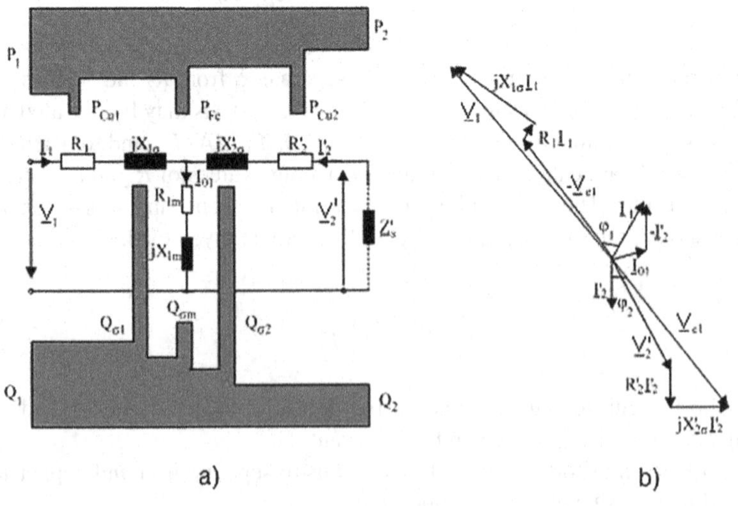

FIGURE 2.31 (a) Active and reactive power balance and (b) phasor diagram under load.

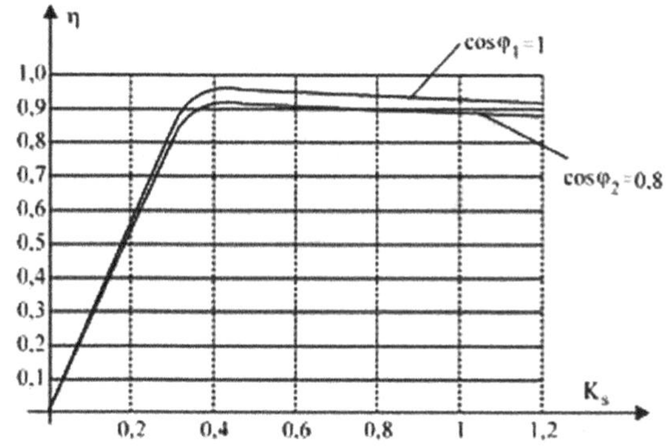

FIGURE 2.32 Efficiency versus load factor.

The maximum efficiency is obtained for $\partial\eta/\partial K_s = 0$, for K_{sk}:

$$K_{sk} = \sqrt{\frac{p_{0n}}{p_{scn}}} \tag{2.103}$$

Consequently, for the critical (maximum efficiency) load factor, copper and core losses are equal to each other.

Note: A transformer with full-load operation most of the time has to be designed with K_{sk} close to unity while one with discontinuous and partial load operation (distribution transformer in schools, public institutions, or companies with large human resources, etc.) should be designed with K_{sk} around 0.5. Typical curves of efficiency versus load factor, $K_s = I_1/I_{1n} \approx P_2/P_{2n}$, are shown in Figure 2.32.

For $I_1/I_{1n} > 0.2$, the magnetization current, I_{01}, may be neglected and thus $I_2 = -I_1$ and thus the equation (Equation 2.82) becomes

$$-\underline{V}_2' = \underline{V}_1 - \underline{I}_1\underline{Z}_{sc} \tag{2.104}$$

Equation 2.104 leads to the simple equivalent scheme in Figure 2.33a and its corresponding phasor diagram (Figure 2.33b).

The secondary voltage variation, ΔV, under load is

$$\Delta V = V_1 - V_2' \approx \overline{AB} \tag{2.105}$$

From Figure 2.33b, $\overline{AB}$ is

$$\Delta V \approx I_1 R_{sc}\cos\varphi_2 + I_1 X_{sc}\sin\varphi_2 = K_s V_{1scn}\cos\left(\varphi_{sc} - \varphi_2\right) \tag{2.106}$$

As, in general, $\varphi_{sc} = (80\text{–}85)^\circ$, the voltage regulation (drop) may be positive if $\varphi_2 \geq 0$ (resistive–inductive load) or negative for $\varphi_2 < \varphi_{sc} - 90^\circ$ (Figure 2.34), in general, for resistive–capacitive load such as overhead long power transmission lines without load.

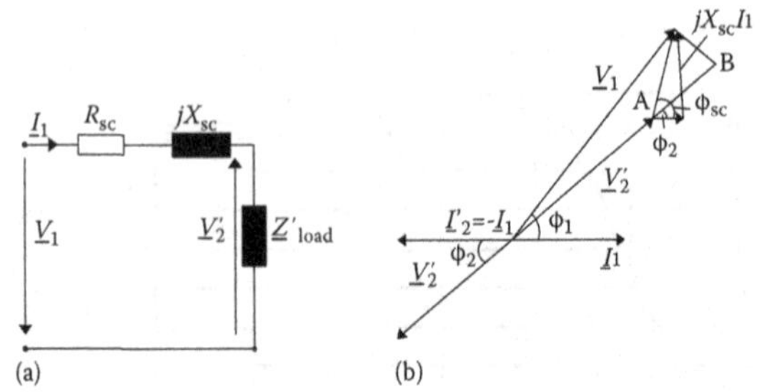

FIGURE 2.33 (a) Simplified equivalent circuit and (b) phasor diagram with neglected magnetization current ($I_{01} = 0$).

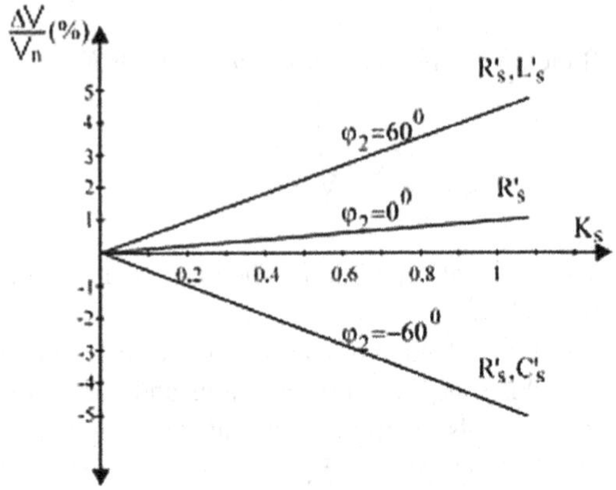

FIGURE 2.34 Voltage regulation versus load factor, $K_s = I_1/I_{1n}$.

Example 2.4: Transformer on Load

For the transformer in Examples 2.3 and 2.4, with p_{on} =22 kW, S_n =10 MVA, V_{1nsc} =1400 V, cos φ_{sc} = 0.25 (with $R_{sc} \approx 1.20\ \Omega$, $X_{sc} = 4.7\ \Omega$), and V_{1n}/V_{2n} =35 kV/6 kV, calculate the rated efficiency, voltage regulation and secondary voltage at pure resistive, inductive, and capacitive rated load.

Solution:

Pure resistive, inductive, and capacitive loads means φ_2 = 0, +90°, −90°

The efficiency for rated load, ($K_s = K_{sn} = 1$), and resistive load, ($\varphi_2 = 0$), is, from Equation 2.102,

$$\eta = \frac{10 \cdot 10^6 \cdot 10 \cdot 10}{10 \cdot 10^6 \cdot 10 \cdot 10 + 22 \cdot 10^3 + 100 \cdot 10^3} = 0.9879$$

For $\varphi_2 = \pm 90°$ (pure inductive and capacitive load), the active power output, P_2, is zero and thus the efficiency is zero.

The voltage (for rated current) regulation, ΔV_n, (from Equation 2.106) is

$$\begin{aligned}\Delta V &= \left(V_1 - V_2'\right) = V_{1nsc} \cdot \cos\left(\varphi_{sc} - \varphi_2\right) \\ &= 1400 \cdot 0.25 = +350 \text{ V}; \quad \varphi_2 = 0 \\ &= 1400 \cdot \cos\left(85° - 90°\right) = +1394 \text{ V}; \quad \varphi_2 = 90° \\ &= 1400 \cdot \cos\left(85° - \left(-90°\right)\right) = -1394 \text{ V}; \quad \varphi_2 = -90°\end{aligned}$$

The secondary voltage is

$$V_{2'} = \left(V_1 - \Delta V_n\right) \cdot \frac{V_{2n}}{V_{1n}} = 5940 \text{ V}, 5761 \text{ V}, 6239 \text{ V}$$

The no-load secondary voltage is 6000 V.

Example 2.5: Dual Output Voltage one-phase Residential Transformers

In the United States and some other countries, dual 120-V and 240-V output voltages are used for residential power supplies (Figure 2.35).

Let us consider the case of a 15-kVA, 240/240/120-V, 60-Hz transformer. The load 1 absorbs 1.5 kW at 120 V and 0.867 PF leading, while load 3 absorbs 4 kW at 0.867 PF lagging. What is the maximum admissible load for the load 2 at unity PF and 120 V and the currents, I_1, I_2, I_3 (Figure 2.35)?

Solution:

The current in load 2 is approximately the RMS value of I_1 (for load 1), that is,

$$I_1 \approx \frac{P_1}{V_1 \cdot \cos\varphi} = \frac{1500}{120 \cdot 0.867} = 14.41 \text{ A}$$

$$\underline{I_1} = I_1\left(\cos\varphi - j\sin\varphi\right) = 14.41 \cdot \left(0.867 - j0.5\right)$$

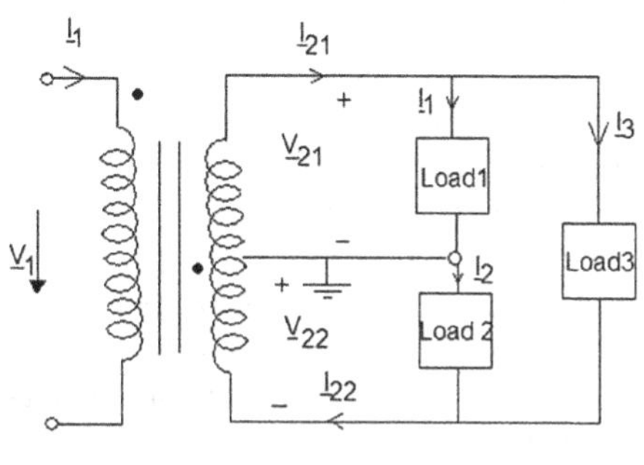

FIGURE 2.35 Dual voltage single-phase residential transformer.

Similarly,

$$\underline{I_3} \approx \frac{P_3}{V_3 \cdot \cos\varphi} = \frac{4000}{240 \cdot 0.867} = 19.22 \text{ A}$$

$$\underline{I_3} = I_3(\cos\varphi + j\sin\varphi) = 19.22 \cdot (0.867 + j0.5)$$

We still cannot calculate $\underline{I_2}$.

For the 120-V output circuit load 2, the power factor is unity, so only active power P_{2max} is delivered.

Now, the apparent total power should be equal to $S_n = 5$ kVA:

$$S_n = \sqrt{(P_1 + P_{2max} + P_3)^2 + (Q_1 + Q_3)^2} = 10.000 \text{ VA}$$

where

$$Q_1 = P_1 \cdot \tan\varphi_1 = 1500/\sqrt{3} \text{ VA}; P_1 = 1500 \text{ W}$$

$$Q_3 = P_3 \cdot \tan\varphi_3 = 4000/\sqrt{3} \text{ VA}; P_3 = 4000 \text{ W}$$

Finally,

$$P_{2max} = 10^3 \cdot \sqrt{15^2 - (1.5+4)^2/3} - 5.5 \cdot 10^3 = 9.160 \text{ kW}$$

So, the current in load 2, I_{2max} (purely active), is

$$I_2 = I_{2max} = \frac{P_{2max}}{120 \text{ V} \cdot \cos\varphi_2} = \frac{9.160}{120 \cdot 1} = 76.33 \text{ A}$$

We end the one-phase transformer steady-state analysis here, to proceed with the three-phase transformer steady state, by starting with the three-phase connections and their order number.

2.11 THREE-PHASE TRANSFORMERS: PHASE CONNECTIONS

The phase connections pertaining to either primary or secondary (or tertiary) three-phase windings may be

- Star (Y) connections (Figure 2.36a)
- Delta (Δ) connections (Figure 2.36b)
- Zig-zag (Z) connections (Figure 2.36c)

For the star connection, the beginnings or the ends of the three phases are connected to a neutral point (o) which may be available or not.

Capital letters (ABC) are used here to denote high voltages side and lower case letters (abc) for lower voltage side phase beginnings and XYZ and xyz for phase ends. (Some national standards use other denotations).

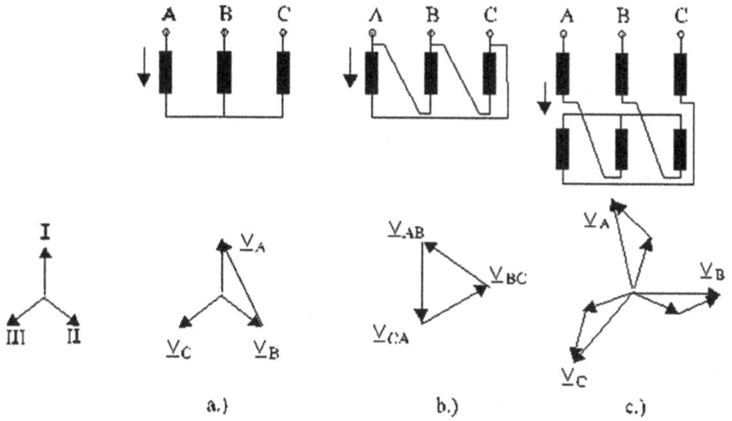

FIGURE 2.36 Three-phase connections: (a) Star (Y), (b) delta (Δ), and (c) zig-zag (Z).

For the star connection (Figure 2.36a), the line voltage $V_{AB} = V_A - V_B$; similarly, $V_{BC} = V_B - V_C$ and $V_{CA} = V_C - V_A$. For symmetric line voltages (equal amplitudes and 120° phase shift), the line voltage V_{lY} is

$$V_{lY} = V_{phY} \cdot \sqrt{3}; \quad I_{lY} = I_{phY} \tag{2.107}$$

For the delta (Δ) connection (Figure 2.36b), the line currents are: $I_{AB} = I_A - I_B$; $I_{BC} = I_B - I_C$; $I_{CA} = I_C - I_A$; the line and phase voltages are equal to each other. For symmetrical phase currents (same amplitude and 120° phase shift), the line current, $I_{l\Delta}$, is

$$I_{l\Delta} = I_{ph\Delta} \cdot \sqrt{3}; \quad V_{l\Delta} = V_{ph\Delta} \tag{2.108}$$

Now, the apparent three-phase power, S, has the same formula for star and delta connections (for symmetric voltages and currents):

$$S_{3ph} = 3 \cdot V_{ph} \cdot I_{ph} = \sqrt{3} \cdot V_l \cdot I_l \tag{2.109}$$

RMS values are considered (Equations 2.106 through 2.108) for all sinusoidal variables. It should be noticed that the Y connection implies zero current summation but the Δ connection allows for a circulation (homopolar, or 3 ν harmonic component) current between phases, without affecting the external power source.

This occurs also when the load is unbalanced in the secondary of the transformer, but at fundamental frequency.

The Z connection (Figure 2.36c) used for the transformer secondary has each phase made of two half-phases. Half-phases of different transformer limbs are connected in opposite series so that each phase voltage is made of the vectorial difference between the voltages of the two component half-phases and thus $V_a = V_{sb} \cdot \sqrt{3}$ (instead of $2V_{sb}$ in the case of Y or Δ connection).

To obtain the same no-load secondary voltage, the Z connections require $2/\sqrt{3}$ times more turns (and more copper losses) than the star or delta connections. But for a Z connection, with an available neutral point, single-phase load is possible without asymmetrization of the unloaded phases voltages, because the neutral circulating current produces emfs in phase on all half-phases and they cancel each other in each phase.

Connection schemes combine $Y(Y_0)$, Δ in the primary or secondary or Z (in secondary) to obtain combinations such as $Y\Delta$, Yy_0, Δy_0, Δy, Yz. A scheme (combination) of connections is characterized by the phase-shift angle, β, between the primary and the secondary homologous line voltages (V_{AB}, V_{ab}), (V_{BC}, V_{bc}), with the secondary line voltage lagging the primary line voltage.

The connection scheme order $n = \beta°/30°$ and is an integer.

It may be demonstrated that when the primary and secondary connections are the same (Yy or $\Delta\Delta$), n is an even number (0, 2, 4, 6, 8, 10), and, when they are different (Δy), n is an odd number (1, 3, 5, 7, 9, 11). Let us consider the connection scheme, Δy, in Figure 2.37 and try to find its order, n.

First we choose a positive direction for $\int \underline{dl} \cdot \underline{E}$ along the transformer columns I, II, and III. We also suppose that the primary line voltages are symmetric. The emfs produced by the primary and secondary windings placed around the same transformer limb are in phase if they are coiled and travel in the same direction, because they are flown by the same flux.

The primary line voltage, say V_{AB}, for the Δ connection is obtained from the emf of phase A placed on a single limb, while V_{ab} is obtained by subtracting emfs from windings on two limbs (Figure 2.37).

The lagging angle, β, between V_{ab} and V_{AB} is 150°, so $n = 150/30 = 5$. It is evident why n is odd: the two line voltages are composed differently. To connect transformers in parallel and avoid circulating current, the connection scheme order has to be the same.

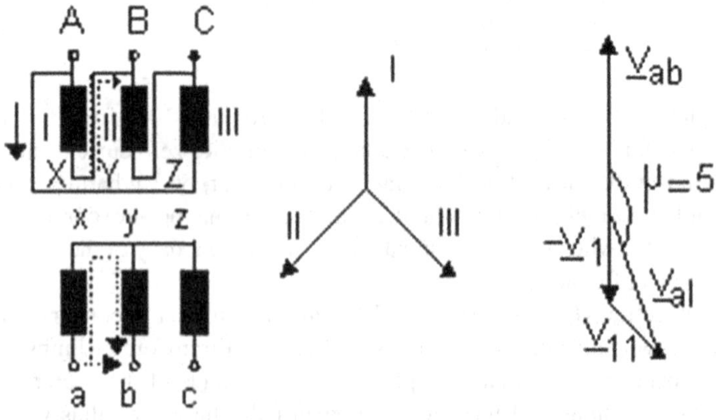

FIGURE 2.37 Connection scheme, Δy5.

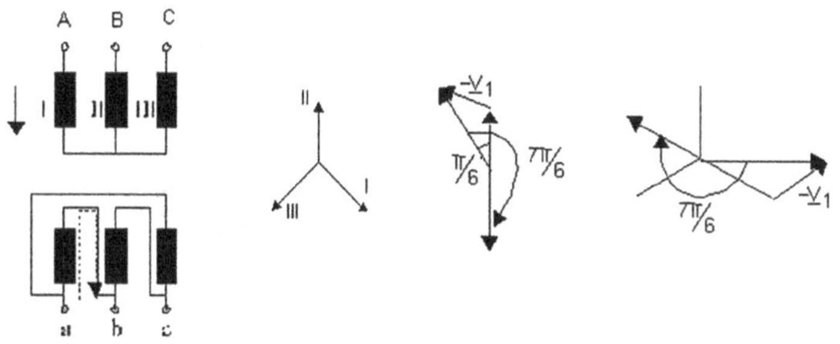

FIGURE 2.38 Connection scheme, YΔ7.

Example 2.6

Connection Scheme with Given Order, n (YΔ7)

Solution:

Building a certain connection scheme is the typical problem at the design stage. First we draw the phase windings, connected only in the primary but not the secondary, with marked beginnings and ends (Figure 2.38).

By setting the position of the three limbs (emfs) we may first obtain V_{AB} from two-phase voltages, V_A and V_B, and then draw V_{ab} phasor lagging V_{AB} by $7 \times 30° = 210°$.

It is now evident that V_{ab}, obtained from a single-limb winding (Δ connection), corresponds to limb I traveled in the negative (vertical) direction. This is how we can place the letters a and b, and then c is completed univoquely. We should mention that n should be valid for all three line voltage pairs. We should then verify it, for one more pair (V_{BC}, V_{bc}).

It may be demonstrated, however, that *n* is met for all three line voltage pairs, after being set for one pair, if the order of phases is A.B.C in the primary and a.b.c or b.c.a or c.a.b in the secondary.

2.12 PARTICULARS OF THREE-PHASE TRANSFORMERS ON NO LOAD

The no-load steady-state operation of three-phase transformers is heavily influenced by the core configuration (3 limb, 5 limb, or 3 × 1 phase) and by the connection scheme.

We treat here only the asymmetry of the no-load phase currents produced by the three-limb core and the no-load current waveform in the presence of magnetic saturation in Yy0 connection scheme.

2.12.1 No-Load Current Asymmetry

The no-load current asymmetry occurs mainly by three-limb/three-phase transformers because the magnetic reluctances corresponding to magnetic fluxes in phases A and C (Figure 2.39a) are larger than that in phase B. Not so is the case in 5-limb cores or in 3 × one-phase transformer groups.

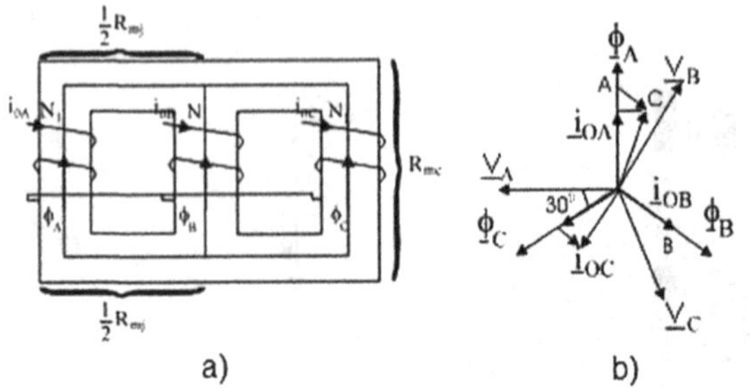

FIGURE 2.39 Three-limb transformer on no load: (a) The core and (b) phasors at no load.

If we neglect the voltage drops along the primary impedance, Z_1, for sinusoidal voltages, the emfs and thus the magnetic fluxes in the three limbs, Φ_A, Φ_B and Φ_C, are sinusoidal in time. So, $\Phi_A + \Phi_B + \Phi_C = 0$ and they are symmetric (equal in amplitude, with 120° phase shift) for symmetric voltages, V_A, V_B, V_C. Let us consider the sum of the currents to be zero, that is, a star connection ($I_{0A} + I_{0B} + I_{0C} = 0$).

Applying Kirchoff's second theorem to the magnetic circuits in Figure 2.39a we obtain

$$N_1 \cdot \left(\underline{I}_{0A} - \underline{I}_{0B}\right) = \left(R_{mc} + R_{my}\right) \cdot \underline{\Phi}_A - R_{mc} \cdot \underline{\Phi}_B \tag{2.110}$$

$$N_1 \cdot \left(\underline{I}_{0A} - \underline{I}_{0C}\right) = \left(R_{mc} + R_{my}\right) \cdot \left(\underline{\Phi}_A - \underline{\Phi}_C\right) \tag{2.111}$$

With the summation of core fluxes and currents being zero, we may find the currents:

$$\begin{aligned}
\underline{I}_{0A} &= \left(\frac{R_{mc} + R_{my}}{N_1}\right) \cdot \underline{\Phi}_A + \frac{1}{3} \cdot \frac{R_{my}}{N_1} \cdot \underline{\Phi}_B \\
\underline{I}_{0B} &= \left(R_{mc} + \frac{R_{my}}{3}\right) \cdot \frac{\underline{\Phi}_B}{N_1} \\
\underline{I}_{0C} &= \left(\frac{R_{mc} + R_{my}}{N_1}\right) \cdot \underline{\Phi}_C + \frac{1}{3} \cdot \frac{R_{my}}{N_1} \cdot \underline{\Phi}_B
\end{aligned} \tag{2.112}$$

So, only the current in phase B, placed in the middle column, is in phase with the corresponding flux, ϕ_B, and it is the smallest of all the three currents ($I_{0A} = I_{0C} > I_{0B}$). The phasor diagram of Equation 2.112—Figure 2.39b—with voltage phasors (V_A,V_B,V_C) at 90° ahead of ϕ_A, ϕ_B, ϕ_C illustrates the fact that only I_{0B} is purely reactive (magnetization) while I_{0A} and I_{0C} have an active component—one positive and the other negative. This simply means a circulation of active power between phases A and C.

A few consequences of this situation are as follows:

- For three-limb core, three-phase transformers, the active power, P_0, related to core losses, is calculated by algebraically adding up the active power on all phases (it is negative on one lateral limb).

- The standardized no-load current, $I_{10n} = (I_{0A} + I_{0B} + I_{0C})/3$.
- So far, magnetic saturation was neglected. It will come up next.

2.12.2 Y Primary Connection for the Three-Limb Core

The Y connection is generally used in the primary, for protection through neutral-point voltage monitoring. Because the summation of currents is zero, in the presence of magnetic saturation (which is notable in practical transformers), the third time harmonic of no-load current can not occur. But it shows up in the core flux, which departs from a sinusoidal waveform, characteristic to the saturated single-phase transformer on no load, and thus produces a third order ($3f_1$-frequency) flux harmonic in the core flux (Figure 2.40a). The third harmonic means a $3 \cdot 120° = 360°$ phase shift between phases.

The third time harmonic of magnetic flux may not close path through the three-limb core because it has to observe Gauss's law (zero flux on a closed area around the central limb). Consequently, the magnetic flux at $3f_1$ frequency (150, 180 Hz for power grid supply) closes through the oil tank walls, inducing notable eddy current losses. To reduce these additional rather severe no-load losses, a tertiary winding with small rating, with series phase connection, is added (Figure 2.40b).

The current induced in the short-circuited winding severely reduces the resultant mmf at $3f_1$ frequency, and thus the eddy current losses in the oil tank are practically zero (for dry transformers there is no oil tank). The electromagnetic interference of the $3f_1$ flux lines in the air is reduced. It is important to note that, for the 3 × one-phase transformer group, the third time harmonic of flux closes paths in the core and thus no additional eddy current losses occur in the oil thank. The situation is almost similar in the 5-limb three-phase transformers.

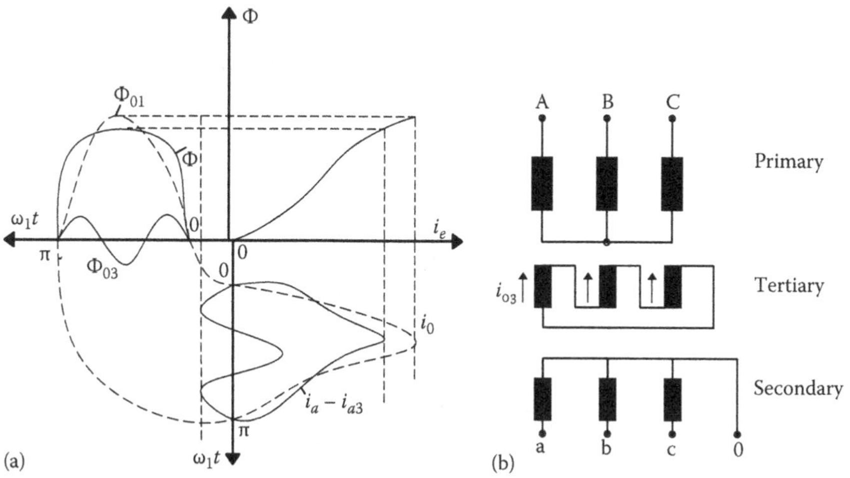

FIGURE 2.40 (a) Third flux time harmonics for the no load of a three-limb three-phase transformer with Yy0 connection and (b) tertiary winding.

The tertiary series-connected winding will also prove useful for unbalanced load, to destroy rated frequency zero sequence (homopolar) secondary current field for Y primary connections (such as in Yy0 connection scheme).

2.13 GENERAL EQUATIONS OF THREE-PHASE TRANSFORMERS

For the 3×one-phase transformer group, we only have to put together the equations of each of the three one-phase transformers as there is no magnetic coupling between them (they have separate cores).

But as most transformers have three-limb cores, with a few with 5-limb cores, there is scope to analyze the coupling between phases, at least to measure correctly the three-phase transformer parameters and study the transients for the general (any) case.

Three-phase three-limb core transformer phase voltages and currents.

Figure 2.41 shows the three-limb core three-phase transformer with its phase voltages and currents.

Again, the primary is sink and the secondary is source. Consequently, in phase coordinates,

$$\frac{d\Psi_{A,B,C}}{dt} = V_{A,B,C} - R_1 \cdot i_{A,B,C}; \quad \frac{d\Psi_{a,b,c}}{dt} = -V_{a,b,c} - R_2 \cdot i_{a,b,c} \tag{2.113}$$

There is magnetic coupling from each of the six windings. Let us consider that interphase coupling occurs only through main (core) flux. So,

$$\begin{aligned} \Psi_A &= L_{1l} \cdot i_A + L_{AAm} \cdot i_A + L_{ABm} \cdot i_B \\ &\quad + L_{ACm} \cdot i_C + L_{Aam} \cdot i_a + L_{Abm} \cdot i_b + L_{Acm} \cdot i_c \\ \Psi_a &= L_{aAm} \cdot i_A + L_{aBm} \cdot i_B \\ &\quad + L_{aCm} \cdot i_C + L_{2l} \cdot i_a + L_{aam} \cdot i_a + L_{abm} \cdot i_b + L_{acm} \cdot i_c \end{aligned} \tag{2.114}$$

Similar expressions for phases, B, b, C, and c can be obtained. And thus,

$$\left|\Psi_{A,B,C}\right| = \left|L_{ABCabc}\right| \cdot \left|i_{ABCabc}\right|^T \tag{2.115}$$

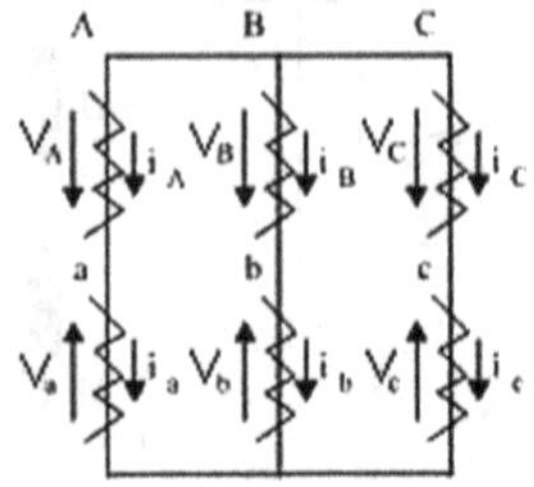

FIGURE 2.41 Three-phase three-limb core transformer phase voltages and currents.

L_{ABCabc} is a 6 × 6 matrix with some of the terms equal to each other such as $L_{ABm} = L_{BCm}$, $L_{abm} = L_{bcm}$, $L_{AAm} = L_{CCm}$, $L_{aam} = L_{ccm}$, etc.

Adding the consumer equations and considering the primary voltages as inputs and flux (Ψ_{ABCabc}) as the variable vector and (i_{ABCabc}) as the dummy variable vector (to be eliminated through Equation 2.114), the transformer equations can be solved by numerical methods for any steady state or transient (with symmetric or asymmetric input voltages and balanced or unbalanced load). However, the transformer parameters have to be known, either from design or through measurements.

2.13.1 Inductance Measurement/Lab 2.4

Let us first consider that $i_A + i_B + i_C = 0$ and $i_a + i_b + i_c = 0$ and also, approximately, that $L_{ABm} = L_{ACm}$, $L_{abm} = L_{acm}$, and $L_{Abm} = L_{aBm}$. In this case, from Equation 2.113, we get

$$\begin{aligned}\Psi_A &= L_{1l} \cdot i_A + \left(L_{AAm} - L_{ABm}\right) \cdot i_A + \left(L_{aam} - L_{abm}\right) \cdot i_a \\ \Psi_a &= L_{2l} \cdot i_a + \left(L_{aam} - L_{abm}\right) \cdot i_a + \left(L_{Aam} - L_{Abm}\right) \cdot i_A\end{aligned} \tag{2.116}$$

Let us call

$$L_{1mc} = L_{AAm} - L_{ABm}; \quad L_{2mc} = L_{aam} - L_{Abm}; \quad L_{12mc} = L_{Aam} - L_{Abm} \tag{2.117}$$

The cyclic inductances, L_{1mc}, L_{2mc}, and L_{12mc}, are valid when all currents exist and current summation is zero. In this particular case, the influence of other phases in one-phase flux is "hidden"' in the cyclic inductances. To measure all inductances in Equation 2.116 we have to supply only one phase, with the other phases open and measure the current on the supplied phase, i_{A0} (RMS), and the voltages on all phases, V_{A0}, V_{B0}, V_{C0}, V_{a0}, V_{b0}, V_{c0} (RMS):

$$\begin{aligned}L_{1l} + L_{AAm} &\approx \frac{V_{A0}}{\omega_1 \cdot I_{A0}}; \quad L_{ABm} = \frac{-V_{B0}}{\omega_1 \cdot I_{A0}}; \quad L_{ACm} = \frac{-V_{C0}}{\omega_1 \cdot I_{A0}} \\ L_{Aam} &= \frac{V_{a0}}{\omega_1 \cdot I_{A0}}; \quad L_{Abm} = \frac{-V_{b0}}{\omega_1 \cdot I_{A0}}; \quad L_{Acm} = \frac{-V_{c0}}{\omega_1 \cdot I_{A0}}\end{aligned} \tag{2.118}$$

To find the cyclic inductances, L_{1mc}, L_{2mc}, and L_{12mc}, directly, we supply first all primary phases, with secondary phases open, and then do the same test with all secondary phases active and the primary phases open:

$$L_{1l} + L_{1mc} = \frac{V_{A03}}{\omega_1 \cdot I_{A03}}; \quad L_{12mc} = \frac{V_{a03}}{\omega_1 \cdot I_{A03}}; \quad L_{2l} + L_{2mc} = \frac{V_{a03}}{\omega_1 \cdot I_{a03}} \tag{2.119}$$

For balanced input voltages and balanced load, in steady state and transients, the single-phase transformer equations (with cyclic inductances) may be applied safely.

However, under unbalanced input voltages or loads in the secondary, the general equations of three-phase transformer, with measured inductances are to be used and solved numerically. Magnetic saturation has to be considered, especially if the input voltage and (or) the frequency vary as in power electronics associated applications (oil rings underground pump drives).

For steady-state unbalanced load, the method of symmetrical components proved to be very practical and intuitive.

2.14 UNBALANCED LOAD STEADY STATE IN THREE-PHASE TRANSFORMERS/LAB 2.5

In three-phase power distribution systems, often, single-phase consumers are served. For electric a.c. locomotives, induction furnaces, and institutional and residential consumers, if care is not exercised at three-phase feeding transformer connections scheme, corroborated with magnetic core configuration, the phase voltages of secondary and primary may become unbalanced. First, the neutral potential departs from earth (ground) potential; second, the less-loaded secondary phase voltage may be large and thus endanger the remaining voltage-sensitive consumers on the respective phase. The method of symmetrical components proves very instrumental to treat such situations.

In essence, the voltages and currents in the primary and secondary are decomposed into direct, inverse, and zero sequences (+, −, 0) (Figure 2.42):

$$\begin{vmatrix} \underline{I}_{a+} \\ \underline{I}_{a-} \\ \underline{I}_{a0} \end{vmatrix} = \frac{1}{3} \cdot \begin{vmatrix} 1 & a & a^2 \\ 1 & a^2 & a \\ 1 & 1 & 1 \end{vmatrix} \cdot \begin{vmatrix} \underline{I}_a \\ \underline{I}_b \\ \underline{I}_c \end{vmatrix}; \quad \begin{aligned} &\underline{I}_{b+} = \underline{I}_{a+} \cdot a^2; \quad \underline{I}_{c+} = \underline{I}_{a+} \cdot a \\ &\underline{I}_{b-} = \underline{I}_{a-} \cdot a; \quad \underline{I}_{c-} = \underline{I}_{a-} \cdot a^2 \\ &\underline{I}_{b0} = \underline{I}_{c0} = \underline{I}_0; \quad a = e^{j \cdot \frac{2\pi}{3}} \end{aligned} \tag{2.120}$$

The inverse transformation is

$$\begin{vmatrix} \underline{I}_a \\ \underline{I}_b \\ \underline{I}_c \end{vmatrix} = \begin{vmatrix} 1 & 1 & 1 \\ a^2 & a & 1 \\ a & a^2 & 1 \end{vmatrix} \cdot \begin{vmatrix} \underline{I}_{a+} \\ \underline{I}_{a-} \\ \underline{I}_{a0} \end{vmatrix} \tag{2.121}$$

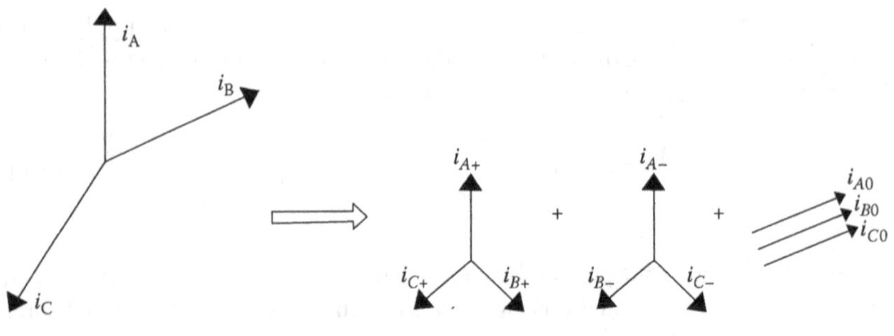

FIGURE 2.42 Symmetrical components superposition.

If the magnetization current +/– components are neglected:

$$
\begin{aligned}
I_{A+,B+,C+} &= -\frac{N_2}{N_1} \cdot i_{a+,b+,c+} \\
I_{A-,B-,C-} &= -\frac{N_2}{N_1} \cdot i_{a-,b-,c-}
\end{aligned}
\tag{2.122}
$$

The direct and inverse components of secondary currents—due to unbalanced load—are reflected in the primary as in Equation 2.122 because the transformer is insensitive to the sequence (order) of phases; it behaves basically through the short-circuit impedance, $Z_{sc} = Z_+ = Z_-$, as known from previous paragraphs. The secondary zero sequence current component, $I_{0a} = I_{0b} = I_{0c}$—due to single-phase unbalanced loads—typical for Y_0, Δ, Z_0 secondary connection cannot occur in the Y-connected primary unless a tertiary series connected winding is added. If the zero sequence current can occur in the primary, the transformer behavior towards it is the same as for +/– sequences. So, for example, in such a case a single-phase load (in the secondary) means no unbalance in the phase voltages in the primary and secondary, and no neutral point potential deviation (such as in x_0, y_0 connection).

On the contrary, if the zero sequence current cannot occur in the primary, then, from its point of view, the transformer is open in the primary and the entire secondary zero sequence mmf, $N_2 \cdot i_{a0} = N_2 \cdot i_{b0} = N_2 \cdot i_{c0}$, produces zero sequence fluxes in the cores. The level of this flux, Φ_{a0}, depends on the core configuration of the three-phase transformer (Figure 2.43a through c).

So, for the three-limb core (Figure 2.43a) the uncompensated zero sequence secondary produces close flux paths through the oil tank walls. The transformer has a reasonably low zero sequence impedance, Z_0, due to the predominant air zone of flux lines.

On the contrary, for 3 × one-phase transformer groups, the uncompensated secondary zero sequence mmfs close flux paths entirely through iron and the

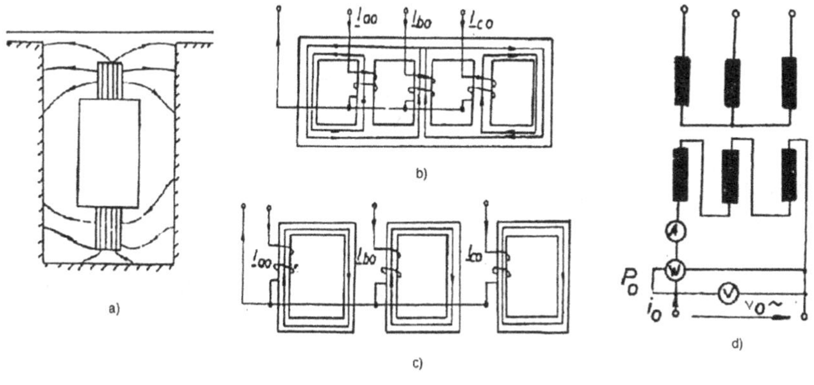

FIGURE 2.43 Zero sequence secondary mmf uncompensated flux lines. (a) three-limb transformer, (b) 5-limb transformer, (c) 3 × one-phase transformer group, and (d) zero sequence impedance measurement.

transformer impedance is the no-load impedance, $Z_0 = Z_{\text{no load}} = R_1 + R_{1m} + j(X_{1l} + X_{1m})$, treated in a previous Section.

The secondary neutral point potential moves by a value, V_{a0} (same for all phases):

$$V_{a0} = -I_{a0} \cdot Z_0 \tag{2.123}$$

The larger the zero sequence impedance for a given load unbalance degree (I_{a0}), the larger is the unbalancing of phase voltages (by $V_{a0} = V_{b0} = V_{c0}$).

So, if the load in the secondary has a single-phase notable component (y0—connection scheme) and the 3 × one-phase transformer group is used, a tertiary series winding is required if a Y connection is used in the primary. Not so is the case for Δ connection in the primary which allows for zero sequence currents, and thus $Z_0 \rightarrow Z_{sc}$. The three-limb transformer, instead, may operate with Yy0 connection scheme and limited zero sequence secondary current because $Z_{sc} < Z_0 \ll Z_{\text{no load}}$. For symmetrical input voltages, and neglecting the voltage on the short-circuit impedance, the voltages will equal the emfs. Consequently, for the Yy0 connection scheme (and no tertiary series winding), $V_{A-} = 0$ and thus the A, a phases voltages will be

$$\begin{aligned} \underline{V}_A &= \underline{V}_{A+} + \underline{V}_{A0} = -\left(\underline{V}_{ea+} - \underline{Z}_0 \cdot \underline{I}_{a0}\right) \cdot \frac{N_1}{N_2} \\ \underline{V}_a &= \underline{V}_{a+} + \underline{V}_{a0} = \underline{V}_{ea+} - \underline{Z}_0 \cdot \underline{I}_{a0} \end{aligned} \tag{2.124}$$

We should mention that the zero sequence impedance, Z_0 (Equation 2.123), should be measured (calculated) in the secondary (Figure 2.43d) with secondary phases supplied in series from a single-phase a.c. source, and the primary open. Now, the primary currents "reflect" only the + and – sequence secondary current (Yy0 without tertiary series winding):

$$\begin{aligned} \underline{I}_A &= \underline{I}_{A+} + \underline{I}_{A-} \approx -\frac{N_2}{N_1} \cdot \left(\underline{I}_{a+} + \underline{I}_{a-}\right) \\ \underline{I}_B &= a^2 \cdot \underline{I}_{A+} + a \cdot \underline{I}_{A-} \approx -\frac{N_2}{N_1} \cdot \left(a^2 \cdot \underline{I}_{a+} + a \cdot \underline{I}_{a-}\right) \\ \underline{I}_C &= a \cdot \underline{I}_{A+} + a^2 \cdot \underline{I}_{A-} \approx -\frac{N_2}{N_1} \cdot \left(a \cdot \underline{I}_{a+} + a^2 \cdot \underline{I}_{a-}\right) \end{aligned} \tag{2.125}$$

With Z_0 secondary connection, even for single-phase load, when large zero sequence currents occur, their mmfs produced by the half-phases cancel effects on each core limb and thus the zero sequence flux, $\phi_{a0} = \phi_{b0} = \phi_{c0} = 0\left(\underline{Z}_0 = 0\right)$.

The Yz0 connection scheme may be single-phase fully loaded without neutral point potential displacement from ground.

Example 2.7: Yy0 Connection with Pure Single-Phase Load

A three-phase transformer with Yy0 connection scheme and line voltages, V_{1nl}/V_{2nl} = 6000/380 V, is purely resistive loaded only on phase a (in the secondary) at I_a =

120 A ($I_b = I_c = 0$). The homopolar impedance is approximated by a reactance, X_0. Calculate the current distribution between the two phases and the neutral potential in the secondary, (V_{a0}) and primary (V_A) and the phase voltages in the secondary, V_b, V_c, for $X_0 = 1\ \Omega$.

Solution:

Making use of Equation 2.123, with $Z_0 = jX_0$,: $V_{ea+} = V_{2ln}/\sqrt{3} = 380/\sqrt{3} = 220$ V

The phase a current, I_a, is in phase with V_a.

From Equation 2.120 with

$$\underline{I}_b = \underline{I}_c = 0;\ \underline{I}_{a+} = \underline{I}_{a-} = \underline{I}_{a0} = \frac{\underline{I}_a}{3} = \frac{120}{3} = 40 \text{ A}$$

Now $N_2/N_1 = V_{2phn}/V_{1phn} = \dfrac{380}{6000} = 0.0633$, and, thus, from Equation 2.125,

$$\underline{I}_A = \underline{I}_{A+} + \underline{I}_{A-} = 2 \cdot \underline{I}_{A+} = -2\frac{N_2}{N_1} \cdot I_{a+} = -2 \cdot 0.0633 \cdot 40 = -5.064 \text{ A}$$

$$\underline{I}_B = a^2 \cdot \underline{I}_{A+} + a \cdot \underline{I}_{A-} = \left(a + a^2\right) \cdot \underline{I}_{A+};\quad \underline{I}_{A+} = \frac{-I_A}{2} = +2.532 \text{ A}$$

$$\underline{I}_C = a \cdot \underline{I}_{A+} + a^2 \cdot \underline{I}_{A-} = \left(a + a^2\right) \underline{I}_{A+} = +2.532 \text{ A} = \underline{I}_B$$

So, the current of phase A in the primary is divided into two equal currents in phases B and C, opposite in phase, while in the secondary there is current only in phase a. From the rectangular triangle in Figure 2.44, we get

$$V_a = V_{eat}^2 - X_0^2 \cdot I_{a0}^2 = 220^2 - (1 \cdot 40)^2 = 216 \text{ V}$$

Solving the triangles in Figure 2.44, we get $V_b = 258$ V, $V_c = 191$ V.

So the secondary phase voltages are unbalanced but not very severely, a sign that the transformer has a three-limb core.

The secondary null point displacement from ground, $V_{a0} = i_{a0} \cdot X_0 = 1 \cdot 40 = 40$ V.

For the neutral point of primary, $V_{A0} = N_1/N_2 \cdot V_{a0} = \dfrac{6000}{380} \cdot 40 = 631.58$ V.

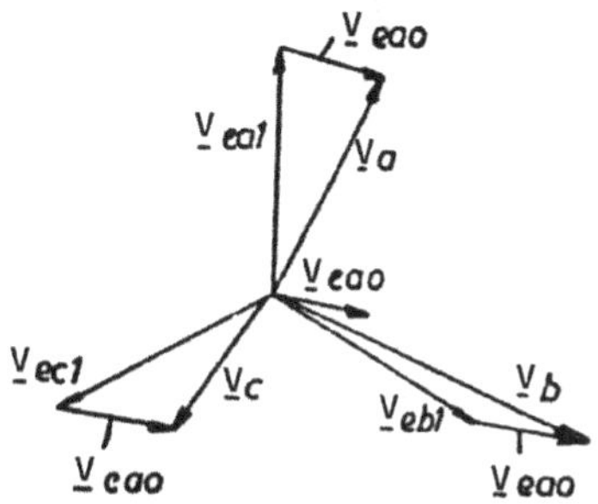

FIGURE 2.44 Secondary voltages phasor diagram for a single-phase resistive load and Yy0 connection scheme.

The other way around, any measured neutral point potential is an indication of unbalanced load or of non-symmetric phase voltages. But this latter case is left out due to lack of space.

2.15 PARALLELING THREE-PHASE TRANSFORMERS

The increase in input electric power of a company is handled through an additional transformer, to be connected in parallel with existing one(s).

The homologous terminals are to be connected together, first in the primary and then, after checking that safe paralleling is feasible, the homologous terminals of the secondary are connected together.

In essence, optimal parallel operation takes place when

- There is no circulation current between transformers at no load
- Each transformer is loaded in the same proportion to the rated condition (same load factor, $K_{S1} = K_{S2}$)
- The currents in all transformers in parallel are in phase
- It is intuitive to consider the following paralleling conditions:
- The rated line voltages of all transformers are the same both in the primary and secondary
- The transformer ratio, N_1/N_2, is the same
- The connection scheme has the same order, n, so that the homologous secondary line voltages are in phase ($n_1 = n_2$)

The above conditions guarantee zero circulating current at no load.

To see that the loading factor, K_s, is the same for all transformers, we have to take up the simplified equivalent circuit that represents the transformer as a short-circuit impedance, Z_{1sc}, but as seen from the secondary (Figure 2.45), $Z_{sc2} = \frac{N_2^2}{N_1^2} \cdot \underline{Z}_{1sc}$.

We may now calculate simply the current in the secondary of the first transformer, I_{2a}, as a function of load current, I_{2t}:

$$\underline{I}_{2a} = \frac{-V_1 \cdot \left(\frac{N_2}{N_1} - \frac{N_2'}{N_1'} \right)}{\underline{Z}_{asc2} + \underline{Z}_{bsc2}} + \frac{\underline{I}_{2t} \cdot \underline{Z}_{bsc2}}{\underline{Z}_{asc2} + \underline{Z}_{bsc2}} \tag{2.126}$$

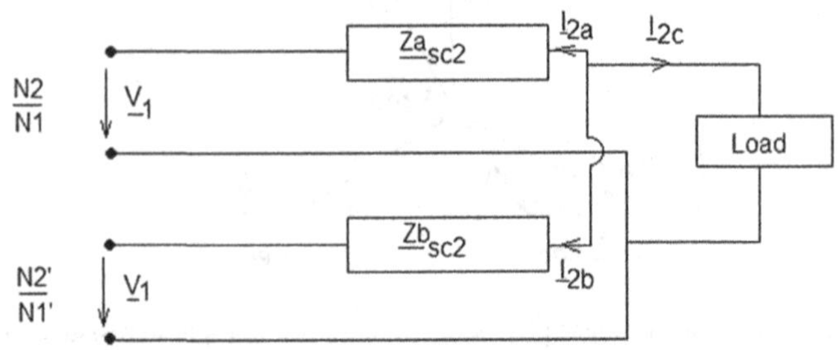

FIGURE 2.45 Transformers in parallel, seen from the secondary.

Also, for the second transformer,

$$\underline{I}_{2b} = \frac{+V_1 \cdot \left(\frac{N_2}{N_1} - \frac{N_2'}{N_1'}\right)}{\underline{Z}_{asc2} + \underline{Z}_{bsc2}} + \frac{\underline{I}_{2t} \cdot \underline{Z}_{asc2}}{\underline{Z}_{asc2} + \underline{Z}_{bsc2}} \quad (2.127)$$

While the second components in Equations 2.126 and 2.127 represent the contribution to the load current, the first terms (equal in amplitude and opposite in phase) represent the circulation current apparent from no load (when $I_{2t} = 0$). The circulating current is due to different transformer ratios. The standard transformer ratios may be different with at most 1% error to avoid notable circulating currents (the denominators in Equation 2.126 contain the short-circuit impedances, which are small).

Now, to provide the same loading factor,

$$\frac{\underline{I}_{2a}}{I_{2an}} = \frac{\underline{I}_{2b}}{I_{2bn}} \quad (2.128)$$

By making use of Equations 2.126 and 2.127, with $N_1 / N_2 = N_1' / N_2'$, Equation 2.128 becomes

$$\underline{Z}_{bsc2} \cdot I_{2bn} = \underline{Z}_{asc2} \cdot I_{2an} \quad (2.129)$$

But this means, in fact, that:

* the short-circuit power factor is the same for both transformers:

$$\cos \varphi_{sca} = \cos \varphi_{scb}$$

* the secondary (and primary) rated short-circuit voltages are the same.

$$V_{nsca2} = V_{nscb2}, \ (V_{1nsca} = V_{1nscb})$$

If paralleling is required on short notice and the rated short-circuited voltages and their power factors have errors above +10%, then, if feasible, additional series impedances are added to the transformer with lower rated short-circuit voltage to come close to fulfilling (2.129).

Example 2.8

Heating Test through Identical Transformers in Parallel on No Load/Lab 2.6

Let us consider two identical power distribution (say residential) transformers with provision for 5% of rated voltage variation. Calculate in relative values (per unit or pu) the circulating current between the two transformers in parallel, without load, one connected on the +5% tap and the other on the −5% tap (Figure 2.46), if the rated short-circuit voltage is 5%.

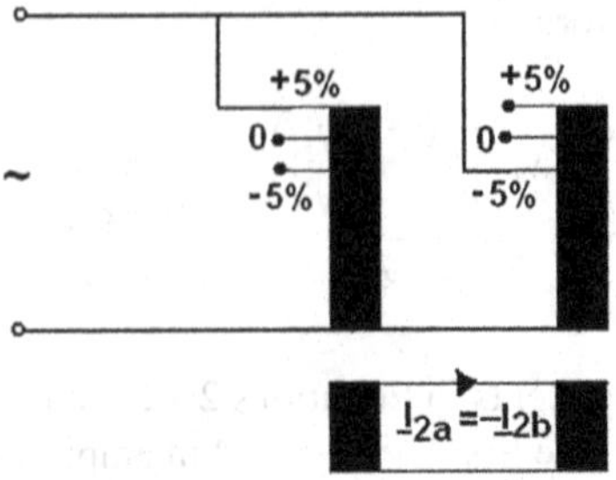

FIGURE 2.46 Tapped identical transformers in parallel on no-load for heating test.

Solution:

From the connection in Figure 2.46 it is clear that the transformer rated turn ratio, K_n, is modified to $1.05 \cdot K_n$ for the first transformer and to $0.95 \cdot K_n$ for the second transformer ($K_n = N_{1n}/N_{2n}$).

Now applying Equations 2.126 and 2.127, we are left only with the circulating current, $I_{2a}(I_{2t} = 0)$:

$$\underline{I}_{2a} = -\underline{I}_{2b} = \frac{-\left(\dfrac{1}{1.05 \cdot K_n} - \dfrac{1}{0.92 \cdot K_n}\right) \cdot V_1}{\underline{Z}_{asc2} + \underline{Z}_{bsc2}}$$

But, seen from the secondary, and with the number of turns slightly changed:

$$\underline{Z}_{asc2} = Z_{1sc} \cdot \frac{1}{(1.05 \cdot K_n)^2}$$

$$\underline{Z}_{bsc2} = Z_{1sc} \cdot \frac{1}{(0.95 \cdot K_n)^2}$$

So, with $I_{2n}/K_n \approx I_{1n}$ and $Z_{1sc} \cdot I_{1n} = V_{1nsc}$,

$$I_{2a} \approx \frac{\dfrac{0.1}{K_n} \cdot I_{2n}}{\dfrac{Z_{1sc}}{K_n^2} \cdot I_{2n} \cdot \left(\dfrac{1}{1.05^2} + \dfrac{1}{0.95^2}\right)} \approx \frac{0.1 \cdot I_{2n}}{2 \cdot \dfrac{V_{1nsc}}{V_1}} = \frac{0.1}{2 \cdot 0.05} \cdot I_{2n} = I_{2n}$$

So, for the case in point, the provoked circulating current is the rated current and thus rated heating of both transformers is obtained. Only the losses of two identical transformers are absorbed from the power grid.

The magnetic saturation conditions are not identical and the rated losses may be considered the average losses in our case. If v_{1nsc} (in percent) $\neq 5\%$, then the provoked circulating current is different from the rated value, but by a known ratio.

2.16 TRANSIENTS IN TRANSFORMERS

Input voltage or load variations are accompanied by intervals of time when the amplitudes of the voltages and currents vary. The time intervals when the voltages or currents are not only sinusoidal may be assimilated with transients.

The inrush current of a transformer which is on no load and is directly connected to the power grid, a sudden short circuit at the secondary terminals, and atmospheric or power electronics steep front voltage pulses, all produce transients.

For slow transients (up to a few kilohertz), the model developed so far for the transformer may be extrapolated but above that (from tens of kilohertz to microsecond voltage pulses of modern power electronics), the capacitors between the winding turns and the oil tank (earth) play a key role. We will use the term electromagnetic transients for slow transients (wherein only resistances and inductances describe the transformer equations and equivalent circuits) and electrostatic transients for fast transients (wherein transformer series and parallel parasitic capacitors play a key role).

2.16.1 Electromagnetic (R,L) Transients

We return to Equations 2.75 and 2.79—valid for electromagnetic transients—to obtain, for single-phase transformer:

$$\begin{aligned} i_1 \cdot R_1 - V_1 &= -\left(L_{1l} + L_{11m}\right) \cdot \frac{di_1}{dt} - L_{11m} \cdot \frac{di_2'}{dt} \\ i_2' \cdot R_2' + V_2' &= -L_{11m} \cdot \frac{di_1}{dt} - \left(L_{2l}' + L_{11m}\right) \cdot \frac{di_2'}{dt} \end{aligned} \tag{2.130}$$

Replacing d/d t by s (Laplace operator), we obtain in matrix format:

$$\begin{vmatrix} R_1 + s \cdot \Big(\underbrace{L_{1l} + L_{11m}}_{L_1}\Big) & s \cdot L_{11m} \\ s \cdot L_{11m} & R_2' + s \cdot \Big(\underbrace{L_{2l}' + L_{11m}}_{L_1'}\Big) \end{vmatrix} \cdot \begin{vmatrix} i_1 \\ i_2' \end{vmatrix} = \begin{vmatrix} V_1 \\ -V_2' \end{vmatrix} \tag{2.131}$$

This leads to two eigenvalues of s from the characteristic equation:

$$\begin{aligned} s^2 \cdot \left(L_1 \cdot L_2' - L_{11m}^2\right) + s \cdot \left(L_1 \cdot R_2' + L_2' \cdot R_1\right) \\ + R_1 \cdot R_2' = 0 \end{aligned} \tag{2.132}$$

With $R_1 = R_2' = R_{sc}/2$, $L_{1l} = L_{2l}'\left(L_1 = L_2'\right)$ and $L_{1l} \cdot L_{2l}$ neglected, we get, from Equation 2.132:

$$s^2 \cdot L_{sc} \cdot L_{11m} + s \cdot L_1 \cdot R_{sc} + \frac{R_{sc}^2}{4} = 0 \tag{2.133}$$

$$s_{1,2} = \frac{\left(-L_1 \pm \sqrt{L_1^2 - L_{sc} \cdot L_{11m}}\right) \cdot R_{sc}}{2 \cdot L_{sc} \cdot L_{11m}} \approx \left\langle \begin{matrix} -\frac{R_{sc}}{L_{sc}} \\ \frac{-R_1}{L_{11m}} \end{matrix} \right. \tag{2.134}$$

So, the electromagnetic transients of transformers are always stable and are characterized by a fast time constant, $T_{sc} = L_{sc}/R_{sc}$, of tens of milliseconds, corresponding

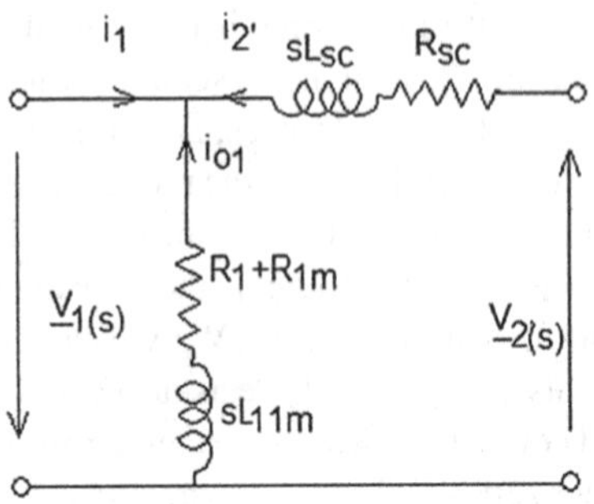

FIGURE 2.47 Transformer structural diagram for electromagnetic transients.

to the load variation process, and a large time constant, $T_m = L_{11m}/R_1$, corresponding to magnetization (voltage variation) as pictured in Figure 2.47.

We are now investigating two particular electromagnetic transients with important practical consequences.

2.16.2 Inrush Current Transients/Lab 2.6

A single-phase transformer with open secondary is connected to the power grid:

$$V_{1t} = V_1\sqrt{2}\cos\left(\omega_1 t + \gamma_0\right) = R_1 \cdot \frac{\Psi_{10}}{L_1\left(i_{10}\right)} + \frac{d\Psi_{10}}{dt}; \quad \Psi_{10} = L_1\left(i_{10}\right)\cdot i_{10} \tag{2.135}$$

At time zero, $\Psi_{10(t=0)} = \Psi_{1rem}$, where Ψ_{1rem} is the remnant flux in the transformer core, corresponding to the material hysteresis loop. As seen from Equation 2.135, only the parallel branch in Figure 2.47 is active $\left(i_2^{'} = 0\right)$. With $\Psi_{10}(i_{10})$ a nonlinear function, due to magnetic saturation, Equation 2.135 may be solved either numerically or graphically to get realistic results.

By considering $L_1(i_{10})$ constant, only in the first right-side term of Equation 2.135, we may solve Equation 2.134 analytically to get

$$\Psi_{10}\left(t\right) = \Psi_{1m}\cdot\left[\sin\cdot\left(\omega_1 t + \gamma_0 - \varphi_0\right) - \sin\cdot\left(\gamma_0 - \varphi_0\right)\cdot e^{\frac{-t}{T_0}}\right] + \Psi_{1rem}$$
$$\varphi_0 = \tan^{1}\cdot\left(\omega_1\cdot\frac{L_{10}}{R_1}\right) \tag{2.136}$$

With $\Psi_{1m} = V_1\sqrt{2}/\omega_1$ and $T_0 = L_{10}/R_1 \approx T_m$, as the rather large time constant in the previous paragraph. Now, adding the nonlinear $\Psi_{10}(i_{10})$ function and neglecting hysteresis, we may graphically find the inrush current waveform (Figure 2.48).

As Equation 2.136 points out, there is no transient (exponential) current component if $\gamma_0 = \varphi_0$, but for $\gamma_0 - \varphi_0 = \pi/2$ the maximum transient current is expected as the peak flux amplitude almost doubles ($\approx 2\Psi_{1m}$) with respect to rated value (already placed in the magnetic saturation zone).

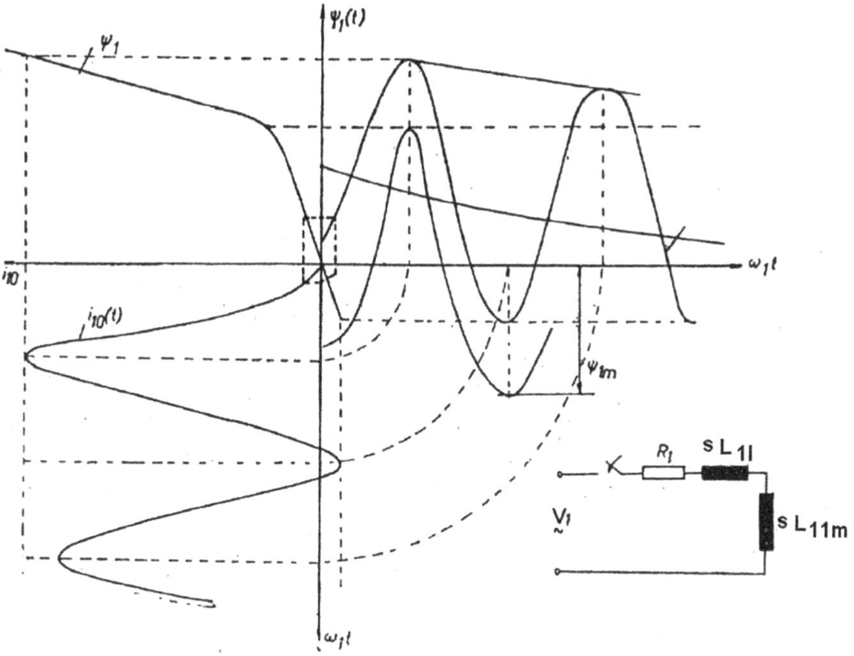

FIGURE 2.48 Inrush current waveform—a graphical solution.

So the peak inrush current may be up to 5–6 times the rated value, though the steady-state no-load current is less than $0.02 \cdot I_{1\,\text{rated}}$. To avoid overcurrent protection tripping, a short duty resistor is connected in series, to limit the current below the tripping value. After the transients (a few seconds) are gone, the resistance is short-circuited by a switch.

2.16.3 Sudden Short Circuit from No Load $(V_2' = 0)$/Lab 2.7

In this case, the transformer is represented only by R_{sc} and sL_{sc} (series impedance) in Figure 2.47.

The current initial value corresponds to no load, i_{10}, and may be neglected this time ($i_{10} \approx 0$). So the transformer equation becomes

$$V_1\sqrt{2}\cdot\cos(\omega_1 t+\gamma_0) = R_{sc}\cdot i_1 + L_{sc}\frac{di}{dt};\quad (i_1)_{t=0} = 0 \tag{2.137}$$

The solution is straightforward as magnetic saturation plays no role:

$$i_{1sc}(t) = \frac{V_1\sqrt{2}}{Z_{sc}}\cdot\left[\cos(\omega_1 t+\gamma_0-\varphi_{sc}) - \cos(\gamma_0-\varphi_{sc})\cdot e^{\frac{-t}{T_{sc}}}\right] \tag{2.138}$$

where φ_{sc} is the short-circuit power factor angle

$$\cos\varphi_{sc} = \frac{R_{sc}}{Z_{sc}}; \quad Z_{sc} = \sqrt{R_{sc}^2 + \omega_1^2 \cdot L_{sc}^2}; \quad T_{sc} = \frac{L_{sc}}{R_{sc}} \tag{2.139}$$

Again, there is no transient for $\gamma_0 - \varphi_{sc} = \pi/2$.

This time, the transient phenomenon is fast and thus short lived (1–7 periods). The peak value of the i_{1sc} may be found for $\partial i_{isc}/\partial t = 0$ and is expressed as

$$i_{sk} = \frac{V_{1n}\sqrt{2}}{Z_{sc}} \cdot K_{sc}; \quad \frac{i_{sk}}{I_{1n}} = \frac{V_{1n}\sqrt{2} \cdot K_{sc}}{Z_{sc} \cdot I_{1n}} = \frac{K_{sc} \cdot \sqrt{2}}{\frac{V_{1nsc}}{V_{1n}}} \tag{2.140}$$

The short-circuit coefficient, K_{sc} = 1.2–1.8, is larger for larger transformers, in general. So, with V_{1nsc}/V_{1n} = 0.04–0.12 and K_{sc} = 1.5, the $i_{sk}/I_{1n} = \dfrac{2}{0.04 \div 0.12} = 50 \div 17$. The short-lived peak short-circuit current may damage the windings through mechanical deformation by huge electrodynamic forces.

Because the steady-state short-circuit current is $\dfrac{I_{1scn}}{I_{1n}} = \dfrac{V_{1n}}{Z_{sc}I_{1n}} = \dfrac{V_{1n}}{V_{1nsc}} = 25 - 8.5$, even the steady-state short-circuit at rated voltage may produce rapid winding over heating, and if by any reason the temperature protection does not trip the transformer in due time, the transformer is designed to withstand the short circuit until the winding temperature reaches 250°C. The corresponding time varies from 5 to 8 s in small transformers and from 20 to 30 s and more in large transformers.

2.16.4 Forces at Peak Short-Circuit Current

Due to the very large peak short-circuit current, i_{sk}, the neighboring primary and secondary phases exert on each other electrodynamic forces (as between parallel conductors).

The leakage field of primary in the zone of secondary (calculated with zero secondary current) interacts with the secondary current to produce these forces whose general *iBl* formula is

$$\mathrm{d}\overline{F} = i\left(\mathrm{d}\overline{l} \times \overline{B}\right) \tag{2.141}$$

The leakage flux density has two components: one which is axial (vertical in Figure 2.23), B_a, and one which is radial, B_r, which shows up only close to the yokes. So the force increment has two components: one radial $\mathrm{d}\overline{F}_r$ and one axial $\mathrm{d}\overline{F}_a$:

$$\mathrm{d}\overline{F}_r = i\left(\mathrm{d}\overline{l} \times \overline{B}_a\right); \quad \mathrm{d}\overline{F}_a = i\left(\mathrm{d}\overline{l} \times \overline{B}_r\right) \tag{2.142}$$

As B_r changes sign over the axial coordinate, F_a acts with two opposite forces, each acting on half the windings.

The radial forces squeeze the low-voltage winding to the core and elongate the high-voltage winding.

Approximate design expressions of radial and axial forces may be obtained from leakage magnetic energy variation along the desired direction:

$$F_r = -\frac{\partial W_{ml1}}{\partial a_{1r}}; \quad F_a = \frac{1}{2}\frac{\partial W_{ml1}}{\partial L_c}; \quad W_{ml1} = L_{11}\cdot\frac{i_{sk}^2}{2} \tag{2.143}$$

According to leakage inductance expressions derived in previous sections

$$\frac{\partial L_{11}}{\partial a_{1r}} = \frac{-L_{11}}{a_{1r}}; \quad \frac{\partial L_{11}}{\partial L_c} = \frac{-L_{11}}{L_c} \tag{2.144}$$

Consequently,

$$F_r = \frac{L_{11}\cdot i_{sk}^2}{2\cdot a_{1r}}; \quad F_a = \frac{L_{11}\cdot i_{sk}^2}{4\cdot L_c} \tag{2.145}$$

So

$$\frac{F_r}{F_a} = \frac{2\cdot L_c}{a_{1r}} \geq 1, \quad \text{in general} \tag{2.146}$$

By design, the transformer windings have to withstand the peak short-circuit current forces without mechanical deformation.

Example 2.9: Peak Electrodynamic Forces

An 1-MVA, 60-Hz, single-phase core transformer has the limb (window) height, $L_c = 1.5$ m, the cylindric (multilayer) windings radial thickness, $a_1 = a_2 = 0.1$ m, and the insulation layer in between, $\delta = 0.01$ m. The rated short-circuit (peak transient overcurrent) factor, $K_{sc} = 1.6$. Calculate the radial and axial forces exerted on the two windings, F_r and F_a, for the peak sudden short-circuit current, with the height of the two windings identical and same rated $V_{1nsc}=0.04V_{1n}$.

Solution:

The radial variable,

$$a = \frac{\delta}{2} + \frac{a_1}{3} = 0.01 + \frac{0.1}{3} = 0.0433 \quad \text{m}$$

According to Equation 2.145, both F_r and F_a may be calculated, with peak short-circuit current, i_{sk}, from Equation 2.140 $i_{sk} = \dfrac{K_{sc}\sqrt{2}}{\frac{V_{1nsc}}{V_{1n}}}\cdot I_{1n}$. To solve the problem with the given data, we need to assume that $L_{1l} = L_{2l} = L_{sc}/2$:

Generally,

$$F_r = \frac{1}{2}\frac{\omega_1 L_{sc}}{2\omega_1}\cdot\left(\frac{K_{sc}\sqrt{2}}{\frac{V_{1nsc}}{V_{1n}}}\right)^2\cdot I_{1n}^2\cdot\frac{1}{a_{1r}} \approx \frac{1}{4}\frac{V_{1n}I_{1n}K_{sc}^2}{\omega_1\frac{V_{1nsc}}{V_{1n}}}\frac{1}{L_{a1}} = \frac{S_n\cdot K_{sc}^2}{4\cdot\omega_1\cdot a_{1r}\cdot\frac{V_{1nsc}}{V_{1n}}} \tag{2.147}$$

$$= \frac{10^6\cdot 1.6^2}{4\cdot pi\cdot 60\cdot 0.0433\cdot 0.04} = 3.62\cdot 10^6 \quad \mathrm{N} = 3620\ \mathrm{kN!}$$

The axial electrodynamic force, F_a, is much smaller:

$$F_a = F_r\cdot\frac{a_{1r}}{2L_c} = F_r\cdot\frac{0.0433}{2\cdot 1.5} = F_r\cdot 0.01443 = 3620\ \mathrm{kN}\cdot 0.01443 \approx 52.25\ \mathrm{kN}$$

Note: As in power distribution transformers winding, tappings of at least ±5% are provided (±10% for generator transformers), notably larger axial forces are expected if the primary and secondary windings are left with unequal vertical lengths.

2.16.5 Electrostatic (C,R) Ultrafast Transients

Steep front overvoltage pulses with microsecond-long fronts may occur due to atmospheric discharges (thunders) and commutation by electromagnetic or power electronic fast power switches. For such ultrafast voltage pulses, the power transformer may not be represented any longer by its electromagnetic model developed so far.

With microsecond front voltage pulses, equivalent frequencies of 10–10^3 kHz may be considered, and thus at least in the first few microseconds after the transients' initiation, the resistances and inductances of the transformer may be neglected. It follows that for super high frequencies the transformer model is represented by its inter-turn, inter-winding and turn-to-earth capacitors. In a real transformer, there are very many parallel (C) and series (K) parasitic capacitors, interconnected to form a very complicated net in a multilayer cylindrical or alternate winding three-phase transformer.

Only for a single-layer winding a simple representation such as in Figure 2.49 is feasible.

Let us now lump all capacitors into only the equivalent capacitor, C_e, that represents the transformer for microsecond front voltage atmospheric wave from a thunder, propagating along an overhead transmission line toward a transformer (Figure 2.50).

Let us consider V_d, the direct voltage wave, V_r, the reflected one, due to transformer equivalent capacitor, $C_e = (10^{-7}–10^{-10})F$, I_d the direct wave current, I_r, the reflected wave current, I_t, the transformer current, and Z, the transmission line impedance (about 500 Ω). From Figure 2.50 it follows that:

$$V_d = ZI_d;\quad V_r = ZI_r;\quad i_t = i_d - i_r;$$
$$V_t = V_d + V_r = \frac{1}{C_e}\cdot\int i_t\,dt \tag{2.148}$$

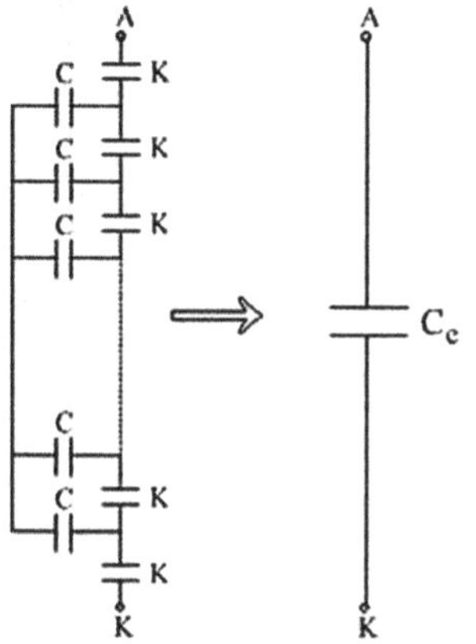

FIGURE 2.49 Super high-frequency (electrostatic) equivalent circuit for a single-layer winding transformer.

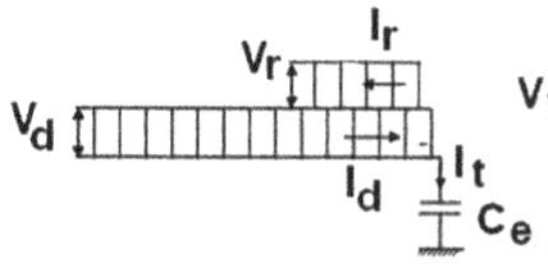

FIGURE 2.50 Microsecond front voltage propagation to a transformer.

Eliminating i_d, i_r, V_d, and V_r from Equation 2.148 and solving the resulting first-order differential equation in V_t, for zero initial transformer voltage ($((V_t)_{t=0} = 0)$), we obtain

$$V_t = 2 \cdot V_d \cdot \left(1 - e^{\frac{t}{C_e Z}}\right)$$
$$i_t = \frac{2V_d}{Z} \cdot e^{\frac{-t}{C_e Z}} \tag{2.149}$$

The time constant of the process, $T_e = (50\text{–}0.05)\ \mu s$, in general.

So the transformer representative capacitor, C_e, charges at double voltage pulse level, $2V_d$, within microseconds, when the resistances and inductances in the transformer do not participate yet.

Now the problem is how this double voltage pulse is distributed along the winding from terminal to the neutral point (star connection). This process also takes place very quickly and thus we still neglect the resistances and inductances in the transformer but now we need to apply the distributed capacitor model (Figure 2.51).

According to Figure 2.51, we may write

$$V(x) + dV - V(x) = q(x) \cdot \frac{dx}{K}$$
$$q(x) + \Delta q - q(x) = V(x) C dx \tag{2.150}$$

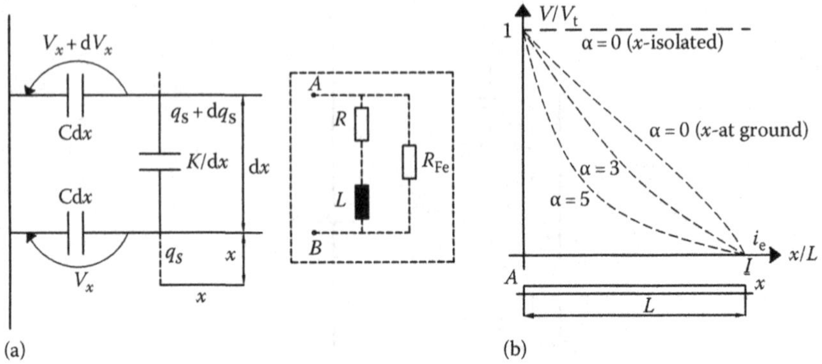

FIGURE 2.51 (a) Distributed capacitor model of the transformer and (b) "initial" voltage distribution along the winding.

or

$$\frac{dV(x)}{dx} = \frac{q}{K}; \quad \frac{dq(x)}{dx} = C \cdot V(x) \tag{2.151}$$

Eliminating q (the electric charge local value) from Equation 2.150 yields:

$$\frac{d^2V(x)}{dx^2} - \frac{C}{K}V(x) = 0 \tag{2.152}$$

This is a wave equation with x, the distance from the neutral point to phase terminal (entry):

$$V(x) = A\sinh(\alpha \cdot x) + B\cosh(\alpha \cdot x); \quad \alpha = \sqrt{\frac{C}{K}} \tag{2.153}$$

With the boundary conditions:

$$\begin{aligned} &V = 0 && \text{for} \quad x = 0 && \text{for earthed neutral point} \\ &\frac{dV}{dx} = 0 && \text{for} \quad x = 0 && \text{for isolated neutral point} \\ &V = V_{max} = 2V_d && \text{for} \quad x = l && \text{for the phase entry} \end{aligned} \tag{2.154}$$

So

$$\begin{aligned} &V(x) = V_{max} \cdot \frac{\sinh(\alpha x)}{\sinh(\alpha L)}; \quad \text{earthed neutral point} \\ &V(x) = V_{max} \cdot \frac{\cosh(\alpha x)}{\cosh(\alpha L)}; \quad \text{isolated neutral point} \end{aligned} \tag{2.155}$$

For the single-phase transformer, the neutral point is replaced by the phase winding end, which may be, again, earthed or isolated.

It is now evident (Figure 2.51b) that if $\alpha = \sqrt{\frac{C}{K}} > 5$, the initial voltage distribution for both neutral situations is highly nonlinear along the winding length, and thus the first 10% of winding may experience more than 50% of the initial voltage, which, if large in itself, will stress heavily the first 10% of the winding. The phenomenon is also common for power electronics–supplied transformers and electric machines, when repeated (by pulse width modulation) such voltage pulses may lead not only to voltage doubling but tripling, with the same nonuniform voltage distribution along the winding length. Special measures are needed to deal with the situation.

Now, after this "initial" distribution installs itself—in a few microseconds or so—the resistances and inductances, including R_{iron} (core loss resistance) show up in the equivalent circuit (Figure 2.51a), and thus resonance phenomena may occur which may lead to even 20%–40% more overvoltage. Antiresonant measures have to be taken to avoid this phenomenon.

2.16.6 Protection Measures Against Overvoltage Electrostatic Transients

There are external (to transformer) measures to reduce the impact of $V_{max} = 2V_d$ overvoltage level and thus implicitly reduce electrostatic stress in the windings of the transformer. Spark gaps are typical of such measures on the current bushings for medium- or low-voltage transformers. In essence, the spark gaps suffer ionization at high-voltage levels, and thus the electric charges flow to the earth through the spark gap, and not through the transformer. Better external protection is obtained through close-to-terminal surge arresters with short ground connectors.

Internal measures, to reduce initial overvoltage stress on the first 10% of transformer winding, have evolved from thinner and taller cables in the first 10% turns (to reduce C and increase K), a metallic protection ring at the phase entry and close to the oil tank, and a cylindrical vertical metallic screen at phase entry and interleaved first 10% turns, to random wound coils in electric machines and ladders of ZnO (zinc oxide) varistor elements around the sensitive (say voltage regulating) parts of the windings.

2.17 INSTRUMENT TRANSFORMERS

To reduce the sensed a.c. voltages and currents to lower levels, acceptable for available instruments (transducers, or sensors), potential (voltage) transformers and, respectively, current transformers are used.

They also provide conduction isolation from high-voltage lines or medium-voltage power sources for transformers and electric machines. Voltmeters range to 600 V (or 1 kV) while ammeters are built up to 5 A.

If a 34-kV line voltage is to be measured, a 350:1 potential transformer works under no load (open secondary), so,

$$V_1 = V_{20}\frac{N_1}{N_2} \tag{2.156}$$

However, due to voltage drop along R_1 and X_{1l}, V_{20} has a small amplitude error and a small phase error with respect to the measured voltage, $V_1 \cdot N_2/N_1$. They should be minimized by design, but a correction factor may be given in precision-sensitive applications. The primary is designed for no-load current and the secondary for no load. So the voltage transformer is small in volume.

The current transformer has, in general, a round-shaped low loss magnetic core with a multiturn secondary winding connected over a small resistance (shunt) whose voltage is proportional to the secondary current. Now the primary is the high current carrying cable which may be twisted to go through the inside of the round core 2, 3, N_1 times ($N_1 < 5$ in general). The current transformer operates basically under short-circuit conditions in the secondary but current fed (and limited) in the primary. Still, if the core has high permeability, the magnetization current, i_{10}, is negligible.

$$N_1 \cdot i_1 + N_2 \cdot i_2 = N_1 \cdot i_{10} \approx 0; \quad i_1 = \frac{-N_2 \cdot i_2}{N_1}; \quad N_2 > N_1 \tag{2.157}$$

The current, $i_2 \leq 5$ A, is, in fact, measured. For I_1 =1 kA and I_2 = 5 A, we would need $N_2/N_1 = 1000/5 = 200/1$. Again, the current transformer has to show small winding losses, besides small magnetization current, to secure small phase and amplitude errors between i_1 and i_2.

The current transformer is designed to a small voltage rating and is small in volume.

Now being heavily underrated in current, the voltage transformer is to be protected against short circuit, while the current transformer, being underrated in voltage, should be protected against open secondary, when, in addition, the large uncompensated mmf in the primary would produce excessive core losses and heavily saturate the core.

2.18 AUTOTRANSFORMERS

For step up and step down of voltages in ratios of up to 2:1 (or 1:2), as needed in long transmission lines, to compensate for voltage drop due to long lines reactance, a transformer with a single but tapped (Figure 2.52) winding (called autotransformer) is used, to cut transformer costs and losses. The autotransformer is used also as lower-cost variable voltage supply.

Neglecting the magnetization current, we may write

$$\left(N_1 - N_2\right) \cdot \underline{I}_1 - N_2 \cdot \underline{I}_s = 0; \quad K_T = \frac{N_1}{N_2} \tag{2.158}$$

where I_s is the secondary current and I_1 the primary current. The load current part of the winding, I_2, is

$$\underline{I}_2 = \underline{I}_1 + \underline{I}_s \tag{2.159}$$

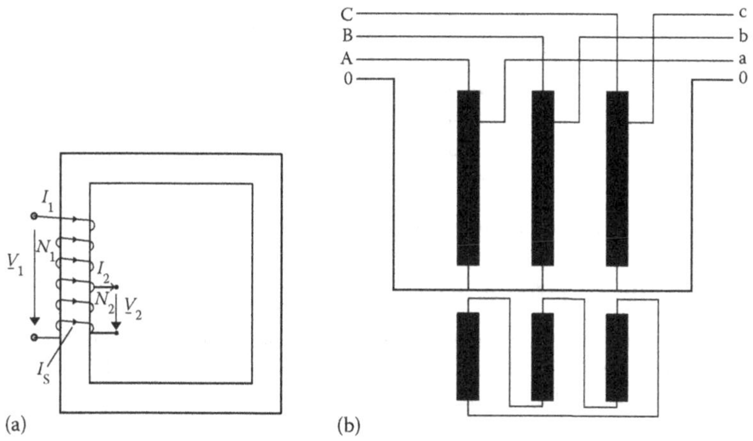

FIGURE 2.52 The step-down autotransformer: (a) Single phase and (b) three phase (Y0y0) with tertiary winding.

From Equation 2.158,

$$\underline{I}_s = \underline{I}_1 \cdot (K_T - 1); \quad \underline{I}_2 = \underline{I}_s \cdot \frac{K_T}{K_T - 1} \tag{2.160}$$

If all losses are neglected, the input and output powers are identical:

$$V_1 \cdot I_1 = V_2 \cdot I_2 \tag{2.161}$$

Consequently, the electromagnetic power, S_e, transmitted through electromagnetic induction (through the core) is

$$S_{en} = V_2 \cdot I_s = V_2 \cdot I_2 \cdot \left(\frac{K_T}{K_T - 1}\right)^{-1} = S_n \cdot \left(\frac{K_T}{K_T - 1}\right)^{-1} \tag{2.162}$$

For $K_T = 2$, $S_e = S_n/2$ and $I_s = I_1$, so only half of the power is transmitted through the core and thus the transformer core design rating for design is 50% of that of a corresponding 2-winding transformer. The rest of the power is transmitted directly—by wire—from the primary to the load.

It follows that the autotransformer is notably less costly; as expected, the core losses, and especially the copper losses, are smaller, and so is the short-circuit rated voltage, V_{1scn}. So the voltage regulation is smaller than that in a 2-winding transformer of the same rating. But the short-circuit current will be larger and thus a stronger over current protection system is required.

Though we treated only the step-down autotransformer, the step-up autotransformer may be treated in a similar way and has similar performance.

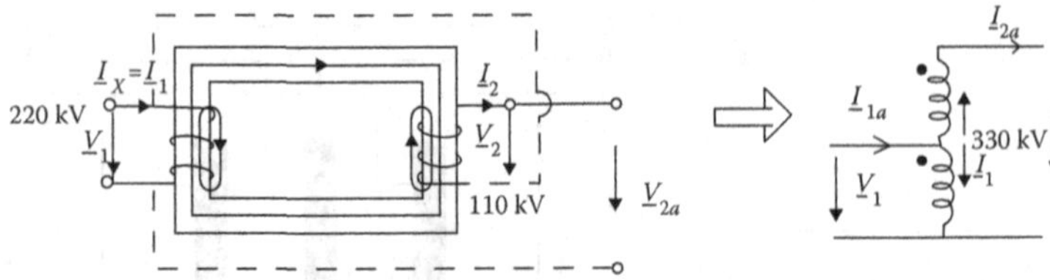

FIGURE 2.53 From transformer to step-up autotransformer.

Example 2.10: From Transformer to Autotransformer

Figure 2.53 shows how to connect a transformer into a step-up autotransformer. Let us consider the transformer ratings: 220/110 kV, 50 MVA. Find the autotransformer rated output voltage and power.

Solution:

The rated input current, I_x, is

$$I_1 = I_x = \frac{S_n}{V_1} = \frac{50 \cdot 10^6}{220 \cdot 10^3} = 227 \text{ A}$$

$$I_{2a} = I_2 = \frac{S_n}{V_2} = \frac{50 \cdot 10^6}{110 \cdot 10^3} = 454 \text{ A}$$

By scalar addition, the input current, I_{1a}, and the output voltage, V_{2a}, of the autotransformer connections are

$$I_{1a} = I_1 + I_2 = 227 + 454 = 681 \text{ A}$$
$$V_{2a} = V_1 + V_2 = (220 + 110) \cdot 10^3 = 330 \text{ kV}$$

So the total delivered power of the autotransformer, S_{an}, is

$$S_{an} = V_1 \cdot I_{1a} = V_{2a} \cdot I_{2a} = 220 \cdot 10^3 \cdot 681 = 150 \text{ MVA}$$

So, as expected in this case, three times more power is available in the step-up autotransformer connection. However, we should notice that, because the input current is three times larger, the feeding cable has to be designed accordingly. Also, the output voltage is three times larger, so the insulation in the secondary winding has to be enforced. But, as the rated losses are the same and the power is tripled, the efficiency of the autotransformer is notably increased in comparison with the original transformer.

2.19 TRANSFORMERS AND INDUCTANCES FOR POWER ELECTRONICS

Power electronics manages to change voltage/current waveforms (amplitude and frequency) or, in other words, electric power parameters, through fast static-power semiconductor-controlled switches.

The modern static power–controlled switches perform on/off commutation within microseconds.

In power electronics, electric power fast processing, by semiconductor- controlled power switches (or rectifiers: SCRs), electric energy storage elements such as inductors and capacitors are used. For galvanic separation of a.c. voltage, large step-up or step-down ratios and high-frequency electric transformers are used for 1–100 kHz frequency range and power in the kilowatt to hundreds of kilowatt and more range. In more distributed electric power systems, power electronics is used to eliminate current or voltage harmonics and to compensate reactive power or voltage drop along power transmission lines. Moreover, high-voltage d.c. power line ties are also used to make the standard a.c. power line more flexible and efficient.

In all these applications, transformers are used together with IGCT power electronics in the tens and hundreds of MVA for switching frequency up to 1 kHz.

Boost d.c.–d.c. converter: (a) The equivalent circuit and (b) storage inductance voltage and current.

Figure 2.54 presents the storage inductance (or inductor), L, used in a d.c.–d.c. converter with d.c. voltage boost.

The inductor, L, is connected to the input d.c. power source for the interval, T_{on}, through the IGBT (insulated gate bipolar transistor) power switch and then, when the inductor charging circuit is opened (during time T_{off}), the output voltage, V_{out}, is

$$V_{out} = V_{in} - L\frac{di}{dt} \tag{2.163}$$

If the switching period, $T_s > T_{on}$, a zero inductor current interval occurs.

The pulsations in the inductor current during the ideal zero current time interval (Figure 2.54b) are due to the reactor parasitic capacitor, C, that acts at high commutation frequencies of the IGBT: $(1/T_s) = (10–20)$ kHz.

Such a reactor, Figure 2.55, is typical for a boost d.c.–d.c. converter for 60 kW power (220–500 Vdc), switched at 10 kHz, in a full hybrid electric vehicle (HEV).

As seen in Figure 2.55, the core has to have multiple airgaps to allow for large currents without heavy magnetic saturation, and should be made of very thin (0.1

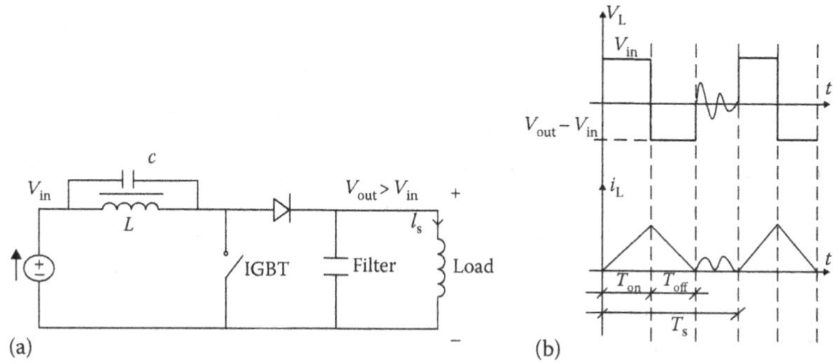

FIGURE 2.54 Boost d.c.–d.c. converter: (a) The equivalent circuit and (b) storage inductance voltage and current.

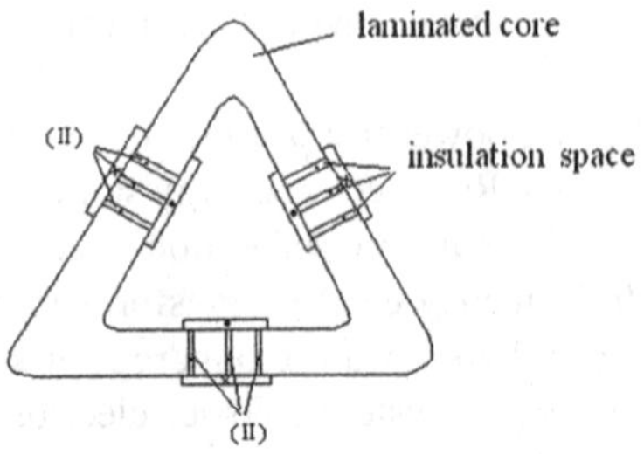

FIGURE 2.55 Storage reactor with multiple gaps in a d.c.–d.c. boost converter.

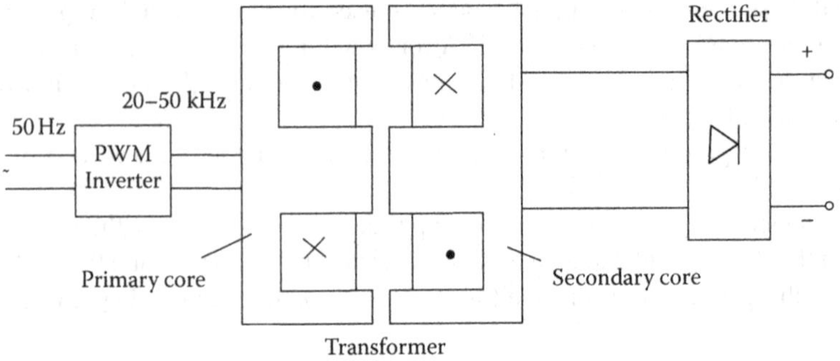

FIGURE 2.56 Battery charger made of high-frequency inverter + transformer + high-frequency diode rectifier.

mm thick) silicon steel laminations, to reduce the core losses at the rather large switching frequency of 10 kHz. Alternatively the core may be made of soft magnetic composites (SMC).

A low-volume (and low weight) electric welding apparatus contains a rectifier an inverter and a 20–100 kHz step-down transformer and a fast-diode rectifier.

Battery charger made of high-frequency inverter + transformer + high-frequency diode rectifier.

A contact-less battery charging system for an HEV (Figure 2.56) or the contact-less power transfer to the rotor of some a.c. machines [10,11] or a mobile phone battery (in situ) charging system imply also the use air-core transformers with small or large airgap at 20–50 kHz [11].

At the other end of the scale, to compensate the voltage drop along long a.c. transmission lines, a two stage a.c.–a.c. power electronics converter and a series-connected transformer is used. In this case, the transformer fundamental frequency is 50(60) Hz but, due to the converter fast switching, there are harmonics, etc.

All the above examples of transformers with power electronics suggest that special materials and configurations and models are needed which are tied to the application power and frequency range [1].

2.20 PRELIMINARY TRANSFORMER DESIGN (SIZING) BY EXAMPLE

By design, we mean here dimensioning the transformer for given specifications.

So, by design we mean synthesis, while calculating performance for a given geometry is called analysis.

The design (synthesis) uses the analysis iteratively. Analysis implies a mathematical model.

The mathematical models may be of field distribution (finite element) type or of circuit type. As FE models are computation-time prohibitive, they are, in most cases, used for performance validation, after design optimization based on analytical models. Here, we will approach the preliminary (general) transformer design by way of a case study and making full use of transformer parameters rather realistic expressions of the circuit model, already derived in this chapter.

2.20.1 Specifications

- Transformer rated kVA: $S_n = 100$ kVA
- Number of phases: 3
- Connections scheme: Yz0
- Magnetization current, $i_{01} < 0.015I_n$
- Frequency, $f_1 = 50$ Hz
- Line voltages, V_{1l}/V_{2l}: 6000/380 V
- Rated short-circuit voltage, $V_{1scn} < 0.045 \cdot V_1$
- Rated current density, $j_{co} = (3\text{–}3.5)$ A/mm^2
- Cylindrical (multi-layer) windings, three-limb iron core

2.20.2 Deliverables

- Core geometry
- Winding design
- Resistance, reactances, and the equivalent circuit
- Losses and efficiency
- Rated no-load current and rated short-circuit voltage

2.20.3 Magnetic Circuit Sizing

The three-limb magnetic core with its main geometrical variables is shown in Figure 2.57.

The magnetization curve (B_m (H_m)) of the laminated silicon steel sheet is given in Table 2.3.

The core losses at $B_m = 1.5$ T, at 50 Hz, $p_{iron} = 1.12$ W/kg and depend on $(B_m/1.5)^2$.

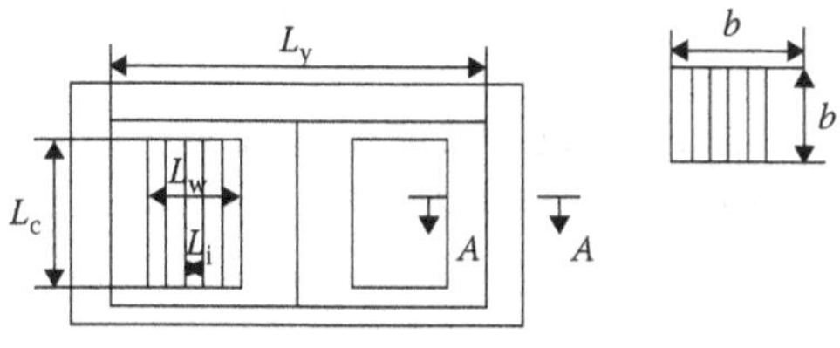

FIGURE 2.57 The three-limb magnetic core.

TABLE 2.3
Silicon Steel Sheet Magnetization Curve

B (T)	0.1	0.15	0.2	0.3	0.4	0.5	0.6	0.7	0.9	1.4	1.5	1.6
H (A/m)	35	45	49	65	76	90	106	124	177	760	1340	2460

We start the magnetic circuit design by setting the limb and yoke flux density values, $B_c = B_y = 1.4$ T. We may find, from the magnetization curve, through interpolation, the magnetic field in the column, $H_c = H_y$.

Based on the magnetic circuit law:

$$H_c \cdot \left(L_c + L_w + \pi \frac{D_c}{4} \right) = N_1 \cdot I_{01} \tag{2.164}$$

With L_w being the window length and D_c being the core column external diameter.

The phase emf, V_{e1}, is almost equal to the phase voltage, $V_{1l}/\sqrt{3}$ (star connection in the primary):

$$V_{e1} = \pi\sqrt{2} \cdot f_1 \cdot B_c \times A_c \times N_1 = K_e \cdot \frac{V_{1n}}{\sqrt{3}}; \quad K_e = 0.985 - 0.95 \tag{2.165}$$

2.20.4 Windings Sizing

The window room of the core has to be enough to place the primary and secondary windings of two phases plus an insulation space, L_i, 10 mm < L_i < 60 mm (for V_{1nl} = 6 kV).

We may assume $N_1 I_1 = N_2 I_2$, and, thus,

$$\left(L_c - b\right) \cdot \left(L_w - L_i\right) = \frac{2 \cdot 2 N_1 I_{1n}}{j_{Co} \cdot K_{fill}} \tag{2.166}$$

$K_{fill} \approx 0.5$ is the winding filling factor.

For star connection of phases, we adopt an initial value for the limb core area, $A_c = 10^{-2}$ m^2.

From Equation 2.165 the number of turns, N_1, of primary is

$$N_1 = \frac{6000}{\sqrt{3} \cdot \pi\sqrt{2} \cdot 50 \cdot 1.4 \cdot 10^{-2}} \approx 1085 \text{ turns}$$

On the other hand, standard industrial experience with tens of kVA transformer designs yields the primary emf per turn [7]:

$$E_{turn} = K_E \cdot \sqrt{S_n \left(\text{in kVA}\right)} \tag{2.167}$$

where

$K_E = 0.6$–0.7 for three-phase industrial transformers

$K_E = 0.45$ for three-phase distribution (residential) transformers

$K_E = 0.75$–0.85 for one-phase transformers

For our case

$$E_{turn} = 0.45\sqrt{100} = 4.5 \text{ V/turn}$$

According to this rationale N_1 would be

$$N_1 = \left(\frac{V_{1n}}{\sqrt{3}}\right) \cdot \frac{K_E}{E_{turn}} = \frac{6000 \cdot 0.97}{\sqrt{3} \cdot 4.5} = 778 \text{ turns} \tag{2.168}$$

We stick with $N_1 = 1085$ turns (from Equation 2.165), to be on safety side.

The rated current, I_{1n}, is

$$I_{1n} = \frac{S_n}{\sqrt{3} \cdot V_{1nl}} = \frac{100 \cdot 10^3}{\sqrt{3} \cdot 6000} = 9.6334 \text{ A} \tag{2.169}$$

The number of turns, N_2, in the secondary is

$$N_2 = N_1 \cdot \frac{V_{2n}}{V_{1n}} \cdot \frac{2}{\sqrt{3}} = 1085 \cdot \left(\frac{\frac{380}{\sqrt{3}}}{\frac{6000}{\sqrt{3}}}\right) \cdot \frac{2}{\sqrt{3}} = 80 \text{ turns} \tag{2.170}$$

The $2/\sqrt{3}$ factor is due to the z_0 connection in the secondary.

The winding radial widths in the window, a_1 and a_2, are

$$a_1 = \frac{N_1 \cdot I_{1n}}{j_{Co} \cdot K_{fill} \cdot L_c}; \quad a_2 = \frac{N_2 \cdot I_{2n}}{j_{Co} \cdot K_{fill} \cdot L_c} = \frac{2}{\sqrt{3}} \cdot a_1 \tag{2.171}$$

The average diameters of the turns are

$$D_{av2} = D_{core} + a_2; \quad D_{av1} = D_{core} + 2 \cdot a_2 + \delta + a_1 \tag{2.172}$$

The leakage reactance expressions (derived earlier in this chapter) are

$$\begin{aligned} X_{1l} &= \omega_1 \mu_0 \frac{N_1^2 \cdot \pi}{L_c} \cdot D_{av1} \cdot \left(\frac{a_1}{3} + \frac{\delta}{2}\right) \\ X'_{2l} &= \omega_1 \mu_0 \frac{N_2^2 \cdot \pi}{L_c} \cdot D_{av2} \cdot \left(\frac{a_2}{3} + \frac{\delta}{2}\right) \end{aligned} \tag{2.173}$$

The primary and secondary resistances are straightforward:

$$R_1 = \rho_{\mathrm{Co}} \cdot \frac{\pi \cdot D_{\mathrm{av1} \cdot N_1}}{\dfrac{I_{1n}}{j_{\mathrm{Co}}}}$$

$$R_2' = \rho_{\mathrm{Co}} \cdot \frac{\pi \cdot D_{\mathrm{av2} \cdot N_1}}{\dfrac{I_{1n}}{j_{\mathrm{Co}}}} \cdot \frac{2}{\sqrt{3}} \tag{2.174}$$

The distance between windings, δ, is assigned to: $\delta = 0.012$ m. We may eliminate a_1 and a_2 from Equations 2.171 through 2.174 to remain with only one unknown, the column (limb) height, L_c, in the short-circuit (rated) voltage expression:

$$V_{1scn} = I_{1n}\sqrt{\left(R_1 + R_2'\right)^2 + \left(X_{1l} + X_{2l}'\right)^2} \tag{2.175}$$

For $\delta = 0.012$, it follows that for $L_c = 0.5$ m, $D_{av1} = 0.2028$ m, $D_{av2} = 0.14768$ m, $X_{2l}' = 6.5\ \Omega$, $X_{1l} = 7.245\ \Omega$, $R_2' = 4.666\ \Omega$, $R_1 = 4.8185\ \Omega$.

The short-circuit rated voltage, V_{1scn}, is then

$$V_{1scn} = 160.87\,\mathrm{V} \tag{2.176}$$

Let us now check the short-circuit rated voltage in percents:

$$\frac{V_{1scn}}{V_{1n}} = 4.638\% \tag{2.177}$$

This value is close to the desired 4.5% and thus the value of L_c of 0.5 m holds.

2.20.5 Losses and Efficiency

The rated copper losses P_{copper} is

$$P_{copper} = 3\left(R_1 + R_2'\right)I_{1n}^2 = 2640\ \mathrm{W} \tag{2.178}$$

To calculate the core losses, we first need to finish up the magnetic core geometry, with window length, L_w, computation:

$$L_w = 2\left(2a_1 + 2a_2 + \delta\right) + D \tag{2.179}$$

where D the radial distance between neighboring phases ($D = 0.1$ m is ok for 6 kV).

So, from Equation 2.179, L_w is

$$L_w = 0.27\ \mathrm{m} \tag{2.180}$$

The core weight, G_{iron}, for same limb and yoke cross-section area, A_c, is

$$G_{iron} = A_c\left[3\left(L_c + D_c\right) + 4L_w\right]\gamma_{iron} = 225.49 \text{ kg} \tag{2.181}$$

So the iron losses, P_{iron}, is (at 1.4 T)

$$p_{iron} = G_{iron}\left(P_{iron}\right)_{1.4} \approx 225.44 \cdot 1.0 = 225.44 \text{ W} \tag{2.182}$$

The rated efficiency, η_n, is thus

$$\eta_n = \frac{S_n}{S_n + p_{iron} + p_{copper}} = \frac{100{,}000}{100{,}000 + 2{,}640 + 225.44} = 0.972 \tag{2.183}$$

2.20.6 No-Load Current

The rated no-load current (I_{10n}) equal to the magnetization current, I_{01}:

$$\begin{aligned} I_{10} \approx I_{01} &= \left(H_c\right)_{1.4T} \cdot \frac{L_c + L_w + \frac{\pi}{4}\cdot D_c}{N_1} = 200.0 \cdot \frac{0.5 + 0.27 + \frac{\pi}{4}\cdot 0.1288}{1085} \\ &= 0.16 \text{ A}; \quad \frac{I_{10}}{I_{1n}} = \frac{0.16}{9.6344} = 1.66\% \end{aligned} \tag{2.184}$$

2.20.7 Active Material Weight

The copper weight, G_{copper}, is

$$\begin{aligned} G_{copper} &= 3\cdot\left(\pi\cdot D_{av1} + \pi\cdot D_{av2}\cdot\frac{4}{3}\right)\cdot\frac{I_n}{\rho_{Co}}\cdot N_1\cdot\gamma_{Copper} \\ &= 3\cdot\left(\pi\cdot 0.2028 + \pi\cdot 0.14768\cdot\frac{4}{3}\right)\cdot\frac{9.6344}{3.2\cdot 10^6}\cdot 1085\cdot 8900 \\ &= 109.32 \text{ kg} \end{aligned} \tag{2.185}$$

The total active material weight, G_a, is

$$G_a = G_{iron} + G_{copper} = 225.44 + 109.32 = 334.72 \text{kg} \tag{2.186}$$

The kVA/kg in the transformer is

$$\frac{S_n}{G_a} = \frac{100 \text{ kVA}}{334.72 \text{ kg}} \approx 0.29 \frac{\text{kVA}}{\text{kg}} \tag{2.187}$$

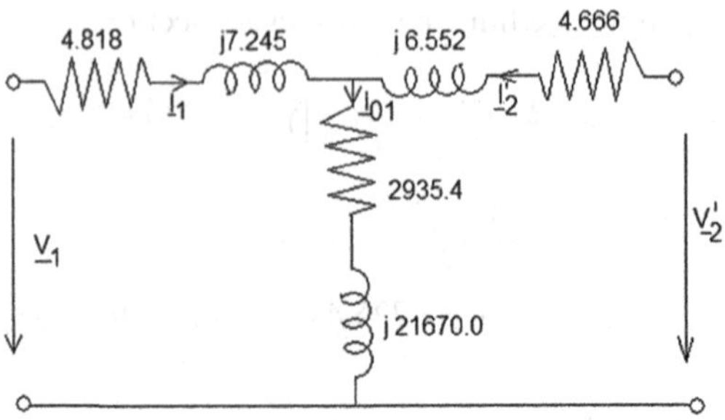

FIGURE 2.58 The equivalent circuit.

2.20.8 Equivalent Circuit

From the equivalent circuit, only the magnetization reactance, X_m, and core loss resistance, R_{1m}, are to be calculated as

$$X_m = \frac{V_{1n}}{\sqrt{3}\cdot I_{10}} - X_{1l} = \frac{6000}{\sqrt{3}\cdot 0.16} - 7.245 = 21{,}676 - 7.245 \approx 21{,}670\,\Omega \quad (2.188)$$

$$R_{1m} = \frac{p_{iron}}{3\cdot I_{10}^2} = \frac{225.44}{3\cdot 0.16^2} = 2935.4\,\Omega \quad (2.189)$$

The equivalent circuit in numbers is pictured in Figure 2.58.

Note: As the results of the preliminary design are reasonable, they could be a good start for the thermal and mechanical design, and for a design optimization code as done elsewhere.

2.21 SUMMARY

- Electric transformers are a set of magnetically coupled electric circuits capable to step up or step down the voltage in a.c. power transmission. They are based on Faraday's law for bodies at standstill. Transformers serve also for galvanic separation.
- The transformer ratio, $K_T = N_1/N_2 = (V_{1n}/V_{2n})_{ph}$, reflects the voltage step up (K_T < 1) or step down (K_T > 1).
- Transformers may be classified as single-phase or 3 (or multiple)-phase configurations. Alternatively, we distinguish power transformers (for power systems, industry and power distribution), measurement transformers, autotransformers, and transformers for power electronics.
- Magnetic circuits of transformers are made of thin sheets of silicon steel at industrial frequency (50(60) Hz) or from soft ferrites or permalloy, etc., at high frequencies as in association with power electronics.

- Magnetic circuits are characterized by magnetic saturation and hysteresis and eddy current losses. The thin sheets provide for low eddy current losses and allow thus for high efficiency.
- Electric circuits are flown by a.c. currents close to the magnetic core and exhibit skin effects which increase the resistance and decrease the leakage inductance. In transformers, skin effect is reduced through strand transposition (up to Roebel bars or Litz wire).
- To study transformer steady state and transients, leakage and main inductances and resistances are defined and calculated to form an equivalent circuit with the secondary reduced to the primary.
- While the large main (magnetization) reactances lead to very low no-load current (up to 2% of rated current), the leakage reactances and resistances determine the short-circuit current, which is large.
- The power, p_0, at no load and rated voltage almost equals the core losses under load. The short-circuit winding losses at rated current, I_{1n} (and at short-circuit rated voltage $V_{1scn} = Z_{sc} \cdot I_{1n} \approx (0.04\text{–}0.12) \cdot V_{1n}$), equal those at rated load. So, from these two tests not only the transformer parameters for the equivalent circuit, but also the losses under specified load factor, $K_s = I_1/I_{1n}$, can be calculated.
- Under load, there is a voltage variation, ΔV_2, of secondary voltage, V_2, from its no-load value, V_{20}, which is called voltage regulation and is proportional to the load factor, K_s, short-circuit rated voltage and to $\cos(\varphi_{sc} - \varphi_2)$; φ_{sc}, φ_2-short-circuit and load power factor angles:

$$\Delta V_2 = \Delta V_1 \cdot \frac{N_2}{N_1} = \frac{N_2}{N_1} \cdot \left(K_s \cdot V_{1nsc} \cdot \cos\left(\varphi_{sc} - \varphi_2\right)\right) = V_{20} - V_2$$

- ΔV_2 should be small for power transmission and distribution to secure pretty constant output voltage with load but it may be intentionally large if the short-circuit current has to be made limited.
- Transformers are often connected in parallel; to share the load fairly they have to have the same primary and secondary rated voltages and connection scheme and order, n, and the same V_{1nsc} and $\cos \varphi_{sc}$ in 3 phase configurations.
- Three-phase transformers sometimes supply unbalanced loads and then the connection scheme and the magnetic core type have to be corroborated wisely to attenuate the zero sequence uncompensated load current emf that unbalances the phase voltages and moves the neutral potential from ground level. Similar aspects occur in three-phase transformers connected to weak power grids with unbalanced voltages.
- Transformers undergo electromagnetic and ultra-high-frequency (electrostatic) transients. Adequate means of transformer protection are needed to avoid transformer thermal or mechanical damage due to severe transients.
- See more on transformers in dedicated books and standards [8,11,12].
- Insulation material system design and testing are crucial to safe operation of transformers and electric machines [9].

2.22 PROPOSED PROBLEMS

2.1 The laminated silicon core of an a.c. coil is made of 0.5 mm thick sheets with an electric conductivity, σ = 1 · 106 (Ω · m) – 1, and a relative permeability, μrel:

$\mu_{rel} = 3000$; for $B < 0.8$ T
$\mu_{rel} = 3000 - (B - 0.8)^2 \cdot 10^3$; for $0.8 \le B < 2$ T

Calculate the eddy current losses per unit volume, in such a core at 60 Hz and at 600 Hz.
Hint: Use Equations 2.27 through 2.29.

2.2 In the open slot of an electric machine there is a single rectangular copper bar with $h = 0.020$ m (height) and $b = 0.005$ m (width). Copper conductivity, $\sigma_{Co} = 5 \cdot 10^7\ (\Omega \cdot \text{m})^{-1}$.

a. Calculate the skin effect resistance and reactance coefficients, K_R and K_x, of the conductor for 60 Hz and for 1 Hz.
b. Replace the single conductor by two conductors with $h' = h/2$ in height, connected in series and calculate again the skin effect coefficients in the same conductor as for (a).

Hint: Use Equations 2.44 through 2.50 with $m = 2$.

2.2 The cylindrical (multilayer) winding of a transformer is characterized by $N_1 = 100$ turns, radial thickness, $a_{1r} = 0.01$ or 0.03 m, and the distance between windings, $\delta_{is} = 0.005$ m, core diameter, $D = 0.1$ m.

a. Calculate the winding leakage inductance for $a_{1r} = 0.01$ and 0.03 m for a column height $L_c = 0.08$m.
b. For the same copper volume and $L_c = 0.15$ m, determine the winding radial thickness, a_{1r}, and, again, the leakage inductance; compare and discuss the results for cases (a) and (b).

Hint: Use Equations 2.61 through 2.63.

2.3 A single-phase 1-MVA V1nl/V2nl = 110/20 kV transformer is characterized by no-load current, I10 = 0.01 · I1n, no-load power factor, $\cos\varphi_0 = 0.05$, short-circuit rated voltage, V1scn = 0.04 · V1n and $\cos\varphi_{sc} = 0.15$. Calculate

a. The rated and no-load currents, I_{1n}, I_{10n}
b. The short-circuit resistance, R_{sc}, and reactance, X_{sc}
c. Rated copper losses, p_{co}
d. Iron losses, p_{iron}
e. Rated efficiency at $\cos\varphi_2 = 1$ and $\cos\varphi_2 = 0.8$ lagging
f. Load factor, $K_{scn} = I_1/I_{1n}$, for maximum efficiency
g. Voltage drop, ΔV, in percent of rated voltage, at rated load and $\cos\varphi_2 = 1$ and $\cos\varphi_2 = 0.867$, lagging and leading
h. The short-circuit rated voltage, V_{1scn}, in Volt

Hint: See Examples 2.2 through 2.4.

2.4 A three-limb core three-phase 60 Hz, 220 V/RMS per phase transformer, with star primary connection, operates on no load. The magnetic reluctances of limbs and yokes are $R_{mc} = R_{my} = 1[\text{H}]^{-1}$.

a. Neglecting all voltage drops in the transformer, express the limb flux phasors $\underline{\Phi}_A$, $\underline{\Phi}_B$, $\underline{\Phi}_C$ with their calculated amplitude
b. Calculate in complex number terms the three-phase currents on no load
c. Determine the active power for each phase, noting that the voltage/flux phase angle is 90°

Hint: See Figure 2.39 and Equations 2.112.

2.5 A three-phase 20/0.38 kV, 500-kVA, Yy0 transformer is loaded at rated secondary current, I_{2n}, only on phase a in the secondary at $\cos\varphi_2 = 0.707$ lagging. Determine
a. The rated phase currents in the primary
b. The current in the phase a of secondary
c. The currents in phases A, B, C of primary for the single-phase load
d. No-load reactance/phase, X_m, if the no-load current is $I_{10} = 0.01 \cdot I_{1n}$
e. The zero sequence emf per phase in the secondary and primary if $X_0 = 0.1 \cdot X_m$ (or the neutral potential to ground)
f. The secondary phase voltages, V_a, V_b, V_c
g. The primary phase voltages, V_A, V_B, V_C

Hint: Check Example 2.7.

2.6 Two three-phase Yy6 transformers of $S_{na} = 500$ kVA and $S_{nb} = 300$ kVA, $V_{1nl}/V_{2nl} = 6000/380$ V, $V_{nsca} = 1.1 \cdot V_{nscb} = 0.04 \cdot V_{1n}$, $\cos\varphi_{sca} = \cos\varphi_{scb} = 0.3$, with same number of turns and transformer ratio, operate in parallel.
a. With the first transformer tapped +5% and second at −5% in the primary, on no load, in parallel, calculate the circulating current between the two transformers in the primary and secondary.
b. With both transformers tapped at 0%, and the first transformer loaded at rated current and the second one in parallel, and both supplying a resistive load, calculate the secondary current in the second transformer and in the load. Discuss the results.

Hint: See Equations 2.120 through 2.127 and Example 2.6.

2.7 A one-phase 60-Hz transformer under no load is connected suddenly to the power grid. The primary resistance, $R_1 = 0.1\ \Omega$, and the primary voltage is 220 V (RMS). The following are required:
a. The approximate total flux-linkage amplitude during steady-state no load, Ψ_{1m0}
b. The magnetization curve is given by $I_0 = a \cdot \Psi_{1m0} + b \cdot \Psi_{1m0}^2$

With $\Psi_{1m0} = 0.95$ Wb for $i_0 = 0.1$ A and $\Psi_{1m0} = 1.8$ Wb for $i_0 = 50$ A. Determine the inductance function, $L_{10}(i_0) = \Psi_{1m0}/i_0$
c. For zero remnant flux and $\gamma_0 - \varphi_0 = 90°$ represent in a graph, the inrush current versus time, for a constant time constant, $T_0 = L_{10}(0.2\text{A})/R_1$, and constant $\varphi_0 = (\tan^{-1}(\omega_1 \cdot T_0))$

Hint: Check Figure 2.48 and Equation 2.136.

2.8 The inter-turn and turn/earth capacitances, K and C per unit winding length (height in meters), of a three-phase transformer are $C = 10\ \mu F = 25$ K.
   a. Calculate the initial distribution of an atmospheric microsecond front voltage of 1 MV along the winding height for an isolated and earthed neutral
   b. After using metal screens connected at primary phase terminal C = K; calculate again the initial voltage distribution along the winding height and discuss the results for the two cases

   Hint: Check Equations 2.150 and 2.155 and Figure 2.51.

2.9 A 10-kVA, 220/110 V single-phase autotransformer is considered. Determine
   a. The electromagnetic rated power, S_{em}
   b. The rated primary, secondary and load currents
   c. Calculate the ratio of copper losses of the autotransformer to transformer with the same voltages and design current density (it means both resistances are proportional to turns squared only)

   Hint: Check Equations 2.158 through 2.162 and Example 2.8.

REFERENCES

1. A. Van den Bossche and V.C. Valchev, *Inductors and Transformers for Power Electronics*, Chapter ch03:chap03, Taylor & Francis, New York, 2004.
2. R.J. Parker, *Advances in Permanent Magnetism*, John Wiley & Sons, New York, 1990.
3. P. Campbell, *Permanent Magnet Materials and Their Application*, Cambridge University Press, Cambridge, U.K., 1993.
4. I.D. Mayergoyz, *Mathematical Models of Hysteresis*, Springer-Verlag, New York, 1991.
5. M.A. Mueller, Calculation of iron losses from time-stepped finite-element models of cage induction machines, *International Conference on EMD, IEEE Conference Publication* 412, 1995.
6. ABB Transformer Handbook, 2005.
7. G. Say, *Performance and Design of AC Machines*, Pitman and Sons Ltd., London, U.K., 1961, p. 143.
8. X.M. Lopez-Fernandez, H.B. Ertan, J. Turowski, "Transformers: Analysis, Design and Measurement" CRC Press, Florida, 2013.
9. T. R. Gaerke, D. C. Hernandez, "Understanding stator insulation in-process testing", *IEEE Trans*, Vol. IA-53, No.2, 2017, pp.1704–1708.
10. J. Dai, D. C. Ludois, "A survey of wireless power transfer and a critical comparison of inductive and capacitive coupling for small gap applications", *IEEE Trans*, Vol. PE-30, No.11, 2015, pp.6017–6029.
11. A. Di. Goia, I. P. Brown, Y. Nie, R. Knippel, D. C. Ludois, J. Dai, S. Hagen, C. Altehed, "Design and demonstration of a wound field synchronous machine for electric vehicle traction with brushless capacitive field excitation", *IEEE Trans, Vol. IA-54*, No.2, 2018, pp.1390–1403.
12. R. Fischer, "Electrical Machines 17th ed. Hanser Verlag Munchen, 2017.

3 Energy Conversion and Types of Electric Machines

3.1 ENERGY CONVERSION IN ELECTRIC MACHINES

Electric machines are sets of magnetically and electrically coupled electric circuits with one (or more) movable element (rotor), which convert electric energy into mechanical energy (motor mode) or vice versa (generator mode). They are based on the law of energy conversion and on Faraday's law for bodies in relative motion. In the following text, we will discuss the energy conversion principle of electric machines and introduce the basic types of electric machines, by making use of the frequency theorem [1].

The energy conversion in electric machines involves energy in four forms:

$$\begin{matrix}\text{Electric}\\ \text{energy}\\ \text{from the}\\ \text{power source}\end{matrix} = \begin{matrix}\text{Mechanical}\\ \text{energy}\end{matrix} + \begin{matrix}\text{Stored}\\ \text{magnetic}\\ \text{energy}\end{matrix} + \begin{matrix}\text{Energy}\\ \text{loss}\\ \text{in the electric}\\ \text{machine}\end{matrix} \qquad (3.1)$$

$$\xrightarrow{\text{Motor}} \qquad \xrightarrow{\text{Motor}}$$

$$\xleftarrow{\text{Generator}} \qquad \xleftarrow{\text{Generator}}$$

There are three main reasons for the loss of energy:

- Magnetic core hysteresis and eddy current losses (as in transformers): p_{iron}
- Winding losses (as in transformers): p_{copper}
- Mechanical losses (windage, bearing, ventilator losses): p_{mec}

Figure 3.1 portrays Equation 3.1 with specified losses.

According to Figure 3.1, the net electric energy converted into magnetic energy, dW_e, is

$$dW_e = (V - Ri)i\,dt \qquad (3.2)$$

To transform the electric energy into magnetic energy (in the electric machine), the coupling magnetic field (of the net of magnetic/electric circuits that make the electric

DOI: 10.1201/9781003214519-3

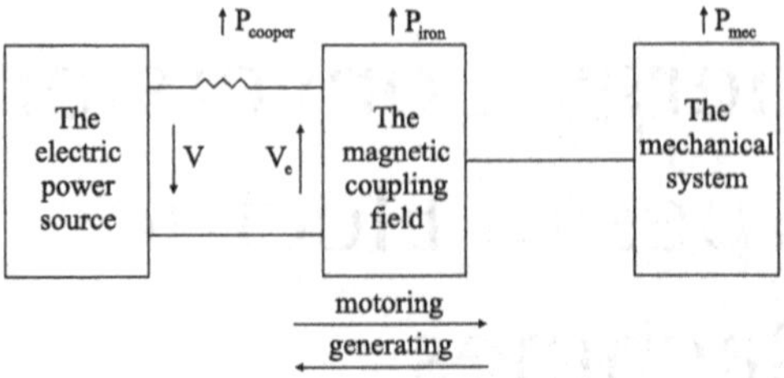

FIGURE 3.1 Energy conversion in electric machines.

machine) has to produce a reaction in the electric circuit, which manifests itself by the electromagnetomotive force (emf), V_e:

$$-V_e = V - Ri; \quad dW_e = (-V_e)i dt \tag{3.3}$$

If the electric energy is transmitted to the coupling magnetic field through a few electric circuits, Equation 3.3 will contain their summation.

3.2 ELECTROMAGNETIC TORQUE

According to Faraday's law, the emf, V_e, is

$$V_e = \frac{-d_s\Psi}{dt} = -\frac{\partial\Psi}{\partial t} - \frac{\partial\Psi}{\partial\theta_r}\frac{d\theta_r}{dt} \tag{3.4}$$

where s refers to the total (substantial) flux-linkage time derivative. V_e contains the pulsational emf and the motion emf (θ_r is the rotor position). It turns out that only the motion emf participates directly in the electro-magneto-mechanical energy conversion and torque production in electric machines. From Equations 3.3 and 3.4,

$$dW_e = i d_s\Psi \tag{3.5}$$

Thus, when the flux linkage is constant ($d_s\Psi = 0$), there is no electric energy transfer between the electric machine and the power source. On the other hand, the mechanical energy increment, dW_{mec}, is defined by the electromagnetic torque, T_e, and the rotor angle increment, $d\theta_r$:

$$dW_{mec} = T_e d\theta_r \tag{3.6}$$

Denoting the stored magnetic energy increment by dW_{mag} and combining the above equations, we obtain

$$dW_e = i d\Psi_s = dW_{mag} + T_e d\theta_r \tag{3.7}$$

Now, if $\Psi = \text{const}$, as stated above, the energy transfer from the electric power source is zero, and thus all of the stored magnetic energy comes from the mechanical energy conversion:

$$T_e = \left(-\frac{\partial W_{mag}}{\partial \theta_r}\right)_{\Psi=\text{const}} \quad (3.8)$$

In practice, such a situation occurs when the mechanical energy of the rotor is converted into magnetic stored energy and finally into iron and copper losses in the electric machine in the generator mode, supplying a braking resistor, for instance, on board of an urban people mover. The electromagnetic torque is negative and brakes the electric generator almost to zero speed.

The torque developed in a PM machine for zero current (zero electric energy transfer from the power source)—(cogging torque)—is due to the variation of PM-produced magnetic energy in the airgap with the rotor position, due to the stator and the core rotor slot openings (or saliencies). It is an almost zero-loss bidirectional conversion of the magnetic energy of PMs to mechanical energy and back. The average torque being zero, no net mechanical energy is produced but torque ripple occurs.

In most cases, $dW_e \neq 0(d\Psi \neq 0)$, so there is an electric energy transfer from (to) the electric power source, so we have to refer to Equation 3.7 in a modified form:

$$dW_{mag} = id\Psi - T_e d\theta_r; \quad I = \frac{\partial W_{mag}}{\partial \Psi} \quad (3.9)$$

Now, Ψ and θ_r are the independent variables. However, the relationship between flux linkages and currents in an electric machine (via inductances) is rather straightforward. Further on, we introduce a new energy function, W'_{mag}, called coenergy:

$$W'_{mag} = -W_{mag} + i \cdot \Psi \quad (3.10)$$

or

$$dW'_{mag} = i \cdot d\Psi + \Psi \cdot di - dW_{mag} \quad (3.11)$$

By substituting Equation 3.11 in Equation 3.9, to eliminate W_{mag}, we obtain

$$T_e = \left(+\frac{\partial W'_{mag}}{\partial \theta_r}\right)_{I=\text{const}}; \quad \Psi = +\frac{\partial W'_{mag}}{\partial i} \quad (3.12)$$

Magnetic energy and coenergy with a single excitation (current) source.

In general, $\Psi(i)$ functions are nonlinear due to the magnetic saturation of magnetic cores in electric machines as in Figure 3.2; from Equations 3.9 and 3.11:

$$W_{mag} = \int_0^{\Psi_m} id\Psi; \quad W'_{mag} = \int_0^{i_m} \Psi di \quad (3.13)$$

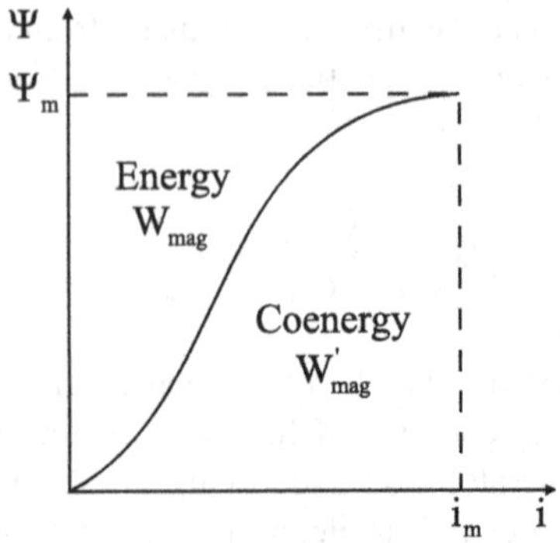

FIGURE 3.2 Magnetic energy and coenergy with a single excitation (current) source.

Equation 3.12 implies nonzero energy transfer from (to) the electric power source ($dW_e = i d\Psi \neq 0$).

The electromagnetic torque is nonzero according to Equations 3.9 and 3.12 only when the magnetic energy (or coenergy) of magnetic fields in the electric machine varies with respect to the rotor (mover) position, θ_r. The application of this general principle has led to numerous practical configurations of electric machines with rotary or linear motion. To classify electric machines, we will subsequently use two principles: with passive or active rotors and with fixed or traveling magnetic fields produced by the fixed part (stator) and/or the rotor in the small air space between them called airgap.

3.3 PASSIVE ROTOR ELECTRIC MACHINES

The passive rotor is made of a soft magnetic material and it does not have any windings or permanent magnets. In order to produce torque (magnetic coenergy variation with the rotor position), it has to have a magnetic saliency, that is, at least one self- or mutual-inductance should vary with the rotor position (Figure 3.3):

$$L(\theta_r) = L_0 + L_m \cos 2\theta_r \tag{3.14}$$

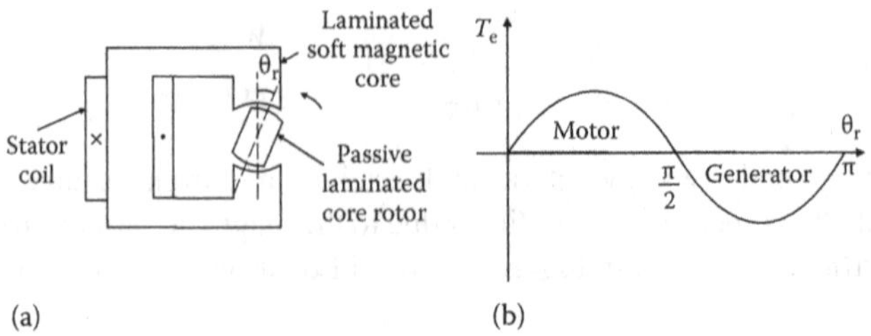

FIGURE 3.3 The single-phase reluctance—passive rotor—electric machine: (a) The configuration and (b) the electromagnetic torque vs. the rotor position, θ_r.

In this primitive configuration, both the stator and the rotor have magnetic saliency. The coenergy is calculated for a single inductance:

$$W'_{\text{mag}} = \int_0^i \Psi di = \int_0^i iL(\theta_r) di = \frac{L(\theta_r)i^2}{2} \tag{3.15}$$

From Equation 3.11, the electromagnetic torque, T_e, is

$$T_{\text{el}} = \frac{\partial W'_{\text{mag}}}{\partial \theta_r} = \frac{i^2}{2}\frac{\partial L}{\partial \theta_r} = -i^2 L_m \sin 2\theta_r \tag{3.16}$$

It is evident that the torque varies with the rotor position and is maximum at $\pi/4$. Left free, the rotor goes back to the $\theta_r = 0$ position. The rotor tends to align the stator field axis and this is the principle of reluctance electric machines. In the single-phase configuration of Figure 3.3, the reluctance machine cannot rotate continuously but may be used for a limited angle motion (ideally 0°–90°). The average torque per revolution is zero.

However, if we place three stators a, b, and c (Figure 3.4a) along the rotor periphery, spatially shifted by 120° and flowed by symmetrical a.c. currents i_a, i_b, and i_c:

$$i_{\text{a,b,c}} = I\sqrt{2}\sin\left(\omega_1 t - (i-1)\frac{2\pi}{3}\right) \tag{3.17}$$

then the average torque per revolution will not be zero and the machine may rotate continuously, as desired:

$$L_{\text{a,b,c}} = L_0 + L_m \cos\left(2\theta_r + (i-1)\frac{2\pi}{3}\right) \tag{3.18}$$

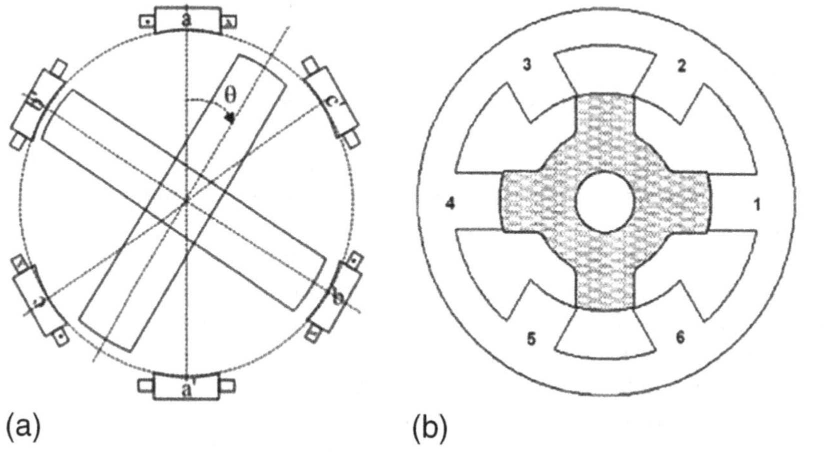

FIGURE 3.4 (a) Elementary three-phase reluctance (passive rotor) machine and (b) switched reluctance machine ($(2p_1)_{\text{rotor}} = 4$).

The rotor still has two salient poles ($2p_1 = 2$). In reality, multiple pole pair ($p_1 > 1$) configurations could be built (Figure 3.4). The machine coenergy formula now shows only three terms as no magnetic coupling between the three stator phases is considered:

$$W'_{\text{mag}} = \sum_{\text{a,b,c}} L_{\text{a,b,c}}(\theta_\text{r}) \frac{i^2_{\text{a,b,c}}}{2} \tag{3.19}$$

with the final electromagnetic torque, T_e:

$$T_{\text{e}3} = \left(\frac{\partial W'_{\text{mag}}}{\partial \theta_\text{r}} \right)_{i=\text{const}} = \frac{3I^2 L_\text{m}}{2} \cos(2\theta_\text{r} - 2\omega_1 t) \tag{3.20}$$

For constant angular rotor speed, ω_r:

$$\theta_\text{r} = \int \omega_\text{r} \text{d}t = \omega_\text{r} t + \theta_{\text{r}0} \tag{3.21}$$

So, finally, Equation 3.20 becomes

$$T_\text{e} = 3I^2 \frac{L_\text{m}}{2} \cos 2\left[\theta_0 + (\omega_\text{r} - \omega_1)t\right] \tag{3.22}$$

Only for $\omega_\text{r} = \omega_1$, the average torque is nonzero and there are idealy no torque pulsations:

$$T_{\text{e}3}(t) = 3I^2 \frac{L_\text{m}}{2} \cos 2\theta_0 \tag{3.23}$$

Thus, the stator current angular frequency, ω_1, and the rotor angular speed should be equal to each other to obtain, in a 2-pole, three-phase a.c. machine, a constant instantaneous (ideal) torque. This is the principle of three-phase (multiphase) a.c. machines with passive rotors (reluctance machines).

Again, multiple pole pair configurations are feasible, but then $\omega_\text{r} = p_1\Omega = 2\pi n p_1$ (n in rps).

Note: In practical reluctance synchronous machines ($\omega_1 = \omega_r$), distributed windings are used in the stator- and mutual (interphase)-inductances are nonzero and bring in more torque.

It is feasible to supply the three phases of the salient pole stator with sequences of the same polarity current pulses such that only one phase produces the torque at a time, as triggered by the adequate rotor position (when its torque is positive, Figure 3.3).

In this case, the stator is also magnetically salient (6, 12 poles) and uses concentrated coils (multiples of two per phase) and the rotor may have 4, 8 poles for three-phase machines. These are called switched reluctance machines (Figure 3.4b)—SRMs—and may be built with one, two, three, four, and more phases, working in sequence with controlled current pulses to produce low pulsation torque with the rotor position. Stepper reluctance motors work on the same principle, but their voltage (current) pulses are open-loop (feed-forward) referenced with some oscillations damping, eventually.

The majority of electric machines have an active rotor (with d.c. or a.c. coils or with PMs) but can still retain the magnetic rotor saliency and thus develop a reluctance torque component.

3.4 ACTIVE ROTOR ELECTRIC MACHINES

A primitive active rotor single-phase electric machine is shown in Figure 3.5.

The flux linkages of the stator/rotor circuits are

$$\psi_1 = L_{11}i_1 + L_{12}i_2; \quad \psi_2 = L_{12}i_1 + L_{22}i_2 \tag{3.24}$$

A coupling between the stator and the rotor windings is evident. The magnetic conergy is thus

$$W'_{\text{mag}} = \frac{1}{2}L_{11}i_1^2 + L_{12}i_1i_2 + \frac{1}{2}L_{22}i_2^2 \tag{3.25}$$

with the torque

$$T_e = \frac{\partial W'_{\text{mag}}}{\partial \theta_r} = \frac{1}{2}i_1^2\frac{\partial L_{11}}{\partial \theta_r} + \frac{1}{2}i_1^2\frac{\partial L_{22}}{\partial \theta_r} + i_1i_2\frac{\partial L_{12}}{\partial \theta_r} \tag{3.26}$$

The first two components of the torque are reluctance torques, as explained earlier. The third term is new and is called an interaction torque. We may draw the conclusion that at least the mutual-inductance, if not the stator self-inductances, has to vary with the rotor position to secure a nonzero torque. This case could be extended to three-phase machines also. Now we have to specify what types of stator/rotor currents—a.c. or d.c.—are used.

3.4.1 DC Rotor and AC Stator Currents

Let us consider that self-inductances above are constant (L_{11} and L_{22}) and $L_{12} = L_m \cos \theta_r$ (2 poles) while $i_1 = I\sqrt{2}\sin\omega_1 t$ and $i_2 = I_{20}$ = const. From Equation 3.26, we get

$$T_{e1} = II_{20}\sqrt{2}L_m\frac{1}{2}\left[\cos\left(\omega_1 t + \theta_r\right) - \cos\left(\omega_1 t - \theta_r\right)\right] \tag{3.27}$$

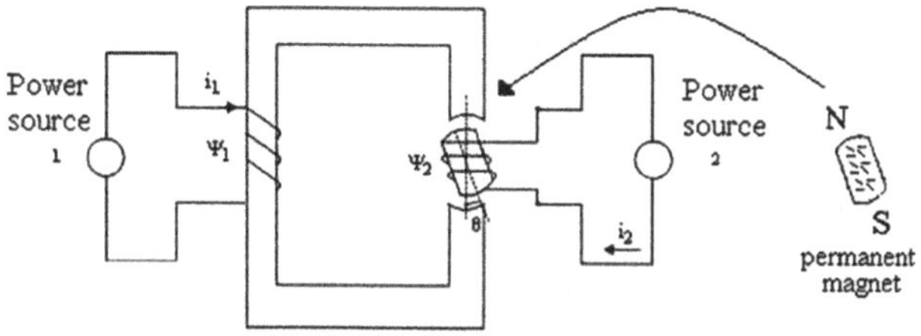

FIGURE 3.5 Primitive active rotor single-phase electric machine.

A nonzero average torque is obtained only if either $\omega_1 = \omega_r$ or $\omega_1 = -\omega_r$ ($\theta_r = \omega_r t$). However, this means that one term in Equation 3.27 is constant while the other one varies with $2\omega_1$. This is typical to the single-phase synchronous ($\omega_1 = \omega_r$) motor with active (I_{20} = const or PM) rotor. To eliminate the pulsating torque in Equation 3.27 with $\theta_r - \omega_1 t = \theta_0$ = const ($\omega_1 = \omega_r$), we may imagine three stators axially shifted by 120° with each other and fed through three-phase a.c. currents $\left(i_{a,b,c} = I\sqrt{2}\sin\left(\omega_1 t - (i-1)(2\pi/3)\right)\right)$. If we neglect the coupling between phases,

$$T_{e3} = I_{20}\left(i_a \frac{\partial L_{ma}}{\partial \theta_r} + i_b \frac{\partial L_{mb}}{\partial \theta_r} + i_c \frac{\partial L_{mc}}{\partial \theta_r}\right) \tag{3.28}$$

with

$$L_{ma,mb,mc} = L_m \cos\left(\theta_r - (i-1)\frac{2\pi}{3}\right); \quad i = 1,2,3 \tag{3.29}$$

the torque becomes

$$T_{e3} = I_{20} I\sqrt{2} L_m \frac{3}{2}\sin\left((\omega_1 - \omega_r)t - \theta_0\right) \tag{3.30}$$

Again, only for $\omega_1 = \omega_r$, the instantaneous torque is constant; this is the case for primitive synchronous three-phase machines with d.c. (or PM) rotor excitation.

Note: We mentioned the PM here directly, but the PM may be modeled by a constant (d.c.) mmf (current) ideal (superconducting) coil, where mmf refers to ampereturns or magnetomotive force, $N_{2x}I_{20} = H_{cx}h_{PM}$; H_c—coercitive field; h_{PM}—PM thickness along magnetization direction;

So speed (ω_r) control may be operated only through frequency (ω_1) control, performed, in general, through power electronics.

3.4.2 AC Currents in the Rotor and the Stator

Let us consider the same single-phase configuration, but the rotor current, i_2, is

$$i_2 = I_2\sqrt{2}\sin\omega_2 t \tag{3.31}$$

The torque from Equation 3.25 becomes

$$T_{e1} = -II_2 \sin\omega_1 t\left(2L_m \sin\omega_2 t\right)\sin\theta_r$$

For constant speed ω_r, $\theta_r = \omega_r t$, it could be demonstrated that the torque may have a nonzero average component when

$$\omega_1 \mp \omega_2 = \omega_r \tag{3.32}$$

Even in this case, the torque has three pulsating additional components.

To eliminate the pulsating components of the torque, we will place three windings on the stator and on the rotor, supplied by a.c. symmetric currents of frequency ω_1 and ω_2, respectively.

This is how we obtain a total torque:

$$T_{e3} = 3L_m I_1 I_2 \sin\left(\left(\omega_1 - \omega_2 - \omega_r\right)t + \gamma\right) \tag{3.33}$$

Again, with $\omega_1 = \omega_2 + \omega_r$ ($\omega_2 \gtrless 0$), the torque is

$$T_{e3} = 3L_m I_1 I_2 \sin\left(\gamma\right) \tag{3.34}$$

where γ is the phase angle between the stator and the rotor currents when represented at the same frequency (ω_1). Now this is, in fact, the so-called doubly fed induction machine where the rotor frequency currents of frequency $\omega_2 = \omega_1 - \omega_r$ should be provided from outside by a PWM inverter.

It is also possible to place short-circuited windings (bars in rotor slots with end rings) and thus "induce" emfs in the rotor at exactly $\omega_2 = \omega_1 - \omega_r$ frequency by motion.

This is the cage–rotor induction machine, the workhorse of the industry.

3.4.3 DC (PM) Stator and AC Rotor

When the stator is d.c.-fed and has $2p_1$ poles surrounded by d.c. coils (or PMs) and the rotor is a.c.-fed through m a.c. coils uniformly placed along the rotor periphery (with $2\pi/2p_1m$ angle span), with currents showing a $2\pi/m$ time lag (Figure 3.6),

$$I_{2i} = I_2\sqrt{2}\sin\left(\omega_2 t - \left(i-1\right)\frac{2\pi}{m}\right) \tag{3.35}$$

The mutual-inductances between the stator winding and the rotor coils are

$$L_{12i} = L_m \cos\left(\omega_r t - \left(i-1\right)\frac{2\pi}{m}\right); \quad i = 1, m \tag{3.36}$$

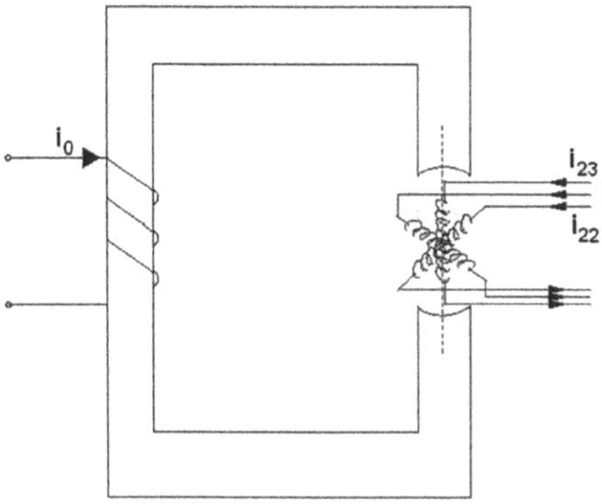

FIGURE 3.6 DC (PM) stator and a.c. multiphase rotor machine.

So the torque T_e is

$$T_e = I_2 I_0 \sqrt{2} L_m \sum_{i=1}^{m} \cos\left(\omega_2 t - (i-1)\frac{2\pi}{m}\right) \cos\left(\theta_r - (i-1)\frac{2\pi}{m}\right) \tag{3.37}$$

Finally,

$$T_{en} = I_2 I_0 \sqrt{2} L_m \frac{n}{2} \sin(\omega_2 t - \theta_r) \tag{3.38}$$

The electromagnetic average torque is nonzero only if $\omega_2 = \omega_r$ (because $\theta_r = \omega_r t + \theta_0$):

$$T_{en} = -I_2 I_0 \sqrt{2} L_m \frac{n}{2} \sin \theta_0 \tag{3.39}$$

So the rotor currents frequency ω_2 is equal to the rotor speed ω_r. This condition is met by the brush–commutator machines that transform d.c. brush currents, through a mechanical commutator, into a.c. currents of $\omega_2 = \omega_r$.

The a.c. currents in the rotor coils of the practical brush–commutator machines are trapezoidal rather than sinusoidal, but the principle above still holds.

The frequency theorem-based classification $\omega_1 = \omega_2 + \omega_r$ has led to the identification of a.c. stator synchronous machines (SMs) with d.c. (PM) rotor excitation, or with a passive (magnetically salient) rotor, to the switched reluctance machines (similar to SMs, but with passive rotors and sequential position-triggered current control pulses) and to induction machines with a.c. stators and a.c. rotors in both configurations: doubly fed or with cage–rotor.

Finally, the brush–commutator machines have been also identified as d.c. (PM) stator and a.c. multiphase current rotor machines.

No machine can apparently escape the frequency theorem principle, so we have them all.

However, we may identify the basic electric machine (brush, synchronous induction ones) also by the type of the magnetic field they show in the airgap: traveling (moving) or fixed.

This time, the condition to obtain an ideal (ripple-less) torque is that the stator- and the rotor-produced airgap magnetic fields should be at a standstill with each other.

3.5 FIX MAGNETIC FIELD (BRUSH–COMMUTATOR) ELECTRIC MACHINES

If we slightly modify the magnetic circuit of the primitive machine in Figure 3.6, and add the brush–commutator, we obtain the contemporary brush–commutator machine (Figure 3.7a and b).

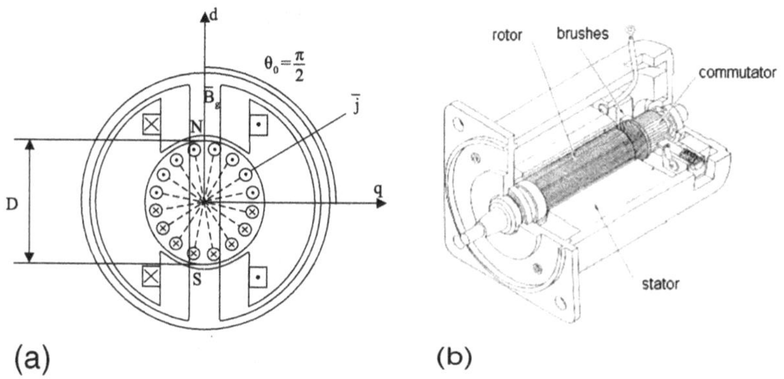

FIGURE 3.7 Two-pole (2_{p1} = 2) electric machine (a) with a fixed magnetic field and (b) without rotor slots.

Through the brush–commutator, the currents in the rotor coils below any stator pole have the same polarity and they alter the polarity when they move under the next stator pole (Figure 3.7a). Now, the stator field is maximum in the stator pole axis; so this is its axis. The rotor currents produce an airgap magnetic field whose maximum mmf lies 90° away from the stator field axis (in the so-called neutral axis). The d.c. (PM) stator magnetic field axis stays the same irrespective of the rotor speed while the rotor current's magnetic field axis is also fixed but 90° away from the stator field axis.

So the two magnetic fields—produced by the stator and the rotor—are both at a standstill and are thus fixed to each other. And this is only due to the brush (mechanical)–commutator.

As the angle between the two field axes is 90°, the interaction between them to produce a torque is optimum. Also, the variation of the rotor current (to vary the torque) does not lead to any emf in the stator d.c. (field) winding.

So the torque (rotor) and the field (stator) current controls are decoupled by machine topology. Permanent magnets may replace the d.c. excitation stator winding, but the principle of fixed magnetic fields interaction still holds.

We may now calculate the torque by the (BIL) tangential force formula:

$$T_e = \frac{D}{2} F_t = \frac{D}{2} B_{gav} \left(A \times \pi D \right) L \tag{3.40}$$

where B_{gav} is the average stator-produced airgap flux density, A is the average rotor current loading in A turns/m, D is the rotor diameter, and L is the laminated core stack length (about the same in the stator and the rotor).

If A would be independent of D and L, with given B_{gav}, the torque would be proportional to the rotor volume. So it is the torque that decides the size of the machine.

3.6 TRAVELING FIELD ELECTRIC MACHINES

Let us reclaim the synchronous machine example, with d.c. (or PM) rotor excitation (Figure 3.8).

First of all, the d.c. or PM rotor heteropolar magnetic field has its axis in the rotor pole axis *d*.

This excitation field becomes a traveling field only when the rotor moves, say at speed ω_r.

But with respect to the rotor, it is (if curvature is neglected)

$$B_F^r(x_r) = B_{Fm}\sin\frac{\pi}{\tau}x_r; \quad \tau = \frac{\pi D}{2p_1}; \quad \tau\text{-pole pitch} \tag{3.41}$$

With respect to the stator (τ - excitation flux density half period (spatially)):

$$x_r = x_s - vt = x_s - \omega_r\frac{\tau}{\pi}t \tag{3.42}$$

at a constant speed ω_r, where x_s is the stator coordinate. So, with respect to the stator (x_s),

$$B_F^s(x_s,t) = B_{Fm}\sin\left(\frac{\pi}{\tau}x_s - \omega_r t\right) \tag{3.43}$$

To the stator, this is truly a traveling field at rotor speed ω_r.

Let us suppose that the stator currents produce a linear current density $A_s(x_s, t)$. Consequently, the torque (as in Equation 3.40) is

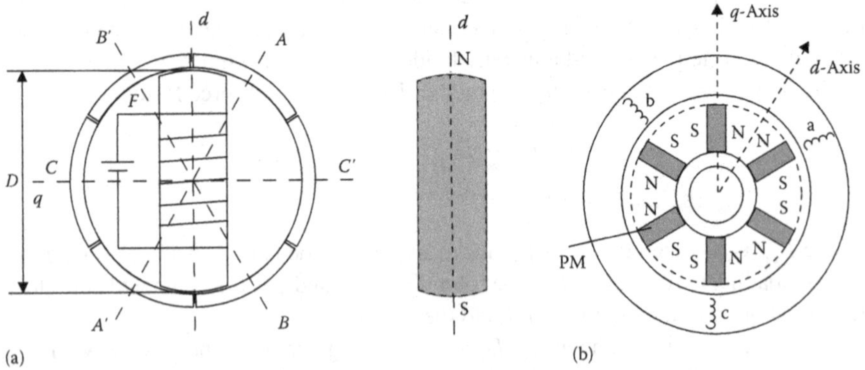

FIGURE 3.8 Traveling field electric synchronous machines: (a) with d.c. rotor excitation and $2p_1 = 2$ and (b) with PM rotor excitation and $2p_1 = 6$.

$$T_e(t) = \frac{D}{2}L\int_0^{2p_1\tau} B_F^s(x_s,t)A_s(x_s,t)\mathrm{d}x \tag{3.44}$$

To obtain a constant instantaneous (ideal) torque (to get rid of time dependence), it is evident that A_s (x_s, t) should be of the form

$$A_s(x_r,t) = A_{ms}\sin\left(\frac{\pi}{\tau}x_s - \omega_r t - \theta_0\right) \tag{3.45}$$

In such conditions,

$$T_e(t) = \frac{\pi D^2 L}{4}B_{Fm}A_{ms}\cos\theta_0 \tag{3.46}$$

So, both the rotor airgap field and the stator current loading (also the stator magnetomotive force) travel at rotor speed under steady state to produce constant instantaneous torque.

As the stator mmf $F_s(x,t) = \int A_s(x_s,t)\mathrm{d}x$, F_s and A_s are 90° phase-shifted. So the maximum torque is obtained at $\theta_0 = 0$ (Equation 3.46), but this means that the excitation rotor field axis and the stator mmf (and field) axis are both running at rotor speed and are 90° phase-shifted; as for the brush–commutator machine, however, both traveling fields are at a standstill (are fixed) with each other.

A similar rationale may be applied to the doubly fed and cage–rotor induction machines whose stator and rotor currents produce magnetic fields traveling at speed ω_1 with respect to the stator, but the frequency of the rotor currents $\omega_2 = \omega_1 - \omega_r \neq 0$. This is why induction machines are also called asynchronous machines.

3.7 TYPES OF LINEAR ELECTRIC MACHINES

For all types of rotary motion electric machines, that have cylindrical or disk-shaped rotors, there is at least one linear version, which is obtained by cutting it longitudinally and spreading it into a plane (flat type) or even rerolling it along an axial axis (tubular types for limited excursion motion)—Figures 3.9 and 3.10.

Three-phase linear induction and synchronous motors have found application in urban people movers on wheels for propulsion and in very rapid interurban transportation (at 400–500 km/h) with magnetic suspension (MAGLEVs) [2].

Single-phase synchronous (PM-mover or stator PM and iron-mover) tubular configurations for linear oscillatory motion are used to drive small-power compressors (e.g., refrigerators) or linear generators (e.g., Stirling engine prime mover)—Figure 3.11a and b. They may be assimilated with PM plunger solenoids connected to the grid or power electronics–controlled.

The loudspeaker/microphone is a typical case of a linear oscillatory stator-PM, coil-mover single-phase synchronous machine (Figure 3.11c). It may be used as an electrodynamic vibrator with frequencies up to around 500 Hz.

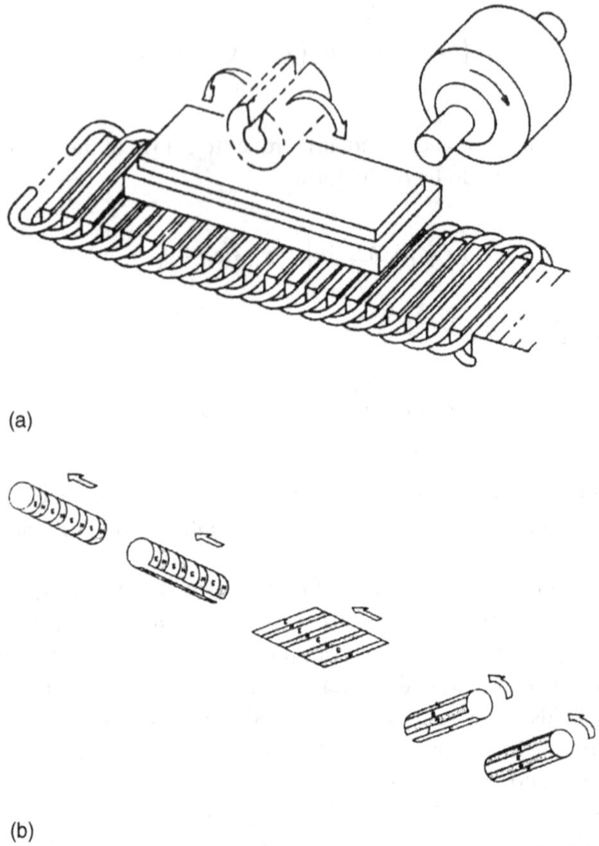

FIGURE 3.9 Obtaining a three-phase linear induction machine (with traveling field) from a rotary machine: (a) Flat and (b) tubular.

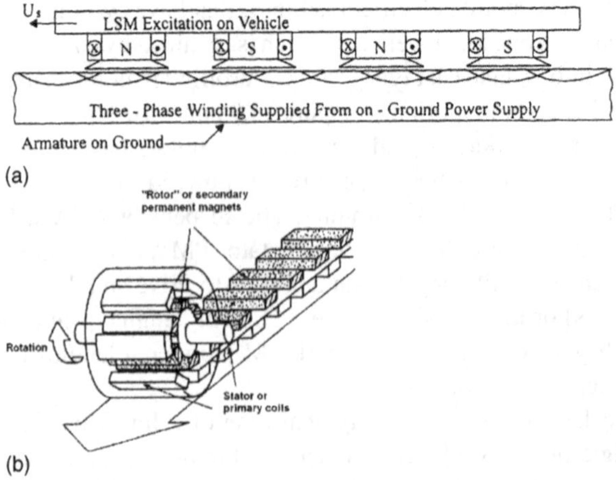

FIGURE 3.10 Three-phase linear flat synchronous machines (with traveling field): (a) With d.c. heteropolar excitation (NSNS) and (b) with PM excitation (NSNS).

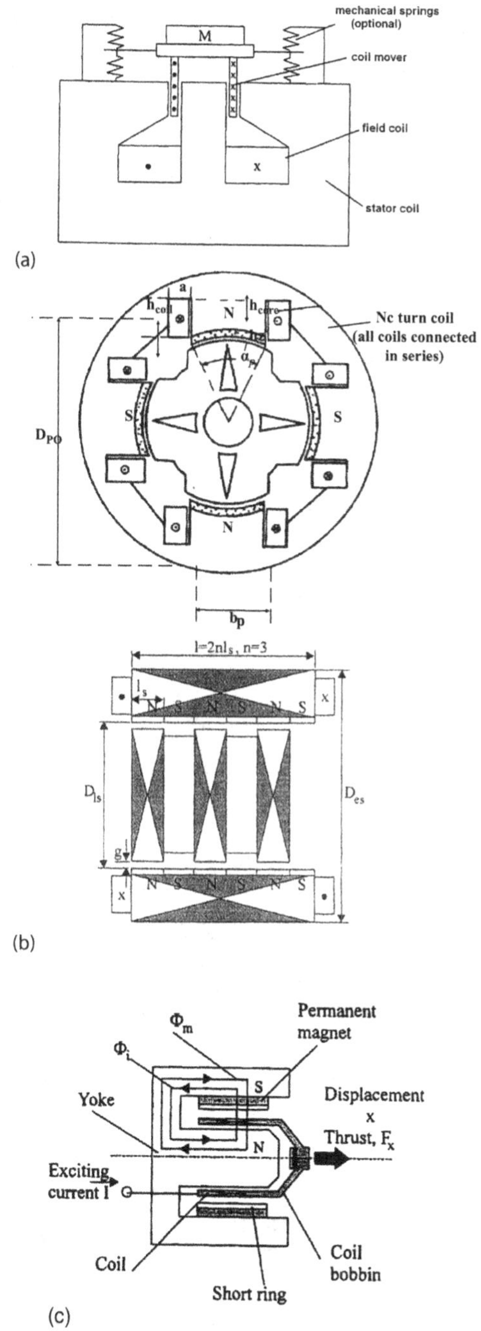

FIGURE 3.11 Linear oscillatory motor/generator: (a) With PM-mover, (b) with stator-PM and iron (reluctance)-mover, and (c) with stator-PM and coil-mover (the microphone/loudspeaker).

Example 3.1: The Loudspeaker/Microphone as a Linear PM Motor

Let us consider the tubular configuration in Figure 3.12 that contains

- A tubular inner and an outer soft magnetic composite shell of soft ferrite or Somaloy 550 stator core, etc.
- A tubular PM, radially magnetized, placed also on the stator
- A nonmagnetic shell holds the tubular copper multiturn coil, which constitutes the mover, connected by flexible terminals to an a.c. (controlled or uncontrolled) power source

The PM flux paths close axially through the two shells and radially through the PM and the coil-mover to produce a unipolar magnetic field B_{PM} in the coil. Then, when the coil carries a current i, a BIL force is developed. The force changes the sign when the current changes the polarity, and thus an oscillatory motion is produced.

To increase the efficiency, the energy necessary to accelerate and decelerate the mover at stroke ends is stored in the mechanical springs, which, in the compressor drives, may be designed to be mechanically resonant at the imposed electric frequency of the currents in the coil ($f_m = f_e$). Very good efficiencies could be obtained down to 20 W power or less in this situation.

The emf V_e in the coil comes from the $B_{PM}Ul$ formula (l—mean coil turn length, U—linear speed):

$$V_e = B_{PM}\pi D_{arc}N_c \frac{dx}{dt} \tag{3.47}$$

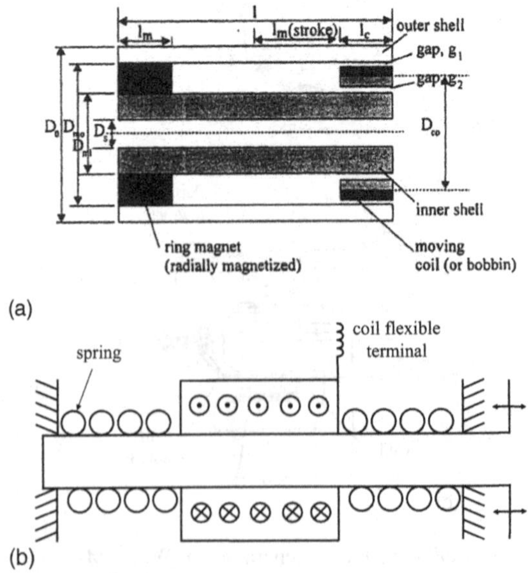

FIGURE 3.12 The loudspeaker/microphone as (a) a linear PM machine and (b) a mover in the middle with springs.

F_e springs from the BIL formula:

$$F_e = B_{PM}\pi D_{arc} i_c N_c \tag{3.48}$$

where N_c is the turns in series per coil.

As the oscillatory motion is quasi-sinusoidal,

$$x = x_1 \cos \omega_r t \tag{3.49}$$

It follows from Equation 3.47 that the emf is sinusoidal:

$$V_e(t) = -B_{PM}\pi D_{mar} N_c \omega_r x_1 \sin \omega_r t \tag{3.50}$$

As the coil inductance does not vary with the mover position L_c = const, the motion and the voltage equations are

$$M_{mover}\frac{dU}{dt} = F_e - F_{load} - K_{spring}\left(X - l_{stoke}/2\right) \tag{3.51}$$

$$\frac{dx}{dt} = U$$

$$i_c R_c + L_c \frac{di_c}{dt} = V_c(t) - V_e(t) \tag{3.52}$$

Under mechanical resonance conditions,

$$\omega_r = \sqrt{K_{spring}/M_{mover}} = \omega_1; \quad F_{load} \approx K_{load} U(t) \tag{3.53}$$

If $V_e = V_{em} \cos \omega_r t$, then the current, under harmonic (steady-state sinusoidal) motion, will have the same frequency, ω_r, and we have a synchronous single-phase PM machine.

This time, the emf is sinusoidal in time due to the harmonic motion imposed by the mechanical springs.

As the mechanical springs move back and forth, they store the mover kinetic energy at stroke ends, and thus the electromagnetic force is responsible mainly for startup and then to cover the load force (for the compressor, this load may be considered proportional to the speed ω_r: $F_{load} \approx K_{load}\, U(t)$).

So, for steady-state harmonic motion,

$$\begin{aligned} \underline{V_1} &= V_0\sqrt{2} e^{j(\omega_r t + \gamma)} \\ \underline{U_1} &= j\omega_r \underline{X_1} \end{aligned} \tag{3.54}$$

and with $\frac{d}{dt} = j\omega_r$ in Equations 3.51 and 3.52, we obtain in complex members,

$$j\omega_r \underline{I_1} = \frac{\underline{V_1} - R_c \underline{I_1} - j\omega_r K_{PM}}{L_c}; \quad K_{PM} = B_{PM}\pi D_{arc} N_c \tag{3.55}$$

(Continued)

Example 3.1: (Continued)

$$-\omega_r^2 \underline{X_1} = \frac{K_{PM}\underline{I_1} - K_{spring}\underline{X_1} - j\omega_r C_{load}\underline{X_1}}{M_{mover}} \tag{3.56}$$

with

$$K_{spring} = M_{mover}\omega_r^2 \tag{3.57}$$

and mechanical resonance conditions,

$$\underline{I_1} = \underline{X_1} j\omega_r \frac{C_{load}}{K_{PM}} = \underline{U_1}\frac{C_{load}}{K_{PM}} \tag{3.58}$$

From Equation 3.55, $\underline{I_1}$ may be calculated and from Equation 3.58, $\underline{X_1}$ may be calculated as sinusoidal current and position phasors (amplitudes and phases included), respectively.

For sinusoidal resonant motion, with the load force proportional to speed, Equation 3.58 shows the current $\underline{I_1}$ in phase with linear speed $\underline{U_1}$, that is, with the emf V_{e1}, and thus the maximum force per current is obtained (best efficiency).

Example 3.2: Linear Compressor Coil-Mover PM Linear Motor

In a numerical example with $P_n = 125$ W, 120 V, 60 Hz compressor load with C_{load} = 86.8N s/m, l_{stroke} = 0.01 m, R_c = 6.3Ω, L_c = 91.6 mN, K_{PM} = 78.6 Wb/m, and M_{mover} = 0.57kg, at resonance conditions (K_{spring} = M_{rotor} $\omega_r^2 = 0.57(2\pi 60)^2$ = 76772N/m), we get the motion amplitude X_1 from Equations 3.55 and 3.58:

$$X_1 = 4.8 \cdot 10^{-3} \text{ m} \quad \text{for } I_n = 1.46 \text{ A (RMS)}$$

The copper losses $p_{copper} = R_c I_n^2 = 6.3 \times 1.46^2 = 13.4$ W; with 5 W more for iron and mechanical spring loss, the efficiency is

$$\eta_n = \frac{P_n - \Sigma p}{P_r} = \frac{120 - 13.4 - 5}{120} = 0.877$$

The power factor

$$\cos\varphi_n = \frac{P_n}{V_n I_n \eta_n} = \frac{120}{120 \times 1.46 \times 0.877} = 0.817$$

This performance is quite satisfactory as the maximum speed $|U| = \omega_r X_1 = 2 \times \pi \times 60 \times 4.8 \times 10^{-3} = 1.76$ m/s.

- The above performance has been secured by the spring's job of converting the mover kinetic energy to the spring's potential energy to relieve the electromagnetic force and source from handling the mover acceleration and deceleration through the oscillatory motion.

- If the load decreases, the motion amplitude (X_1) decreases, as expected, but the efficiency remains good.
- Designed with an electric frequency equal to mechanical eigen (resonance) frequency, the electric frequency, ω_1, should be kept constant or varied slightly through power electronics to track the small mechanical resonance frequency variations due to temperature and wearing (K_{spring} changes) and thus maintain a high efficiency.

More information on linear electric machines can be found in [3].

3.8 FLUX-MODULATION ELECTRIC MACHINES: A NEW BREED

So far in this chapter, we did present the principles of d.c. brush and a.c. traveling field rotary machines that are now in wide use in various industries. Also we did describe in principle their linear motion machines counterparts.

To complete the picture here we present briefly the so-called flux-modulation electric machines which reunite quite a few "novel" electric machine topologies proposed in the last decades but which did not reach yet widespread industrialization in high torque density low-medium speed electric motor/generator drives.

Among these novel entries we mention here [4–9]:

- the switched reluctance machine (doubly salient)/single rotor/simply of doubly fed (Figure 3.13)
- the flux–switching stator PM machine (doubly salient)/single rotor (Figure 3.14)
- the flux-reversal stator and rotor PM machine (doubly salient)/single rotor (Figure 3.15)
- the Vernier rotor PM doubly salient machine/single rotor (Figure 3.16)
- brushless doubly-fed reluctance machine (simply salient)/single rotor (Figure 3.17)

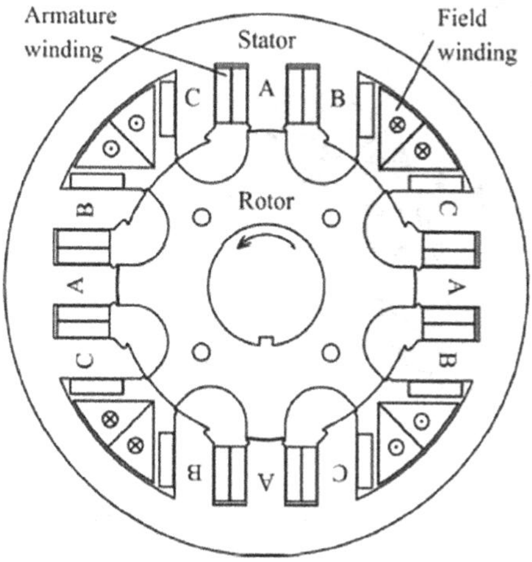

FIGURE 3.13 Dual fed three phase switched reluctance motors.

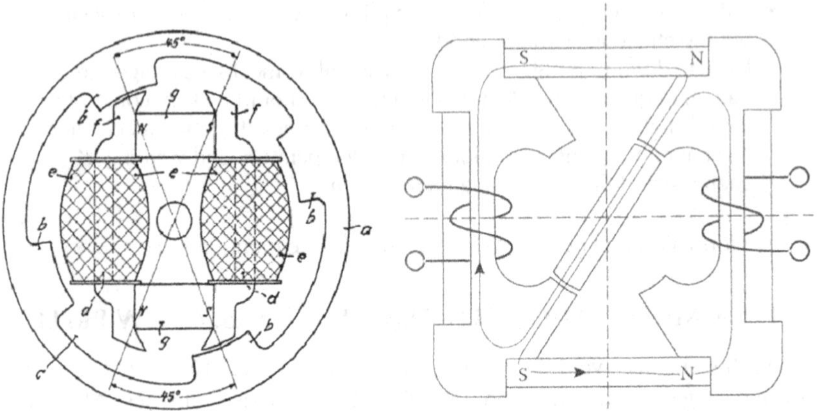

(a) (b)

FIGURE 3.14. Typical flux-switching PM motor.

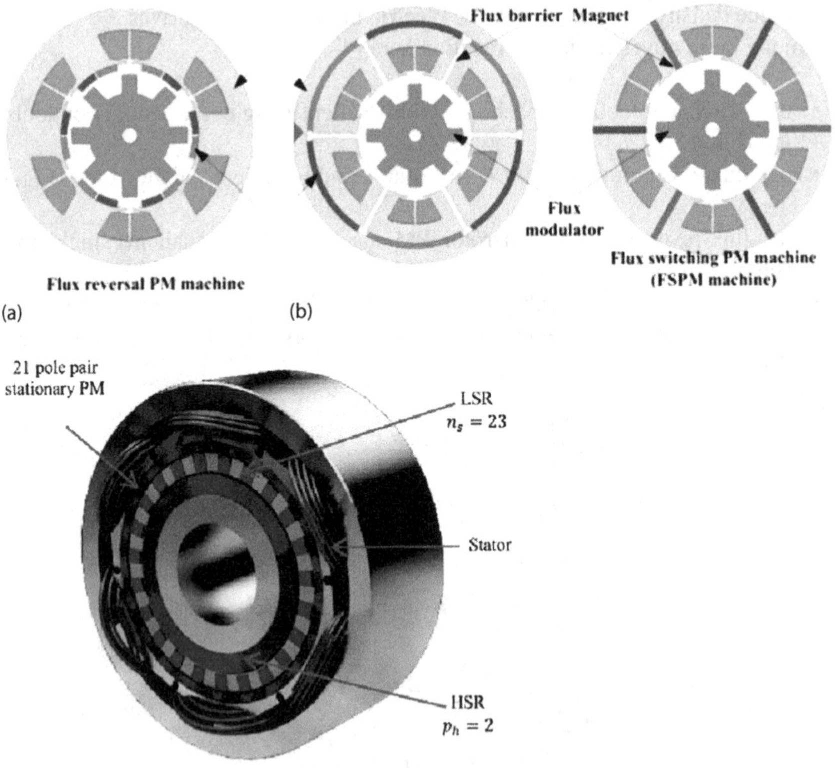

(a) (b) (c)

FIGURE 3.15 Flux reversal PM motor: (a) with stator PM, (b) with rotor PM, (c) magnetically geared.

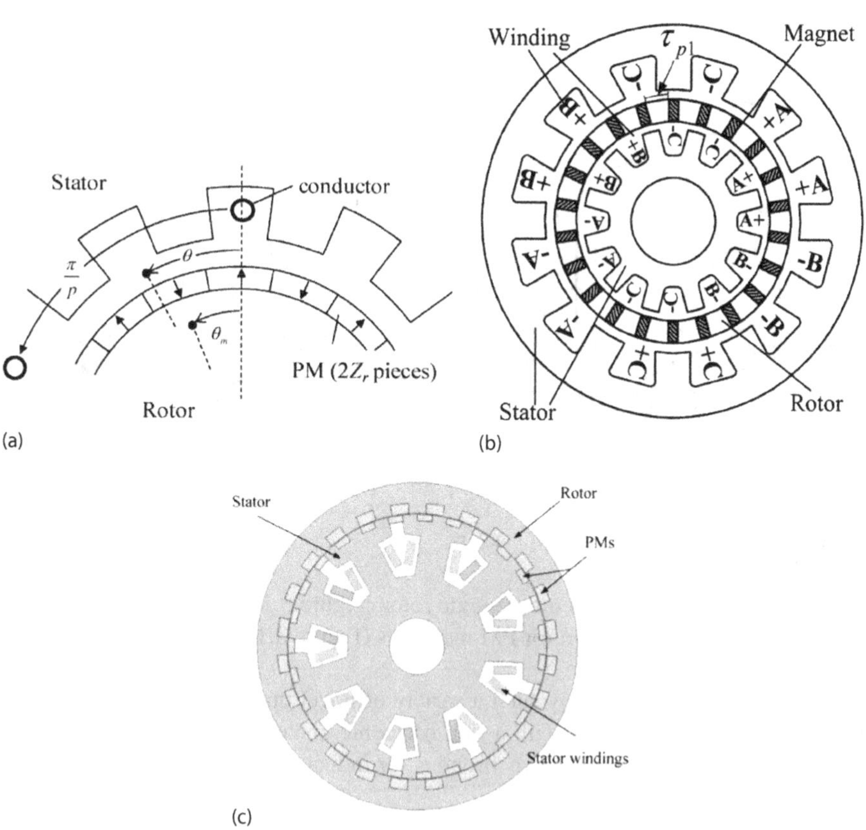

FIGURE 3.16 Vernier rotor PM doubly salient motor: (a) with surface PMs, (b) with spoke PMs, (c) with stator and rotor PMs.

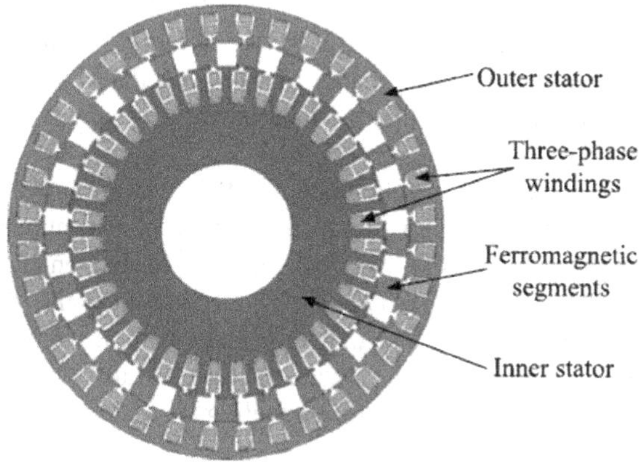

FIGURE 3.17 Brushless doubly fed reluctance machine.

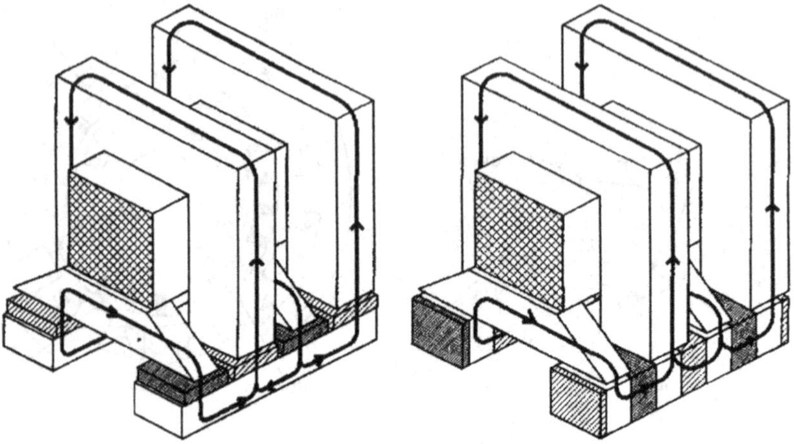

(a) (b)

FIGURE 3.18 Transverse flux PM machine.

- transverse flux PM machines/single rotor (Figure 3.18)
- magnetic-geared dual-rotor PM machines (Figure 3.19)

One way to approach unitarily such a variety of novel machines is to find periodic mmf and airgap permeances with a rich harmonic contents which, in some conditions, reach synchronism as in synchronous (or induction) machines where the fundamental mmfs are in synchronism.

A kind of flux modulation occurs and the equivalent number of poles and frequency increase, for smaller speed and high torque density is obtained by a kind of torque magnification process where variable magnetic reluctance plays a crucial role.

It may be asked why we can not only increase the number of poles of tooth-wound PMSMs to obtain similar results. The answer is: because there is not enough room for tooth-wound coils to secure high torque density. But higher mmf per pole in "flux-modulation" machines comes with the price of larger "armature reaction" and thus, in general, lower power factor. That is, a larger kVA PWM converter is required. To briefly present the principle of flux modulation we treat here briefly the magnetic gear (Figure 3.20) [5].

The magnetic gear has basically three parts (all or two are movable). The two—outer (slow) and inner (fast)—rotors are provided with $p_1 > p_2$ number of pole pairs of permanent magnets. Here (Figure 3.20) they are placed on the rotor surfaces (it could be of spoke shape, also). The part in the middle is the "flux-modulator" made of laminated (or SMC) p_{FM} ferromagnetic poles with the same number of non-magnetic inter poles. A variable reluctance structure is thus obtained. The two PM rotors, when rotated at Ω_{ro} and Ω_{ri}, produce traveling (fundamental) mmfs F_{o1} and F_{i1}:

$$F_{o1} = F_{o1m}\cos\left(p_1\theta - k_1 p_1 \Omega_{ro} + \gamma_1\right) \tag{3.59}$$

$$F_{i1} = F_{i1m}\cos\left(p_2\theta - k_2 p_2 \Omega_{ri} + \gamma_2\right) \tag{3.60}$$

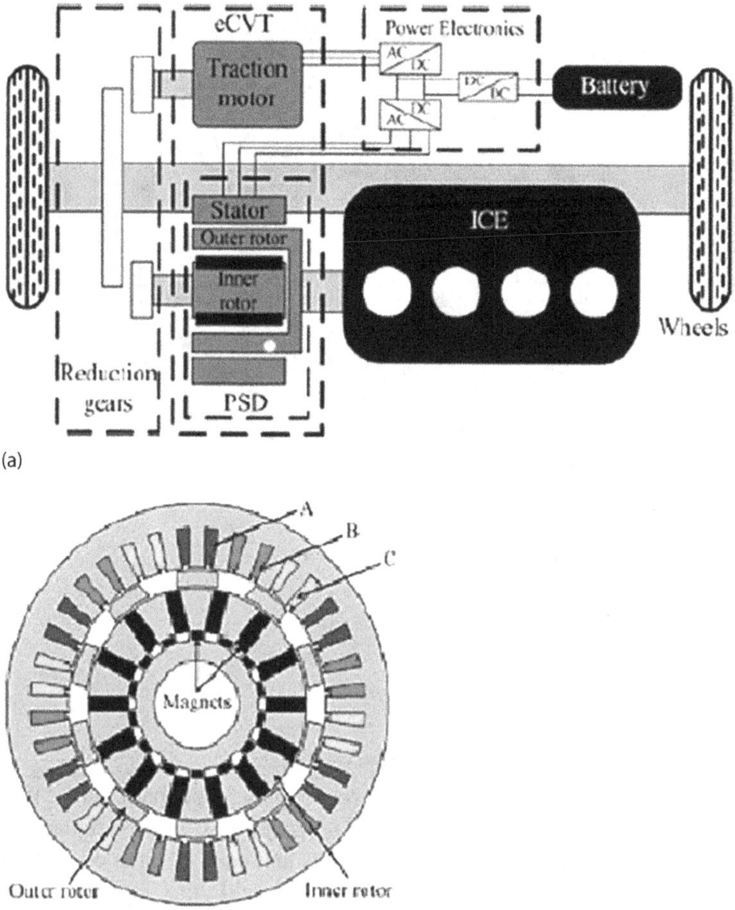

FIGURE 3.19 Magnetic geared dual rotor PM machine: automotive application, (a) and basic structure, (b).

The coefficients k_1, k_2 = ±1 refer to the motion direction (trigonometric or anti-trigonometric). On the other hand, the "flux-modulator" produces a periodic magnetic permeance λ_{FM1}:

$$\lambda_{FM1} = \lambda_{ave} + \lambda_1 \cos\left(p_{FM}\theta - p_{FM}\Omega_{FM} + \Phi_{FM}\right) \tag{3.61}$$

As in general $p_1 \neq p_2 \neq p_{FM}$, the average airgap magnetic permeance λ_{ave} does not produce torque. Both mmfs F_{o1} and F_{i1} interact with airgap magnetic permeance fundamental in (3.61) and produce in some conditions synchronous airgap flux density and mmf waves:

$$p_{FM}\Omega_{FM} = p_2\Omega_2 - p_1\Omega_1 \tag{3.62}$$

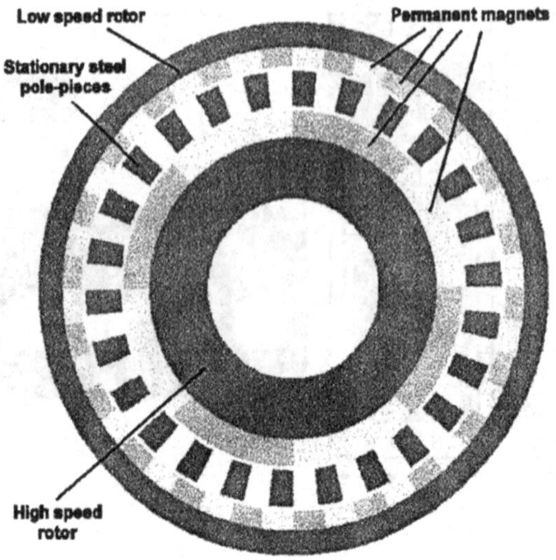

FIGURE 3.20 Standard magnetic gear (with two rotors)

in other words for:

$$p_{FM} = p_2 - p_1 \tag{3.63}$$

such synchronization conditions are met.

Example: p_1 = 22-pole pairs (outer PM rotor), p_2 = 3 pole pairs (inner PM rotor) and p_{FM} = 22. Each of the two PM rotors of magnetic gear may be replaced by distributed (or rotor wound) a.c. (or d.c.) windings and each of them could be kept at standstill while the flux-modulator may move, to yield "flux modulation" electric machines.

The FM may be assembled in the outer part with distributed (or rotor wound) winding fed at frequency f_1, when the flux modulator may be eliminated and the machine may have a stator and a rotor (Vernier machine, Figure 3.16). In the flux-switching (stator PM) machine a stationary outer-part winding is a.c. fed to produce a mmf at $\Omega_1 = 2\pi f_1/p_1$ speed. The PMs on the inner part are "relocated" in the stator and the modulator rotor moves with the speed $\Omega_r = \Omega_{FM}$ in (3.62) (Figure 3.14). The flux reversal machine is similar to the flux-switching machine even when the PMs are placed in the rotor. Now the TF-PMSM (Figure 3.18) armature reaction is homopolar ($p_1 = 0$), the rotor parts have p_2 PM pole pairs while the stator core plays also the role of the flux-modulator: $p_{FM} = p_2$; thus the machine acts as a synchronous motor.

Now the brushless dual stator winding reluctance machine (Figure 3.17) has the outer and inner part with stationary a.c. windings (in one or two stators)—fed at f_1, f_2 frequencies—while the rotor is in fact the flux modulator. In this case again:

$$\Omega_{FM} = \Omega_r = 2\pi\left(f_1 \pm f_2\right)/p_r;\ p_r = p_1 \pm p_2 \tag{3.64}$$

A magnetic-geared FM machine (Figure 3.19) with two rotors as shown follows again (3.62) but this time the torques on the stator T_s, on outer (slow) rotor $T_{FM} = T_{or}$ and on inner rotor T_{ir} (in lossless conditions) are related as:

$$T_{FM} = T_{or} = \frac{-p_{or}}{p_{FM}} T_{em}; \quad T_{ir} = \frac{p_{ir}}{p_{FM}} T_{em} \tag{3.65}$$

in this case (Figure 3.19) the inner rotor is provided with spoke-shape PMs.

In the flux-reversal magnetic geared machine (Figure 3.15c) the stator has only 6 (or multiple of 6) large stator teeth and coils with an mmf with p_1 = 2-pole pairs, the inner high-speed PM rotor has p_2 = 2-pole pairs. There are also 21 PM pole pairs on the stator $p_1 = 21$ and $p_{FM} = 23$ pole pairs (segments) on the outer rotor (the flux modulator). The magnetic gear part has a transmission ratio $G_r = p_{FM}/p_1 = 23/2 = 11.5$. As seen in Figure 3.15c the inner (faster) PM rotor is coupled to the load. In fact, a 2-pole pair stator mmf and a 2-pole pair inner plus a magnetic gear make this machine a high torque density one. The large torque at low speed will cost a surplus of PMs (on the stator). It is even feasible to add magnets between the ferromagnetic poles of the flux-modulator. Consequent pole PMs on both of the FM (outer rotor) and on inner rotor are also feasible [8].

Note. For each rotary FM machine a linear counterpart is feasible [9].

3.9 SUMMARY

- This book deals only with electromagnetic machines that use magnetic energy storage.
- There are also electrostatic machines with electrostatic energy conversion, but they are used in sub mm (10^{-3} m) diameter micromachines. We will not discuss them here (see *IEEE Transactions on Microelectromechanical Systems, for knowledge acquisition*).
- There are also piezoelectric (traveling) field machines for a large torque (Nm or more) at very small speeds—rotary and linear (see [10,11]). However, they are not discussed in this book.
- In this chapter, the electromagnetic torque of electric machines is derived from the stored magnetic energy (coenergy), based on the generalized force/torque concept.
- The main types of electric machines are derived, with respect to passive and active (magnetically or electrically) rotors, based on the frequency theorem $\omega_1 = \omega_2 + \omega_r$ (with ω_1—stator electric frequency, ω_2—rotor electric frequency, and ω_r—rotor mechanical frequency in electric terms ($\omega_r = p_1\Omega_1$; p_1—pole pairs or electric periods per revolution)); three-phase and single-phase machines are introduced. Multi-phase (m > 3) machines are similar.
- For $\omega_2 = 0$ (d.c. rotor excitation), synchronous machines are obtained while brush–commutator machines correspond to $\omega_1 = 0$ (d.c. stator excitation). Finally, for induction machines, $\omega_2 \neq 0$ is positive for motoring and negative for generating.

- The same practical (commercial) machines are classified into fixed (brush–commutator) and traveling (a.c. synchronous and induction) magnetic field machines.
- The linear electric machines are counterparts of the rotary machines in all configurations (principles).
- The case of the loudspeaker as a linear oscillatory-motion PM machine is discussed in detail for steady state, in a small compressor application, to envisage the general energy conversion details of all other electric machines.
- "Flux-modulation" reluctance machines with one or two rotors illustrate a new breed of EMs suitable for high torque density and efficiency medium-low speed applications, in full R&D swing today [4].
- As this is an introductory chapter, only one proposed problem is included.

3.10 PROPOSED PROBLEMS

3.1 A U-shaped plunger solenoid (Figure 3.21) with a soft composite material core is needed to activate an internal combustion engine (ICE) valve. The travel starts from 0.5×10^{-3} m airgap and ends at 8.5×10^{-3} m. For the geometrical data in Figure 3.13 and an ideal magnetic core (infinite permeability):

a. Derive the expression of thrust (from the energy formula) as a function of airgap, for a given coil current i_c, number of turns/coil N_c, and active area at the airgap A:
   - $W_w = 20 \times 10^{-3}$ m
   - $d = 20 \times 10^{-3}$ m
   - Stack depth $L = 0.05$ m

b. Calculate the coil ampere turns $N_c i_c$ for the same thrust $F_x = 500$ N for $x_{min} = 0.5 \times 10^{-3}$ m and $x_{max} = 8.5 \times 10^{-3}$ m.

Hint: First calculate the coil inductance, noticing that there are two airgap magnet circuit branches in parallel: $L_c = \left(\mu_0 N_c^2 / 2x\right) dL/2$ and $F_x = \left(i_c^2/2\right)\left(\partial L_c / \partial x\right)$.

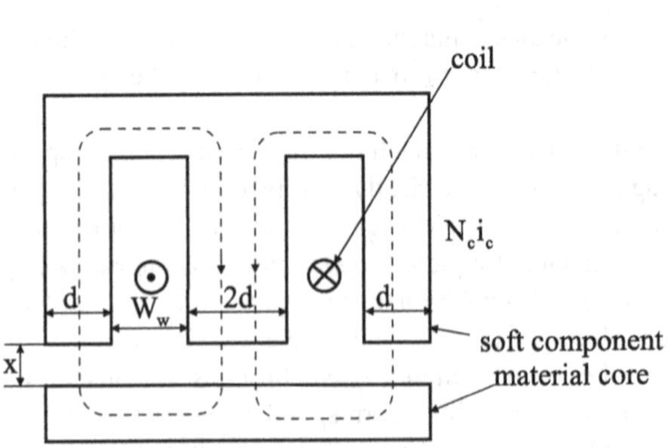

FIGURE 3.21 Tubular plunger solenoid.

REFERENCES

1. I. Boldea and S.A. Nasar, *Electric Machines Dynamics*, Chapter 1, MacGraw Hill, New York, 1986.
2. I. Boldea and S.A. Nasar, *Linear Motion Electromagnetic Systems*, John Wiley & Sons, New York, 1985.
3. I. Boldea and S.A. Nasar, *Linear Motion Electromagnetic Devices*, Chapter 7, Taylor & Francis Group, New York, 2001.
4. I. Boldea, L. Tutelea, "Reluctance Electric Machines: Design and Control", CRC Press, Taylor & Francis, New York, 2018.
5. Y. Chen, W. Fu, Xu Weng, "A concept of general flux-modulated electric machines based on a unified theory and its application to developing a novel doubly-fed dual-stator motor", *IEEE Trans*, Vol. IE-64, No. 12, 2017, pp. 9914–9923.
6. L. Sun, M. Cheng, M. Tong, "Key issues in design and manufacture of magnetic-geared dual-rotor motor for hybrid vehicles", *IEEE Trans*, Vol. EC-32, No. 4, 2017, pp. 1492–1501.
7. M. Bouheraoua, J. Wang, K. Atallah, "Rotor position estimation of a pseudo direct—drive PM machine using EKF", *IEEE Trans*, Vol. IA-53, No. 2, 2017, pp. 1088–1095.
8. Y. Wang, Sh. Niu, W. Fu, "Sensitivity analysis and optimal design of a dual mechanical port bidirectional flux-modulated machine", *IEEE Trans*, Vol. IE-65, No. 1, 2018, pp. 211–220.
9. Y. Shen, Q. Lu, H. Li, J. Cai, X. Huang, Y. Fang, "Analysis of a novel double-sided yokeless multitooth linear switched-flux PM motor", *IEEE Trans*, Vol. IE-65, No. 2, 2018, pp. 1837–1845.
10. T. Sashida and T. Kenjo, *An Introduction to Ultrasonic Motors*, Oxford University Press, Oxford, U.K., 1993.
11. M. Bulo, *Modeling and control of traveling piezoelectric motors*, PhD thesis, EPFC, Lausanne, Switzerland, 2005.

4 Brush–Commutator Machines *Steady State*

4.1 INTRODUCTION

Brush–commutator electric machines are commercially also called d.c. machines, but today any machine can be supplied from a d.c. source, provided that a power electronic converter is available.

Also, besides d.c. brush–commutator machines, there is the a.c. brush–commutator series (universal) motor still in use in many home appliances (e.g., vacuum cleaners, home robots, and hair dryers), for construction (vibration) tools, up to 1 kW at 30,000 rpm. They operate with fixed magnetic fields (a.c. stator and a.c. rotor currents, Chapter 3).

Though considered a "doomed species," due to brush–commutator scintillation and wearing limitations, faced with faster power electronic (static or brushless) commutation in traveling field machines, the brush–commutator PM small motors and for a few megawatts, low-speed ones (less than 150 rpm) for special applications (automotive and metallurgy drives, respectively), will die hard, for single quadrant variable speed drives, due to overall lower PWM converter cost.

We will give preference to PM small d.c. motors and to a.c. brush–commutator (universal) motors because of their potential for the future.

The d.c. brush motors still in use for rail, urban, or marine transportation or metallurgy will be discussed only briefly because most probably these motors will not be in use in the next decades or so. A d.c. brush–commutator PM motor with its main parts is shown in Figure 4.1.

4.1.1 Stator and Rotor Construction Elements

PM d.c. brush machines (which contain a stator and a rotor as main parts) may be built with [1–4]

- Radial airgap (or cylindrical rotor/stator) (Figure 4.1)
- Axial airgap (or disk-shaped rotor/stator)
- Slotted thin-sheet silicon steel rotor core (Figure 4.2)
- Slotless rotor core
- Surface PM $2p_1$ poles in the stator (Figure 4.2a)
- Interior PM $2p_1$ poles in the stator (Figure 4.2b)
- Interior rotor (Figure 4.2a,b)
- External rotor (Figure 4.2c)

DOI: 10.1201/9781003214519-4

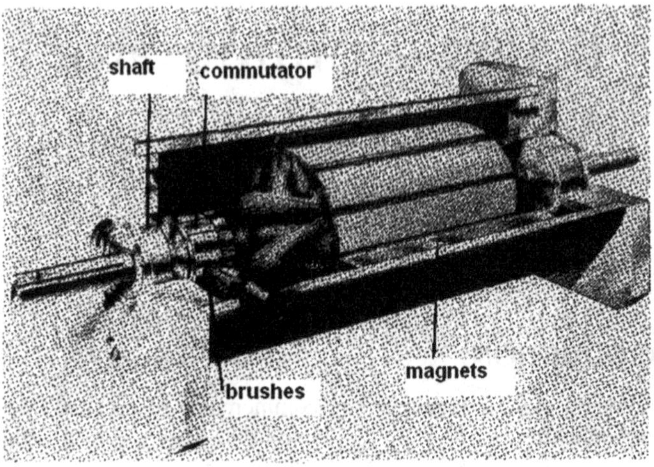

FIGURE 4.1 Typical d.c. brush–commutator PM motor with its mains parts.

The machine has a stator and a mover (rotor) with an air layer in between: the airgap.

The axial-airgap configuration with a disk-shaped rotor is most often (with rotor windings in the airgap) used (with $2p_1 \geq 4$) to reduce the axial length, volume, and rotor electric time constant, and thus obtain an ultrafast rotor-current (torque) control with power electronics at moderate costs.

The stator contains either a laminated (or solid iron) back iron and radially magnetized surface strong PM stators (Figure 4.2a) or laminated thin-sheet silicon steel poles with radially deep PMs (with tangential magnetization provided by "easy-to-demagnetize" ALNICO PMs [$B_r = 0.8T$, $H_c = 80$ kA/m] or Ferrite PMs [$B_r = 0.4T$, $H_c = 350$ kA/m]) (Figure 4.2b).

In slotless rotors or in rotor silicon steel sheet cores, there are uniform slots to hold identical coils that span π/p_1 geometrical radians and are all connected in series through the brush–commutator insulated copper sectors. They are called the armature winding. In slotless rotors with radial airgap, there is a laminated silicon iron back core in the rotor to complete the PM magnetic flux path.

For automobile engine–starter motors, hybrid surface PM/reluctance stator poles with cylindrical slotted rotors are used to secure a high starting torque and a large torque up to 300–400 rpm where the engine ignites at low ambient temperatures (Figure 4.2d) (see [4] for more details).

The d.c. (or a.c.) electromagnetic excitation stator with laminated silicon steel core (made of 0.5 mm thick sheets) shows salient poles and concentrated coils that produce the excitation field (Figure 4.3).

A.C.-excited stators, where excitation coils are connected in series at commutator brushes in the universal motor, lack the interpoles in general and have, again, salient poles with concentrated coils placed in a laminated silicon steel (or soft composite material) stator core (Figure 4.4).

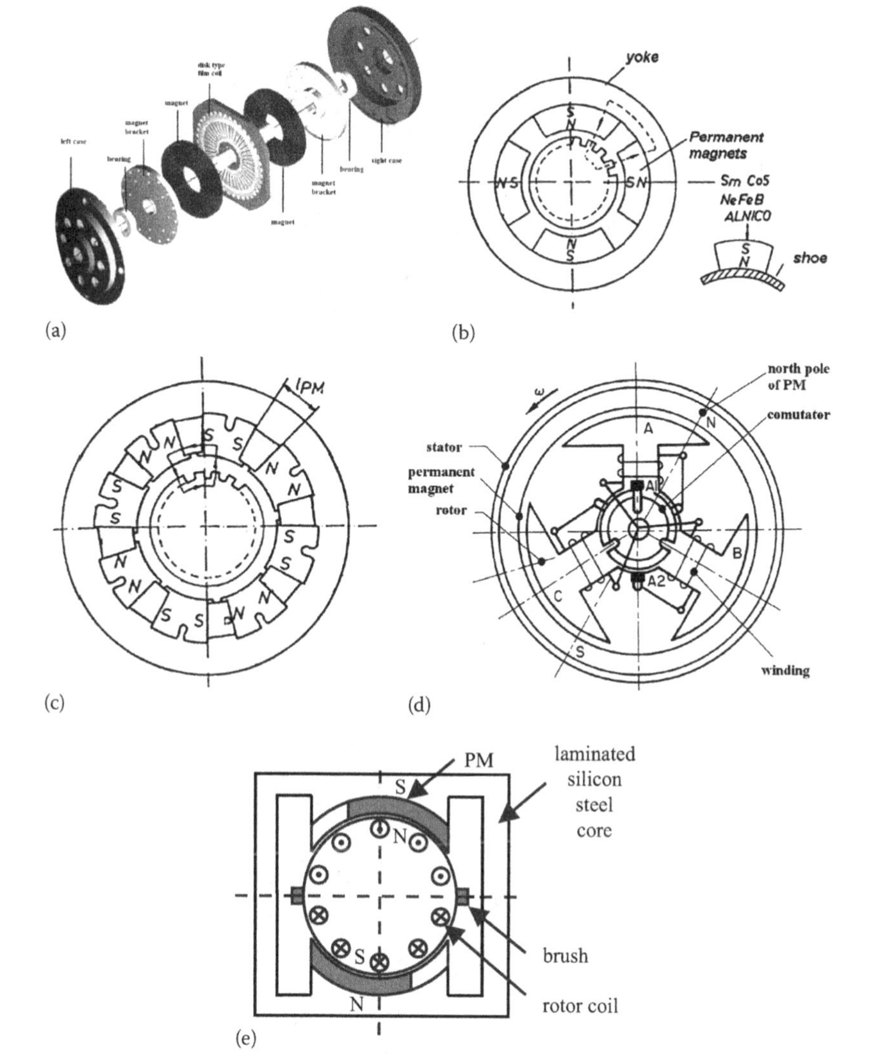

FIGURE 4.2 PM d.c. brush machines: (a) with surface PMs, (b) with interior PMs, (c) with external rotor (for ventilators), and (d) hybrid PM–iron (reluctance) stator poles.

4.2 BRUSH–COMMUTATOR ARMATURE WINDINGS

There are two types of armature windings that are placed in the uniform slots of the rotor silicon steel sheet core, made of lap coils or wave coils (Figure 4.5).

The span of the coils $y_b \approx \tau$ (τ = pole pitch; $\tau = \pi D/2p_1$), to embrace all the stator pole flux and thus to produce maximum electromagnetic force (emf).

The step of the coils at the commutator y_c is

$$y_c = m; \quad m = 1,2 \text{—lap winding} \tag{4.1}$$

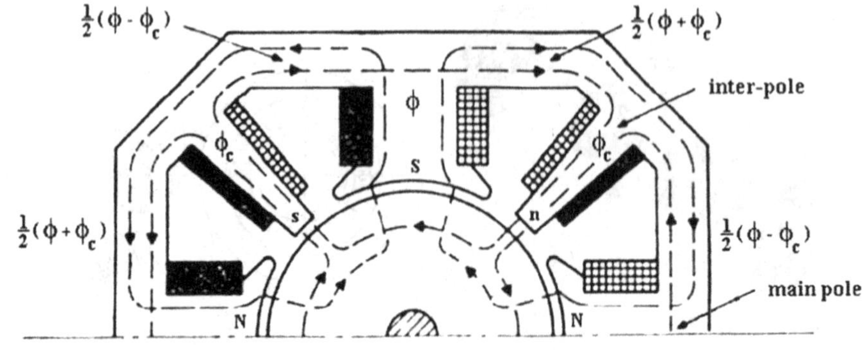

FIGURE 4.3 Cross section of a d.c. brush–commutator machine with interpoles to ease the commutation process.

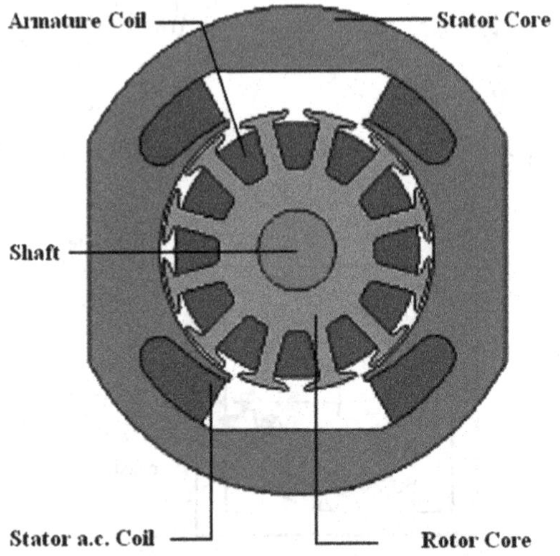

FIGURE 4.4 Cross section of a universal (a.c. brush–commutator single-phase) motor.

$$y_c = \frac{k-m}{p_1}; \quad m = 1,2\text{—wave winding} \tag{4.2}$$

K is the number of commutator copper sectors along the periphery. For $m = 1$, we obtain simple windings; for $m = 2$, we obtain what are called double windings.

The coils may have 2, 4, or even 6 ends, i.e., they may be simple, double, or triple coils (Figure 4.6).

For multiple end coils, it is possible that part of the turns be placed in adjacent slots to ease the commutation process (Figure 4.6c).

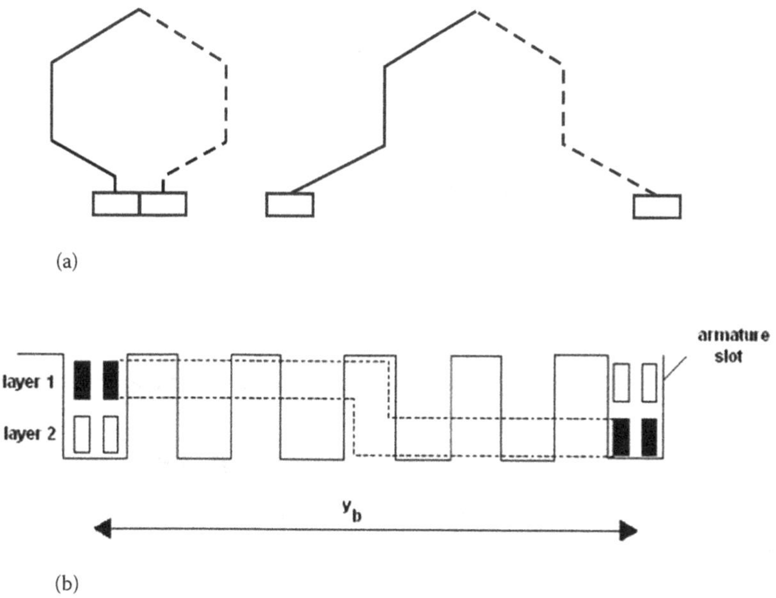

FIGURE 4.5 (a) Lap and wave coils and (b) their placement in two layers in slots.

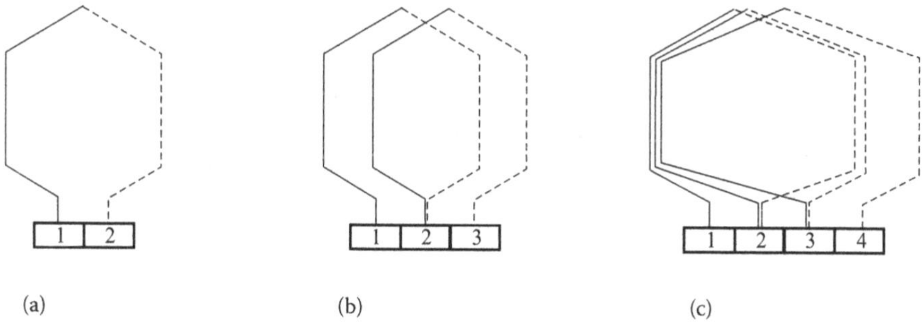

FIGURE 4.6 Lap coils: (a) simple, (b) double, and (c) triple in steps.

The emfs induced in the coils sides, placed in certain rotor slots, are a.c. and trapezoidal in time, but we consider here only the fundamental.

In this latter case, each slot emf is characterized by an electric phase angle α_{ec}:

$$\alpha_{ec} = \frac{2\pi}{N_s} p_1 \tag{4.3}$$

where

p_1 is the pole pairs in the stator and
N_s is the number of slots on the rotor.

If the winding is fully symmetric, after every two poles (a period), the emfs are in phase. In general, however, t emfs are in phase:

$$t = \mathrm{LCD}\left(N_s, p_1\right) \le p_1 \tag{4.4}$$

so the number of distinct phase emfs in slots is N_s/t and they form a regular polygon with N_s/t sides with a phase shift angle α_{et}:

$$\alpha_{et} = \frac{2\pi}{N_s} t \tag{4.5}$$

If all the slot emfs are placed as phasors one after the other, we end up with t polygons.

4.2.1 Simple Lap Windings by Example: $N_s = 16$, $2p_1 = 4$

Let us proceed directly with an example.

Consider a simple lap winding for $N_s = 16$, $2p_1 = 4$. According to Equations 4.3 through 4.5,

$$\alpha_{ec} = \frac{2\pi}{N_s} p_1 = \frac{2\pi}{16} 2 = \frac{\pi}{4}$$
$$\alpha_{et} = \frac{2\pi}{16} t = \frac{2\pi}{16} 2 = \frac{\pi}{4} = \alpha_{ec}$$

Consequently, the order of the slots in the emf polygon is the same with the physical order of slots along the rotor periphery, and the polygon has $N_s/t = 16/2 = 8$ sides. So, there are two polygons that overlap completely because $t = p_1$ (Figure 4.7).

A side of the polygon contains the forward and backward sides of a coil situated under the neighboring excitation (PM) poles.

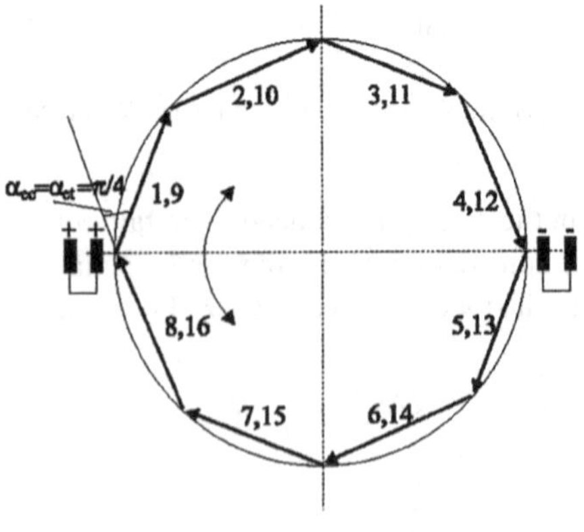

FIGURE 4.7 Elementary emf polygons: $N_s = 16$, $2p_1 = 4$, $m = 1$ (simple lap winding), and $u = 1$ (two-end (simple) coils).

The brush–commutator, made up of insulated copper sectors has $K = uN_s = 1 \cdot N_s = 16$ sectors, connects in series all coils, with a span y equal to the pole pitch $\tau = N_s/2p_1 = 16/2 \cdot 2 = 4$ slot pitches (Figure 4.7).

Both the coils and the commutator sector move with the rotor while the stator poles are fixed (Figure 4.8).

To complete the commutator, brushes that are fixed are added and they make mechanical contact to the commutator sectors to input or collect the d.c. current into (from) the rotor coils.

The brushes are placed such that the coil which is momentarily short-circuited by the brushes—which undergoes commutation—should have the sides between the stator poles (in the neutral axis) where the excitation field is zero. For symmetric coils (Figure 4.8), physically the brushes end up being located, axially seen, in the middle of the stator pole. Only for asymmetric (Siemens) coils, the brushes are physically in the neutral axis.

The distance between (+) and (−) brushes is one pole pitch (Figure 4.7) to collect, diametrically, the maximum available emf. The coils in series are all placed with forward sides under one pole and with the backward sides in the neighboring stator pole (of opposite polarity). They form a current path. In Figures 4.7 and 4.8, there are in all $2a = 4$ current paths, so the current at the brush I_{brush} is divided $2a$ times to get the coil current I_{coil}:

$$I_{\text{brush}} = 2aI_{\text{coil}} \tag{4.6}$$

For lap windings, the number of current path $2a$ is

$$2a = 2p_1 m \tag{4.7}$$

For simple lap windings, $m_1 = 1$, so $2a = 2p_1 = 4$ in our case. The number of brushes equals the number of poles $2a = 2p_1$ (for double lap windings, the brushes span two commutator sectors).

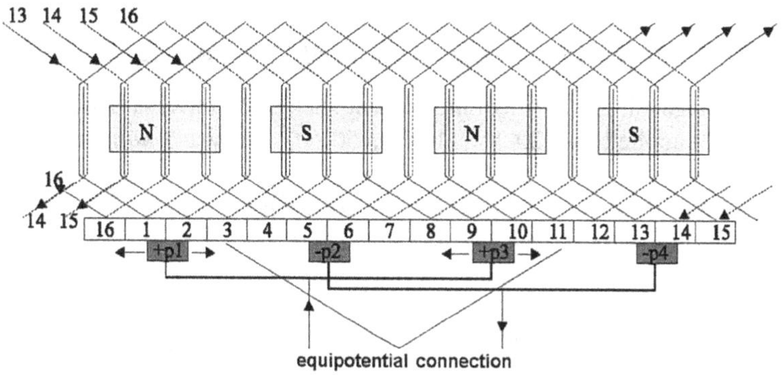

FIGURE 4.8 Simple ($m = 1$) lap winding layout: $N_s = 16$, $2p_1 = 4$.

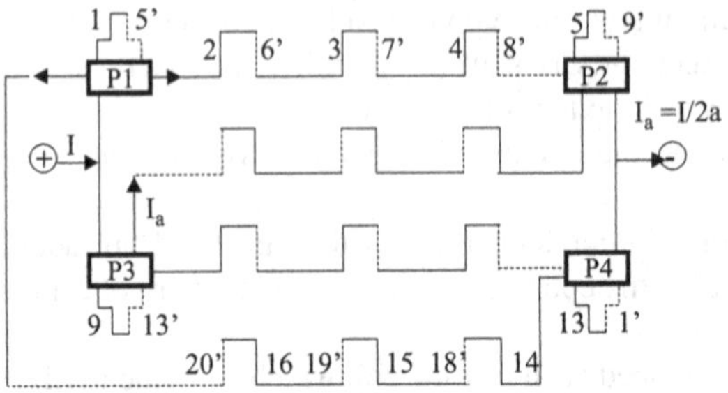

FIGURE 4.9 The current paths composition for simple lap winding: $N_s = 16$, $m = 1$, $2p_1 = 4$, and $2a = 4$.

So the lap windings are suitable for low-voltage, large-current (automotive) motors where the large number of current paths allows for the usage of thin-wire coils, which are easier to manufacture and placed in slots and experience lower skin effect.

We should also note that in any time instant on each current path, one coil is short-circuited (it commutates, that is changes current polarity while the rotor moves with one commutator sector; or the coil switches from one (+) coil path to the next (−)).

So out of $4(N_s/2a)$ coils per current path, only $3 = ((N_s/2a) - 1)$ produce emfs in series (Figure 4.9). The commutating coils are 1–5′, 5–9′, 9–13′, and 13–1′. The nonuniformity of the airgap between various stator poles and the rotor, due to manufacturing imperfections, may produce emfs/current paths which differ from each other. As all the current paths are in parallel at the brushes, circulating currents may occur which then circulate through the brush–commutator contact. To divert these currents, equipotential connections at the collector side are made between all (or most) polygon corners (Figure 4.7) as seen in Figure 4.8.

Note: For the double lap windings ($m = 2$), in essence, two simple lap windings are built: one for the even number of slots and one for the odd number of slots and then they are connected in parallel by doubling the brushes' span. So we end up with $2a = 2p_1m$ current paths but still with $2p_1$ brushes, albeit of double span.

4.2.2 Simple Wave Windings by Example: $N_s = 9$, $2p_1 = 2$

For large voltage and small (medium) current such as in universal motors for some applications, supplied at 220 V (110 V), 50(60) Hz, wave windings are more appropriate.

Although $2p_1 = 2,4$ poles, $2p_1 = 2$ is more common.

Let us consider here the case of $2p_1 = 2$.

The coil step at the commutator y_c is (Equation 4.1):

$$y_c = \frac{k-m}{p_1} = \frac{9-1}{1} = 8; \quad k = N_s = 9,\ m = 1$$

Now the coil span y_c = integer $(N_s/2p_1)$ = integer $(9/2) = 4$ slot pitches.

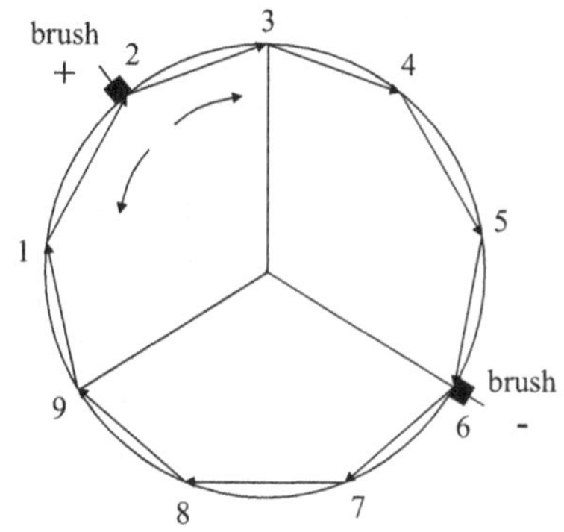

FIGURE 4.10 Elementary polygon for simple wave winding ($N_s = 9$, $2p_1 = 2$).

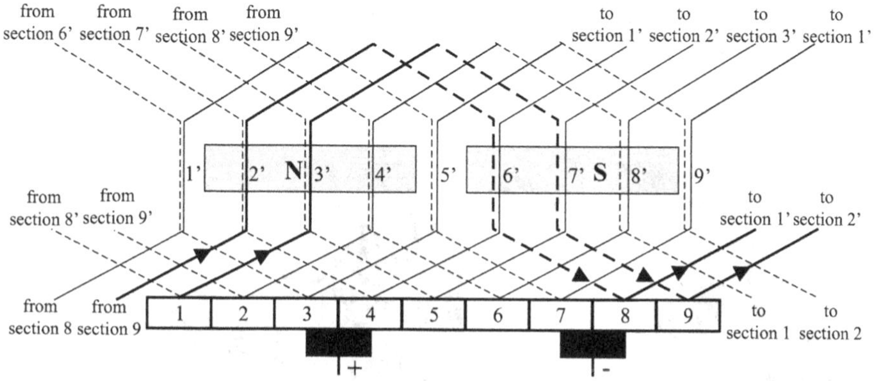

FIGURE 4.11 Simple wave winding: $N_s = 9$, $2p_1 = 2$, and $m = 1$.

This time, $t = \text{LCD}\,(9,1) = 1$; all slot emfs have distinct phase angles, so, still, the order of polygon sides corresponds to the physical order of the slots (Figure 4.10).

For $2p_1 = 4$, this is not the case.

The number of current path $2a = 2$ (it is $2a = 2$ even for $2p_1 = 4$ poles; for double wave winding ($m = 2$), however, $2a = 2m$).

The winding layout is shown in Figure 4.11.

We should notice that now the coil 6–9′ is short-circuited at brush (+) and coil 1–5′ at brush (−).

So one current path contains three active slot (coil) emfs in series and the other contains four. This is very important in a realistic design when calculating the average emf per path V_{ea}.

So there is always some circulation current between current paths through the brush–commutator which has to be taken care of for the design. Also, the emf/path (no-load voltage) has notable time pulsations (N_s of them per revolution).

To improve commutation, it is possible to increase the number of commutator sectors by adopting double coils ($u = 2$, $K = N_s\,u = 9 \times 2 = 18$).

Note: The $N_s = 3$, $2p_1 = 2$, combination used for the micromotor in Figure 4.2c is an extreme case when one coil commutates (is short-circuited) all the time.

As seen already in Figure 4.2c, the short-circuited (commutating) coil does not have all the sides exactly in the neutral axis.

The brushes are moved away from their ideal position to improve commutation in small machines with a preferred direction of motion.

For more on windings of heavy duty (in transportation or metallurgy) brush–commutator motors, which have combined simple lap/double wave windings, see [3].

Airgap (slotless) windings are built in quite a few configurations which show notable peculiarities [1–3,5].

4.3 THE BRUSH–COMMUTATOR

The brush–commutator of small machines consists of K hard-drawn or silver copper wedge-shaped segments (sectors) insulated from one to another and connected to the armature coil ends (Figure 4.12a and b). The commutator segments are insulated also from the shaft by a die-cast resin holder, which is fixed on the motor shaft. The

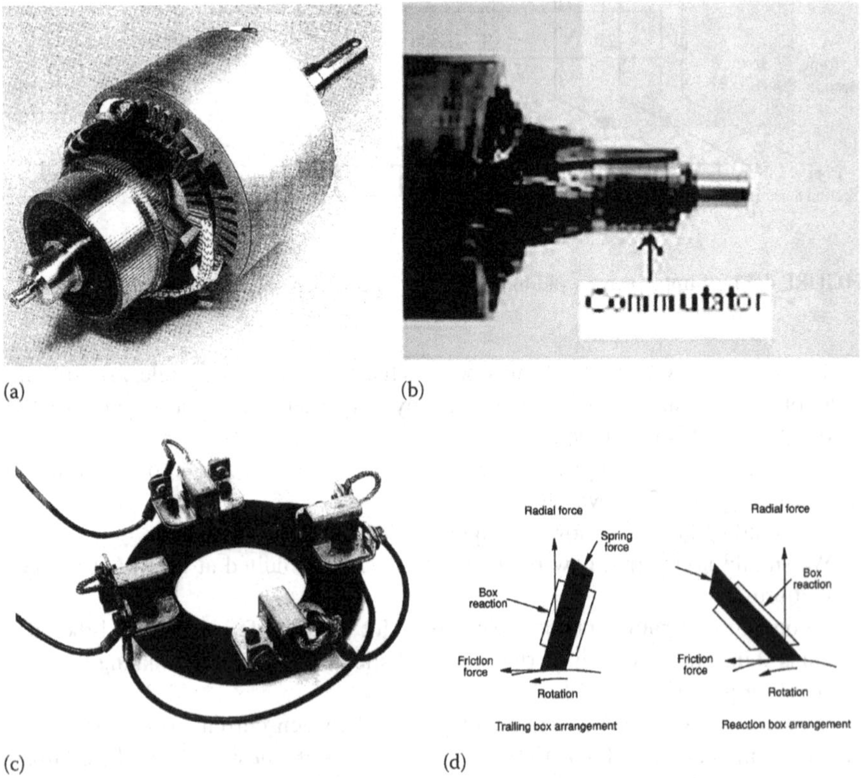

FIGURE 4.12 (a) Brush–commutator motor, (b) rotor, (c) the brush gear, and (d) brush positioning.

silver–copper segments can survive the flood soldering of armature-coil ends to the tows at 300°C.

Spacers between copper segments are made from shellac-bound mica splitting (90° mica) or from epoxi-resin-bound fine mica (samicanite). They should wear as slowly as copper segments and be mechanically hard and elastic.

The brush gear consists of $2a$ brush holders, fitted to a yoke of insulating material which houses brushes of suitable electric conductivity and hardness (Figure 4.12c).

Good sliding friction quality and adequate (large) electric conductivity characterize good brushes. Brushes are made of:

- Natural graphite (good for large-voltage, small motors)
- Hard carbon (low cost, used for fractional power and low-speed machines)
- Electrographite (good conductivity, good for industrial and traction motors)
- Metal graphite (high conductivity, good for low-voltage motors such as automobile actuators, etc.)

The brushes are pressed to the commutator by mechanical springs and are placed radially for bidirectional motion (inclined by 30°–40° for narrow brushes) or in a trailing box or a reaction (inclined by 10°–15°) box, for a preferred direction of motion (Figure 4.12d).

4.4 AIRGAP FLUX DENSITY OF STATOR EXCITATION MMF

When the armature current is zero (no load) and the machine is excited, through the stator heteropolar excitation mmf ($N_F i_F$/per pole), the flux lines flow through the stator pole radially, then through the rotor teeth, then below the slots through the back iron, and then back to the airgap, stator poles, and stator back iron (Figure 4.13a).

The excitation airgap flux density $B_{gF}(x)$ distribution (Figure 4.13b) reveals ripples due to rotor slot openings. In fact, slotting leads to a reduction in the average value of the airgap flux density which is calculated by considering an increase in the equivalent airgap g_e (from g) by $K_C > 1$. K_C is the so-called Carter coefficient (derived initially by conformal transformation to determine magnetic field distribution in the airgap with open rotor slots):

$$K_c \approx \frac{\tau_s}{\tau_s - \gamma b_s}; \quad \gamma \approx \frac{(b_s / g)^2}{5 + b_{s/g}}; \quad g_e = K_c g \tag{4.8}$$

where

τ_s is the rotor slot pitch (in meters),
b_s is the rotor slot opening span (in meters),
g is the airgap (in meters), and
g_e is the equivalent airgap (in meters).

The open-slot configuration, used for preformed armature coils insertion in slots, may increase the equivalent airgap by as much as 20%–30%, so it has to be considered in any practical design.

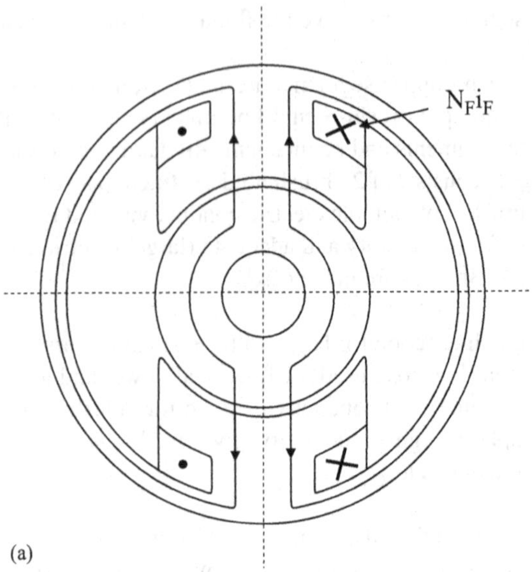

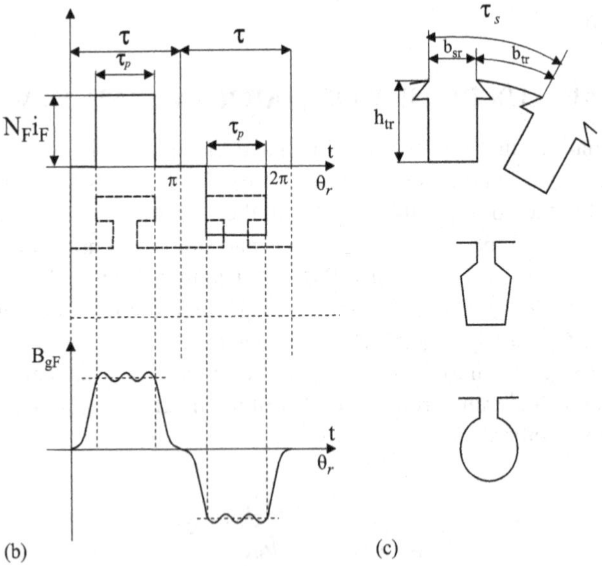

FIGURE 4.13 (a) Flux line of stator excitation field ($2p_1 = 2$ poles), (b) excitation mmf and flux, and (c) rotor slots.

4.5 NO-LOAD MAGNETIZATION CURVE BY EXAMPLE

The relationship between the excitation (or PM) magnetic pole-flux, ϕ_{pole}, and the pole mmf N_Fi_F, at zero armature current, is called the no-load magnetization curve and is a crucial design target.

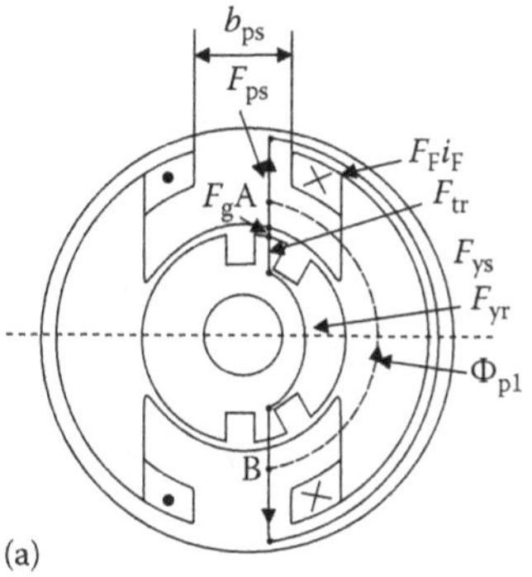

(a)

B_m	H_m
0.1	35
0.2	49
0.3	65
0.4	76
0.5	90
0.6	106
0.7	124
0.8	148
0.9	177
1.0	220
1.1	237
1.2	356
1.3	482
1.4	760
1.5	1,340
1.6	2,460
1.7	4,800
1.8	8,270
1.9	15,220
2.0	34,000

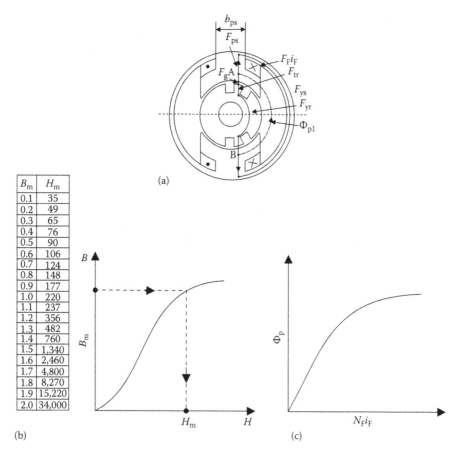

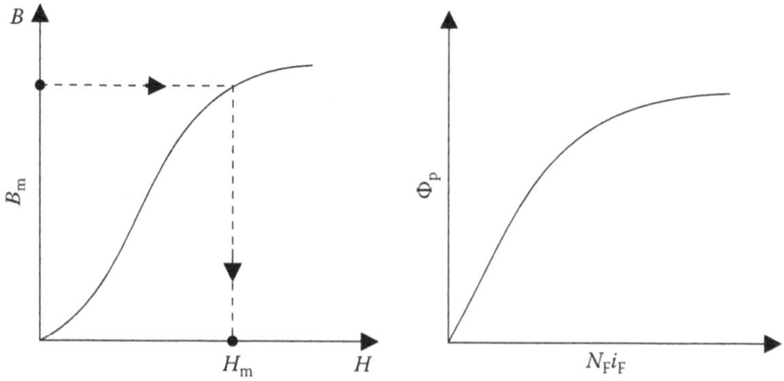

(b) (c)

FIGURE 4.14 (a) Excitation flux line and its mmf contributions, (b) magnetic core $B(H)$ curve and table, and (c) no-load magnetization curve ϕ_p (N_Fi_F).

The magnetic flux line path in Figure 4.14a is decomposed in a few components that can be represented by unique flux density/magnetic field values: the two airgaps, the yoke and the poles in the stator and the two teeth and the yoke in the rotor, characterized by their corresponding mmf contributions: $2F_g$, F_{ys}, F_{ps}, F_{tr}, and F_{yr} (Figure 4.14).

Ampere's law along the flux line yields

$$2N_Fi_F = 2F_g + 2F_{tr} + F_{yr} + 2F_{ps} + F_{ys} \tag{4.9}$$

Let us start with a given airgap flux density B_{gF}. Then the pole flux ϕ_p is

$$\phi_p \approx \tau_p L_e B_{gF}; \quad \frac{\tau_p}{\tau} = 0.65 - 0.75 \tag{4.10}$$

where

τ_p is the stator pole shoe,

τ is the pole pitch ($\tau = \pi D/2p_1$),

L_e is the equivalent stack length, and

$L_e = K_{fill}L$, where $K_{fill} > 0.9$ is the lamination filling factor and L is the measured stack length

Let us consider an example with $B_{gF} = 0.5$T, stack length $L_e = 0.05$ m, rotor diameter $D = 0.06$ m, $2p_1 = 2$, airgap $g = 1.5 \times 10^{-3}$ m, $K_c = 1.2$, and $\tau_p/\tau = 0.7$; $(\phi_{p\,g})_{0.5\,T}$ is

$$\left(\phi_{p_g}\right)_{0.5T} = 0.5 \times \frac{\pi 0.06}{2} \times 0.7 \times 0.05 = 1.648 \times 10^{-3}\ \text{Wb}$$

What interests us from now on is only the airgap flux density $B_{gF} = 0.5$ T.

The airgap mmf: $F_g = gK_c H_g = gK_c B_{gF}/\mu_0 = 1.5 \times 10^{-3} \times 1.2 \times 0.5/1.256 \times 10^{-6} =$ 716 A turns. The rotor teeth flux density B_{tr} is calculated by equalizing the flux along a rotor tooth pitch to the one through the rotor tooth (b_{tr}-width):

$$B_{gF}\tau_s = B_{tr}b_{tr} \tag{4.11}$$

As in rectangular (open) slot rotors, the tooth width, b_{tr}, varies radially with the magnetic flux density, B_{tr}, H_{tr} determined at the airgap, H_{t0}, at slot middle, H_{tm}, and at slot bottom, H_{t1}, are considered and averaged. Thus, we obtain H_{tav} as

$$H_{tav} = \frac{1}{6}\left(H_{t0} + 4H_{tm} + H_{t1}\right) \tag{4.12}$$

The rotor tooth mmf F_{tr} is

$$F_{tr} = h_{sr}H_{tav} \tag{4.13}$$

For our case, $b_{tr}/\tau_s = 0.4$ (0.40–0.55 in general); also the rotor diameter D per rotor slot height h_{tr} is large.

So B_{tr} is constant along the rotor tooth height (not so, in general).

So from Equation 4.12, $B_{tr} = B_{gF}\tau_s/\tau_{tr} = 0.5/0.4 = 1.25$ T.

From the table in Figure 4.14, $H_{tr}(B_{tr} = 1.25$ T) is $H_{tr} = 417$ A/m.

For a rotor slot height $h_{tr} = 0.012$ m, F_{tr} is

$$F_{tr} = 417 \times 0.012 = 5.3\ \text{A turns}$$

For the rotor yoke, we first need an average thickness, h_{yr}:

$$h_{yr} = \frac{\left(D_{rotor} - D_{shaft}\right)}{2} - h_{tr} = \frac{(0.06 - 0.01)}{2} - 0.012 = 0.025 - 0.012 = 0.013\ \text{m} \tag{4.14}$$

So the rotor yoke average flux density B_{yr} is (conservatively):

$$B_{yr} \approx \frac{\left(B_{gF} \times (\tau/2)\right)}{h_{yr}} = \frac{(0.5 \times 0.0471)}{0.013} = 1.81\ \text{T}$$
$$\tau \approx \pi \frac{D_r}{2p_1} = \pi \frac{0.06}{2} = 0.0942\ \text{m} \tag{4.15}$$

The value of B_{yr} is close to the limit in practice because the armature current mmf will add additional yoke flux.

From the table in Figure 4.14, $H_{yr} = 8300$ A/m.

The average magnetic field path in the rotor back iron l_{yr} is

$$l_{yr} = \frac{\pi\left(D - 2h_{sr} - h_{yr}\right)}{2} = \frac{\pi(0.06 - 2\times 0.012 - 0.013)}{2} = 0.0361\ \text{m} \tag{4.16}$$

So the rotor yoke mmf F_{yr} is

$$F_{yr} = l_{yr} \times N_{yr} = 0.0361 \times 8310 = 300\ \text{A turns} \tag{4.17}$$

As seen in Figure 4.14a, there is some leakage flux ϕ_{pe}, which closes the paths directly between the stator poles (or PMs), which is proportional to the total mmf F_{pp} between the stator poles through the rotor (points AB):

$$F_{pp} = 2F_g + 2F_{tr} + F_{yr} = 2\times 716 + 2\times 5.3 + 300 = 1742\ \text{A turns} \tag{4.18}$$

As the first approximation, we may consider

$$\phi_{pe} = \phi_{pg} \frac{F_{pp} - 2F_g}{2F_g} = 1.648 \times 10^{-3} \times \frac{1742 - 1432}{1432} = 0.3574 \times 10^{-3}\ \text{Wb} \tag{4.19}$$

So the total stator pole flux, ϕ_{PF}, is

$$\phi_{PF} = \phi_{pg} + \phi_{pe} = (1.648 + 0.3574) \times 10^{-3} = 2.01 \times 10^{-3}\ \text{Wb} \tag{4.20}$$

Now the flux densities in the stator pole shoe and pole body are

$$B_{pshoe} = B_{gF} \times \frac{\phi_{pF}}{\phi_{pe}} = 0.5 \times \frac{2.01 \times 10^{-3}}{1.648 \times 10^{-3}} = 0.61\ \text{T} \tag{4.21}$$

$$B_{pbody} = B_{pshoe} \times \frac{\tau_p}{b_{pstator}} = \frac{0.61}{0.5} = 1.22\ \text{T} \tag{4.22}$$

(with $\tau_p = 0.7\tau$ in our case)

The stator pole body width $b_{pstator}/\tau_p \approx 0.4$–$0.55$ to leave enough room for the excitation coils, while avoiding too heavy magnetic saturation (we take here $b_{pstator}/\tau_p = 0.5$).

Now it is reasonable to assume that the excitation coil radial height h_{cF} is equal to or smaller than the rotor slot depth h_{tr}.

From the table in Figure 4.14, at $B_{pbody} = 1.22$ T, we find $H_{pb} = 360$ A/m and with pole body height $h_{cF} = h_{tr} = 0.012$ m (and neglecting the mmf in the pole shoe because B_{pshoe} is small (0.61 T)), the pole body mmf F_{ps} is

$$F_{ps} = h_{tr} \times H_{pbody} = 0.012 \times 360 = 5 \text{ A turns} \tag{4.23}$$

Finally, the flux density in the stator back iron B_{ys} is

$$B_{ys} = B_{gF} \times \frac{(\tau_p / 2)}{h_{ys}} \tag{4.24}$$

The stator core is designed for $B_{ys} \leq 1.4$ T to leave room for armature flux contribution before a too heavy magnetic saturation is reached.

With $\tau_p = 0.7\ \tau = 0.7 \times 0.0942 = 0.066$ m, $B_{gF} = 0.5$ T, and $B_{ys} = 1.3$ T, we obtain the stator yoke, h_{ys}:

$$h_{ys} = \frac{B_{gF}}{B_{ys}} \frac{\tau_p}{2} = \frac{0.5}{1.3} \times 0.066 = 0.02536 \text{ m} \tag{4.25}$$

Now for $B_{ys} = 1.3$ T from the table in Figure 4.14a, $H_{ys} = 482$ A/m.

The length of the flux path in the back iron, l_{ys}, is approximately

$$l_{ys} = \frac{\pi\left(D_{rotor} + 2g + 2h_{pshoe} + 2h_{pbody} + h_{ys}\right)}{2p_1} \tag{4.26}$$

$$h_{pshoe} \approx \frac{\left(\tau_p - b_{pbody}\right)}{2\sqrt{3}} = \frac{(0.066 - 0.0662)}{2\sqrt{3}} = 0.0095 \text{ m} \tag{4.27}$$

The $\sqrt{3}$ factor stands for a 30° angle of the pole shoe geometry.

$$l_{ys} = \frac{\pi(0.06 + 2\times 0.0015 + 2\times 0.00095 + 2\times 0.12 + 0.02536)}{2\times 1} = 0.206 \text{ m}$$

The external diameter of the machine D_{out} is

$$D_{out} = D_{rotor} + 2g + 2h_{pshoe} + 2h_{pbody} + 2h_{ys} = 0.15672 \text{ m} \tag{4.28}$$

The ratio $D_{rotor}/D_{out} = 0.06/0.15672 = 0.3828$. This is too low a value for a close to optimal design. For two poles, this design ideally recommends $(D_{rotor}/D_{out})_{2p1=2} \approx 0.45–0.55$.

The mmf in the stator yoke F_{ys} is

$$F_{ys} = h_{ys} \times H_{ys} = 0.206 \times 482 = 99.292 \text{ A turns} \tag{4.29}$$

Now the total field (excitation) A turns per pole $N_F i_F$ is (Equation 4.10):

$$N_F i_F = F_g + F_{tr} + F_{pr} + \left(F_{yr} + F_{ys}\right)/2 = 716 + 5.3 + 5 + \left(300 + 99.292\right)/2 = 926.3 \text{ A turns/pole} \tag{4.30}$$

The total contribution of iron may be assessed through a saturation coefficient $K_s > 0$:

$$1 + K_s = \frac{N_F i_F}{F_g} = \frac{926.3}{716} = 1.294 \tag{4.31}$$

Values of $K_s = 0.2–0.4$, or even more, are considered feasible.

The computation sequence above may be easily mechanized by building a computer code where the airgap flux density is set at 10(20) values to finally yield the entire $N_F i_F$ (ϕ_{pg}) curve, that is the no-load magnetization curve, which is used in design and performance assessment.

Note that in this paragraph, a preliminary machine sizing was, in fact, done.

The rotor diameter D_{rotor} and stack length L have been given, but they may be assigned starting values based on tangential force f_{tn}; $L/D_r = 0.5–2.5$.

Let us consider here the inverse process, that is, to calculate a feasible rated torque for the case study here. For a rated tangential specific force $f_t = 0.7$ N/cm^2, the rated torque would be

$$T_e = f_t \pi D_r L_e \frac{D_r}{2} = 7 \times 10^3 \pi 0.06^2 \frac{0.05}{2} = 1.978 \approx 2 \text{ N m} \tag{4.32}$$

We may even calculate the slot mmf required to produce this torque (from the BIL formula [Chapter 1]):

$$T_e \approx B_{gF} \frac{\tau_p}{\tau} L \frac{D_r}{2} \pi D_r A \tag{4.33}$$

$$A = \frac{2 \times 2.2}{0.5\pi 0.06^2 \times 0.7 \times 0.05} = 22242 \text{ A turns/m} \tag{4.34}$$

This is a rather large value for $D_r = 0.06$ m, but let us see what it requires in terms of current density if the slot depth $h_{tr} = 0.012$ m and the tooth/slot (b_{tr}/b_{sr}) ratio is unity:

$$A = \frac{2N_c i_c N_s}{\pi D_r}; \quad 2N_c i_c = K_{fill} \frac{\pi D_r}{N_s} \frac{1}{2} h_{tr} j_{corotor} \tag{4.35}$$

where K_{fill} is the rotor slot copper fill factor $K_{fill} = 0.4–0.6$; $K_{fill} > 0.45$ only for preformed coils. The slot width $b_{tr} \approx (\pi D_r/N_s)*1/2$ (half of rotor slot pitch).

So from Equation 4.35, we can determine the rated current density $j_{corotor}$, then the A turns per coil $N_c i_c$ (there are two coils in every rotor slot because the armature winding has two layers), and then the number of slots, which has to be an even number for simple lap windings (low voltage) and an odd number for simple wave windings (large voltage):

$$j_{corotor} = \frac{2A}{K_{fill} h_{tr}} = \frac{2 \times 22242}{0.45 \times 0.012} = 8.2377 \times 10^6 \text{ A/m}^2 = 8.2377 \text{ A/mm}^2 \tag{4.36}$$

Forced air cooling is required to secure thermally safe operation at this current density.

The number of slots N_s is our choice, and we should consider a commutation that favors large N_s, but then the commutator geometry limitations have to be considered.

For our case ($D_{rotor} = 0.06$ m), we may safely choose $N_s = 12$, 16, or 18.

4.6 PM AIRGAP FLUX DENSITY AND ARMATURE REACTION BY EXAMPLE

Both surface PMs and interior PMs (Figure 4.2) have been used, but today strong surface PMs of NdFeB or of hard Ferrites (for micromotors) are preferred. The main reason is improved brush commutation, because the total magnetic airgap includes the PM radial (axial, in the case of axial airgap configurations) height h_{PM} (Figure 4.15a).

The PM may exhibit radial or parallel magnetization and both have merits and demerits. Rounded PM corners—to be studied by FEM—are provided to reduce PM flux density gradients at their ends and thus reduce the PM (cogging) torque at zero current, due to slot openings. The PM span τ_{PM} per pole pitch τ may now be larger than for d.c. excitation: $\tau_{PM}/\tau = 0.66–0.85$.

Larger values of this ratio lead to more torque per given current (more of the periphery is active) but the leakage flux between the stator poles now takes more from the useful PM flux density, B_{gPM}:

$$B_{gPM} = \frac{B_m}{1 + K_{lPM}} \tag{4.37}$$

where B_m is the flux density in the PM.

The leakage factor for surface PMs is in general $K_{lPM} \approx 0.15–0.3$ and may be determined with good precision by numerical methods such as FEM.

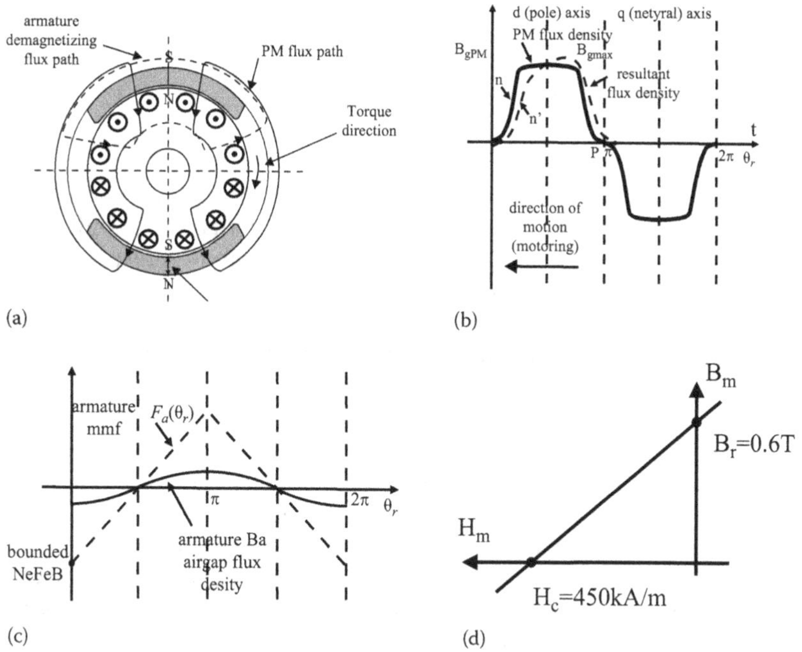

FIGURE 4.15 (a) PM airgap flux density, (b) its distribution, (c) armature mmf and flux density, and (d) PM demagnetization curve.

A few remarks are in order:

- The current polarity in all coils beneath a PM pole is the same due to the brush–commutator operation if the brushes are in the ideal position (neutral axis of PM flux).
- The mmf of rotor currents rises stepwise in each slot (approximately linearly) from the *d* (pole(PM)) axis position (Figure 4.15c). And so does its flux density:

$$F_a(\theta_r) = F_{am}\left(\frac{2\theta_r}{\pi} - 1\right); \quad 0 \le \theta_r \le \pi, \text{ in electric radians} \tag{4.38}$$

The maximum armature mmf F_{am} corresponds here to half the pole pitch:

$$F_{am} = \frac{2N_c i_c N_s}{4\pi} \tag{4.39}$$

For contemporary magnets, the recoil permeability $(\mu_{PM})_{pu}$ = 1.05–1.2.

The armature flux density B_a in the airgap is

$$B_a(\theta_r) \approx \frac{4F_a(\theta_r)\cdot\mu_0}{gK_c + h_{PM}\cdot(\mu_{PM})_{pu}} \tag{4.40}$$

$(\mu_{PM})_{pu}$ is the relative value of PM recoil permeability.

Now, for the surface PM stator along the entire periphery, the $B_a(\theta_r)$ is linear (Equation 4.40 and Figure 4.15) since the total magnetic airgap, $gK_c + h_{PM}$ * $(\mu_{PM})_{pu}$, is all along the same (uniform).

In contrast, for excited stators or interior PM stators, the magnetic airgap becomes much larger between the poles, and thus in Equation 4.39, F_{am} is reduced by τ_p/τ and the airgap is increased for $0 < \theta_r < 1 - (\pi/2)(\tau_p/\tau)$ and $\pi > |\theta_r| > \pi(\tau_p/\tau)$ (between poles).

- The armature (rotor mmf) flux density in the airgap brings up additional flux per trailing half a pole and a reduction on the entry half a pole for motoring (Figure 4.15a and b).

  In the surface PM stator case, it is unlikely that magnetic saturation in the rotor teeth of the trailing half pole will occur. This is in contrast to the excited stator, where this highest rotor tooth flux density, $B_{trmax} = B_{gmax}(\tau_s/b_{tr})$, becomes a major design variable (limitation).
- As expected, the large magnetic airgap of the surface PM stator will lead to a notably smaller rotor commutating coil (leakage) inductance, L_a, which in turn will ease the commutation process.
- For the PM stator, the excitation flux density is assigned a value—for given PMs and machine geometry—and thus the no-load magnetic curve losses its meaning. But sizing the PM to produce the calculated PM airgap flux density B_{gPM} is necessary.

Let us consider the same rotor example as in the previous Section 4.5, now with $B_{gPM} = 0.5$ T.

As a rule, for surface PM design, $B_{gPM} \approx (0.5–0.8)B_r$, where B_r is the remnant flux density of PMs (lower values are adequate for slotless (in air) armature windings).

Let us consider bonded (lower cost) NdFeB magnets with $B_r = 0.8$ T and $H_c = 650$ kA/m:

$$(\mu_{rPM})_{pu} = \frac{B_r}{(\mu_0 H_c)} = 1.05$$

The PM pole may be replaced by a coil in air with an mmf $\theta_{PM} = H_c h_{PM}$, whose radial thickness is equal to the magnet thickness.

The pole mmf is replaced by the PM mmf $h_{PM} \times H_m$, where H_m is the actual magnetic field in the PM corresponding to B_m, but with positive sign.

Now as the flux density in the PM, B_m (with leakage flux coefficient $K_{lPM} = 0.3$) from Equation 4.36,

$$B_{\rm m} = B_{\rm gPM}\left(1+K_{\rm lPM}\right) = 0.5\left(1+0.3\right) = 0.65\ {\rm T}$$
$$\left|H'_{\rm m}\right| \approx \frac{B_{\rm m}}{\mu_0} = \frac{0.65}{1.25\times10^{-6}} = 0.5175\times10^6\ {\rm A/m} \tag{4.41}$$

So the mmf in the PMs is

$$F_{\rm PM} = H'_{\rm m}\times h_{\rm PM} = 0.5175\times10^6\times h_{\rm PM} \tag{4.42}$$

Now for the case in point, the balance of mmf is

$$2H_c\times h_{\rm PM} = 2\theta_{\rm PM} = 2F_{\rm g} + 2F_{\rm tr} + F_{\rm yr} + F_{\rm ys} + 2F_{\rm PM} \tag{4.43}$$

For $B_{\rm gPM} = 0.5$ T as in Section 4.5., $F_{\rm g} = 716$ A turns, $F_{\rm tr} = 5.3$ A turns, and $F_{\rm yr} = 300$ A turns.

The above interpretation of $B_{\rm m}\left(H'_{\rm m}\right)$ linear curve through origin (Equation 4.41) for PMs is valid if the PM is first replaced by a coil in air with an mmf of $H_c h_{\rm PM}$. But F_{ys} (mmf in the stator yoke) is not known because the PM height is not known. Even in Section 4.5., it was moderate ($F_{\rm ys} = 99$ A turns). As the PM height $h_{\rm PM}$ is to be smaller, $F_{\rm ys}$ will be even smaller (smaller flux line length in the stator yoke).

So the only unknown in Equation 4.43 is the magnet height $h_{\rm PM}$, but the equation has to be solved iteratively.

For a conservative solution, let us keep $F_{\rm ys} = 99$ A turns and, consequently, from Equation 4.43

$$2h_{\rm PM}\left(0.650\times10^6 - 0.5175\times10^6\right) = 2\times716 + 2\times5.8 + 300 + 99 = 920.8 \tag{4.44}$$

So the magnet thickness is

$$h_{\rm PM} = 3.4747\times10^{-3}\ {\rm m}$$

Note that the mechanical airgap $g = 1.5$ mm.

As the total stator yoke thickness $h_{\rm ys} = 0.025$ m (because the total flux per pole holds), the external diameter of the stator is

$$D_{\rm out} = D_{\rm rotor} + 2g + 2h_{\rm PM} + 2h_{\rm ys}$$
$$= 0.06 + 2\times0.0015 + 2\times0.00347 + 2\times0.025 \approx 0.12\ {\rm m} \tag{4.45}$$

(instead of 0.156 m as it was with d.c. excitation)

So the surface PM stator leads to a reduced stator outer diameter for the same rotor, same airgap flux density, same torque, and same rotor with same copper losses.

So, the PM d.c. brush motor is not only smaller but also with larger efficiency as the excitation losses are considered zero.

Note: To be fair, we should add the PM magnetization energy losses instead of excitation losses. However, the PM magnetization energy losses are so small when spread over the operational life of the PM machine that they can be neglected.

4.7 THE COMMUTATION PROCESS

Commutation can be defined as a group of phenomena related to the rotor coils current polarity reversal, when each one of them passes through the zero excitation (PM) flux density (neutral) axis.

In fact, this polarity reversal of current reduces to the coil switching from the (+) current path to the subsequent (–) current path.

During this interval, the respective rotor coil is short-circuited by the brush (brushes), and this phenomenon is known as brush commutation.

This is very similar to the soft (slow) power electronic commutation which means also applying a zero voltage to the respective coil.

The commutation is good if there is no visible scintillation and the commutator surface remains clean and undeteriorated for continuous duty at maximum rotor current and speed for which the machine was designed.

The behavior of the copper brush contact phenomenon with speed, current, etc. is a science in itself, and so it warrants separate treatment.

This is beyond our scope here.

Also, the brush span here is considered equal to the copper segment span at the commutator, and the intersegment insulation thickness is neglected.

The unfolding of commutation process in time is illustrated in Figure 4.16.

Let us denote the rotor coil resistance and inductance as R_c and L_c, respectively, and the brush–segment contact resistance corresponding to copper segment 1 and 2 as R_{b1} and R_{b2}, respectively:

$$R_{b1} = R_b \frac{A_b}{A_{b1}} = R_b \frac{1}{1 - t/T_c}; \quad R_{b2} = R_b \frac{A_b}{A_{b2}} = R_b \times T_c/t \tag{4.46}$$

A_b—brush (segment) total area.

Now the commutating coil circuit (voltage) equation is (Figure 4.16)

$$R_c i + R_{b1}\left(i_a + i\right) - R_{b2}\left(i_a - i\right) = -L_c \frac{di}{dt} + V_{er} + V_{ec} = \sum V_{ei} \tag{4.47}$$

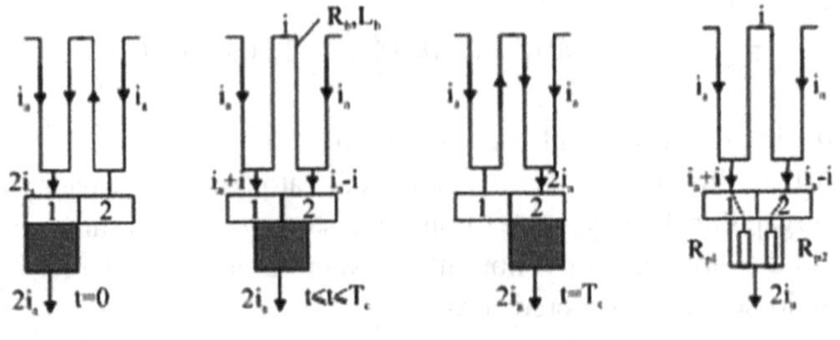

FIGURE 4.16 Rotor coil current reversal during commutation.

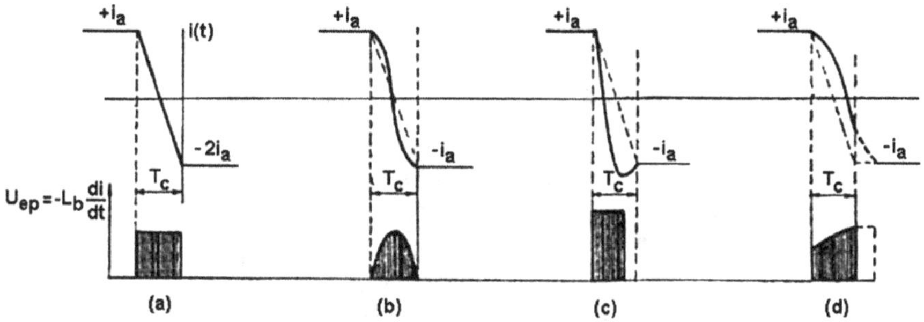

FIGURE 4.17 Commutation and self-emf ($-L_c(di/dt)$) during commutation: (a) linear, (b) resistive, (c) early, and (d) late.

$-L_c\,(di/dt) > 0$ is the self-induced coil voltage

$V_{er} > 0$; $V_{er} = B_{amax}\,\pi D_r n L N_c$ is the motion-induced voltage by the remaining (non-commutating) armature coils

$V_{ec} < 0$ is the motion-induced voltage by interpoles (Figure 4.3) on stator with their coils in series at brushes (PM motors lack interpoles). V_{ec} has opposite sign in comparison with V_{er} and $-L_c\,(di/dt)$ to effect total emf cancellation and thus provide for safe, in time, resistive commutation (Figure 4.17b).

In fact, the interpole mmf/pole $F_{interpole} = N_{cinterpole}\, i_{brush}$ > armature mmf per half a pole (Equation 4.39) to secure $|V_{ec}| > V_{er}$ for brushes placed in the neutral electric axis.

Note that the interpole current is the current at the brushes: $i_{brush} = 2aI_c$. For an intuitive interpretation of current commutation, let us consider a few particular simplified cases:

Linear commutation: $\sum V_{ei} = 0$ and $R_c = 0$.

In this case, the current $i(t)$ solution from Equation 4.46 with Equation 4.47 is

$$i(t) = i_c\left(1 - 2t/T_c\right) \tag{4.48}$$

The current variation is linear (Figure 4.17a) and the self-induced emf, $e = -L_c\,(di/d) = L_c\,2i_c\,nK$, is constant, where n is the speed in rps and K is the number of commutator copper segments (sectors). The zeroing of the resultant emf presupposes, in fact, the existence of interpoles.

Resistive commutation: $\sum V_{ei} = 0$ but $R_c \neq 0$.

Again, from Equation 4.46 with Equation 4.47, it follows that (Figure 4.17b)

$$i(t) = \frac{i_c\left(1 - 2t/T_c\right)}{1 + \dfrac{R_c}{R_b}\dfrac{t}{T_c}\left(1 - \dfrac{t}{T_c}\right)} \tag{4.49}$$

For resistive commutation, total zero emf is needed, so interpoles are necessary (in excited stators).

But, again, the commutation current variation from $+i_c$ to $-i_c$ is completed exactly within commutation time $T_c = 1/(nK)$; for $n = 60$ rps and $K = 24$, $T_s \approx 0.7 \times 10^{-3}$s.

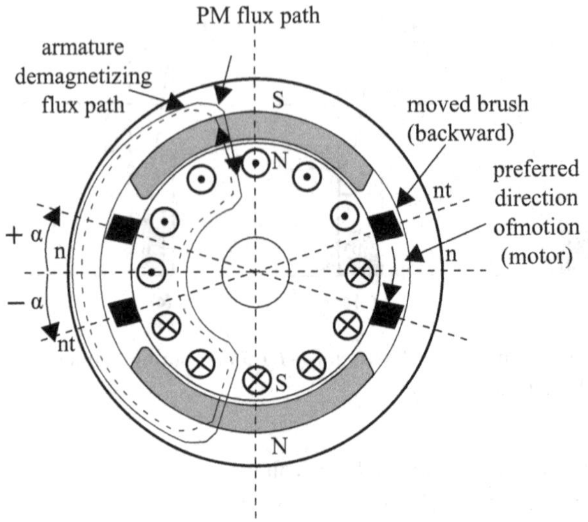

FIGURE 4.18 Brushes shifted countermotion-wise by α from neutral axis for better commutation.

Early commutation means that the interpoles are too strong, and thus total emf becomes negative at some point (Figure 4.17c); an increase of interpoles' airgap solves the problem.

Late commutation means a leftover emf between adjacent segments (Figure 4.17d), and thus scintillation occurs. The solution is to enforce the interpoles or reduce the airgap of interpoles.

For the PM motors, which do not have interpoles, and a single direction of motion, it is feasible to move the brushes in countermotion direction-wise by a small angle (−α) to produce a negative PM flux density in the commutating coil (see n, n', Figures 4.15b and 4.18).

This produces a negative V_{er} to cancel $-L_c$ (di/dt) and produces again zero total emf in the commutating coil short-circuited by brushes. The α/π part of rotor (armature) mmf now "demagnetizes" the magnets (Figure 4.18). This is the "price" paid for better commutation.

4.7.1 The Coil Commutation Inductance

The commutating inductance is placed with its sides in excitation (PM) neutral axis, and thus its inductance L_c refers to the d axis magnetic reluctance main field, L_{cm}, and the leakage inductance, L_{cl}:

$$L_c = L_{cl} + L_{cm} \tag{4.50}$$

The coil leakage inductance L_{cl} is having the same formula as in a.c. rotary machines and includes the slot leakage component L_{cls} and the coil end-connection component L_{cle}, with L_{cls} from Chapter 2 (Equation 2.44):

$$L_{cls} = 2\mu_0 N_c^2 \left(\frac{h_{rs}}{3b_{rs}} + \frac{L_{end}}{L_e} \ln \frac{L_{end}}{r_{end}} \right) \tag{4.51}$$

L_{cm} is

$$L_{cm} \approx \frac{N_c^2}{R_{gm}} = \frac{\mu_0 N_c^2 \tau_p L_{stack}}{g_m}; \quad g_m = gK_c + h_{PM}$$

where

h_{rs} is the slot height,
b_{rs} is the rotor slot average width (valid for open slots),
L_{end} is the coil end-connection length at one machine end,
r_{end} is the radius of coil end-connection bundle,
L_e is the magnetic machine stack length, and
R_{gm} is the airgap magnetic reluctance per pole.

It is thus evident that the surface PM stator, with a much larger magnetic airgap (due to $h_{PM} > g$), has a smaller commutating coil inductance.

4.8 EMF

As already discussed, the emf at the brushes corresponds to one of the $2a$ current paths in parallel and thus collects coil emfs (V_{ec}) from all coils, on a current path, in series.

But, again, the regular flux density in the airgap is nonuniform due to the armature reaction (see Section 4.7) as shown in a different form in Figure 4.19.

The motion emf in a coil, produced by excitation, from BIL formula, is

$$V_{ec}(x) = 2N_c \left(B_a(x) + B_{gF}(x) \right) \times L_{stack} \times \pi \times D_r \times n \tag{4.52}$$

where

$B_a(x)$ is the local armature flux density in the airgap and
$B_{gF,PM}$ is the local excitation (PM) flux density in the airgap.

Factor 2 comes from the fact that there are two active sides for each coil, and they are considered here exactly one pole-pitch aside (diametrical coils).

As evident from Figure 4.19, the local resultant flux density varies while the coil moves under the pole; so its emf, which is the voltage drop between neighboring commutator segments, is nonuniform.

This voltage nonuniformity may be large in non-PM motors, and thus it may endanger the life of insulation layer between the commutator segments; the former has to stay below 25 V.

Compensating windings, placed in stator pole slots and connected, again in series with the armature (at brushes), destroy the armature reaction under stator poles for all rotor currents and thus eliminate the problem (Figure 4.19b).

In addition, compensation windings, used only in very heavy duty transportation motors, reduce the magnetic saturation produced by the armature-reaction field.

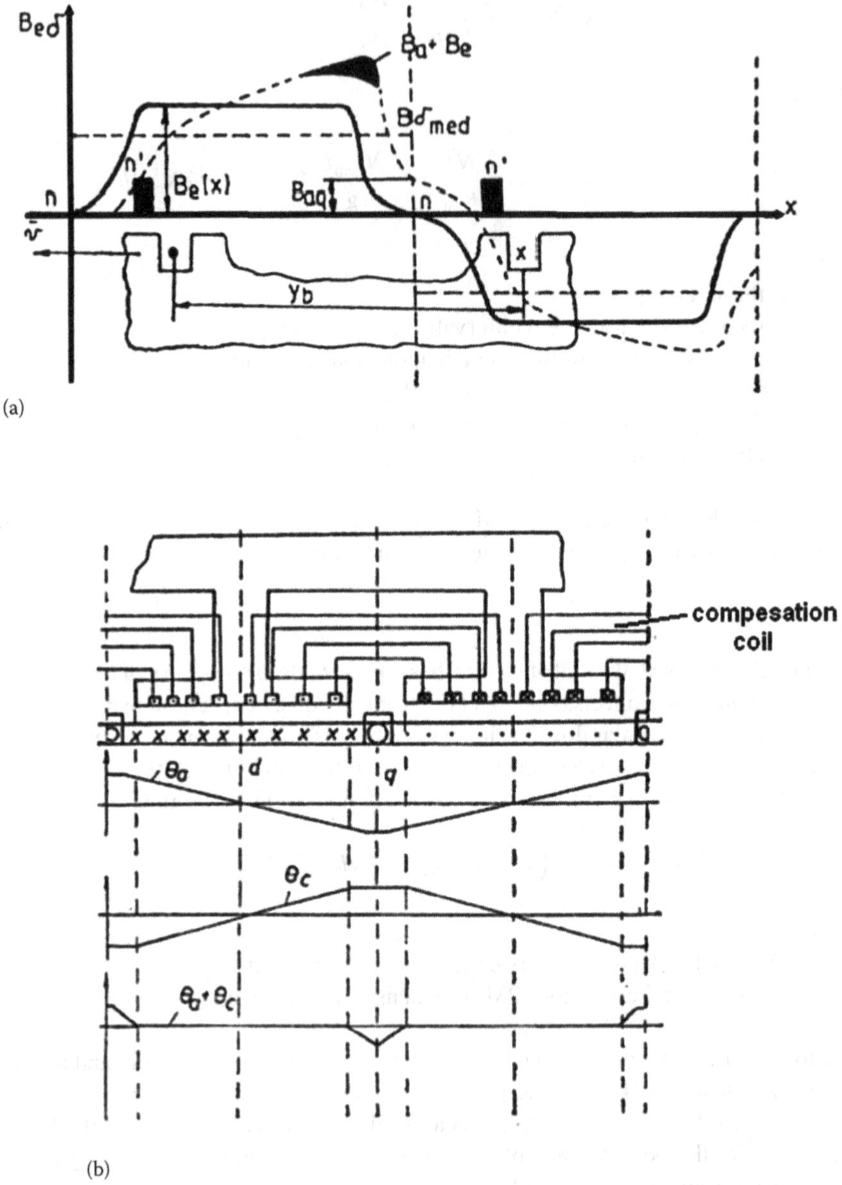

FIGURE 4.19 (a) Resultant nonuniform flux density in the airgap and (b) compensation winding.

Also, the armature field that produces V_{ec} (positive) in the commutating coil is much smaller as the active armature mmf per half a pole is reduced only to the uncompensated part located outside the stator poles (Figure 4.19b). So commutation is also notably improved or lighter interpoles are needed.

Now we may define an average total airgap flux density B_{gan}:

$$B_{gan} = \phi_p / (\tau L_e) \tag{4.53}$$

where L_e is the magnetic stack length ($L_e = L_{stack} K_{filliron}$).

There are ideally $N_s/2a$ (in reality $\approx N_s/2a - 1$) active coils in series per current path, and the brush emf V_{ea} is

$$V_{ea} = K_e n \phi_p \tag{4.54}$$

$$K_e = \frac{Np_1}{a}; \quad N = N_s 2N_c \tag{4.55}$$

where N is the total slot conductors per rotor. So, the no-load (average) voltage V_{ea} (at brushes) is proportional to speed n (rps) and total flux/pole through a proportionality coefficient which depends on the number of rotor slots N_s, number of pole pairs p_1, and the inverse of current path pairs a.

It is now evident why lap windings (with larger $2a = 2p_1m$) are preferred for low-voltage, high-current applications.

4.9 EQUIVALENT CIRCUIT AND EXCITATION CONNECTIONS

As we have settled that stator excitation (or PM) and armature mmfs and airgap magnetic fields have orthogonal electric axes (d and q; $\alpha_{edq} = 90° = p_1\alpha_{gdq}$), the equivalent circuit is straightforward, especially for separate or PM excitation (Figure 4.20).

C is the compensation winding and K is the interpole winding.

The commutation (K) and compensation (C) stator windings, both in the armature windings axis, are lumped into the armature winding. To eliminate the excitation power source, shunt or series excitation connection is used (Figure 4.20b and c). Mixed excitation is used today only for a few remaining d.c. generators in vessels or in some old Diesel–electric locomotives (Figure 4.20d).

In the following sections, we will consider only the separate (and PM) excitation in some detail for both motor and generator modes and, in short, d.c. series and a.c. series brush motors.

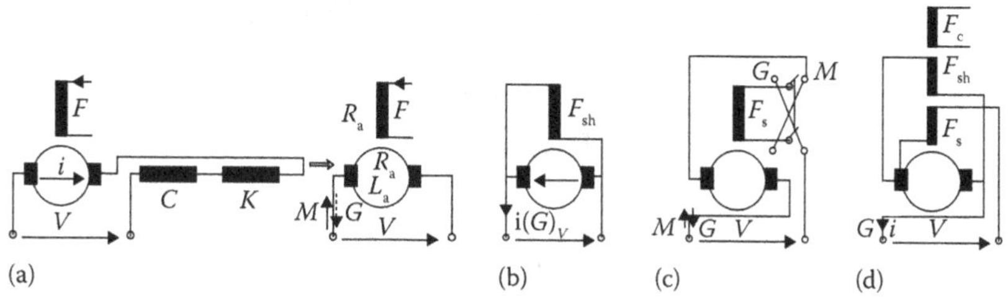

FIGURE 4.20 Equivalent circuits and excitation connection options: (a) separate excitation (or PM), (b) shunt excitation, (c) series excitation, and (d) mixed excitation.

4.10 D.C. BRUSH MOTOR/GENERATOR WITH SEPARATE (OR PM) EXCITATION/LAB 4.1

Let us consider a separate excitation d.c. brush machine (Figure 4.21) [4].

The machine is fed with a d.c. source in the stator and then in the rotor, separately, in this sequence. This is the motor mode. The general equation is

$$V_a = R_a i_a + L_a \frac{di_a}{dt} + V_{ea} + \Delta V_{brush} \tag{4.56}$$

For steady state (Equation 4.54),

$$V_{ea} = K_e n \phi_p; \quad i_a = \text{const} \tag{4.57}$$

The brush voltage drop ΔV_{brush} (around 1 V) is not negligible for low-voltage (automotive) motors.

Multiplying by the armature current, i_a, we obtain the power balance:

$$V_a i_a = R_a i_a^2 + V_{ea} i_a + \Delta V_{brush} i_a \tag{4.58}$$

The electromagnetic power $V_{ea} i_a$ is equal to the product of electromagnetic torque T_e and mechanical speed ($2\pi n$):

$$V_{ea} i_a = T_e 2\pi n \tag{4.59}$$

Making use of Equation 4.57, for V_{ea},

$$T_e = K_e \phi_p i_a / 2\pi \tag{4.60}$$

The electromagnetic torque is proportional to pole flux ϕ_p and brush (input (armature)) current i_a. And, with brushes in neutral electric axis, any change in the armature current i_a will be effect-less on the excitation circuit current (due to 90° electric phase

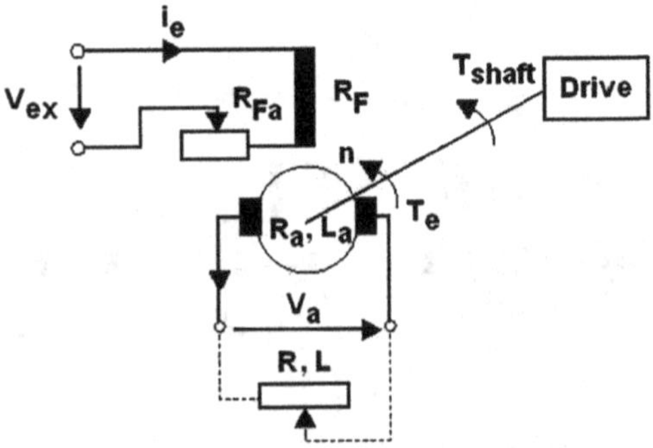

FIGURE 4.21 D.C. brush machine with separate excitation.

shift between armature and excitation circuits), if magnetic saturation is neglected (or small).

So the excitation circuit equation is

$$V_F = R_F i_F + L_{Ft}\frac{di_F}{dt} \tag{4.61}$$

$L_{Ft}(i_F)$ is the transient inductance (Chapter 2) in the presence of magnetic saturation:

$$L_{Ft}\left(i_F\right) = L_F\left(i_F\right) + \frac{\partial L_F}{\partial i_F} i_F \tag{4.62}$$

For the PM excitation, ϕ_p is constant in general if the magnetic saturation in the machine magnetic cores is negligible. Otherwise, a small pole flux decrease with armature current occurs. This may lead to motor self-overspeeding with load if the supply voltage stays constant. However, this does not happen if closed-loop torque (speed) control is performed through power electronics; it is still important for refined machine design (sizing).

Now, if we add the mechanical losses, p_{mec}, and the rotor iron losses (the stator excitation (or PM) field produces, in any point of the rotor core, a traveling field at frequency $\omega_r = 2\pi n p_1$, and thus both hysteresis and eddy current core losses occur), we get the complete power balance (Figure 4.22). Generating mode (Figure 4.22a) is used in contemporary motor drives for regenerative braking with bidirectional power electronics supply.

It is generator braking because the torque is negative ($i_a < 0$) and the input power is negative ($i_a < 0$) (Equation 4.59).

The excitation losses are traditionally left out just because they are obtained from a different source, but the total efficiency for generating is

$$\eta_{tg} = \frac{P_{2electrical}}{P_{1mechanical}} = \frac{V_a i_a}{V_a i_a + p_{mec} + p_{iron} + p_{copper} + p_{brush} + p_{excitation} + p_s} \tag{4.63}$$

For the motor mode (Figure 4.22b),

$$\begin{aligned}\eta_{tm} &= \frac{P_{2mechanical}}{P_{1electrical} + p_{excitation}} \\ &= \frac{T_{shaft}2\pi n}{T_{shaft}2\pi n + p_{mec} + p_{iron} + p_{copper} + p_{brush} + p_{excitation} + p_s} \\ &= \frac{T_{shaft}2\pi n}{V_{1a} i_a + p_{excitation}}\end{aligned} \tag{4.64}$$

Torque T_{shaft} is

$$T_{shaft} = T_e \mp \frac{p_{mec} + p_{iron}}{2\pi n} \tag{4.65}$$

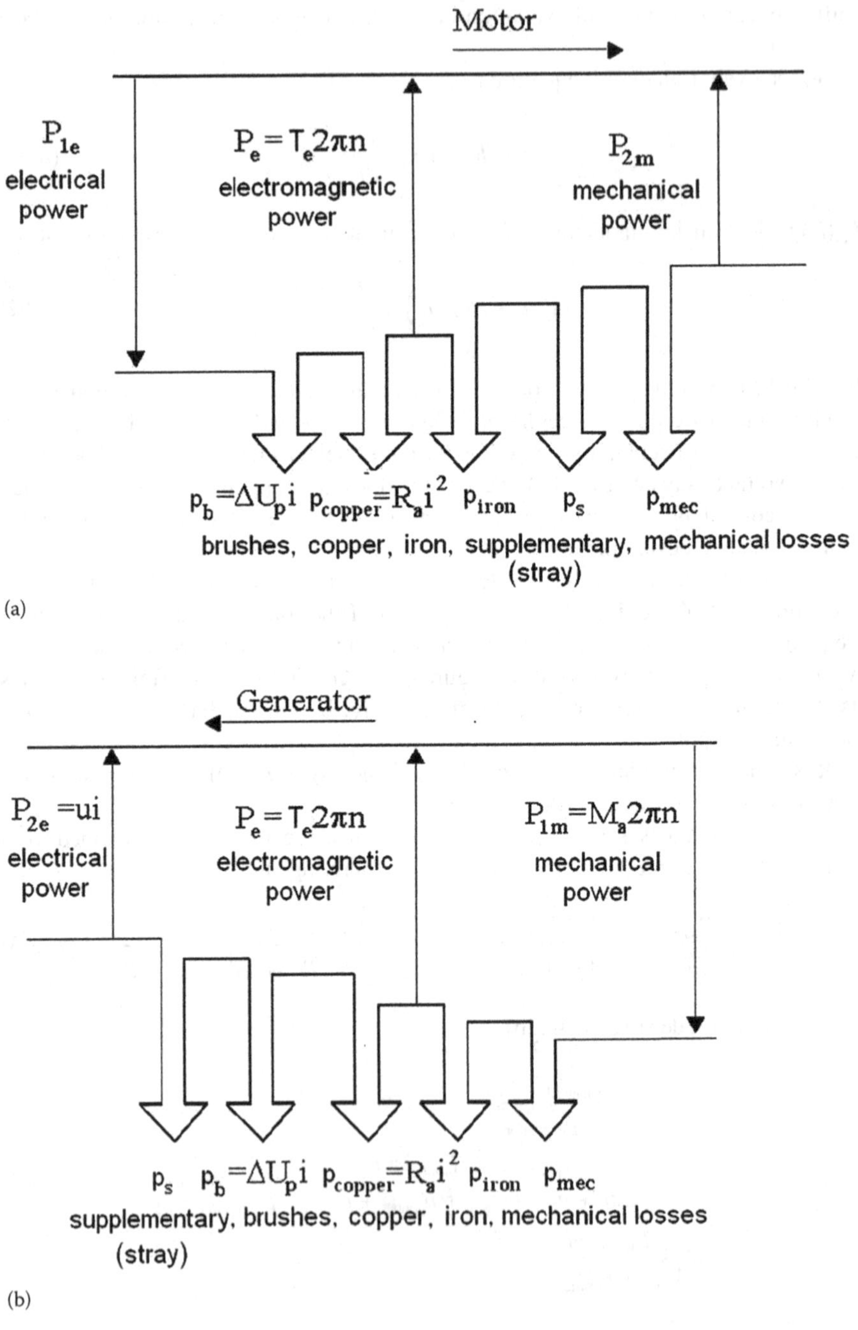

FIGURE 4.22 D.C. brush machine power balance: (a) for motoring and (b) for generating.

where

p_s is the stray losses,
p_{mec} is the mechanical losses in W,
and n is the speed in rps, and
Sign (−) is for motor and sign (+) is for generator.

For the generator (Equation 4.63),

$$P_{1\text{mechanical}} = T_{\text{shaftgen}} 2\pi n \tag{4.66}$$

4.11 D.C. BRUSH PM MOTOR STEADY-STATE AND SPEED CONTROL METHODS/LAB 4.2

D.c. brush PM (or any d.c.(a.c.)) motor is characterized at steady state by

Speed/current: $n = f_i(i_a)$; i_F and V_a = const;
Speed/torque curve: $n = f_{Te}(T_e)$; i_F and V_a = const;
Efficiency versus torque curve: $\eta_m = f_n(T_e)$; i_F or V_a = const.

The current/speed curve is obtained directly from the armature circuit. Equation 4.56 at steady state ($di_a/dt = 0$)

$$i_a = \frac{V_a - K_e\phi_p n - \Delta V_{\text{brush}}}{R_a + R_{\text{add}}} \tag{4.67}$$

Similarly, the speed/torque curve is obtained from Equation 4.66 and torque formula 4.60

$$n = n_{0i} - \frac{2\pi\left(R_a + R_{\text{add}}\right)T_e}{\left(K_e\phi_p\right)^2} - \frac{\Delta V_{\text{brush}}}{K_e\phi_p} \tag{4.68}$$

$$n_{0i} = \frac{V_a}{K_e\phi_p} \tag{4.69}$$

The speed n_{0i} for zero armature current ($i_a = 0$) is called the ideal no-load speed (in rps).

The ideal no-load speed is proportional to the d.c. armature supply voltage V_a. When V_a is changed, n_{0i} is changed, and, consequently, speed n is also changed almost in proportion.

It is also possible to reduce (vary) the current by adding a resistance, R_{add}, in series with R_a (Equations 4.67 and 4.68).

All these are shown in Figure 4.23.

The short-circuit torque, T_{esc}, corresponds to the short-circuit ($V_a = 0$) current, i_{sc}:

$$T_{esc} = K_e\phi_p i_{sc} \tag{4.70}$$

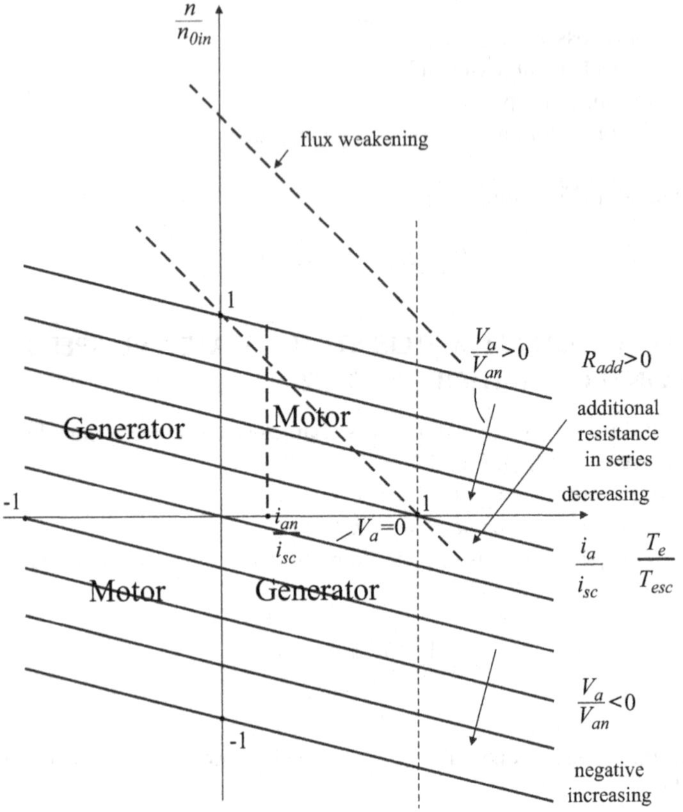

FIGURE 4.23 Per unit (pu) speed, n/n_{0in}, versus current, i_a/i_{an}, or torque, T_e/T_{en}.

$$i_{sc} = \frac{V_a}{R_a} \tag{4.71}$$

obtained at rated voltage V_a. The ideal rated no-load speed is

$$n_{0in} = \frac{V_{an}}{K_e \phi_p} \tag{4.72}$$

$$\frac{i_a}{i_{an}} = 1 - \frac{n}{n_{0in}}; \quad \frac{i_{sc}}{i_{an}} > \frac{20}{1} \tag{4.73}$$

$$\frac{n}{n_{0in}} = 1 - \frac{T_e}{T_{esc}} \tag{4.74}$$

The four quadrant operation is shown in Figure 4.23. The fact that the speed/current and speed/torque in (pu) are identical (Equations 4.73 and 4.74) indicates that torque control is the same as current control.

When the current is positive, it is motor (or generator) in forward (backward) motion direction (quadrants 1 and 4) mode, depending on the armature voltage polarity (+) or (−).

For motor/generator operation in reversal motion, the voltage is negative (quadrant 3) or positive (quadrant 2).

From this, we may derive motor starting and speed control methods for d.c. brush PM motors.

4.11.1 Speed Control Methods

Speed may be controlled for a given torque (current) (Equation 4.67)—positive or negative—for motoring and regenerating by controlling the armature voltage V_a (positive or negative): a four-quadrant a.c.–d.c. or d.c.–d.c. static converter (chopper) is required (Figure 4.23).

This easy voltage-control method of speed has secured the d.c. brush PM motor a future in niche applications. This method maintains low losses when controlled speed decreases, so it is efficient in energy conversion.

Speed for a given torque (current) may also be reduced by an additional resistance R_{add} in series to armature, but this will lead to an increase of losses in R_{add}, so it is energy intensive and it should be used only in short-duty small power motors to reduce the initial motor costs.

Note: For separate excitation, there is one more speed control method, above the rated speed, i.e., by weakening the pole flux ϕ_p (by reducing the field current i_F), and thus increasing the no-load ideal speed (Figure 4.23).

This method would be good for maintaining constant electromagnetic power, $P_e = T_e \times 2\pi n$, when speed increases above n_{0in} for rated voltage; 2 to 1 to 3 to 1 constant power speed range (above the rated base speed) is practical in urban transportation drives with d.c.–d.c. converter supply.

All of the above are explained in numerical Example 4.1.

Example 4.1: D.C. Brush PM Motor/Generator/Countercurrent Braking

Let us consider a small automotive d.c. brush PM motor with rated voltage V_{dc} = 42 V, rated power P_n = 55 W, and rated efficiency of η_n = 0.9 at n_n = 30 rps (1800 rpm). The brush voltage drop at rated current is ΔV_{brush} = 1 V and the core losses p_{iron} and mechanical losses (at rated speed) p_{mecn} are $p_{iron} = p_{mecn} = 0.01\ P_n$. Calculate the following:

a. The rated armature current, I_{an}
b. The copper rated losses, p_{con}
c. The armature resistance, R_a
d. The rated emf V_{en} and the rated electromagnetic torque, T_{en}
e. The rated shaft torque, T_{shaft}
f. Ideal no-load speed, n_{0in}
g. Calculate the input power and efficiency at half rated speed $n_n/2$ and rated current (torque) and the input (reduced) voltage V_a'; notice that

(Continued)

Example 4.1: (Continued)

core losses and mechanical losses are proportional to the required speed n squared.

h. Calculate the voltage required at $n_n/2$ and rated torque in the regenerative braking
i. For the same $n_n/2$ and T_{en} (i_{an}), calculate the required additional resistance R_{add} and efficiency
j. At zero speed, determine the voltage V_a for rated but braking torque T_{en} (negative), called the countercurrent braking
k. For n_n and generator mode on resistive load at armature terminals and for rated voltage V_{an}, calculate the load R_{load}. This may be a designated d.c. generator mode or it may be a braking regime called dynamic braking because the machine takes the energy from the load machine (kinetic energy) which decelerates gradually
l. Draw the characteristics for investigated motor (generator braking) modes

Solution:

a. The rated i_{an} springs from the input electric power $P_{1e} = V_{an} i_{an}$.

$$P_{1e} = \frac{P_n}{\eta_n} = \frac{55\text{ W}}{0.9} = 61.11\text{ W}$$

$$I_{an} = \frac{P_{e1}}{V_{an}} = \frac{61.11}{42} = 1.455\text{ A}$$

b. The rated copper loss is the only unknown component of all losses $\sum_p$:

$$\sum_p = P_{e1} - P_n = 61.11 - 55 = 6.11\text{ W}$$

$$P_{coppern} = \sum_p - p_{iron} - p_{mec} - p_{brushes} = 6.11 - 0.01 \times 2 \times 55$$

$$-1 \times 1.455 = 3.555\text{W}$$

c. The armature resistance R_a is

$$R_a = \frac{P_{coppern}}{I_{an}^2} = \frac{3.555}{1.455^2} = 1.68\ \Omega$$

d. For the rated emf V_{ean}, we make use of Equation 4.56 with $di_a/dt = 0$:

$$V_{ean} = V_{an} - R_a I_{an} - \Delta V_{brush} = 42 - 1.68 \times 1.455 - 1 = 38.556\text{ V}$$

The torque T_{en} is (Equation 4.59)

$$T_{en} = \frac{V_{ean} I_{an}}{2\pi n_n} = \frac{38.556 \times 1.455}{2\pi \times 30} = 0.2977\text{ N m}$$

e. The shaft rated torque T_{shaftn} (Equation 4.66) is

$$T_{shaftn} = T_{en} - \frac{(p_{mecn} + p_{iron})}{2\pi n_n} = 0.2977 - \frac{0.02 \times 55}{2\pi \times 30}$$
$$= 0.2918\ \text{N m} = \frac{P_n}{2\pi \times n_n}$$

f. The ideal no-load speed $n_{0n} = V_{an}/(K_e\ \phi_p)$ (4.72), but with $V_{ean} = K_e\ \phi_p\ n_n$ from Equation 4.54, it follows that

$$n_{0ni} = \frac{V_{an}}{V_{ean}} n_n = \frac{42}{38.556} \cdot 30 = 32.679\ \text{rps} \approx 1960\ \text{rpm}$$

g. At half rated speed $n_n/2 = 15$ rps (900 rpm) and rated motor torque T_{en} = 0.2977 N m, the required voltage, from Equation 4.57, is

$$V_a = R_a I_{an} + V_{ean} \frac{n_n/2}{n_n} + \Delta V_{brush} = 1.68 \times 1.455 + 38.556/2 + 1 = 22.722\ \text{V}$$

As both core losses and mechanical losses are reduced $((n_n/2)/n_n)^2 = 4$ times, we may calculate the total losses for this case $\sum_p$:

$$\left(\sum p\right)_{n_n/2} = p_{coppern} + (p_{iron} + p_{mecn}) \left(\frac{n_n/2}{n_n}\right)^2 + \Delta V_{brush} \cdot I_{an}$$
$$= 3.555 + 2 \times 0.01 \times 55 \times 1/4 + 1.0 \times 1.455 = 5.286\ \text{W}$$

Now the input electric power $P_{1e} = V_a I_{an} = 22.722 \times 1.455 = 33.06$ W. So the efficiency $\eta_{motor} = (P_{1e} - \sum p)/P_{1e} = (33.06 - 5.286)/33.06 = 0.84$. This is still a very good value.

h. For the regenerative braking, the torque $T_e = -T_{en} = -0.2977$ Nm and the current $I_a' = -I_{an} = -1.455$ A; $\Delta V_{brush} = 1$ V. Now from Equation 4.56

$$V_a' = R_a I_a' + V_{ean} \frac{n_n/2}{n_n} - \Delta V_{brush}$$
$$= 1.68 \times (-1.455) + \frac{38.556}{2} - 1 = 15.8336\ \text{V}$$

The losses are the same as for motoring at $n_n/2$ and $+T_{en}$ torque, so the efficiency is

$$\eta_{gen}' = \frac{V_a' |I_a'|}{V_a' |I_a'| + \sum p} = \frac{15.8336 \times 1.455}{15.8336 \times 1.455 + 5.286} = 0.813$$

The shaft torque T_{shaft} is larger than the electromagnetic torque:

(Continued)

Example 4.1: (Continued)

$$\left(T_{shaft}\right)_{regen} = T_e - \frac{p_{mec} + p_{iron}}{2\pi n/2} = -0.2977 - \frac{0.02 \times 55}{2\pi 30/2}\frac{1}{4} = -0.3006 \text{ Nm}$$

i. At $n_n/2$ and rated torque T_{en} and current I_{an}, with additional series resistance R_{add} at rated voltage V_{an}, Equation 4.56 becomes

$$V_{an} = \left(R_a + R_{add}\right)I_{an} + \frac{1}{2}V^*_{ean} + \Delta V_{brush}$$

so

$$R_{add} = \left(42 - \frac{1}{2} \times 38.556 - 1\right)/1.455 - R_a = 13.249\ \Omega$$

Now total losses are

$$\left(\sum_p\right)R_{add}* = \left(\sum_p\right)R_a, I_{an}, n_n/2 + R_{add}i^2_{an} = 5.286$$
$$+13.249 \times 1.455^2 = 33.33 \text{ W}$$

So, the efficiency is

$$\left(\eta_n\right)_{n_n/2, R_{add}, I_{an}} = \frac{V_{an}I_{an} - \left(\Sigma p\right)_{R_{add}}}{V_{an}I_{an}} = \frac{61.11 - 33.43}{61.11} = 0.4529$$

This is not an acceptable value for sustained operation.

j. At zero speed, the voltage Equation 4.56 degenerates into

$$V_a'' = R_a I_a + \Delta V_{brush}$$

For rated braking, torque $T_e = -T_{en} = -0.2977$ Nm, the current $I_a' = -I_{an} = -1.455$ A, and thus the voltage $(V_a'')_{n_n=0,\,T_e} = 1.68 \times (-1.455) + 1 = -1.444$ V. The four-quadrant converter has to be able to deliver negative current at negative voltage (third quadrant).

This is called the countercurrent braking (which is traditionally performed with $-V_{an}$ and a large additional R_{add} to limit the current, at the price of very large losses at low (or zero) speed).

The method corresponds to motoring in the reverse direction at zero speed and may be used as a controllable loading machine at zero speed for the convenient testing of other variable speed drives near zero speed.

k. The generator mode on a resistive load R_{load} at n_n makes use of Equation 4.57:

$$R_{load}I_{an} = \left(V_{an}\right)_{gen} = R_a\left(-I_{an}\right) + V_{ean} - \Delta V_{brush}$$
$$= 1.68 \times \left(-1.455\right) + 38.556 - 1 = 35.116 \text{ V}$$

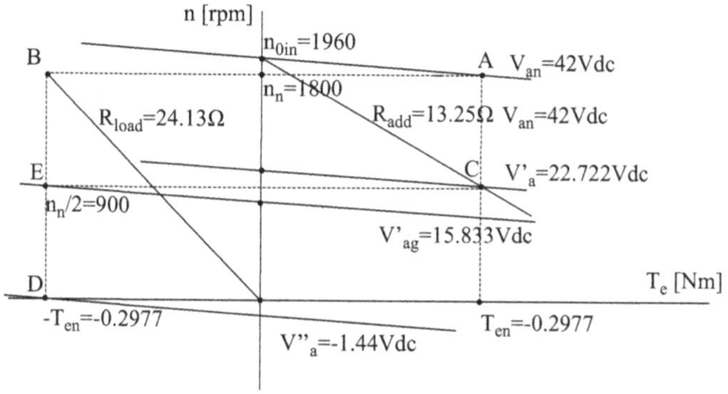

FIGURE 4.24 Motor/generator/braking speed/torque curves of d.c. brush PM machine.

So, the load resistance $R_{load} = (V_{an})_{gen}/I_{an} = 35.116/1.455 = 24.13\ \Omega$.

The output electric power in the load resistance $P_{2e} = (V_{an})_{gen}\ I_{an} = 35.116 \times 1.455 = 51.09$ W.

The fact that at rated current there is a voltage drop from no-load voltage V_{ean} to full generator load, V_{an}, ΔV:

$$\Delta V = \frac{V_{ean} - (V_{an})_{gen}}{V_{ean}} = \frac{38.556 - 35.116}{38.556} = 0.0892$$

shows that the generator design has to be dealt with separately if given rated load voltage or rated voltage regulation is to be met.

1. The mechanical speed/torque curves for the instances explained above are shown in Figure 4.24.
   A—motor/rated; *B*—generator/rated (on R_{load}); *C*—motor/rated torque at 50% rated speed, either produced by voltage V_a or with additional resistance R_{add}; *D*—counter-rated current braking at zero speed with reduced negative voltage V_a''; and *E*—regenerative braking at rated torque, 50% rated speed and reduced voltage V_a'.

Note: The separately excited d.c. brush machine has flux weakening (reducing ϕ_p) as an additional method for speed control above the rated speed and constant electromagnetic power; up to 3:1 max speed/rated speed ratios are feasible. In such cases, if the d.c.-excited brush motor is ruled out, a.c. machines with power electronics control are most suitable for such cases.

Example 4.2: Derivation of Rotor Parameters R_a and L_a

The rotor resistance and leakage inductance are essential for machine design. R_a contains the brush–commutator component. So the rotor coils resistance R_{ar}, considering the ideal number of coils in series per current path $N_s/2a$, is

$$R_{ar} \approx \left(\rho_{Co}\frac{l_{coil}N_c}{A_{copper}}\frac{N_s}{2a}\right)\frac{1}{2a}; \quad l_{coil} = 2\left(L_{stack} + L_{end}\right) \tag{4.75}$$

(Continued)

Example 4.2: (Continued)

ρ_{c0}—copper electric resistivity in Ohm meter, L_{end}—coil end-connection length at one machine end, N_s—number of rotor slots, and A—copper wire (cable) cross-section:

$$A_{copper} = \frac{I_{an}}{2aj_{cor}}; \quad i_c = \frac{I_a}{2a} \tag{4.76}$$

where

j_{cor} is the design current density in the rotor,

I_{an} is the rated current at the terminals,

The rotor inductance, L_a, has to consider the magnetic energy of the armature, and flux produced by the rotor current in the airgap and in the iron of the machine. The axis of this flux is along the neutral axis of the stator poles.

An exact value may be obtained by FEM, but at least for surface PM rotor with the total magnetic airgap $g_m = K_{cg} + h_{PM} * (\mu_{rPM})_{pu}$ ($(\mu_{rPM})_{pu} = 1.05 - 1.2$), its main component L_{am} comes from the armature magnetic energy in a cylindrical large airgap g_m as produced by a triangular mmf variation (Figure 4.15c and Equations 4.39 and 4.40):

$$W_{ma} = \frac{L_a i_a^2}{2} = \frac{1}{2} H_0 L_e 4p_1 \int_0^{\pi/2} g_m \left(\frac{2N_c i_c N_s}{4p_1 g_m} \frac{2\theta}{\pi} \right)^2 \frac{D_r}{2} d\theta_r \tag{4.77}$$

The integral is taken only for a half a pole during which the armature mmf varies from zero to its maximum and then the result is multiplied by $4p_1$ (the number of half poles per periphery). Finally:

$$L_{am} \approx \frac{\mu_0 N_c^2 N_s^2}{24a^2} \cdot \frac{L_e \tau}{g_m} \tag{4.78}$$

The total rotor (armature) inductance L_a also contains the leakage inductance, corresponding to the slot leakage field and the end-connection leakage field L_{al}.

L_{al} is related to one coil leakage inductance L_{cl} Equation 4.51, with $N_s/2a$ coils in series and with $2a$ paths in parallel:

$$L_{al} = \left(L_{cl} \frac{N_s}{2a} \right) \frac{1}{2a} \tag{4.79}$$

Finally,

$$L_a = L_{am} + L_{al} \tag{4.80}$$

For surface PM stators, L_{am} and L_{al} may have comparable values.

Note: For airgap (slotless) windings, the total magnetic gap g_m extends over the windings depth h_{tr}, while the slot leakage component in L_{cl} (first term in Equation 4.79) is taken out.

For excited brush machines with interpoles and compensation poles, the expression of L_a takes a more involved mathematical form.

4.12 D.C. BRUSH SERIES MOTOR/LAB 4.3

With the excitation circuit in series with the armature (at brushes), the d.c. brush series motor is convenient for wide constant power speed range; so it is needed in traction applications or ICE starters (Figure 4.25).

This machine is used mainly as a motor, and it may be used for self-excitation regenerative braking only after switching the terminals of the field winding first (to reverse the excitation field), Figure 4.20.

Alternatively, the excitation may be separated first and then supplied separately (at low voltage) for regenerative braking, as done routinely in standard diesel–electric or electric locomotives. (Recent and new traction drives use induction machines with full power electronics control.)

We will deal here only with the motor mode for which the voltage equation is

$$V_a = \left(R_a + R_{Fs}^e\right)i_a + \left(L_a + L_{Fs}\right)\frac{di_a}{dt} + V_{ea} \tag{4.81}$$

$$V_{ea} = K_e\phi_p\left(i_{Fs}\right)n; \quad R_{Fs}^e = \frac{R_{Fs}R_{Fad}}{R_{Fs} + R_{Fad}}; \quad i_{Fs} = i_a\frac{R_{Fad}}{R_{Fad} + R_{Fs}} \tag{4.82}$$

For steady state, $di_a/dt = 0$.

Now the pole flux is produced by the field current i_{Fs}, which is proportional (Equation 4.82) to the armature current (and equal to it when no flux weakening is performed by adding R_{Fad} in parallel with the field winding R_{Fs}).

The torque T_e is

$$T_e = \frac{V_{ea}i_a}{2\pi n} = \frac{K_e\phi_p\left(i_{Fs}\right)i_a}{2\pi} \tag{4.83}$$

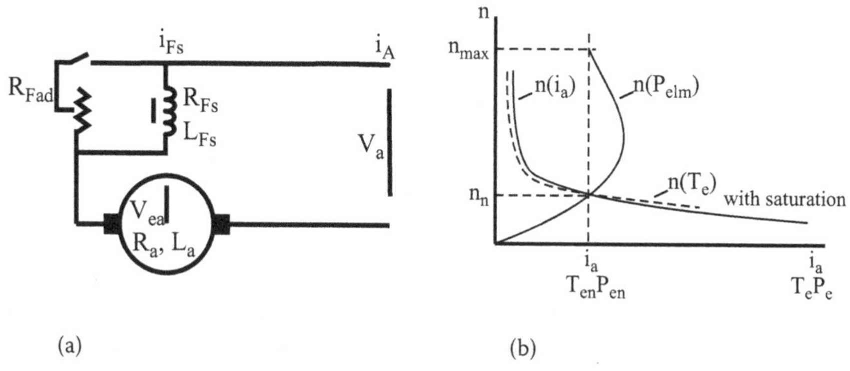

FIGURE 4.25 D.C. brush series machine: (a) equivalent circuit and (b) natural characteristic curves for rated voltage and no R_{Fad}.

The d.c. brush series motor may be described by the curves:

- $n(i_a)$
- $n(T_e)$; $n(P_{em})$
- $\eta(P_{em})$

$$\eta = \frac{P_{em} - p_{mec} - p_{iron}}{P_{em} + p_{copper} + p_{brush}} d \tag{4.84}$$

Magnetic saturation occurs above a certain i_a value and, from then on, the pole flux ϕ_p stays rather constant, and thus the characteristics at high current degenerate into those of d.c. brush separate excitation motors. Let us neglect the saturation and consider a linear relationship between i_{Fs} and ϕ_p:

$$\phi_p \approx K_\phi i_{Fs} \tag{4.85}$$

From Equations 4.81 through 4.84, we derive

$$i_a = \frac{V_a}{R_a + \frac{R_{Fs}R_{Fad}}{R_{Fs} + R_{Fad}} + K_e K_\phi \frac{R_{Fad}}{R_{Fad} + R_{Fs}} n} \tag{4.86}$$

$$T_e = K_e K_\phi \frac{R_{Fad}}{R_{Fad} + R_{Fs}} i_a^2; \quad P_{elm} = T_e 2\pi n \tag{4.87}$$

Graphical representation of $n(i_a)$, $n(T_e)$, and $n(P_{em})$ is shown in Figure 4.25b.

It is evident from Equations 4.86 and 4.87 that the electromagnetic (developed) power is maximum at a certain speed for constant voltage V_a (Figure 4.25b).

Also, the ideal no-load speed is $n_{0i} \rightarrow \infty$ because at zero current, the flux tends to zero (in fact to a permanent small value) ϕ_{prem}. So, a d.c. brush series motor should not be left without a load, a condition always fulfilled in traction drives.

$n(i_a)$ and $n(T_e)$ are mild characteristics, allowing wide speed variation, with appropriate control, to provide rigorously constant electromagnetic power over an $n_{max}/n_n = 3{:}1$ speed range.

4.12.1 Starting and Speed Control

Two effective methods of starting (limiting the current) and speed control are (for given torque) evident from Equations 4.86 and 4.87:

- Voltage control (by voltage reduction below V_{an}): $n < n_n$

  The torque at a given speed is proportional to the voltage squared (with neglected saturation), and thus $n(i_a)$ and $n(T_e)$ curves fall below those at V_{an}, if $V_a < V_{an}$, to produce the desired current (torque) at any speed below the rated speed n_n (Figure 4.26a). For example, constant torque up to the base (rated) speed may be provided this way (Figure 4.26b).
- Field weakening speed control: $n_n < n < n_{max}$

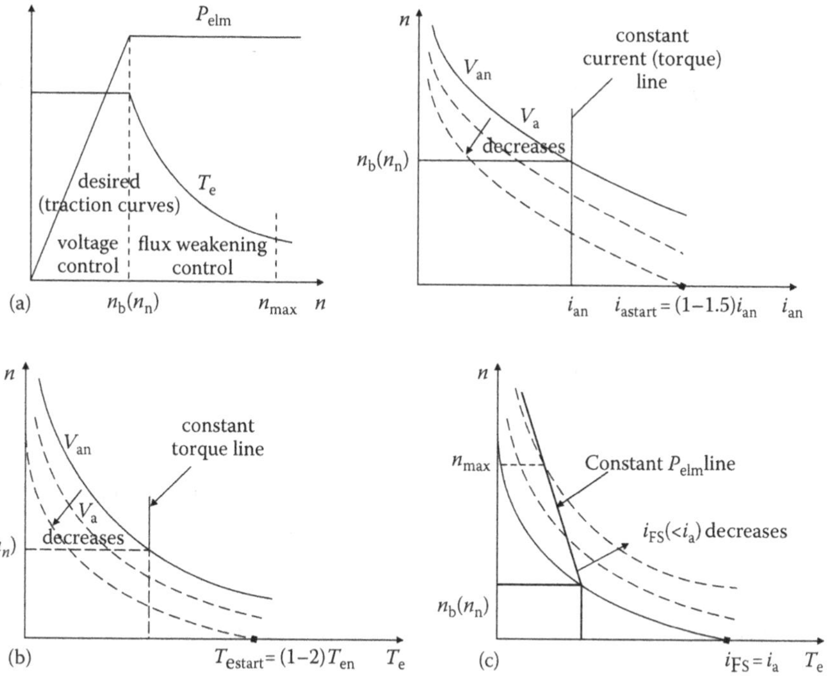

FIGURE 4.26 Speed control of d.c. brush series motor curves: (a) Traction motor torque (power) speed envelops, (b) voltage reduction control under base speed, n_b, for constant torque limit ($n(i_a)$ and $n(T_e)$ curves), and (c) flux weakening above base speed, n_b, for constant P_{elm} limit.

By modifying stepwise the resistance R_{Fad} in parallel with the field winding R_{Fs}, the field current i_{Fs} becomes smaller than the armature current i_a, and thus flux weakening occurs.

Above the base speed, we may modify R_{Fad} (Figure 4.25a) to maintain $P_{elm} = P_{en}$, if so desired, up to the maximum speed n_{max} (Figure 4.26c).

4.13 A.C. BRUSH SERIES UNIVERSAL MOTOR

As already discussed earlier, the universal motor is a.c. series–excited at constant (grid) frequency, and, supplied at a variable voltage amplitude, is used for speed control. For home appliances or construction tools (below 1 kW), a triac a.c. voltage changer (soft starter) produces the required voltage. For 2(3)-fixed-speeds simpler appliances, winding tapping or additional series resistance is used.

Right after 1900, universal motors in the tenths of kilowatts, with variable output voltage tapping of winding under load transformer supply, in configurations provided with interpoles and series or short circuit compensation (Figure 4.27), have been used in Switzerland, for example, in the mountain railroad locomotives at 16.33 Hz; some of them are apparently still in use today!

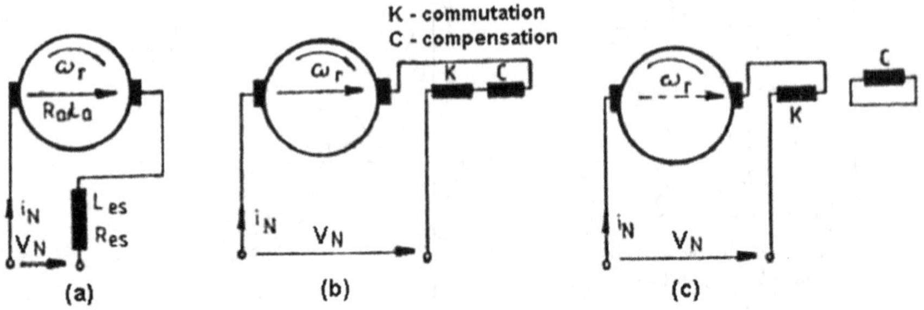

FIGURE 4.27 A.C. brush series motor: (a) low power, (b) medium power with series compensation *C*, and (c) medium power with short-circuit compensation *C*; *K*—interpole (commutation) winding.

The machine voltage equation should take into consideration the fact that the a.c. excitation induces a pulsational V_{ep} and a motion emf V_{er} into the rotor circuit:

$$\left(R_a + R_{Fs}\right)i - V = V_{ep} - V_{er} \tag{4.88}$$

The pulsational (transformer-like) self-induced voltage (emf) V_{ep} is

$$V_{ep} = -\left(L_a + L_{Fs}\right)\frac{di}{dt} \tag{4.89}$$

The motion emf V_{er} is, as in d.c., with neglected magnetic saturation:

$$V_{er} = K_e n \phi_p \approx K_e K_\phi n i \tag{4.90}$$

The electromagnetic torque T_e, as for d.c., is

$$T_e\left(t\right) = \frac{V_{er} i}{2\pi n} \approx \frac{K_e K_\phi i^2}{2\pi} \tag{4.91}$$

The torque expression is similar to the formula for the d.c. brush series motor, but the current is a.c. and, during steady state, at stator (supply) frequency ω_1.

For steady state (constant speed and constant torque load),

$$V\left(t\right) = V_1\sqrt{2}\cos\omega_1 t; \quad i\left(t\right) = I_1\sqrt{2}\cos\left(\omega_1 t - \varphi_1\right) \tag{4.92}$$

Current i in the stator and at the rotor brushes is indeed at frequency ω_1, but in the rotor, one more frequency, $\omega_r = 2\pi n p_1$, occurs as in the case of d.c. brush machines.

For rated speed, $\omega_r = (3\text{–}6)\omega_1$, in order to secure a large motion emf, V_{er} (which, being in phase with the current, leads to a very good power factor).

Now with Equation 4.92 in Equation 4.91, the instantaneous steady-state torque, $T_e(t)$, is

$$T_e(t) = \frac{K_e K_\phi I_1^2}{2\pi}\left(1 - \cos 2(\omega_1 t - \varphi_1)\right) \tag{4.93}$$

As expected, from a single-phase winding a.c. machine (of any kind), the torque pulsates at $2\omega_1$ and care must be exercised in damping the frame vibration because of this.

As with a.c. sinusoidal terminal voltage and current circuits (leaving out the true $\omega_1 + \omega_r$ frequency in the rotor), phasors may be used:

$$v \to \underline{V} = V_1\sqrt{2}e^{j\omega_1 t}; \quad i \to \underline{I} = I_1\sqrt{2}e^{j(\omega_1 t - \varphi_1)} \tag{4.94}$$

With Equation 4.94 in Equations 4.88 through 4.90, we obtain

$$\underline{V} = (R_a + R_{Fs})\underline{I} + j\omega_1(L_a + L_{Fs})\underline{I} + K_e K_\phi n \underline{I} \tag{4.95}$$

With $R_a + F_{Fs} = R_{ae}$, with $\omega_1(L_a + L_F) = X_{ae}$, and by introducing a series resistance R_{iron} to account for iron losses (both in the stator and the rotor), Equation 4.95 becomes

$$\underline{V} = \underline{I}(R_{ae} + jX_{ae} + R_{iron} + K_e K_\phi n) \tag{4.96}$$

The average torque T_{eav} (from Equation 4.93) is

$$T_{eav} = \frac{K_e K_\phi I^2}{2\pi}; \quad \cos\varphi = \frac{(R_{ae} + R_{iron} + K_e K_\phi n)I}{V} \tag{4.97}$$

With current from Equation 4.96 in Equation 4.97, the torque speed curve is

$$T_{ean} = \frac{K_e K_\phi}{2\pi} \frac{V^2}{(R_{ae} + R_{iron} + K_e K_\phi n)^2 + X_{ae}^2} \tag{4.98}$$

The $n(T_{ean})$ curve (Figure 4.28c) resembles the d.c. brush series motor and speed control is handled by amplitude voltage control easily through a triac soft starter (with an adequate power filter to attenuate harmonics current in the rather weak (residential) power sources (transformers)) [6].

As already explained in the paragraph on commutation, a.c. current commutation is more difficult because of the additional a.c. excitation emf in the commutating coil present at all speeds. This a.c. emf has to be reduced below 3.5 V at start and to 1.5–2.5 V at full speed running.

Frequency reduction is another way to improve commutation, which is the main design limiting factor.

Due to overall reduced cost, the universal motor is still heavily used in home appliances, from hair dryers to vacuum cleaners and some washing machines, and for handheld tools with dual or variable speed.

One contemporary automotive small d.c. brush PM motor and one handheld tool universal motor are shown in Figure 4.29.

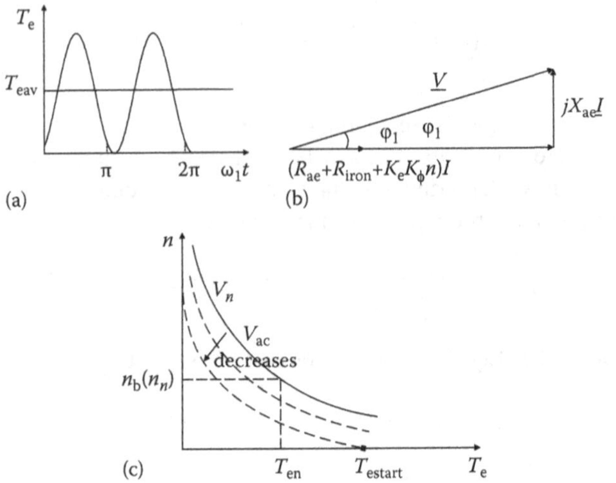

FIGURE 4.28 A.C. brush series (universal) motor characteristics: (a) instantaneous torque pulsations at $2\omega_1$ frequency, (b) the phasor diagram at rated speed: $\cos\varphi_1 > 0.9$ for $(p_1n_n/f_1) \approx (3 \div 6)$, and (c) speed control by a.c. voltage reduction: $n(T_e)$ curves.

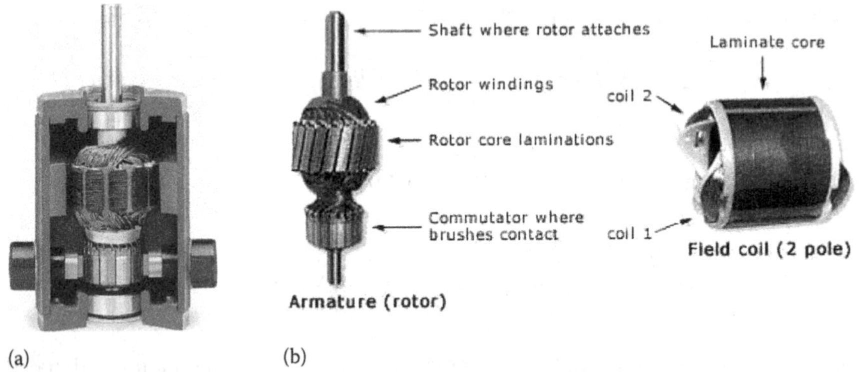

FIGURE 4.29 (a) Automotive small d.c. brush PM motor and (b) handheld universal contemporary motor.

4.14 TESTING BRUSH–COMMUTATOR MACHINES/LAB 4.4

Testing is done for acceptability and performance, and there are international and national standards for testing, such as IEEE or IEC Standards.

Here, we concentrate on modern methods to assess the performance and for heating (rated duty loading) test.

It all starts with the efficiency definition, say, for motor mode:

$$\eta_{motor} = \frac{P_{2mechanical}}{P_{1electrical}} = \frac{P_{1electrical} - \sum p}{P_{1electrical}} \quad (4.99)$$

Let us first consider the d.c. brush PM motor.

4.14.1 D.C. Brush PM Motor Losses, Efficiency, and Cogging Torque

First we should note that the PM field is always there and so are the iron losses due to it, situated mainly in the rotor at frequency $f_r = p_1 n$, variable with speed.

Let us suppose that we are able to attach to the shaft a driving machine (an identical twin motor, for example) with a torque-meter or at least an encoder with speed estimation down to very low speeds (1–2 rpm) (Figure 4.30a and b).

With a complete test rig having twin motors, a torque-meter, and an encoder, we can run M_2 as the motor and M_1 as the generator, and thus first at rated voltage V_{an} and rated current I_{a2n}, machine 2 is loaded.

We can measure rather precisely the electric input power in both machines P_{11el} and P_{12el}; their difference is twice the total losses of each of the twin machines:

$$2\left(\sum p\right) = P_{12el} - \left|P_{11el}\right| \tag{4.100}$$

$P_{12el} = V_{an} i_{a1n}$, $P_{11el} = V_{an} i_{a2}$, and $i_{a2} \neq i_{a1n}$.

While driving machine M_2, at zero current ($i_{a2} = 0$), by machine M_1 in the speed closed-loop control mode at very low speed (1–20 rpm), we may record the rotor position θ_r simultaneously with the torque from the torque-meter (precise enough for 5% rated torque).

This is the cogging torque of the machine M_2: $T_{cogg}(\theta_r)$ (Figure 4.31).

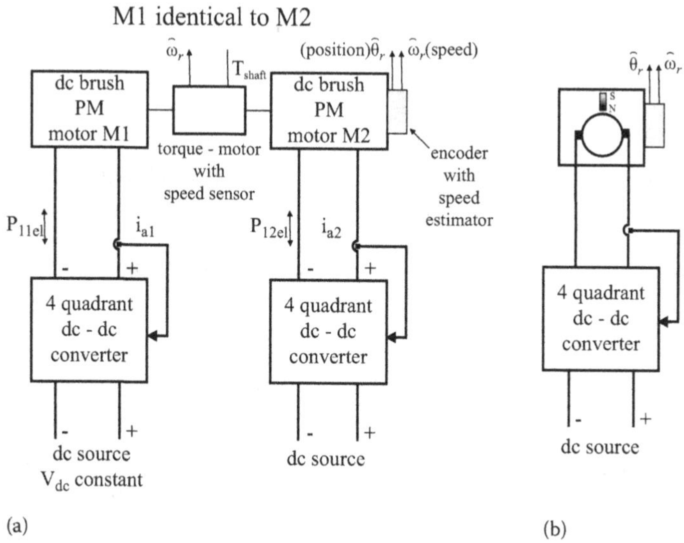

FIGURE 4.30 (a) Testing rig for small d.c. brush PM motors: complete, with twin motors M_1 and M_2 and (b) minimal.

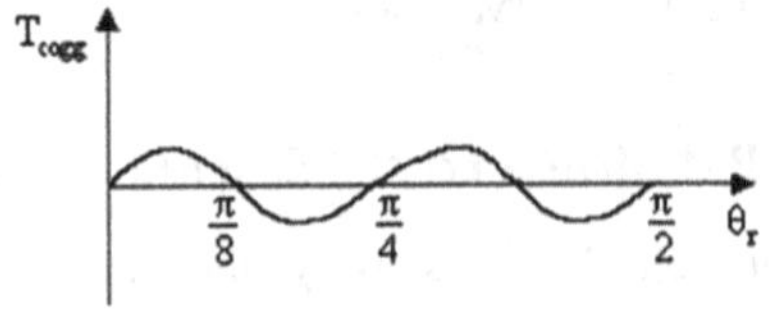

FIGURE 4.31 Typical cogging torque, T_{cogg} ($i_a = 0$), versus position for $N_s = 8$ rotor slots and $2p_1 = 2$; LCM (8,2) = 8.

The number of periods of cogging torque is the lowest common multiplier (LCM) of the stator PM poles $2p_1$ and the rotor slot number N_s. The larger the LCM, the smaller the cogging torque.

The shaping of the PM pole ends might also reduce the peak cogging torque value considerably.

The cogging torque produces speed pulsations, frame noise, etc., and it has to be reduced in most, if not all, the drives used today.

PM pole flux ϕ_{PM} may also be determined under no load by measuring the no-load voltage V_{ea2} and speed of M_2 (on no load) when driven by M_1:

$$K_e\phi_{PM} = \frac{V_{ea2}(V)}{n(rps)} \tag{4.101}$$

The same schematic in Figure 4.30a may be used to program the loading (through the current i_{a1} in M_1) according to real duty cycles and thus perform the heating or endurance tests. Ony the losses in the two machines are absorbed from the power grid.

The minimal test rig (Figure 4.30b) requires apriori specific tests, according to the voltage equation of M_2:

$$V_{a2} = R_a I_{a2} + V_{ea2} \tag{4.102}$$

R_a is measured in advance, with brushes on commutator (it contains part of brush voltage drop contribution), at rated current, quickly (by stalling the motor with a wrench), to avoid overheating influence on R_a.

Then the motor running on-load, with V_{a2} and n measured (Equation 4.102), delivers V_{ea2}. Thus,

$$K_e\phi_p = \frac{V_{ea2}}{n}$$

Now we may pulsate the input voltage up and down with a certain amplitude and frequency until the motor cannot follow, and thus speed pulsations are small.

The machine switches from the motor mode to the generator mode at input voltage a.c. frequency, with the total average power from the converter equal to the total losses in the machine (Figure 4.32).

This is known as artificial loading.

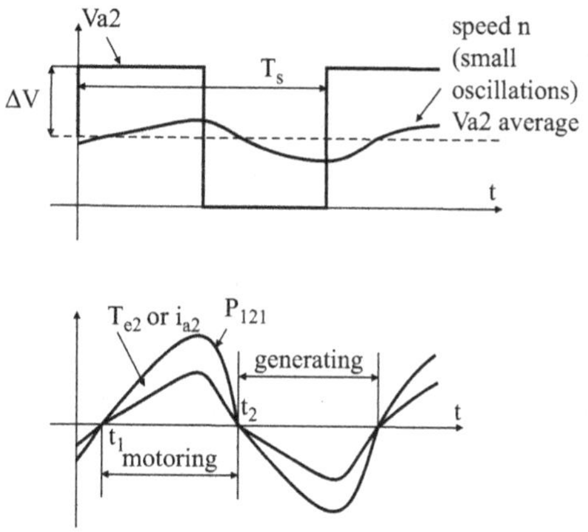

FIGURE 4.32 Artificial loading of d.c. brush PM motor.

$$\sum p = \frac{1}{T_s} \int_0^{T_s} P_{12e} \mathrm{d}t \tag{4.103}$$

$$i_{a2}(\mathrm{RMS}) = \sqrt{\frac{1}{T_s} \int_0^{T_s} i_{a2}^2 \mathrm{d}t} \tag{4.104}$$

$$\left(P_{an12e}\right)_{motoring} = \frac{1}{T_s} \int_{t_1}^{t_2} P_{e12e}(t) \mathrm{d}t \tag{4.105}$$

Now the efficiency is (with Equations 4.103 through 4.105)

$$\eta_{motor} = \frac{\left(P_{an12e}\right)_{motor} - \sum p}{\left(P_{an12e}\right)_{motor}} \tag{4.106}$$

If the endurance (heating) test at the rated duty cycle is desired, then the voltage amplitude ΔV pulsation amplitude is the output of a closed-loop current regulator which monitors the $(i_{a2})_{RMS}$ current. Only the voltage V_{a2}, current i_{a2}, and speed n are measured ($P_{12e} = V_{a2} \cdot i_{a2}$) and processed using Equations 4.103 through 4.105.

The $K_e\phi_p$ coefficient in the emf V_{ea2} may be also measured by free deceleration of M_2 with V_{ea2} and speed n measured ($K_e\phi_p = V_{ea2}/n$).

The cogging torque may not be measured with this minimal (but contemporary) test rig.

Testing of universal motor is somewhat similar but artificial loading is not feasible. For full loading, the complete test rig (Figure 4.30a), but with a d.c. brush PM loading machine (M_1) and a universal motor (M_2), supplied through an a.c. triac (soft starter), is necessary.

4.15 PRELIMINARY DESIGN OF A D.C. BRUSH PM AUTOMOTIVE SMALL MOTOR BY EXAMPLE

As for any electric machine design, we start with specifications such as

- Rated voltage: $V_{dc} = 36$ V
- Rated continuous torque: $T_{en} = 0.5$ Nm, constant to maximum speed, n_{max}
- Maximum speed: $n_{max} = 3600$ rpm = 60 rps
- Peak torque at maximum speed $(T_{emax})_{nmax} = 0.6$ Nm
- Short duty cycle: 14% self-ventilation, $j_{comax} \approx 4.5$ A/mm²
- Surface PM stator

Solution:
The design items are

- Rotor diameter, D_r, and stack length, L_{stack}
- PM stator design
- Rotor slot and armature winding design
- R_a, L_a, K_e ϕ_p, rated losses, and efficiency

Given the small torque, a 2-pole motor is chosen for start.

The tangential specific force is chosen at $f_{tmax} = 0.3 \times 10^4$ N/m² (for low current density), and thus (Equation 4.13),

$$T_{emax} = f_{tmax}\pi D_r L_{stack}\frac{D_r}{2} = f_{tmax}\frac{\pi D_r^3}{2}\frac{L_{stack}}{D_r} \tag{4.107}$$

With $\lambda_s = L_{stack}/D_r = 1.2$, we obtain the rotor diameter:

$$D_r = \sqrt[3]{\frac{2T_{emax}}{f_{tmax}\pi\lambda_s}} = \sqrt[3]{\frac{2\times 0.6}{0.3\times 10^4\times\pi\times 1.2}} = 0.0474 \text{ m}$$

The stack length is

$$L_{stack} = D_r\lambda_s = 1.2\times 0.0474 = 0.05685 \text{ m} \approx 0.06 \text{ m} \tag{4.108}$$

4.15.1 PM Stator Geometry

First, we have to consider the PM's material and its remnant flux density, B_r, and coercive field, H_c. Let us consider lower cost, bonded NdFeB, with $B_r = 0.6$ T and $H_c = 450$ kA/m. For surface PMs, let us consider an airgap flux density $B_{gPM} = 0.45$ T, with a fringing (leakage) $K_{lPM} = 0.15$. From Equation 4.25, the PM flux density, B_m, is

$$B_m = B_{gPM}\left(1 + K_{lPM}\right) = 0.45\left(1 + 0.15\right) = 0.5175 \text{ T}$$

For an airgap $g = 1$ mm, let us find the PM thickness, h_{PM}, if the iron saturation factor $1 + K_s = 1.15$ (Equation 4.31):

$$h_{PM}\left(H_c - H_m\right) \approx F_g\left(1+K_s\right); \quad F_g = K_c \cdot g \cdot \frac{B_{gPM}}{\mu_0}; \quad H_m \approx \frac{B_m}{\mu_0 \times 1.05} \tag{4.109}$$

The saturation factor K_s value is low due to the large total (magnetic) airgap $g_m \approx K_c \cdot g + h_{PM}$.

Also, due to large g_m, with semiclosed rotor slots, $K_c \approx 1.03$–1.06; let us consider $K_c = 1.05$.

From Equation 4.109,

$$h_{PM} \approx \frac{F_g}{B_r - B_m} = \frac{1.05\times 10^{-3} 0.45}{0.6-0.5175} = 5.727\times 10^{-3} \text{ m}$$

The PM radial height has been chosen large in order to reduce the rotor coil inductance L_c and thus ease the process of commutation at 3000 rpm. The pole span of the PM is $\tau_p = 0.7\tau$, with $\tau = \frac{\pi D_r}{2p_1} = \frac{\pi 0.0474}{2} = 0.0744$ m.

The stator yoke flux density B_{ys} is

$$B_{ys} = \frac{\left(\tau/2\right)\cdot B_m}{h_{ys}}; \quad B_{ys} = 1.4 \text{ T} \tag{4.110}$$

So, $h_{ys} = ((0.0744 \cdot 0.7)/2)(0.5175/1.4) = 0.0096$ m.

So, the stator outer diameter, D_{out}, is

$$\begin{aligned} D_{out} &= D_r + 2g + 2h_{PM} + 2h_{ys} \\ &= 0.0474 + 2\times 0.001 + 2\times 0.005727 + 2\times 0.0096 = 0.08 \text{ m} \end{aligned}$$

The ratio $D_r/D_{out} = 0.0474/0.08 = 0.5925$, which is in the practical interval for 2-pole motors.

The torque is (Equation 4.60)

$$T_{emax} = \frac{p_1}{2\pi}\frac{N}{a}\phi_p i_a; \quad i_a = 2ai_c; \quad N = N_c N_s \tag{4.111}$$

We first have to settle the rotor yoke depth, h_{yr}, after considering a shaft diameter of 10 mm for 0.6 N m:

$$h_{yr} \approx \frac{\tau_p B_{gPM}}{2B_{yr}} = \frac{0.7\times 0.0747}{2}\frac{0.45}{1.5} = 0.0078 \text{ m}$$

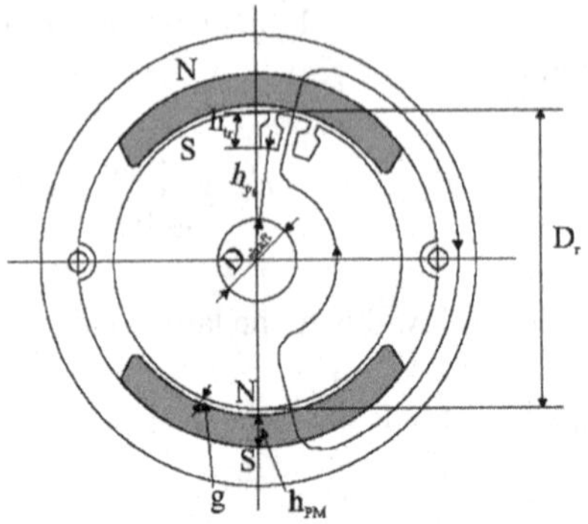

FIGURE 4.33 D.C. brush PM motor cross section.

4.15.2 Rotor Slot and Winding Design

The slot height left, h_{tr} (Figure 4.33), is

$$h_{tr} = \frac{D_r - D_{shaft} - 2h_{ys}}{2} = \frac{0.0474 - 0.01 - 2\times 0.0078}{2} \approx 0.011\ \text{m}$$

From Equation 4.111, the A turns per rotor periphery Ni_a for a simple lap winding ($2a = 2p_1 = 2$) is

$$Ni_{amax} = \frac{T_{emax}\times 2\pi \times a}{p_1 \times B_{gPM}\times l_{stack}\times \tau_p} = \frac{0.6\times 2\pi\times 1}{1\times 0.45\times 0.06\times 0.7\times 0.074}$$
$$= 2680\ \text{A turns/periphery}$$

Now the number of turns should be an even number for a simple lap winding. A slot pitch τ_s easy to get through stamping is $\tau_s = (0.009\text{–}0.012)$ m.

With 16 rotor slots, $\tau_s = \pi D_r/N_s = \pi \times 0.0474/16 = 0.0093$ m. With the rotor teeth area equal to the slot area and a slot fill factor, K_{fill}, the current density is

$$j_{cormax} = \frac{Ni_{amax}}{K_{fill}\times \frac{1}{2}\times\frac{\pi}{4}\left(D_r^2 - \left(D_r - 2h_{h_{tr}}\right)^2\right)}$$
$$= \frac{2680\times 10^6}{0.45\times\frac{1}{2}\times\frac{\pi}{4}\left(47.4^2 - \left(47.4 - 2\times 11\right)^2\right)} = 9.47\times 10^6\ \text{A/m}^2$$

For 14% duty cycle,

$$\left(j_{cor}\right)_e = j_{cormax}\sqrt{0.14} = 9.47\times 10^6\sqrt{0.14}\ \text{A/m}^2 = 3.532 < 4.0\ \text{A/mm}^2$$

The slot active area, A_{slota}, is

$$A_{slota} = \frac{\pi\left(D_r^2 - \left(D_r - 2h_{tr}\right)^2\right)}{2\times 4\times N_s} = \frac{\pi\left(47.4^2 - \left(47.4 - 21.1\right)^2\right)\times 10^{-6}}{2\times 4\times 16}$$
$$= 39.29\times 10^{-6}\,\text{m}^2$$

With the active slot height $h_{tra} = 8$ mm $< h_{tr} = 11$ mm, the average width $b_{tsa} = A_{slota}/h_{tr} = 39.29/8 \approx 5$ mm, with an average tooth width $b_{tra} \approx \tau_s - b_{ta} \approx 9.3 - 5 = 4.3$ mm, which will secure a tooth flux density below 1.2 T on no load ($B_{gPM} \approx 0.45$ T).

It is feasible to further decrease the rotor tooth width in order to increase the slot active area and thus further reduce the peak current density and copper losses. The number of conductors, N, per rotor periphery may be calculated from the emf expression:

$$V_{ean} = \frac{p_1}{a} N\phi_p n_{max} = \left(0.9 - 0.92\right)V_{dcn} = \left(32.4 - 33.12\right)\text{V}$$
$$\phi_p = \tau_p B_{gPM} L_{stack} = 0.7\times 0.0744\times 0.45\times 0.06 = 1.406\times 10^{-3}\ \text{Wb}$$
$$N = \frac{32.4}{1\times 1.406\times 10^{-3}\times 60} \approx 384\ \text{turns}$$

There are 16 slots and 2 coils per slot, so the number of turns per coil is

$$N_c = \frac{N}{2N_s} = \frac{384}{2\times 16} \approx 12.00 = 12\ \text{turns/coil}$$

So, $N = N_c 2N_s = 12 \times 2 \times 16 = 384$ and $i_{amax} = \frac{Ni_{amax}}{N} = \frac{2679}{384} = 6.976$ A.

The coil current $i_{cmax} = \frac{i_{amax}}{2a} = \frac{6.976}{2\cdot 1} = 3.488$ A.

So, the coil wire diameter, d_{c0}, is

$$d_{c0} = \sqrt{\frac{i_{cmax}}{j_{cormax}}\frac{4}{\pi}} = \sqrt{\frac{3.488\times 4}{9.47\times 10^6\pi}} = 0.685\times 10^{-3}\ \text{m}$$

Note: We now proceed with the motor parameters: R_c, L_c, R_a, and L_a. But we will first stress on copper losses:

$$p_{copper} = R_a I_a^2 = \frac{1}{2a}\rho_{c0}\frac{l_{coil}}{\frac{\pi d_{c0}^2}{4}} N_c \frac{N_s}{2a} I_{c\,max}^2$$
$$= \frac{1}{2} 2.1\times 10^{-8}\frac{0.337\times 4}{\pi\times\left(0.685\right)^2\times 10^{-6}}\times 12\times\frac{16}{2}\times 3.488^2 = 33.3\ \text{W}$$
$$l_{coil} = 2\left(l_{stack} + 1.46\tau\right) = 2\left(0.06 + 1.46\times 0.0744\right) = 0.337\ \text{m}$$

The electromagnetic power, P_{elmax}, is

$$P_{elmax} = T_{emax} \times 2 \times \pi \times n_{max} = 0.6 \times 2 \times \pi \times 60 = 266.08\,\text{W}$$

As we can see, the efficiency of this preliminary design is

$$\eta < \frac{226.08}{226.08 + 33.3} = 0.8716$$

Note: We also notice that the stack length is still too small, so the active coil length (0.06 m) is much smaller than the end-connection length (0.11 m).

A longer stack length, with same or slightly lower rotor diameter D_r, should bring smaller mmf Ni_{amax} and, thus, smaller copper losses.

Also, an axial airgap disk-shaped rotor solution might be considered for $2p_1 = 4$ poles.

The above preliminary design, to be finished by calculating R_c, L_c, R_a, and L_a, serves at least as a good start for a detailed optimal design code that accounts for all losses, commutator design, costs, thermal and mechanical limitations, etc.

For more on design, see [7–10].

4.16 SUMMARY

- Electric machines convert electric energy to mechanical energy or vice versa via stored magnetic energy (or coenergy) in a set of magnetically coupled electric circuits with a fixed part (the stator), a movable part (the rotor), and an airgap in between.
- The d.c. brush machines are supplied from a d.c. voltage source at the two terminals of the rotor, but they contain, on the rotor, a mechanical commutator made of electrically insulated copper segments that perform series connection of all identical coils placed in uniform slots of silicon sheet soft iron core of the rotor. Brushes, fixed to the stator, press the commutator segments such that the current in the rotor coils changes polarity when the coils move from one pole to the next.
- A pole pitch, τ, is the circumferential extension of the positive magnetic flux density produced by the $2p_1$ concentrated coils placed on salient poles or by $2p_1$ PM poles on the stator.
- The d.c. excitation on the stator produces a trapezoidal heteropolar flux density distribution with $2\,p_1$ poles in the airgap (p_1 electrical periods); this is how the electrical angle, $\alpha_e = p_1\alpha_m$ (α_m—the mechanical angle), is defined.
- The stator d.c. excitation (or PM) magnetic field axes are fixed, with their maxima along the middle of the $2p_1$ stator pole (*d* axis). The rotor current magnetomotive force has its $2p_1$ maxima along the neutral axis (*q* axis, 90° electrical degrees away from the d axis), irrespective of speed, due to the brush–commutator principle, which acts as the rectifier (for generator mode) or as the inverter (for the motor mode) for the $f_r = p_1 n$ frequency coil emfs and currents, I_c. The brush–commutator machine operates with fix orthogonal magnetic fields [11].

- The lap and wave armature (rotor) windings have $2a$ parallel current paths via the brushes ($2a = 2p_1m$ for the lap windings and $2a = 2m$ for the wave windings; $m = 1,2$ is the winding order), and the $m = 2$ brush span covers two copper segments of the commutator). Lap windings are preferred for larger current and low voltage motors, and wave windings for lower current and large voltage motors.
- Because of motion, the d.c. heteropolar field of the stator produces hysteresis and eddy currents in the laminated core of the rotor at a frequency $f_r = p_1 \cdot n$.
- The voltage drop along the brush copper segment contact varies from 0.5 V (in metal brushes) to 1.5–2 V for carbon brushes.
- The d.c. excitation of the stator may be supplied separately, shunted, or in series to the (+)(−) brushes of the commutator.
- The brushes commutate the coils (one by one) from one current path (+) to the next one (−), when the coils move from under one stator pole to the next.

  The process of current reversal through a short circuit should be terminated in time $T_c = (1/kn)$ (k is the number of commutator segments, n is speed in rps), which corresponds to the rotor angle rotation by one commutator segment.
- Proper commutation is limited by coil inductance, speed, and current level. Commutator wear and scintillation are the major demerits of d.c. brush machines.
- The motor characteristics of a d.c. brush PM motor, for example in automotive applications at low power, relate, speed, n, to, armature (rotor) brush current, i, electromagnetic torque, T_e, and electromagnetic power, and are rather rigid (linear).

  So speed drops only a little with torque.
- D.c. brush PM motors exhibit an LCM (N_s,$2p_1$) period pulsating torque at zero rotor current, which is called *cogging torque*. Cogging torque is a result of PM magnetic energy conversion to mechanical energy and back, for an almost zero average value per revolution. Cogging torque may be reduced by increasing the number of rotor slots, decreasing slot opening width and shaping the PM ends adequately. Cogging torque reduction means less noise, less vibration.
- Speed effective control (and starting) is performed through armature voltage control, or via series additional resistance R_{ad} at brushes; in this latter case the initial cost of R_{ad} is small but the Joule losses are large.
- D.c. and a.c. brush series motors eliminate the need for an excitation power source and are characterized by mild $n(I_a)$, $n(T_e)$ curves and an easy to obtain (through voltage control) 3:1 n_{max}/n_{base} speed ratio at constant electromagnetic power, so typical for traction (or methalurgy) motor, home appliances or hand-held tools.
- The a.c. brush series (universal) motor is a.c.-fed (at f_1), and thus exhibits $2f_1$ steady-state full torque pulsations.
- Testing of brush–commutator machines is done according to the evolutionary standards: here, only two (one complete and one minimal) test rigs, containing four-quadrant d.c.–d.c. converters, are introduced and shown capable to load the machine and measure the losses and parameters (R_a,L_a, and V_{ea}).

4.17 PROPOSED PROBLEMS

4.1 Build a simple lap winding for $2p_1 = 4$ poles and $N_s = 24$ slots, including the computation of t, α_{ec}, and α_{et}, drawing the emf polygon, armature winding overlay with copper segments and brushes, the coils in series in all four current paths, and the commutating four coils at the four brushes.
Hint: See Section 4.2: lap windings.

4.2 Build a simple wave winding for $2p_1 = 4$ poles and $N_s = 17$ slots, revealing all information as in Example 4.1.
Hint: See Section 4.2: wave windings.

4.3 For the d.c.-excited brush–commutator motor in Figure 4.34, calculate the pole excitation A turns $N_F I_F$ for an airgap flux density $B_g = 0.70$ T; note that the pole shoe was eliminated. All dimensions are given in millimeters.
Hint: See Section 4.5.

4.4 For the PM stator brush–commutator geometry in Figure 4.35, calculate the PM radial thickness h_{PM} and the outer stator diameter D_{out} for airgap flux density under PM, $B_{gPM} = 0.8$ T, by using NdFeB PMs with $B_r = 1.2T$ and $H_c = 960$ kA/m. The rotor has $N_s = 16$ slots and $2p_1 = 4$.
Hint: See Section 4.7.

4.5 In Example 4.4, consider a current density $J_{cr} = 5$ A/mm² and a slot filling factor $k_{fill} = 0.45$. Calculate the slot A turns $2N_cI_c$ (N_cI_c—A turns/coil) and the maximum of the armature-reaction flux density. Draw the PM armature and resultant airgap flux density distribution along the rotor's complete periphery.
Hints: See Equations 4.38 through 4.40 and Figure 4.15.

4.6 A 4-pole surface PM stator d.c. brush–commutator motor has the following data:
Turns/rotor coil: $N_c = 15$
Airgap: $g = 1 \times 10^{-3}$ m
Rotor diameter: $D_r = 0.06$ m

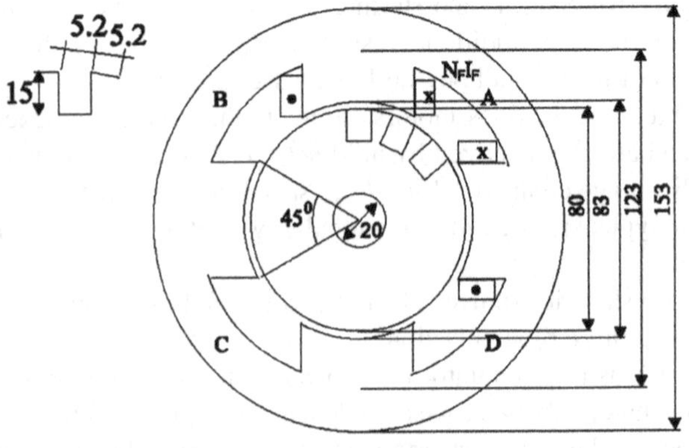

FIGURE 4.34 D.C.-excited brush–commutator motor geometry.

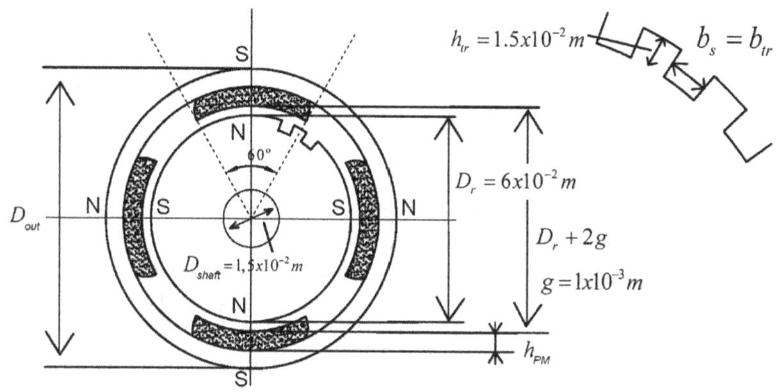

FIGURE 4.35 PM stator brush–commutator motor geometry with $N_s = 16$ slots and $2p_1 = 4$ poles.

Number of slots: $N_s = 16$
Number of poles: $2p_1 = 4$
PM radial thickness: $h_{PM} = 4 \times 10^{-3}$ m
Stack length: $L_e = 0.06$ m
Coil end-connection length: $L_{end} = 0.06$ m
Radius of coil bundle: $r_{end} = 0.01$ m
Rotor open slot height: $h_{rs} = 1.5 \times 10^{-2}$ m
Rotor slot average width: $b_{rs} = 6 \times 10^{-3}$ m
Simple lap winding
Calculate the coil inductance, L_c, and the armature-reaction inductance, L_a.
Hints: See Equations 4.50, 4.51, 4.78, and 4.80.

4.7 A small d.c. brush PM small motor is fed at 12 V d.c. and is rated at $P_n = 25$ W; $\eta_n = 0.8$, $n = 30$ rps, and $\Delta V_{brush} = 0.5$ V, with $P_{iron} = P_{mec}/2 = 0.025\ P_n$.

Calculate

a. The rated input current I_{an} and copper losses p_{con}
b. The armature resistance R_a and the rated emf
c. The ideal no-load speed at rated voltage
d. Rated electromagnetic torque T_{en} and shaft torque T_{shaftn}

Hint: See Example 4.1.

4.8 A d.c. brush series urban traction (streetcar) motor has the data: $V_{an} = 500$ V d.c., $P_a = 50$ kW, $n_n = 1800$ rpm, and $\eta_n = 0.93$; the total copper losses (R_a and R_{Fs}; $R_a = R_{Fs}$) is $p_{cop} = 0.04\ P_n$, $p_{iron} = p_{mec}$, and $\Delta V_{brush} = 2$ V. The magnetic saturation is neglected.

Calculate

a. Rated current and armature and series-excitation copper losses, R_a and R_{Fs}.
b. Iron, brush, and mechanical losses (p_{iron}, p_{mec}, and p_{brush}).

c. The emf, V_{ean}, the rated electromagnetic torque, T_{en}, and electromagnetic power, P_{elm}.
d. For rated torque, calculate the required voltage V_a values at 900 rpm and at standstill.
e. For rated voltage, at $3n_n$ = 5400 rpm and $T_{en}/3$, calculate the required resistance R_{fields} in parallel with the field circuit. Determine the electromagnetic power again.

Hints: See Section 4.13 and Equations 4.81 through 4.87.

4.9 A home appliance (washing machine) 2-pole universal motor has the rated power P_n = 350 W at n_n = 9000 rpm, at 50 Hz and V_n = 220 V (RMS). The core loss $p_{iron} = p_{copper}/2$, $p_{mec} = 0.02\ P_n$, the rated efficiency $\eta_n = 0.9$, and rated power factor $\cos\varphi_n = 0.95$ lagging. Calculate
a. Rated current I_n
b. Total winding resistance R_{ae} and core loss resistance R_{iron}
c. The motion-induced emf V_{er}
d. The total machine inductance L_{ae}
e. The electromagnetic torque
f. The shaft torque
g. Starting current and torque at rated voltage

Hints: See Section 4.14 and Equations 4.96 through 4.101.

REFERENCES

1. I. Kenjo and S. Nagamori, *PM and Brushless D.C. Motors*, Clarendon Press, Oxford, U.K., 1985.
2. K. Vogt, *Electric Machines: Design* (in German), VEB Verlag, Berlin, Germany, 1988.
3. G. Say and E. Taylor, *Direct Current Machines*, Pitman, London, U.K., 1985.
4. S.A. Nasar, I. Boldea, and L.E. Unnewehr, *Permanent Magnet Reluctance and Selfsynchronous Motors*, Chapters 1–5, CRC Press, Boca Raton, FL, 1993.
5. J.F. Gieras and M. Wing, *Permanent Magnet Motor Technology*, 2nd edition, Chapter 4, Marcel Dekker, New York, 2002.
6. A. diGerlando and R. Perini, A model for the operation analysis of high speed universal motor with triac regulated mains voltage supply, *Symposium on Power Electronics and Electric Drives Automation Motion*, Ravello, Italy, 2007, pp. c407–c412.
7. E. Hamdi, *Design of Small Electric Machines*, Chapters 4–6, Wiley, New York, 1994.
8. J.J. Cathey, *Electric Machines*, Chapter 5, McGraw-Hill, Boston, MA, 2001.
9. Ch. Gross, *Electric Machines*, Chapter 9, CRC Press, Taylor & Francis Group, New York, 2006.
10. J. Pyrhönen, T. Jokinen, V. Hrabovcova, "Design of Rotating Electrical Machines", John Wiley & Sons Ltd, UK, 2008.
11. I. Boldea and S.A. Nasar, *Electric Drives*, 3nd edition, Chapter 4, CRC Press, Taylor & Francis Group, New York, 2017.

5 Induction Machines
Steady State

5.1 INTRODUCTION: APPLICATIONS AND TOPOLOGIES

Induction (asynchronous) machines are provided with electric windings that are placed in uniform slots located along the periphery of the stator and the rotor (with slot openings toward an airgap) in silicon–steel sheet (laminated), soft, magnetic cores. Ac currents of different frequencies, f_1 and f_2, flow through the electrical windings in the stator and the rotor ($|f_2| < |f_1|$) [1–3].

Induction machines (IMs), as classified in Chapter 3, are a.c. current stators, a.c. current rotors, and traveling field machines.

The stator full power winding is also called the primary winding and is a one-, two-, or three-phase distributed winding. Most IMs, above 100 W, are three-phase machines. The rotor winding is called the secondary winding.

The rotor is built in two configurations:

- Cage rotor: with uninsulated aluminum (brass or copper) bars, short-circuited by end rings
- Wound rotor: with a three-phase distributed winding connected to slip rings and, via brushes, to an impedance or to a power source of frequency f_2 such that,

$$f_2 = f_1 - np_1 \tag{5.1}$$

Consequently, according to the frequency theorem (Chapter 3), rippleless (ideal) torque at steady state is obtained.

Being reversible, the IM operates as a motor or as a generator. Though it is predominantly used as a motor up to 30 MW for cage rotors, it has been used at 400 MW as a motor and a generator with wound rotor in pump storage hydropower plants [1]. The IM is the "workhorse," but recently it has become the "race horse" of the industry in power electronics variable speed (via variable frequency f_1 or f_2) applications.

The IM may also be built as a flat or as a tubular linear induction motor for applications in urban and industrial transportation on wheels or for magnetic levitation (for clean rooms).

5.2 CONSTRUCTION ELEMENTS

As any electric machine, the IM has a fixed part, the stator (primary), and a moving part, the rotor (secondary) (Figure 5.1).

DOI: 10.1201/9781003214519-5

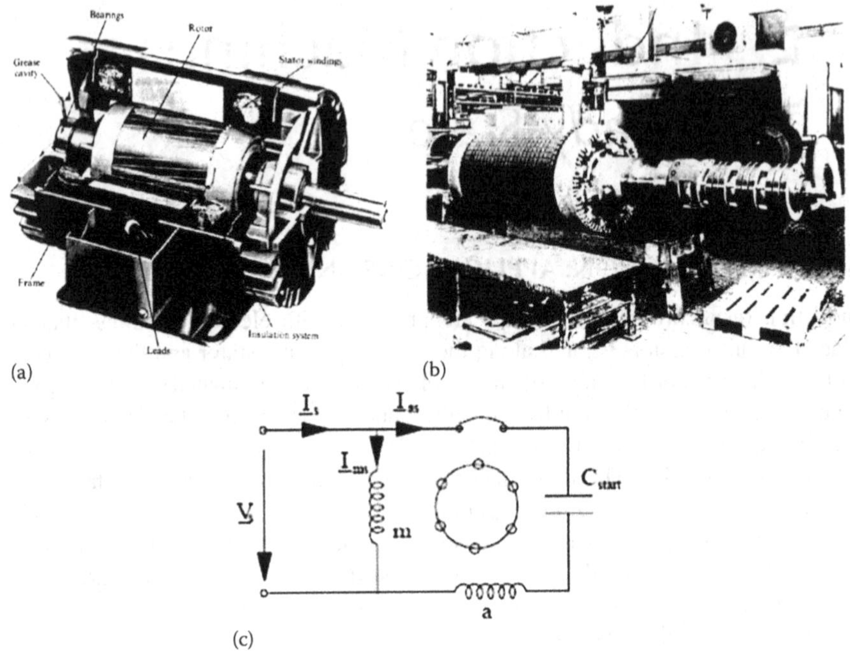

FIGURE 5.1 The induction machine: (a) three-phase with cage rotor and (b) three-phase with wound rotor.

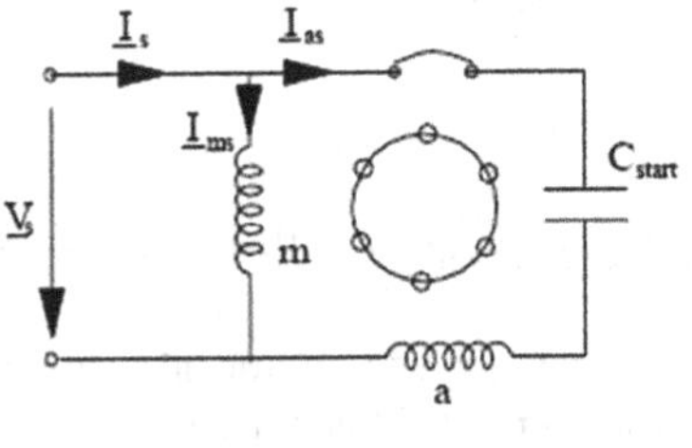

FIGURE 5.2 Single-phase supply capacitor IM.

The main part of the stator is the stator core (stack) made as a soft, *nonoriented grain* silicon–steel sheet (laminated) core. The stator core (stack) is provided toward its interior (airgap) side with uniform slots that will host the a.c.-distributed one-, two-, or three-phase winding supplied at a frequency f_1.

The rotor also has two main parts: the laminated core with uniform slots which host either uninsulated conductors (bars) short-circuited at stack ends by conductor end rings (the cage rotor)—Figure 5.1a—or a three-phase a.c. winding connected to three copper slip rings (the wound rotor)—Figure 5.1b—and via brushes, to either an impedance or to a separate power source at frequency f_2 (according to

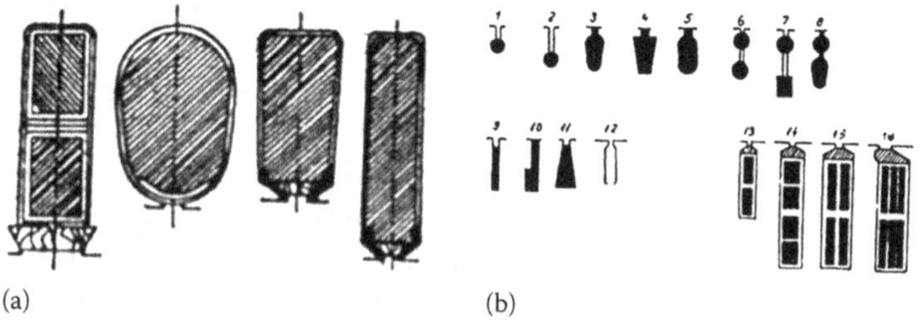

FIGURE 5.3 Typical slots for induction machines: (a) stator and (b) rotor: 1–5 (single cage), 6–8 (double cage), 9–12 (deep bars), 13–16 (wound rotor).

Equation 5.1). Besides three-phase, single-phase supply small-power IMs are built (Figure 5.2).

The airgap—the radial distance between the stator and the rotor—has values ranging from 0.3 mm (at 100 W) to 2.0 (3.0) mm at large powers (thousands of kW). Axial airgap topologies may be feasible too in niche applications.

In general, the stator and rotor slots are semiclosed (Figure 5.3) but for large machines, on one side (either rotor or stator), they are open, to allow preformed coils. For low-power machines the teeth are rectangular, while for high-power machines, the slots are rectangular, to allow preformed coils with a large copper fill factor.

The cage rotors are more robust than the wound rotors. The wound rotor IMs are used only in limited variable speed applications. For small machines, die-cast aluminum cage rotors are typical, but copper round-bar cage rotors may be preferred for low-power applications when high efficiency is top priority (small pumps, refrigerators). For cage rotor large IMs, copper, or brass bars are common. The frame is made of steel but mostly, even up to 500 kW, of aluminum (Figure 5.1a). Ball bearings are used for small- and medium-power IMs, and slip bearings for large-power IMs.

To reduce torque pulsations, noise, and additional rotor core and cage losses, magnetic field harmonics are reduced by reducing slot openings/airgap ratio to less than (4–5)/1, by slot skewing, winding coil chording, or increasing the number of slots.

Only a few combinations of N_s and N_r numbers of slots avoid the so-called synchronous torque pulsations and should be observed in any practical design. The stator (or wound rotor) a.c. distributed windings constitute the main part of an IM (or of a synchronous machine), and thus will be treated in some detail, for practical use, in what follows.

5.3 AC DISTRIBUTED WINDINGS

We use the generic term of a.c. distributed windings to refer to the a.c. full-power windings that produce a traveling airgap field in induction and synchronous machines. They contain interleaved phase coils rather than tooth-wound coils that are used for some permanent magnet synchronous machines, for salient stator, salient rotor steppers, and switching reluctance machines, which are discussed in Chapter 6.

A sinusoidal flux density distribution in the airgap along the rotor periphery is needed. By interleaved phase coils, we mean multiple slot pitch span for coils in the winding. The three-phase case is the most used for high power while two-phase windings (one is main, the other auxiliary (or starting)) are used (below a few kW) as suitable for single-phase residential (or rural) a.c. power grids at 50 (60) Hz.

Designing a.c. windings means assigning coils to various slots and to machine phases, and then connecting them in Y (or Δ) for three-phase machines.

We will first treat the ideal, traveling magnetization force (mmf) by superposition of phase components, and then the mmf of primitive single-layer and two-layer chorded-coil a.c. windings (with integer count of slots per pole per phase, $q = 1–15$).

Further on, the mmf space harmonic content of the integer q windings is calculated with distribution, chording, and skewing winding factor expressions.

Rules for building practical one- and two-layer three-phase a.c. distributed windings are presented and applied for integer and fractional $q \geq 1$. A case study of a $q > 1$ but fractionary slots/pole/phase, intended to reduce mmf space harmonics for $q < 3$, is also presented.

Two-phase (main and auxiliary (or starting)) distributed windings that are similar to a sinusoidal space distribution for small, single-phase a.c. power grid-supplied, a.c. motors are treated in some detail too.

5.3.1 Traveling MMF of AC Distributed Windings

As already discussed in Chapter 3, traveling field a.c. machines require traveling magnetomotive forces:

$$F_{sf}\left(x,t\right) = F_{sfm}\cos\left(\frac{\pi}{\tau}x - \omega_1 t - \theta_0\right) \tag{5.2}$$

where

x is the coordinate along the stator core periphery after unfolding it in a plane,
τ is the spatial half-period (pole pitch) of an mmf fundamental (ideal) wave,
ω_1 is the angular frequency of stator currents, and
θ_0 is the angular stator position of stator mmf with respect to the phase A axis at $t = 0$

We may decompose it into terms:

$$F_{sf}\left(x,t\right) = F_{sfm}\left[\cos\left(\frac{\pi}{\tau}x - \theta_0\right)\cos\left(\omega_1 t\right) + \sin\left(\frac{\pi}{\tau}x - \theta_0\right)\sin\left(\omega_1 t\right)\right] \tag{5.3}$$

This mere trigonometric exercise indicates that we are now having two mmfs at a standstill, with a spatial sinusoidal distribution and sinusoidal currents. Both the space and the time angles between the two mmf components are 90°. So two-phase windings with a sinusoidal spatial distribution and 90° space phase lag, through which a.c. currents flow with 90° time phase lag, can produce a traveling mmf.

Now, according to Ampere's law, the airgap flux density, $B_{gs}(x,t)$, is simply:

$$B_{gs}(x,t) = \frac{\mu_0 F_{sf}(x,t)}{g_m(x)(K_s+1)} \tag{5.4}$$

where g_m is the magnetic (total) airgap which may vary (as in the stator) with the rotor position—for salient pole rotors—and K_s is the magnetic saturation factor $0 < K_s < 0.8$, in general, and accounts for iron mmf requirements as discussed already in Section 4.6.

g_m is a constant (for nonsalient pole, d.c.-excited or surface-PM pole rotors) if the mmf is a pure traveling wave, and so is the airgap flux density wave (Equation 5.4). Thus, the condition for a rippleless ideal torque may be obtained through two orthogonal phases (Equation 5.3). Their arrangement is typical for single-phase a.c. power grid-connected small a.c. (synchronous and induction) motors.

We may decompose Equation 5.2 also into m terms:

$$\begin{aligned} F_{sf}(x,t) = \frac{2}{m} F_{sfm}\Bigg[& \cos\left(\frac{\pi}{\tau}x - \theta_0\right)\cos(\omega_1 t) + \cos\left(\frac{\pi}{\tau}x - \theta_0 + \frac{2\pi}{m}\right) \\ & \times \cos\left(\omega_1 t - \frac{2\pi}{m}\right) + \cdots + \cos\left(\frac{\pi}{\tau}x - \theta_0 + (m-1)\frac{2\pi}{m}\right) \\ & \times \cos\left(\omega_1 t - (m-1)\frac{2\pi}{m}\right)\Bigg] \end{aligned} \tag{5.5}$$

Consequently, m phase windings with a spatial sinusoidal distribution and sinusoidal currents dephased spatially, and in time by $2\pi/m$, also produce a traveling mmf.

For the two-phase case, $m = 4$ in Equation 5.5. The most common case is $m = 3$ phases. The situation for two- and three-phases is illustrated in Figure 5.4.

The pole pitch, τ, calculated at a stator interior diameter, D_{is}, is

$$\tau = \frac{\pi D_{is}}{2p_1} \tag{5.6}$$

The number of pole pairs, p_1 (or mmf electrical periods per one mechanical revolution), leads to the same definition of the electrical angle, α_e, as for brush–commutator machines:

$$\alpha = p_1\alpha_g; \quad \alpha_g, \text{geometrical angle} \tag{5.7}$$

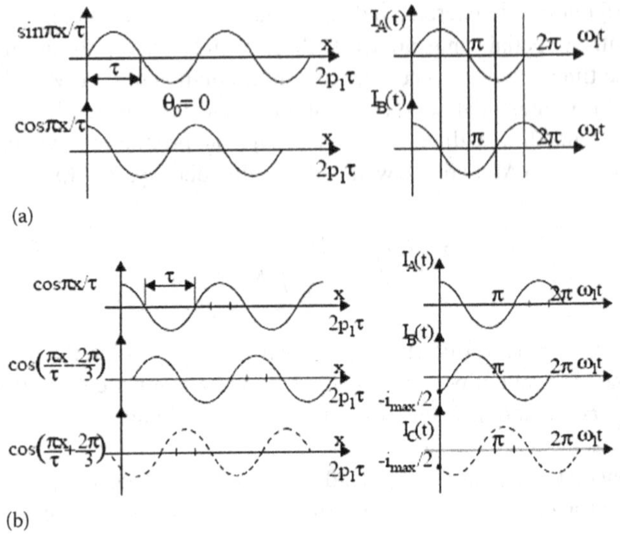

FIGURE 5.4 Ideal multiphase mmfs: (a) two-phase machine and (b) three-phase machine.

A pure sinusoidal distribution of mmf is feasible only with slotless windings of unequal numbers of turns (cosinusoidal variation of turns/coil on the periphery for each phase).

As this is not practical, even for slotless a.c. windings (used for some small-power PMSMs), the a.c. windings are placed in a limited number of slots N_s:

$$N_s = 2p_1qm \tag{5.8}$$

The total number of slots, N_s, has to be divisible by m (number of phases), to retain some symmetry of the phase windings, with an integer q, characteristic of interleaved phase coils in distributed windings:

$$q = \frac{N_s}{2p_1m} = \text{Integer} \tag{5.9}$$

N_s has to be divisible by $2p_1m$; $q = 1,2,\ldots,12$ or even more in turbogenerators or large two-pole IMs. It is also feasible to have $q = a + b/c$, $a \geq 1$, for fractional q distributed windings.

The a.c. distributed windings are made of lap or wave coils as for brush–commutator windings (Chapter 4), Figure 5.5, placed in one or two layers in uniform slots along the stator periphery.

Single-layer windings make use of diametrical coils (full pitch span: $y = \tau$), while two-layer windings often make use of chorded coils ($2\tau/3 \leq y \leq \tau$) to reduce the coil end-connection length (and copper losses) and to reduce some mmf space harmonics, as shown later. Unfortunately, the mmf fundamental amplitude is also reduced by chorded coils.

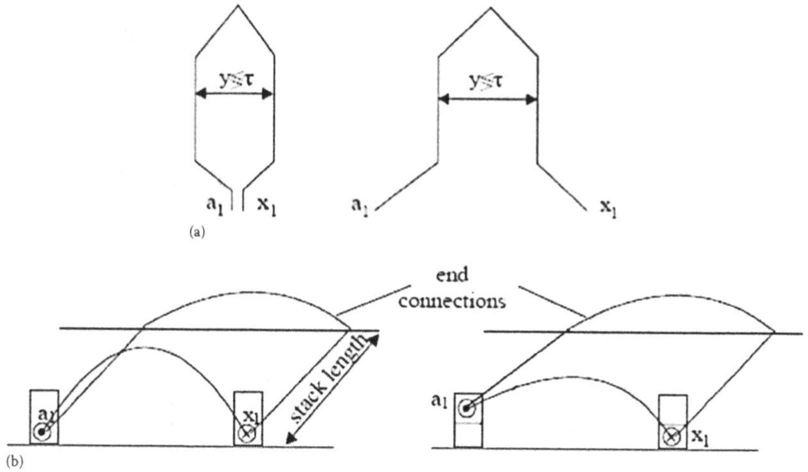

FIGURE 5.5 Lap and wave single-turn (bar) coils placed in (a) one layer in slots and (b) two layers in slots.

5.3.2 Primitive Single-Layer Distributed Windings ($q \geq 1$, Integer)

Let us consider a primitive $2p_1 = 4$ poles three-phase distributed winding with $q = 1$ slot/pole/phase: $N_s = 2p_1mq = 2 \times 2 \times 3 \times 1 = 12$ slots in all. There are four poles, so each pole has $N_s/2p_1 = 12/(2 \times 2) = 3$ slot pitches τ_s, or three slots, one per phase ($q = 1$).

For single-layer windings one coil fully occupies two slots, so there are, in all, six coils; two coils per phase; and four slots per phase, one pole pitch wide coils, τ, or slot pitches, τ_s, apart. For phase A we have slots 1, 4, 7, and 10. Phases B and C are placed in slots by moving 2/3 and 4/3 of a pole pitch from phase A to the right (Figure 5.6a and b). If the slot opening is considered zero, the mmf "jumps" by $n_c I_{A,B,C}$ along the middle of each slot. For $q = 1$, a rectangular heteropolar mmf distribution for each phase is obtained (Figure 5.6b through d).

The coils of a phase may all be connected in series (Figure 5.6e) to form a single current path, $a = 1$, or some of them (in our case all of them) in parallel to form a current paths: $1 < a \leq p_1$ (Figure 5.6f).

The rectangular phase mmf distribution may be used as it is for rectangular current control in PMSMs with $q = 1$, or it may be decomposed into harmonics for each phase:

$$F_{A\nu}(x,t) = \frac{2}{\pi}\frac{n_c I\sqrt{2}\cos\omega_1 t}{\nu}\cos\frac{\nu\pi}{\tau}x \tag{5.10}$$

For the fundamental ($\nu = 1$), we obtain the maximum mmf amplitude, as expected ($q = 1$).

For $q \geq 1$, a multistep rectangular mmf distribution is expected, with lower harmonics content. Two-layer windings with chorded coils further reduce the phase mmf space harmonics content for $q > 1$.

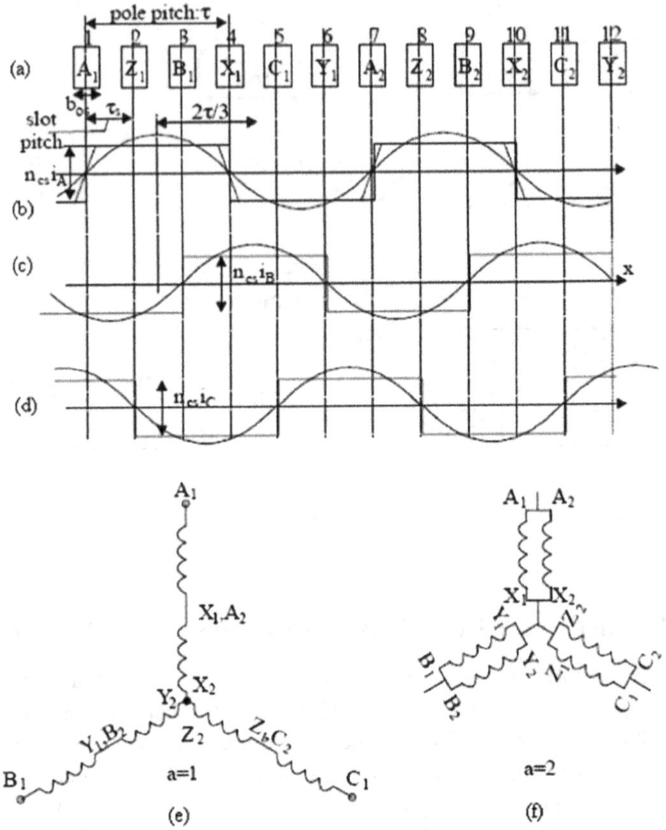

FIGURE 5.6 Single-layer primitive distributed three-phase winding ($2p_1 = 4$, $m = 3$, $q = 1$ slot/pole/phase ($N_s = 12$ slots in all)): (a) phase coil allocation to slots, (b–d) phase mmfs rectangular distribution, (e) series star connection ($a = 1$ current path), and (f) parallel star connection ($a = 2$ current paths).

5.3.3 Primitive Two-Layer Three-Phase Distributed Windings (q = Integer)

Let us consider again $2p_1 = 4$ poles, $m = 3$ phases, but now q changes from 1 to 2 slots/pole/phase: $N_s = 2p_1mq = 2 \times 2 \times 3 \times 2 = 24$ slots. The pole pitch span in slot pitches, τ, is then $\tau = N_s/2p_1 = 24/(2 \times 2) = 6$ slot pitches, τ_s.

The chorded-coil span is taken as $y/\tau = 5/6$ ($y/\tau = 4/6 = 2/3$, is the lowest for q = integer and distributed windings).

Disregarding the second layer, let us allocate phase coils to slots in the first layer, observing that there are $q = 2$ neighboring slots for each phase under each pole; in all $N_s/m = 8$ coils per phase (two layers). The 2/3 pole phase lag of the phase mmf is provided as for the single-layer winding case (Figure 5.7a). Once the allocation of phase coils to slots in the first layer is done, the distribution for the second layer is moved to the left by $(1 - y/\tau)mq$ slots (Figure 5.7a). Then, for the moment when the

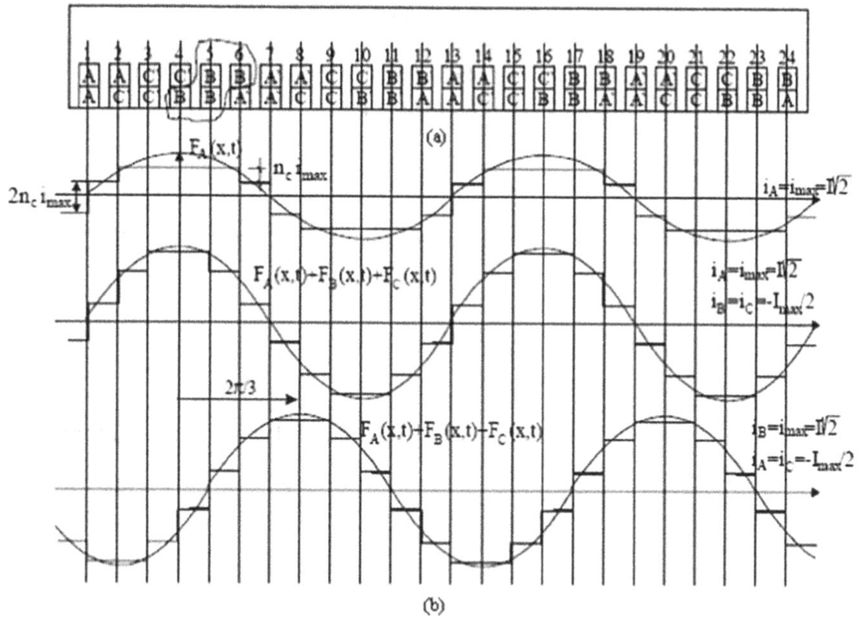

FIGURE 5.7 Two-layer winding ($N_s = 24$, $2p_1 = 4$ poles, $y/\tau = 5/6$): (a) allocation of phase coils to slots and (b) mmf distribution.

phase A current is maximum, its mmf distribution is rectangular but with two steps/polarity (Figure 5.7b). If we add the contribution of all three-phase mmfs for $I_A = I_{max} = -2I_B = -2I_C$ (when I_A is maximum and the currents are symmetric: $I_A + I_B + I_C = 0$), we obtain the three-phase mmf distribution with 3 steps/polarity (Figure 5.7b). For the later moment, when the current in phase B is maximum ($I_B = I_{max} = -2I_A = -2I_C$), the three-phase mmf distribution peaks have moved to the right by $(2/3)\tau$ or $2\pi/3$ electrical radians. So the mmf is traveling, as expected.

5.3.4 MMF Space Harmonics for Integer q (Slots/Pole/Phase)

The geometrical stepwise representation of phase mmfs in Figure 5.7b suggests that generalizing Equation 5.10 is straightforward for $q > 1$ and $y/\tau < 1$ and, for only the fundamental ($\nu = 1$):

$$F_{A1}(x,t) = \frac{2}{\pi} n_c q I \sqrt{2} K_{q1} K_{y1} \cos\left(\frac{\pi}{\tau} x\right) \cos(\omega_1 t) \tag{5.11}$$

with

$$K_{q1} = \frac{\sin\dfrac{\pi}{6}}{q \sin\dfrac{\pi}{6q}} \le 1; \quad K_{y1} = \sin y \frac{\pi y}{2\tau} \le 1 \tag{5.12}$$

K_{q1} is known as the winding distribution factor, and K_{y1} is the chording factor (for $q = 1$ and $y = \tau$; $K_{q1} = K_{y1} = 1$). Keeping the windings symmetric implies $(y/\tau) > (2/3)$, because all phases have the same number of slots per pole (q = integer).

With all coils in series, we have W_1 turns per phase:

$$W_1 = 2p_1qn_c \tag{5.13}$$

where n_c is the turns per coil (for single-layer windings $W_1 = p_1qn_c$ because there is 1 coil/slot). Thus F_{A1} in Equation 5.11 becomes

$$F_{A1}(x,t) = \frac{2}{\pi p_1} W_1 I\sqrt{2}K_{q1}K_{y1} \cos\left(\frac{\pi}{\tau}x\right)\cos\omega_1 t \tag{5.14}$$

Adding the three-phase contributions, with $2\pi/3$ time and space phase shift, yields

$$F_1(x,t) = F_{1m}\cos\left(\frac{\pi}{\tau}x - \omega_1 t\right) \tag{5.15}$$

with

$$F_{1m}(x,t) = \frac{3W_1 I\sqrt{2}K_{q1}K_{y1}}{\pi p_1}(\text{A turns/pole}) \tag{5.16}$$

The derivative of the pole mmf, $F_1(x,t)$, with respect to x is called the linear current density, $A(x,t)$:

$$A_1(x,t) = \frac{\partial F_1(x,t)}{\partial x} = -A_{1m}\sin\left(\frac{\pi}{\tau}x - \omega_1 t\right) \tag{5.17}$$

$$A_{1m} = \frac{\pi}{\tau}F_{1m} \tag{5.18}$$

A_{1m} is the peak value of the linear current density or current loading, a key design factor; A_{1m} = (1,000–50,000) A/m, from small to very large machines and increases with the rotor diameter (and torque).

The space harmonics content of three-phase winding mmf is obtained from Equation 5.10 as

$$F_\nu(x,t) = \frac{3W_1 I\sqrt{2}K_{q\nu}K_{y\nu}}{\pi p_1 \nu}\left[K_{BI}\cos\left(\nu\frac{\pi}{\tau}x - \omega_1 t - (\nu-1)\frac{2\pi}{3}\right)\right.$$
$$\left. -K_{BII}\cos\left(\nu\frac{\pi}{\tau}x + \omega_1 t - (\nu+1)\frac{2\pi}{3}\right)\right] \tag{5.19}$$

with

$$K_{q\nu} = \frac{\sin \frac{\nu\pi}{6}}{q \sin \frac{\nu\pi}{6q}}; \quad K_{y\nu} = \sin \frac{\nu\pi y}{2\tau}; \quad K_{\mathrm{BI}} = \frac{\sin(\nu-1)\pi}{3 \sin(\nu-1)\frac{\pi}{3}};$$
$$K_{\mathrm{BII}} = \frac{\sin(\nu+1)\pi}{3 \sin(\nu+1)\frac{\pi}{3}} \tag{5.20}$$

Due to the full phase symmetry with q = integer, only odd order space harmonics survive in the mmf. For a star connection, at least $3k$ harmonics do not occur as their current summation is zero. So we are left with $\nu = 3k \pm 1 = 5, 7, 11, 13, 17, 19, \ldots$, all prime numbers!

We should notice in Equation 5.20 that for $\nu_{\mathrm{d}} = 3k + 1$ (7,13,19), $K_{\mathrm{BI}} = 1$ and $K_{\mathrm{BII}} = 0$. But the remaining first term in Equation 5.19 represents direct (forward) traveling waves:

$$\frac{\nu\pi}{\tau}x - \omega_1 t = \text{const.}; \quad \frac{\mathrm{d}x}{\mathrm{d}t} = \frac{\omega_1\tau}{\pi\nu} = \frac{2\tau f_1}{\nu}; \quad \omega_1 = 2\pi f_1 \tag{5.21}$$

On the contrary, for $\nu_i = 3k - 1$ (5,11,17,...), $K_{\mathrm{BI}} = 0$ and $K_{\mathrm{BII}} = 1$ and we are left only with the second term in Equation 5.19, which refers to inverse (backward) traveling waves:

$$\frac{\mathrm{d}x}{\mathrm{d}t} = -\frac{\omega_1\tau}{\pi\nu} = -\frac{2\tau f_1}{\nu} \tag{5.22}$$

We should notice that the traveling speed of space harmonics, ν, is ν times lower than that of the fundamental.

The mmf space harmonics, as derived above, refer to $\pi/3$ phase belt sequences, which are predominant in the industry today. In some applications, $2\pi/3$ phase belt sequences are still used (such as in 2/1 ratio pole count changing IM winding).

Example 5.1: MMF Harmonics for Integer q

Let us consider a stator with an interior diameter, $D_{\mathrm{is}} = 0.12$ m; the number of stator slots, $N_s = 24$; $2p_1 = 4$; $y/\tau = 5/6$; two-layer winding, one current path, ($a = 1$); the slot area, $A_{\mathrm{slot}} = 120$ mm^2; total copper filling factor, $k_{\mathrm{fill}} = 0.45$; the rated current density, $j_{\mathrm{con}} = 5$ A/mm^2; and the number of turns per coil, $n_c = 20$. Calculate:

a. The rated RMS current, I_n, and wire gauge
b. The pole pitch, τ, and the slot pitch, τ_s
c. K_{q1}, K_{y1}, and $K_{w1} = K_{q1}K_{y1}$
d. The number of turns/phase, W_1, mmf, and current loading fundamental amplitudes, F_{1m}, A_{1m}

e. K_{q7}, K_{y7}, F_{7m} ($\nu = +7$)

Solution:

a. The slot copper area, A_{cos}, is covered by an mmf of $2n_cI$ (2 coils/slot):

$$A_{cos} = A_{slot}k_{fill} = \frac{2n_cI_n}{j_{con}}$$

So the rated current, I_n, is

$$I_n = \frac{A_{slot}k_{fill}}{2n_c}j_{con} = \frac{120\times10^{-6}\times0.45\times5\times10^6}{2\times20} = 6.75 \text{ A}$$

The copper wire bare diameter, d_{Co}, is

$$d_{Co} = \sqrt{\frac{I_n}{j_{con}}\frac{4}{\pi}} = \sqrt{\frac{6.75\times4}{5\times10^6\times\pi}} = 1.3\times10^{-3} \text{ m}$$

b. The pole pitch, τ, is (Equation 5.6)

$$\tau = \frac{\pi D_{is}}{2p_1} = \frac{\pi\times0.12}{2\times2} = 0.0942 \text{ m}$$

with the slot pitch, τ_s

$$\tau_s = \frac{\tau}{3q} = \frac{0.0942}{3\times2} = 0.0157 \text{ m}$$

c. From Equation 5.12,

$$K_{q1} = \frac{\sin\frac{\pi}{6}}{q\sin\frac{\pi}{6q}} = \frac{0.5}{2\times\sin\frac{\pi}{12}} = 0.9659$$

$$K_{y1} = \sin\frac{y\pi}{2\tau} = \sin\frac{5\pi}{12} = 0.9659$$

$$K_{w1} = K_{q1}K_{y1} = 0.9659\times0.9659 = 0.9329$$

d. The number of turns per phase, W_1 (with $a = 1$ current paths), from Equation 5.13 is

$$W_1 = 2p_1qn_c = 2\times2\times2\times20 = 160 \text{ turns/phase}$$

According to the F_{1m} and A_{1m} expressions (Equations 5.16 through 5.18)

$$F_{1m} = \frac{3W_1 I\sqrt{2}K_{q1}K_{y1}}{\pi p_1} = \frac{3\times 160\times 6.75\sqrt{2}\times 0.9659\times 0.9659}{\pi\times 2}$$
$$= 678 \text{ A turns/pole}$$

$$A_{1m} = F_{1m}\frac{\pi}{\tau} = 678\times\frac{\pi}{0.0942} = 22{,}622.8 \text{ A turns/m}$$

e. From Equation 5.20,

$$K_{q7} = \frac{\sin\frac{7\pi}{6}}{2\sin\frac{7\pi}{6\times 2}} = -0.2588$$

$$K_{y7} = \sin\frac{7\pi}{2}\frac{5}{6} = 0.2588$$

$$F_{7m} = \frac{3W_1 I\sqrt{2}K_{q7}K_{y7}}{\pi p_1\times 7} = \frac{3\times 160\times 6.75\sqrt{2}\times 0.2588^2}{\pi\times 2\times 7}$$
$$= 6.96 \text{ A turns/pole}$$

As seen above, $F_{7m}/F_{1m} = 6.96/678 \approx 0.01$! The chording and distribution factors have contributed massively to this practically good result.

It may be shown that for a 120° phase belt and $q = 2$

$$K_{q1} = \left(\frac{\sin\frac{\nu\pi}{3}}{q\sin\frac{\nu\pi}{3q}}\right)_{\nu=1} = 0.867$$

This is almost 10% smaller than K_{q1} above ($K_{q1} = 0.9569$) for 60° belt windings. This partly explains why 60° belt windings are preferred in the industry.

5.3.5 Practical One-Layer AC Three-Phase Distributed Windings

As already mentioned, windings are made by lap and wave coils. Wave coils are used for single-turn (bar) coils of large a.c. synchronous or IMs (Figure 5.8).

Multi-turn coils are made of lap coils. But the phase-belt lap coils under a single pole may be made of unequal concentric coils (Figure 5.9a) or of identical (chain) coils (Figure 5.9b) for single-layer windings. For double-layer winding flexible coils, made of round wire, the coils are identical and mechanically flexible (for small machines). They may be preformed for rectangular cross-section conductors and twisted to fit with one side in one layer and the other side in another layer.

Rules to build single-layer winding by example are as follows (for $N_s = 24$, $2p_1 = 4$, $m = 3$):

- Calculate the self-induced emf angle shift between neighboring slots:

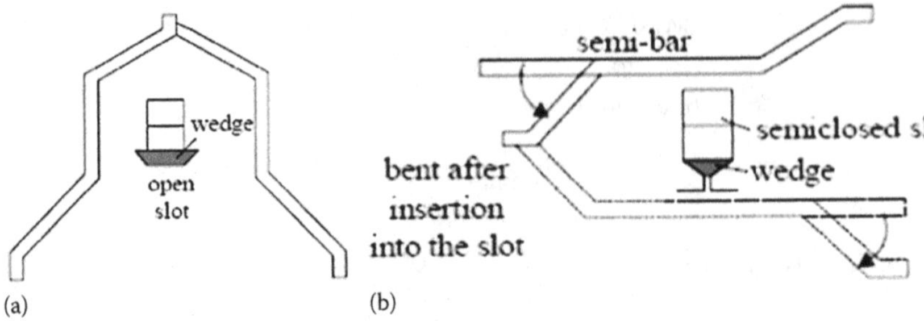

FIGURE 5.8 Bar (single turn) coils: (a) continuous bar and (b) semi-bar.

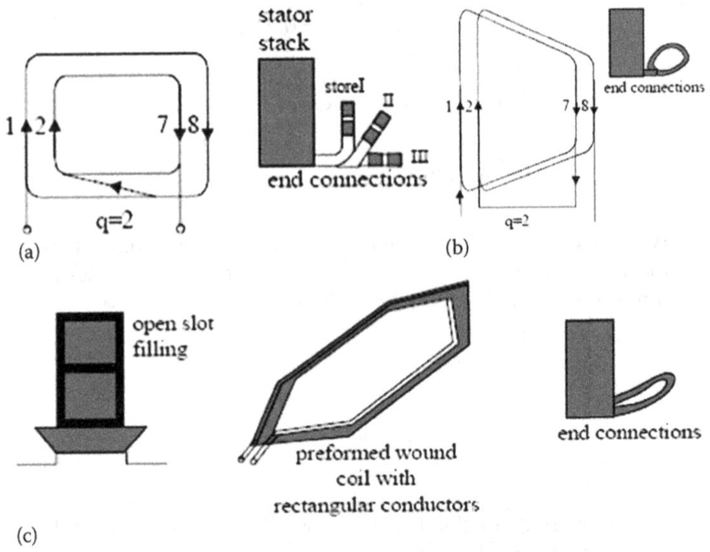

FIGURE 5.9 AC winding practical coils: (a) concentric coil phase belt/single layer, (b) identical (chain) coil phase belt/single layer, and (c) preformed coil/two layer.

$$\alpha_{es} = \frac{2\pi}{N_s} p_1 = \frac{2\pi}{24} 2 = \frac{\pi}{6} \tag{5.23}$$

- Calculate the number of emfs in phase, t:

$$t = \text{LCD}\left(N_s, p_1\right) = \text{LCD}\left(24,2\right) = 2 = p_1 \tag{5.24}$$

- Build a star of N_s/t arrows with an angle among them:

$$\alpha_{et} = \frac{2\pi}{N_s} t = \frac{2\pi}{24} 2 = \frac{\pi}{6} = \alpha_{es} \tag{5.25}$$

so the order of arrows is the natural order of slots.

- Select $N_s/2m = 24/(2 \times 3) = 4$ arrows to represent the entry coil sides of phase A, and take another opposite $N_s/2m$ to represent the exit coil sides for the same phase.
- Move 120° and arrange for phase B slot allocation and then again for phase C (Figure 5.10).

For the two-layer windings, the basic rules by example are

- Consider the case of $N_s = 27$, $2p_1 = 4$.
- Build the slot emf phasors star as for the single-layer case.
- Choose $N_s/m = 27/3 = 9$ arrows for each phase, after dividing them into two, almost (or completely) opposite as phase groups to represent phase A.
- For the case in point, where $t = \text{LCD}(N_s, p_1) = \text{LCD}(27,2) = 1$, $\alpha_{ec} = 2\alpha_{et}$ and thus the order of arrows is not the same as the natural order (Figure 5.11); this is so because $t = 1$ and $q = N_s/2p_1m = 27/(2 \times 2 \times 3) = 2 + 1/4$, that is fractional q.
- Then the next $N_s/m = 9$ arrow group is allocated to phase C and then the last one to phase B (Figure 5.11b).
- The slot allocation to phases in the slot emf phasor star (Figure 5.11a) refers to the first layer. The coil span $y = \text{integer}(N_s/2p_1) = \text{integer}(27/4) = 6$ slot pitches. This is implicitly a chorded-coil winding.
- For bar-coils, either lap (Figure 5.11b) or wave coils (Figure 5.11c) are used. The latter results in shorter additional connection cables between coils in a phase.
- The winding factor, $K_{w1} = K_{q1} \cdot K_{y1}$, may be calculated as such for q = integer. For $q = a + b/c$, $a \geq 1$, which is fractional; the distribution factor, K_{q1}, is calculated as for integer $q' = a.c. + b = 2 \times 4 + 1 = 9$ (in our case), as seen in Figure 5.11a.

$$K_{q1} = \frac{\sin\dfrac{\pi}{6}}{(ac+b)\sin\dfrac{\pi}{6(ac+b)}} \tag{5.26}$$

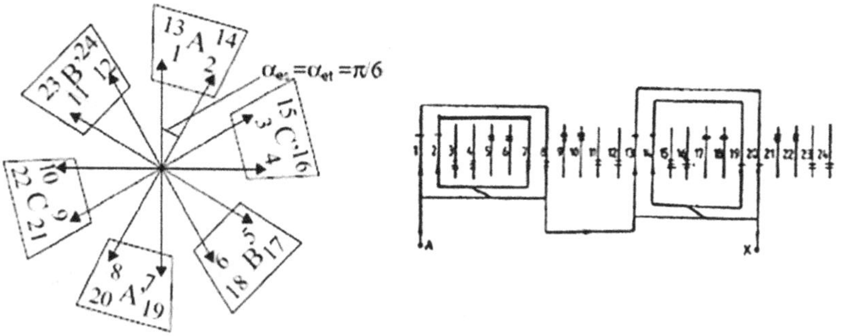

(a) (b)

FIGURE 5.10 Single-layer 24-slot, 4-pole, three-phase winding: (a) emf phasors (star of arrows) and (b) winding layout with only phase A coils shown.

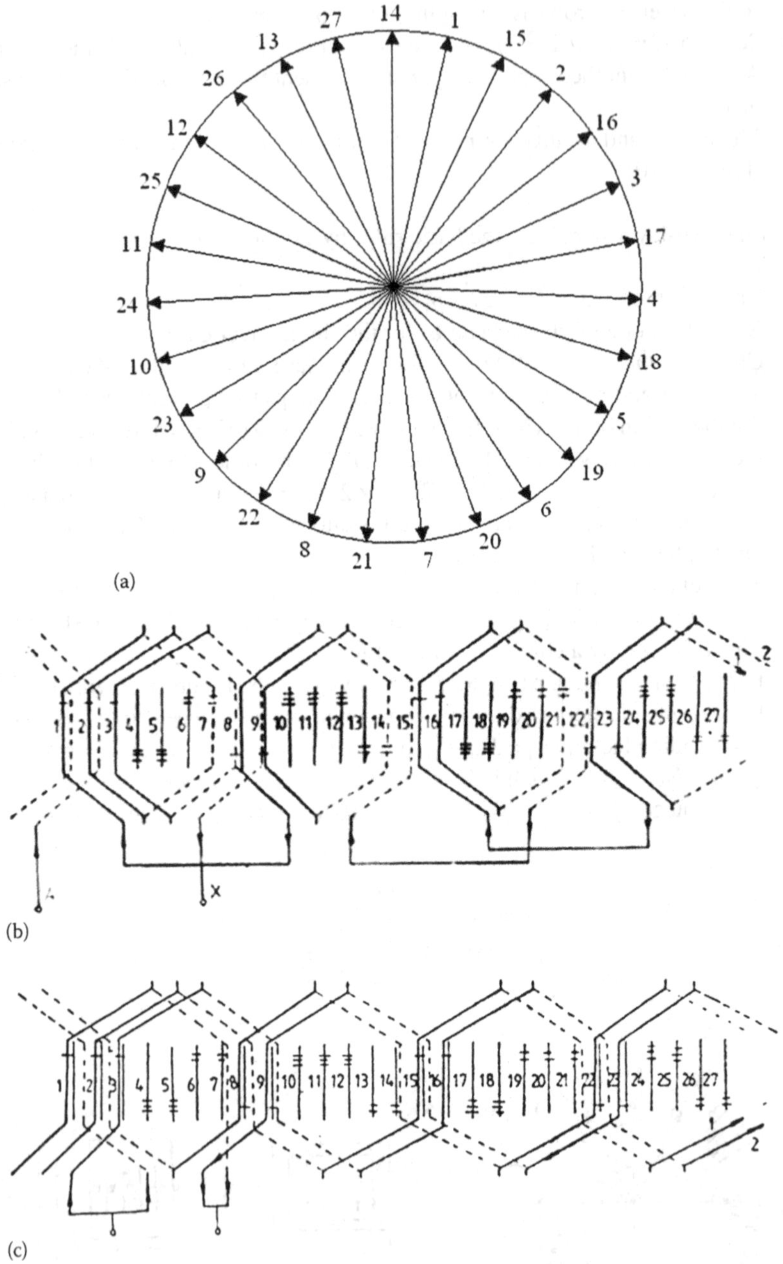

FIGURE 5.11 Distributed fractional three-phase, two-layer windings: $N_s = 27$ slots, $2p_1 = 4$ poles: (a) slot-emf phasor star, (b) winding layout with lap bar coils, and (c) with wave bar coils

- Fractional $q > 1$ is used to increase the order (and amplitude) of the first slot harmonic of the airgap flux density:

$$(\nu_s)_{\min} = (2qkm \pm 1)_{k=1} = 2(ac+b)m \pm 1 \tag{5.27}$$

Note: Pole count ($2p_1$) changing windings, used for IMs to produce two synchronous speeds ($\omega_1 = f_1/p_1$), will be discussed in the next section.

5.3.6 Pole Count Changing AC Three-Phase Distributed Windings

In some applications, two-speed operation (in the 2:1 ratio for example) is required but the total motor cost is paramount (hairdryers or hand-drill-tools).

Let us consider again the $N_s = 24$ slot, $2p_1 = 4$ poles, two-layer winding. Changing the number of poles changes the ideal no load (synchronous or field) speed, $n_1 = f_1/p_1$. To reduce the number of poles from $2p_1 = 4$ to $2p_1' = 2$, each phase winding is divided into two parts ($A_1 - X_1, A_2 - X_2$ for phase A), each part referring to the two neighboring poles of the four-pole winding (Figure 5.12). By reversing the connection of 1-phase half winding the number of poles is reduced from $2p_1 = 4$ to 2 (or vice versa). We now have four terminals for each phase in the terminal box. The pole switching (changing) may be done via an electromechanical power switch or, in variable speed drives to increase constant power speed range, by two twin PWM inverters serving each winding half.

When operated at a constant voltage and frequency power grid, the connection between the two-half windings for $2p_1$ and $2p_1'$ leaves two main alternatives: the same torque or the same power for both speeds. For the case of constant power, the series triangle connection for $2p_1 = 4$ (small speed) is transformed into a parallel star connection for $2p_1' = 2$ (high speed).

Winding utilization is worse at higher speed ($2p_1' = 2$ in our case) as the neighboring low span coil groups form a large span pole. More on pole changing windings may be found in Ref. [4, Chapter 4].

5.3.7 Two-Phase AC Windings

When only a single-phase a.c. supply is available (residential applications), two-phase windings are needed: main winding (m) and auxiliary winding (aux). The latter has either a resistor or a capacitor in series. As already shown in Section 5.3.1, the two windings are displaced by 90° (electrical), and they are symmetrical for a certain speed by choosing a pertinent capacitor, C.

Symmetrization for the start (C_{start}) and separately for the rated speed ($C_{\text{run}} \ll C_{\text{start}}$) is typical for heavy-start, high-efficiency and power factor applications. For light and infrequent starting, a single capacitor, C, may be used, and, for bidirectional motion, the capacitor may be switched from one winding to another (Figure 5.13).

For some single-capacitor two-phase windings, the auxiliary winding is disconnected after starting, so it is designed to occupy only 33.3% of the periphery with the main winding occupying the other 66.6%. For reversible motion, they have to be identical (Figure 5.14).

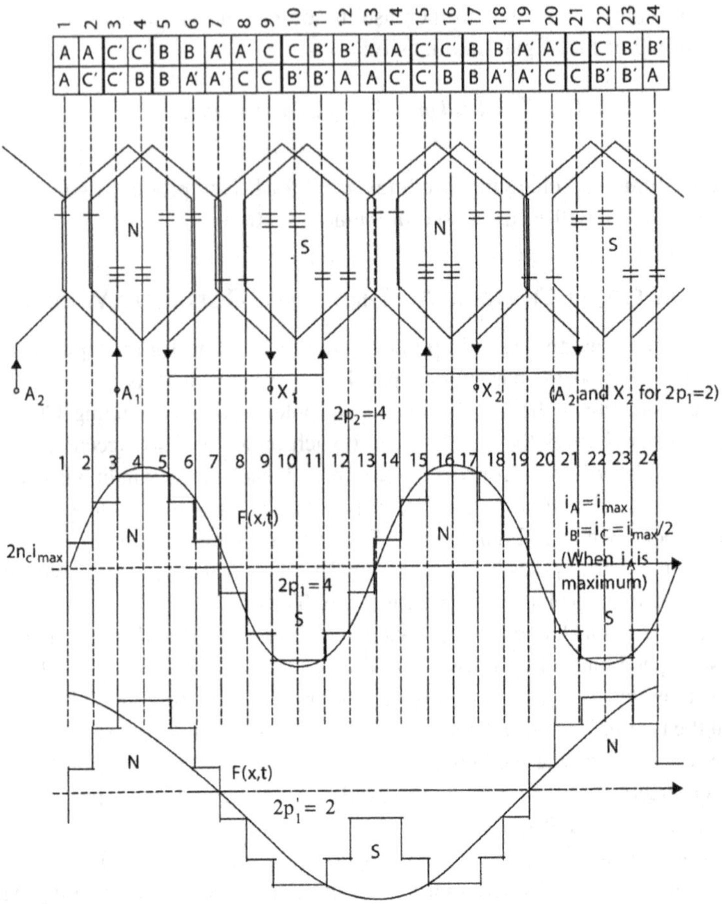

FIGURE 5.12 Pole changing winding: $N_s = 24$, $2p_1 = 4$, $2p_1 = 2$.

To reduce torque pulsations due to mmf space harmonics, the two windings are built with different numbers of turns/coil so as to produce a quasi-sinusoidal mmf spatial distribution.

The pole count changing windings may also be built for two-phase windings [4, Chapter 4].

Three-phase and two-phase distributed windings are characteristic not only of IMs but also of synchronous machine stators.

Note: Typical windings for linear induction motors will be discussed in Section 5.24. Tooth-wound coil ($q \leq 0.5$) design for PM synchronous machines is introduced in Chapter 6.

5.3.8 Cage Rotor Windings

Faultless cage rotor windings are symmetric windings, and thus the time phase shift between currents in neighboring end ring segments, I_i and I_{i+1}, is $\alpha_{er} = 2\pi p_1/N_r$ (Figure 5.15).

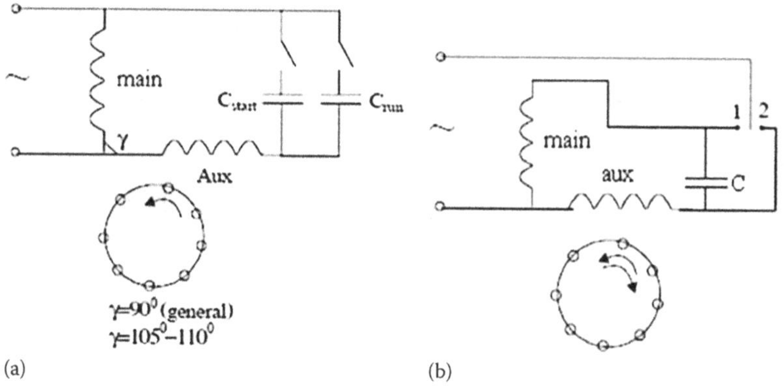

FIGURE 5.13 Two-phase induction motor: (a) unidirectional motion and (b) bidirectional motion: 1—forward; 2—backward.

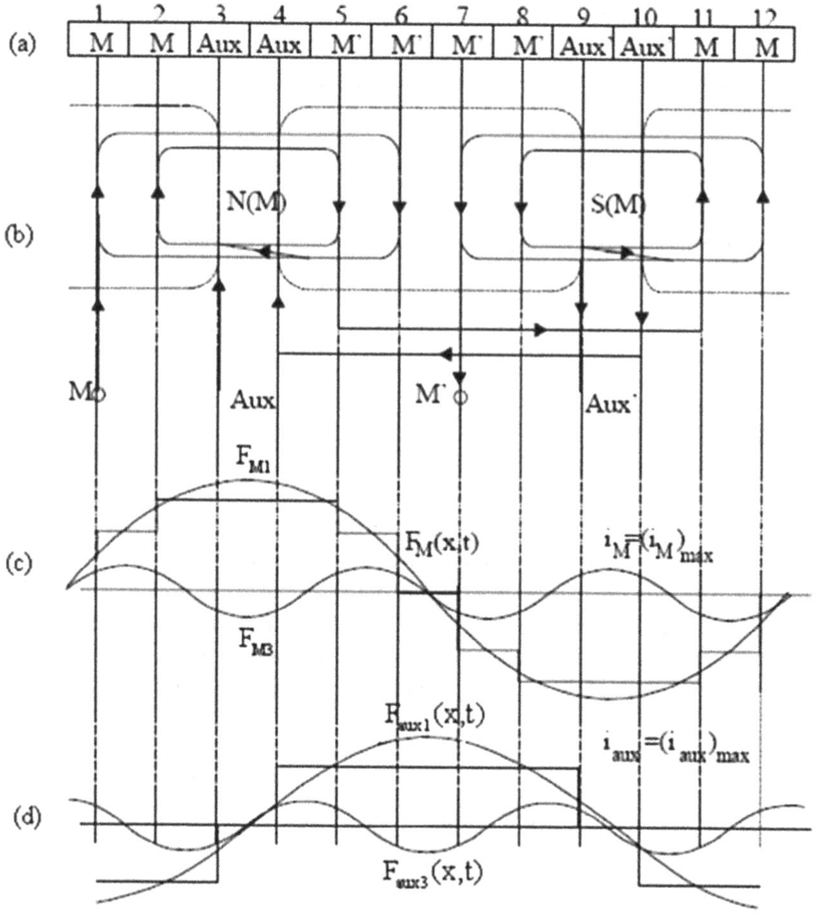

FIGURE 5.14 Capacitor start two-phase winding: $N_s = 12$ slots, $2p_1 = 2$ poles: (a) slot/phase allocation, (b) coil connections, (c) main phase mmf, and (d) auxiliary phase mmf.

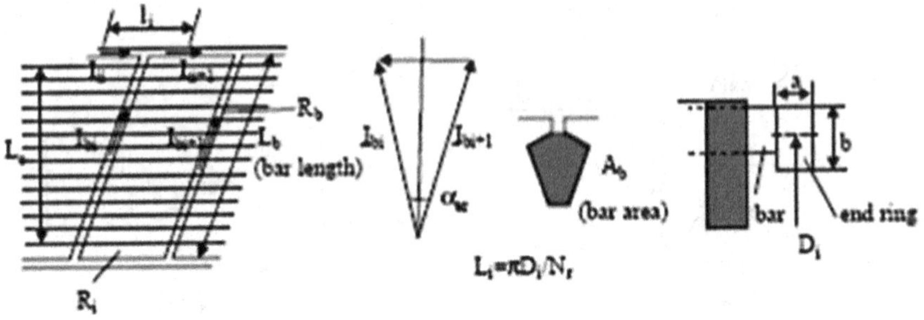

FIGURE 5.15 Rotor cage geometry.

Consequently, the end ring and cage-bar currents, I_r and I_b, are related by

$$I_r = \frac{I_b}{2\sin\dfrac{\alpha_{er}}{2}} \tag{5.28}$$

Let us denote by R_b and R_r the bar and ring segment resistances:

$$R_b = \rho_b\frac{l_b}{A_b};\quad R_r = \rho_r\frac{l_r}{A_r};\quad A_r = a\times b;\quad l_r = \frac{\pi D_{ring}}{N_r} \tag{5.29}$$

We may lump the bar and ring resistances into an equivalent bar resistance, R_{be}:

$$R_{be}I_s^2 = R_bI_b^2 + 2R_rI_r^2 \quad \text{or} \quad R_{be} = R_b + \frac{R_r}{2\sin^2\dfrac{\pi p_1}{N_r}} \tag{5.30}$$

When the number of rotor slots per pole pair (N_r/p_1) becomes small or fractional, the above expressions are less trustworthy.

Let us consider that the rotor d.c. or PM excitation is zero (as in IMs) and the airgap is uniform (non-salient poles) and that the machine stator is three-phase a.c. current fed to produce a sinusoidal mmf, and thus a sinusoidal airgap flux density is marked by mmf and slot harmonics (Figure 5.16).

For low-power machines, the stator or rotor slots may be skewed (Figure 5.17a) by a length c along the rotor axis. The flux density along the rotor axis is phase shifted due to skewing (Figure 5.17b) with an emf reduction/slot by the so-called skewing factor, K_{cv}:

$$K_{cv} = \frac{\overline{AB}}{\overline{AOB}} = \frac{\sin\dfrac{c\nu\pi}{2\tau}}{\dfrac{c\nu\pi}{2\tau}} \tag{5.31}$$

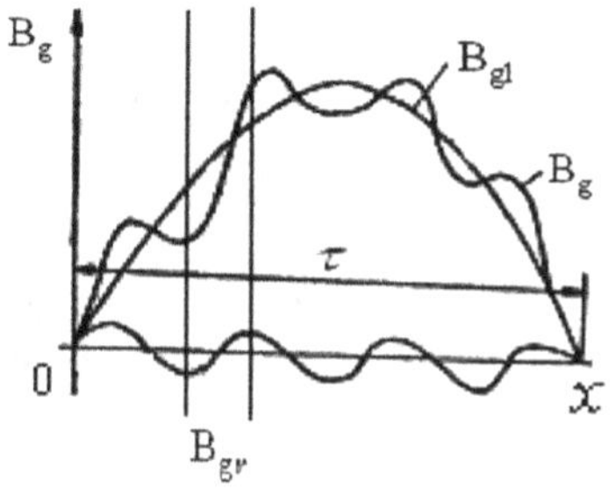

FIGURE 5.16 Airgap flux density of a stator a.c. winding with uniform airgap and a passive iron rotor.

5.3.9 EMF of AC Windings

Thus, the emf is self-induced by the airgap flux density harmonic, $B_{g\nu}$, in a conductor in slot, $V_{econ\nu}$:

$$V_{econv} = B_{gv}\frac{UL_e}{\sqrt{2}}K_{cv} = \frac{1}{2\sqrt{2}}\omega_1 K_{cv}\Phi_v; \quad U = 2\tau_v f_1 \tag{5.32}$$

where Φ_ν is the pole flux for the harmonic, ν.

A coil with span y (Figure 5.17) leads to another coil emf reduction by the already defined chording factor, $K_{y\nu}$:

$$V_{ecoilv} = 2K_{yv}V_{econv}n_c; \quad K_{yv} = \sin\frac{vy}{\tau}\frac{\pi}{2} \tag{5.33}$$

There are n_c turns/coil, q coils/phase belt/pole (Figure 5.17d):

$$V_{eqv} = qK_{qv}V_{ecoilv}; \quad K_{yv} = \frac{\sin\frac{v\pi}{6}}{q\sin\frac{v\pi}{6q}} \tag{5.34}$$

We have just reobtained the distribution factor, $K_{q\nu}$, of the mmf, as expected.

With all $2p_1$ phase belts in series, the phase emf RMS value $V_{e\nu}$ is

$$V_{ev} = \pi\sqrt{2}f_1 W_1 K_{wv}\Phi_v; \quad (\text{RMS}); \quad K_{wv} = K_{qv}K_{yv}K_{cv}; \quad W_1 = 2pqn_c \tag{5.35}$$

$K_{w\nu}$ is the total winding factor for the space harmonic, ν. For $\nu = 1$, the fundamental component is obtained. It is very similar to the emf in a transformer winding with the total winding correction factor, $K_{w\nu}$, and with the polar flux, Φ_ν.

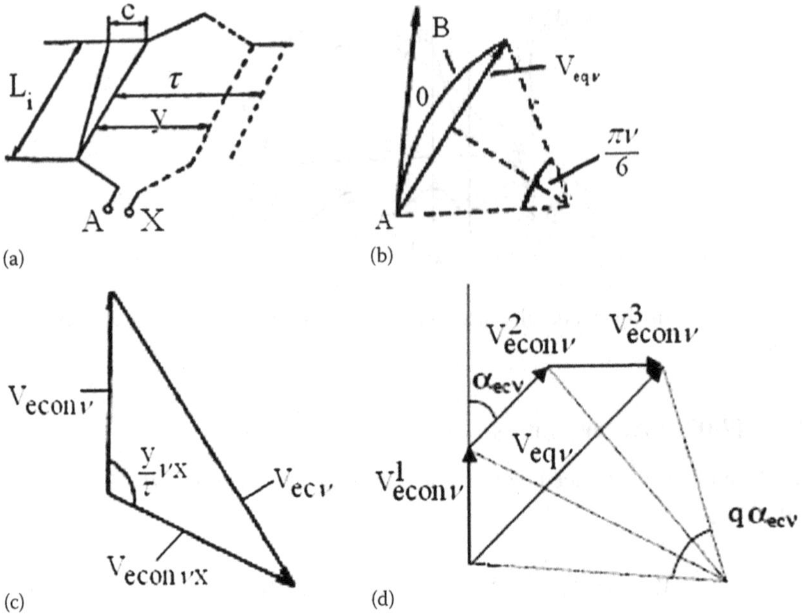

FIGURE 5.17 Winding factor components: (a and b) skewed slots, (c) with chorded coils, and (d) distribution factor $K_{q\nu}$ derivation.

5.4 INDUCTION MACHINE INDUCTANCES

In order to investigate the IMs using the equivalent electric circuit theory (to calculate performance), we need to define the phase self and mutual inductances and phase resistances for the stator and rotor, as previously done for the transformer. The self-inductance of a stator winding considered here is made of two main components:

- Main inductance (L_{ssm} for the stator)
- Leakage inductance (L_{sl} for the stator)L_{ssm} refers to the main magnetic field that crosses the airgap and surrounds both stator and rotor windings, while L_{sl} refers to the leakage flux that surrounds only one winding as in the transformer.

5.4.1 Main Inductance

For a single-phase a.c. distributed winding, L_{ssm} may be calculated either from its flux linkage or from its magnetic energy. Let us use the flux linkage route as in Equation 5.4:

$$B_{g1} = \frac{\mu_0 F_{m1phase}}{g_e}; \quad g_e = K_c g\left(1+K_s\right) \tag{5.36}$$

F_{m1} is the amplitude of a single-phase mmf per pole (Equation 5.10).

$$F_{\text{m1phase}} = \frac{2W_1 I\sqrt{2}K_{q1}K_{y1}}{p_1\pi} \tag{5.37}$$

K_c = (1.1–1.3)—Carter coefficient to account for the slot opening effect on airgap

K_s = (0.2–0.5) (larger for $2p_1 = 2$ poles)—magnetic saturation factor (as in Chapter 4). But the fundamental flux per pole, Φ_1, is

$$\Phi_1 = \frac{2}{\pi}B_{g1}L_e\tau; \quad \Psi_{\text{ssm1}} = \Phi_1 W_1 K_{w1} \tag{5.38}$$

and thus, from Equations 5.36 through 5.38, L_{ssm} is

$$L_{\text{ssm}} = \frac{\Psi_{\text{ssm}}}{I\sqrt{2}} = \frac{4\mu_0}{\pi^2}\left(W_1 K_{w1}\right)^2 \frac{\tau L_e}{p_1 g_e} \tag{5.39}$$

For space harmonics, Equation 5.39 becomes simply

$$L_{\text{ssm}\nu} = \frac{\Psi_{\text{ssm}\nu}}{I\sqrt{2}} = \frac{4\mu_0}{\pi^2}\left(\frac{W_1 K_{w1\nu}}{\nu}\right)^2 \frac{\tau L_e}{p_1 g_e} \tag{5.40}$$

For a three-phase current supply, the so-called cyclic main (magnetization) inductance is $L_{1m\nu}$

$$L_{1m\nu} = \frac{V_{e\nu}}{\omega_1 I} = \frac{6\nu_0}{\pi^2}\left(\frac{W_1 K_{w\nu}}{\nu}\right)^2 \frac{\tau L_e}{p_1 g_e} \tag{5.41}$$

with $V_{e\nu}$ from Equation 5.35.

The harmonics ($\nu > 1$) in $L_{1m\nu}$ are also part of the leakage field as they, in fact, do not couple tightly with the rotor.

For the three-phase connection, the airgap magnetic flux density is

$$\left(B_{g1}\right)_{3\text{ phase}} = \frac{\mu_0 F_{1m}}{K_{cg}\left(1+K_s\right)}; \quad F_{1m} = \frac{3\sqrt{2}w_1 I_0 K_{w1}}{p_1\pi} \tag{5.42}$$

The main inductance field corresponds to the magnetization current, I_0, which is about the same as the no-load current, I_{10} (as for the transformer), that is, with zero rotor currents. Also, we may "build" (calculate) the no-load magnetization curve as for the brush–commutator machine discussed in Section 4.3. We do not repeat the process here but introduce it as a proposed problem. Industrial design experience has synthesized the no load magnetization curve as $B_{g1}(I_0/I_n)$ with $2p_1$ poles as parameter (Figure 5.18) or as l_{1m} (p.u.) versus I_0/I_n; (l_{1m} (p. u.) $= \omega_1 L_{1m} I_n/V_n$).

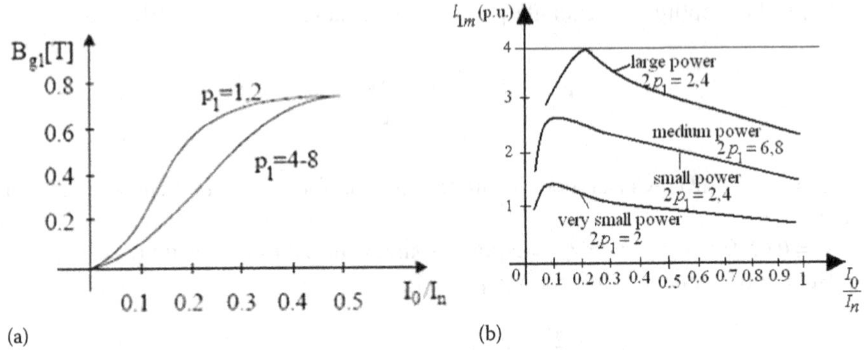

FIGURE 5.18 Magnetization characteristics: (a) typical IM magnetization curves and (b) magnetization inductance.

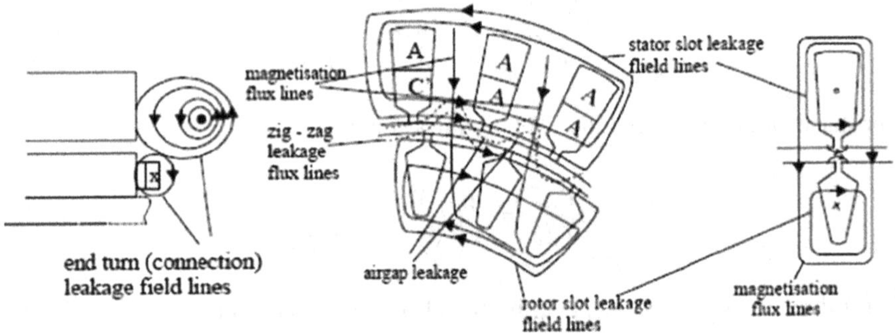

FIGURE 5.19 Classification of leakage flux lines.

5.4.2 Leakage Inductance

The leakage inductance is related to a single phase as no leakage flux coupling between phases is considered. This is not valid for dual windings in the same slots. Leakage flux lines are illustrated in Figure 5.19, and they lead to the following components of phase leakage inductance:

- Slot leakage inductances: L_{slslot}, L_{rlslot} which refer to the flux lines that cross the slot
- Zig-zag leakage inductances: L_{zls}, L_{zlr}
- End-connection leakage inductance: L_{els}, L_{elr}
- Skewing leakage inductance: L_{skewr}
- Differential leakage inductance: L_{dls}, L_{dlr} The last subscript "s" stands for the stator and "r" for the rotor. The differential leakage stator inductance, L_{dls}, refers to the space harmonics, ν, of the main inductance, $L_{1\mathrm{m}}$ (Equation 5.41):

$$L_{\mathrm{dls}} = \sum_{\nu>1} L_{1\,\mathrm{m}\nu} = 2\mu_0 \frac{W_1^2 L_{\mathrm{e}}}{p_1 q} \lambda_{\mathrm{dls}}; \quad \lambda_{\mathrm{dls}} = \frac{3q}{\pi^2 g K_{\mathrm{c}}} \sum^{\nu>1} \frac{K_{\mathrm{w}\nu}^2}{\nu^2} \tag{5.43}$$

for $\nu = km \pm 1$. For single-phase machines, L_{dls} tends to be larger.

Note: As the rotor cage-induced currents may reduce L_{dls} in Equation 5.43, a more detailed assessment of L_{dls} is required for a rigorous design (see Ref. [4, Chapter 6]).

The slot leakage inductance may be calculated, in the absence of skin effect (Chapter 2), by considering a linear magnetic field distribution from the slot bottom to the slot top for rectangular slots (Figure 5.19).

Ampere's law, along the contours in Figure 5.20, leads to

$$H(x)b_s = \frac{n_s I x}{h_s} \quad 0 \le x \le h_s; \quad n_s \text{ — conductors/slot}$$
$$H(x)b_s = n_s I \quad h_s \le x \le h_s + h_{0s} \tag{5.44}$$

The magnetic energy in the slot volume, W_{ms}, leads to the leakage inductance for slot L_{slslot}:

$$L_{slslot} = \frac{2}{I^2} W_{ms} = \frac{2}{I^2}\frac{1}{2} L_e b_s \mu_0 \int_0^{h_s + h_{0s}} H^2(x)\mathrm{d}x = \mu_0 n_s^2 L_e \lambda_s; \quad \lambda_s = \frac{h_s}{3b_s} + \frac{h_{0s}}{b_{0s}} \tag{5.45}$$

λ_s is called the slot geometrical permeance coefficient:

$$\lambda_s = 0.5 - 2.5; \quad \text{for } \frac{h_s}{b_s} < 6; \quad h_{0s} = (1-3)10^{-3} \text{ m} \tag{5.46}$$

which depends on the slot aspect ratio, h_s/b_s; deeper slots lead to larger λ_s.

Now a phase occupies $N_s/m = 2p_1 q$ slots, and thus the phase leakage inductance, L_{sls}, is

$$L_{sls} = L_{slslots} 2 p_1 q = 2\mu_0 \frac{W_1^2 L_e}{p_1 q}\lambda_s \tag{5.47}$$

For trapezoidal or round slots, separate expressions are available (see Ref. [4, Chapter 6]).

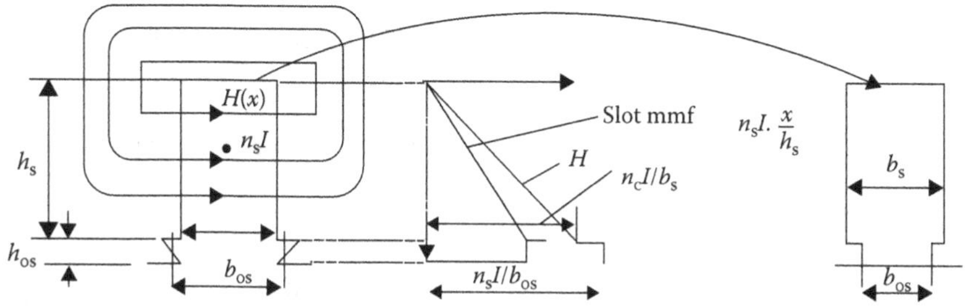

FIGURE 5.20 Rectangular slot leakage field.

The zig-zag leakage inductance, $L_{zls,r}$, is [4]

$$L_{zls,r} = 2\mu_0 \frac{W_s^2 L_e}{p_1 q} \lambda_{zs,r}; \quad \lambda_{zs,r} = \frac{5g\dfrac{K_c}{b_{0s,r}}}{5+4g\dfrac{K_c}{b_{0s,r}}} \frac{3\beta_y+1}{4} < 1; \quad \beta_y = \frac{y}{\tau} \tag{5.48}$$

$\beta_y = 1$ for cage rotors or full-pitch stator coils, and $b_{0s,r}$ is the stator (rotor) slot opening at the airgap.

The end-connection leakage inductance refers to a 3D magnetic field at machine axial ends.

$$L_{els,r} = 2\mu_0 \frac{W_1^2 L_e}{p_1 q} \lambda_{es,r} \tag{5.49}$$

while the geometrical permeance coefficient, $\lambda_{es,r}$, [Ref. 4, Chapter 6] is

$$\lambda_{es} \approx 0.67 \frac{q}{L_e}\left(l_{es} - 0.64\tau\right); \quad \text{single-layer windings} \tag{5.50}$$

where l_{es} is the coil end-connection length.

$$\lambda_{es} \approx 0.34 \frac{q}{L_e}\left(l_{es} - 0.64\tau\right); \quad \text{double-layer windings} \tag{5.51}$$

For cage rotors with end rings close to the rotor stack (core) [4]:

$$\lambda_{ering} \approx \frac{3D_{ring}}{2L_e \sin^2\left(\dfrac{\pi p_1}{N_r}\right)} \log 4.7 \frac{D_{ring}}{a+2b} < \left(1.5-2\right) \tag{5.52}$$

Short stacks (L_e/τ—small) lead to relatively large λ_{es}, and thus larger leakage inductance, besides larger copper losses. This is why the stack length (L_e) per pole pitch (τ) ratio should be larger than unity if possible.

The skewing leakage inductance, L_{skew}, is due to the fact that the stator–rotor magnetic coupling is reduced by the inclination of rotor slots [4]:

$$L_{skew} = \left(1 - K_{skew}^2\right) L_{1m}; \quad K_{skew} = K_{c1} = \frac{\sin \alpha_{skew}}{\alpha_{skew}}; \quad \alpha_{skew} = \frac{c}{\tau}\frac{\pi}{2} \tag{5.53}$$

Skewing increases the leakage inductance, and this will reduce the peak (breakdown) torque as shown later in this chapter.

The total leakage inductance, $L_{sls,r}$, is

$$L_{s,rl} = L_{s,rlslot} + L_{zls,r} + L_{dls,r} + L_{els,r} + L_{skews,r} = 2\mu_0 \frac{W_s^2 L_e}{p_1 q} \sum \lambda_{is,r} \tag{5.54}$$

The rotor leakage inductance (L_{rl}) is already reduced to the stator but let us derive this reduction, as it is necessary for the equivalent circuit of the IM, be it with cage or wound rotor.

5.5 ROTOR CAGE REDUCTION TO THE STATOR

The rotor cage may be considered as a multiphase winding with $m_r = N_r$ (rotor slots) phases, and $W_2 = 1/2$ turns per phase and whose winding factor K_{w2} (for axial slots) is unity.

The reduction to the stator means mmf equivalence with a three-phase equivalent winding, with W_1 turns per phase and I_r' current:

$$\left(F_{1m}\right)_r = \frac{3W_1K_{w1}I_r'\sqrt{2}}{\pi p_1} = N_r\frac{1}{2}\frac{I_b\sqrt{2}}{\pi p_1}K_{skewr} \tag{5.55}$$

So,

$$I_r' = K_iI_b; K_i = \frac{N_rK_{skewr}}{6W_1K_{w1}} \tag{5.56}$$

For loss equivalence,

$$\begin{aligned} R_{be}I_b^2N_r &= 3R_r'I_r'^2 \\ R_r' &= R_{be}\frac{N_r}{3K_i^2} \end{aligned} \tag{5.57}$$

In a similar way, for the cage leakage inductance

$$L_{rl}' = L_{bel}\frac{N_r}{3K_i^2} \tag{5.58}$$

5.6 WOUND ROTOR REDUCTION TO THE STATOR

As for the transformer, the wound rotor three-phase winding may be reduced to the stator by conserving the fundamental mmf, winding losses and magnetic leakage energy, and power.

$$\frac{I_r'}{I_r} = K_i = \frac{V_r}{V_r'}; \quad K_i = \frac{W_2K_{w1r}}{W_1K_{w1s}}; \quad \frac{R_r'}{R_r} = \frac{L_{r1}'}{L_{rl}} = \frac{1}{K_i^2} \tag{5.59}$$

$$\frac{3W_1K_{w1s}I_r'\sqrt{2}}{\pi p_1} = \frac{3W_2K_{w1r}I_r\sqrt{2}}{\pi p_1} \tag{5.60}$$

By this reduction, implicitly, the mutual inductance between the stator and rotor equivalent three phases (L_{srm}) is equal to the main self-inductance, L_{ssm}.

5.7 THREE-PHASE INDUCTION MACHINE CIRCUIT EQUATIONS

The three-phase IM may be represented by three stator and three rotor equivalent windings with magnetic coupling between them (Figure 5.21).

Let us neglect for the time being magnetic saturation, iron losses, and mmf space (or emf time) harmonics. For phase coordinates: stator coordinates for the stator, rotor coordinates for the rotor, there are no motion-induced voltages:

$$\begin{aligned} I_{A,B,C}R_s - V_{A,B,C} &= -\frac{d\Psi_{A,B,C}}{dt} \\ I'_{a,b,c}R'_r - V'_{a,b,c} &= -\frac{d\Psi'_{a,b,c}}{dt} \end{aligned} \tag{5.61}$$

Because the mmf distribution is considered sinusoidal along the stator bore, the mutual inductance between stator and rotor phases varies sinusoidally with the angle between the respective phases (rotor position electrical angle, Θ_{er}). The self-inductance and stator–stator and rotor–rotor mutual inductances do not depend on the rotor position. Consequently, Ψ_A and Ψ'_a flux linkages are

$$\begin{aligned} \Psi_A = L_{sl}I_A + L_{ssm}\left(I_A + I_B\cos\frac{2\pi}{3} + I_C\cos\frac{-2\pi}{3}\right) \\ + L_{ssm}\left(I'_a\cos\Theta_{er} + I'_b\cos\left(\Theta_{er} + \frac{2\pi}{3}\right) + I'_c\cos\left(\Theta_{er} - \frac{2\pi}{3}\right)\right) \end{aligned} \tag{5.62}$$

$$\begin{aligned} \Psi'_a = L'_{ler}I'_a + L_{ssm}\left(I'_a + I'_b\cos\frac{2\pi}{3} + I'_c\cos\frac{-2\pi}{3}\right) \\ + L_{ssm}\left(I_A\cos(-\Theta_{er}) + I_B\cos\left(-\Theta_{er} + \frac{2\pi}{3}\right) + I_C\cos\left(-\Theta_{er} - \frac{2\pi}{3}\right)\right) \end{aligned} \tag{5.63}$$

Let us consider

$$\begin{aligned} I_A + I_B + I_C = 0 \\ I'_a + I'_b + I'_c = 0 \end{aligned} \tag{5.64}$$

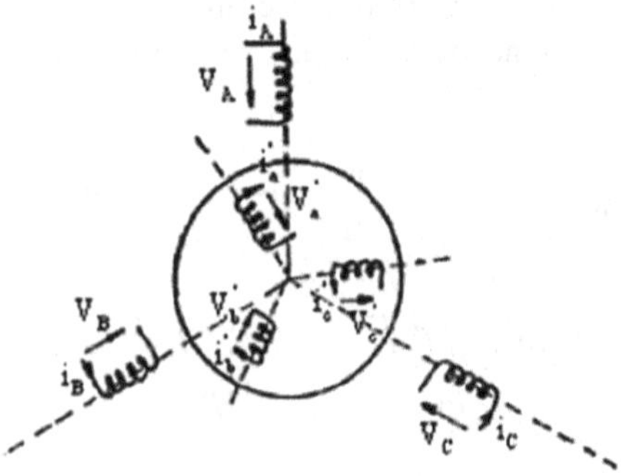

FIGURE 5.21 The three-phase induction machine with reduced rotor windings.

Then, with Equation 5.64, Ψ_A, Ψ'_a become

$$\Psi_A = (L_{sl} + L_{1m})I_A + L_{1m}\left(I'_a \cos\Theta_{er} - \frac{1}{\sqrt{3}}(I'_b - I'_c)\sin\Theta_{er}\right) \tag{5.65}$$

$$\Psi_A = L'_{s1}L'_a + L_{1m}\left(I_A \cos(-\Theta_{er}) - \frac{1}{\sqrt{3}}(I_B - I_C)\sin(-\Theta_{er})\right) \tag{5.66}$$

$L_{1m} = 3L_{ssm}/2$—the cyclic main inductance

To eliminate I'_b, I'_c, I_B, and I_C from the above equations, and for phase segregation we need to observe that for symmetric direct sequence currents (see also [5])

$$jI_{A,a'+} = -\frac{1}{\sqrt{3}}(I_{B,b'+} - I_{C,c'+}) \tag{5.67}$$

while for inverse sequence currents

$$jI_{A,a'-} = \frac{1}{\sqrt{3}}(I_{B,b'-} - I_{C,c'-}) \tag{5.68}$$

So, from Equations 5.65 through 5.68 for direct and inverse sequences, the flux/current relationships becomeok

$$\begin{aligned}\Psi_{A+} &= L_{sl}I_{A+} + L_{1m}\left(I_{A+} + I^{|}_{a+}e^{j\Theta_{er}}\right)\\ \Psi'_{a+} &= L'_{r1}I'_{a+} + L_{1m}\left(I'_{a+} + I_{A+}e^{-j\Theta_{er}}\right)\end{aligned} \tag{5.69}$$

$$\begin{aligned}\Psi_{A-} &= L_{sl}I_{A-} + L_{1m}\left(I_{A-} + I'_{a-}e^{-j\Theta_{er}}\right)\\ \Psi'_{a-} &= L'_{rl}I'_{a-} + L_{1m}\left(I'_{a-} + I_{A-}e^{j\Theta_{er}}\right)\end{aligned} \tag{5.70}$$

Let us consider symmetric (positive sequence) currents in the three phases; drop the "+" subscript and denote

$$I'^{s}_a = I'_a e^{j\Theta_{er}} : \Psi'^{s}_a = \Psi'_a e^{j\Theta_{er}} \tag{5.71}$$

Now from Equations 5.69 and 5.70, we get

$$\begin{aligned}\Psi_A &= L_{sl}I_A + L_{1m}\left(I_A + I'^{s}_a\right)\\ \Psi'_a &= L'_{er}I_a + L_{1m}\left(I'^{s}_a + I_A\right)\end{aligned} \tag{5.72}$$

The total time derivative of Ψ'_a in Equation 5.70 becomes

$$\frac{d\Psi'_a}{dt} = \frac{d}{dt}\left(\Psi'^{s}_a e^{-j\Theta_{er}}\right) = \frac{d\Psi'^{s}_a}{dt}e^{-j\Theta_{er}} - j\frac{d\Theta_{er}}{dt}\Psi'^{s}_a e^{-j\Theta_{er}}; \quad \frac{d\Theta_{er}}{dt} = \omega_r = 2\pi p_1 n \tag{5.73}$$

Finally, the rotor equation in stator coordinates $\left(I'^{s}_{a}, \Psi'^{s}_{a}\right)$ becomes

$$I'^{s}_{a} R'_{r} - V'^{s}_{a} = -\frac{d\Psi'^{s}_{a}}{dt} + j\omega_{r}\Psi'^{s}_{a}; \quad V'^{s}_{a} = V'_{a} e^{j\Theta_{er}} \tag{5.74}$$

Let us add the stator equation:

$$I_{A} R_{s} - V_{A} = -\frac{d\Psi_{A}}{dt} \tag{5.75}$$

A few remarks are in order:

- Equations 5.74 and 5.75 are valid for symmetric (direct) sequence transients and steady state.
- For the negative sequence symmetric currents, ω_r, becomes $(-\omega_r)$.
- In the rotor equation, too, the variables are all reduced to stator coordinates, including the rotor voltage, V'^{s}_{a}
- For steady state, complex number variables may be used because the voltages and currents in Equations 5.74 and 5.75 are all sinusoidal at frequency ω_1:

$$V_{A,B,C} = V_s\sqrt{2}\cos\left(\omega_1 t - (i-1)\frac{2\pi}{3}\right); \quad i = 1,2,3 \tag{5.76}$$

  so $d/dt \rightarrow j\omega_1$.
- Multiplying rotor equation by $3\left(I'^{s}_{a}\right)^{*}$, we notice that only the speed-containing term refers to electromagnetic power:

$$P_{em} = T_e \frac{\omega_r}{p_1} = -\text{Real}\left(3j\omega_1\Psi'^{s}_{a}\left(I'^{s}_{a}\right)^{*}\right) \tag{5.77}$$

The electromagnetic torque of the stator on rotor is then

$$T_e = -3p_1\text{Imag}\left(\Psi'^{s}_{a}\left(I'^{s}_{a}\right)^{*}\right) = 3p_1\,\text{Imag}\left(\Psi_{A} I_{A}^{*}\right) \tag{5.78}$$

Equation 5.78 reflects Newton's principle of action and reaction. From now on, we may abandon the A, a phase subscripts, as the three-phase equations are obtained from three single-phase segregated equations (as for transformers) with the influence of other phases lumped in (via L_{1m}).

5.8 SYMMETRIC STEADY STATE OF THREE-PHASE IMS

By symmetric steady state, we mean symmetric sinusoidal voltages in the stator at frequency, ω_1, eventually also in the rotor, at frequency, ω_2, and constant speed $\omega_r = \omega_1 - \omega_2$. In complex variables,

$$V_{A,B,C} = V_s\sqrt{2}\cos\left(\omega_1 t - (i-1)\frac{2\pi}{3}\right) \tag{5.79}$$

becomes

$$\underline{V}_s = V_s\sqrt{2}e^{j\omega_1 t};\quad \underline{V}_r^{1s} = V_r^{1}\sqrt{2}e^{j(\omega_1 t-\gamma)} \tag{5.80}$$

Also in Equations 5.74 and 5.75, d/d$t \rightarrow j\omega_1$ and thus

$$\begin{aligned}
&\underline{I}_s\underline{Z}_{sl} - \underline{V}_s = \underline{V}_{es}^0;\quad S = \frac{(\omega_1 - \omega_r)}{\omega_1};\quad \underline{Z}_{sl} = R_s + j\omega_1 L_{sl}\\
&\underline{I}_r^{1s}\underline{Z}'_{rl} - \frac{\underline{V}_r'^s}{S} = \underline{V}_{es}^0;\quad \underline{Z}_{rl}'^s = \frac{R_r'}{S} + j\omega_1 L_{rl}\\
&\underline{V}_{es}^0 = -j\omega_1 L_{1m}\underline{I}_{01};\quad \underline{I}_{0s} = \underline{I}_s + \underline{I}_r'^s
\end{aligned} \tag{5.81}$$

If we add the core losses, as for the transformer, via a resistor in series with $j\omega_1 L_{1m}$, the emf V_{er}^0 becomes V_{er}:

$$\underline{V}_{er} \approx -\underline{Z}_{1m}\underline{I}_{01};\quad \underline{Z}_{1m} = R_{iron} + j\omega_1 L_{1m} \tag{5.82}$$

S is called the slip of the IM. It is the p.u. difference of ideal no-load electrical speed, ω_1, and the electrical rotor speed, ω_r, a kind of p.u. speed regulation.

Equations 5.81 resemble those of a transformer (with rotor sink association of signs) but the slip, S, is the new variable directly related to the machine speed. The apparent load, if the rotor is short-circuited $\left(V_r'^s = 0\right)$, is $R_r'(1-S)/S$ and is speed dependent. The equivalent circuit for Equations 5.81 is thus straightforward (Figure 5.22). Note that in the equivalent circuit, all variables are at frequency ω_1. There is no motion "seen" by the equivalent circuit and thus the mechanical power of the real machine is equal to the active power in $R_r'(1-S)/S$.

The electromagnetic torque, from Equation 5.77, is

$$T_e = -3p_1\text{Imag}\left(\underline{\Psi}_r'^s\left(\underline{I}_r'^s\right)^*\right) = 3p_1 L_{1m}\,\text{Imag}\left(\underline{I}_s\underline{I}_r'^*\right) \tag{5.83}$$

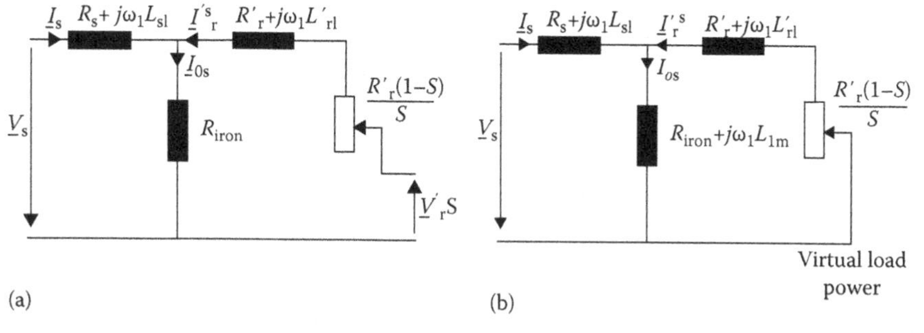

FIGURE 5.22 The IM equivalent circuit for symmetric steady state: (a) with wound rotor (b) with cage rotor.

TABLE 5.1
Operation Modes ($f_1/p_1 > 0$) for Cage Rotor IMs

S	$-\infty\leftarrow$	0	++++	1	$\rightarrow +\infty$
n	$+\infty\leftarrow$	f_1/p_1	++++	0	$\rightarrow -\infty$
T_e	—	0	++++	++++	++++
P_{em}	—	0	++++	++++	++++
Operation mode	Generator		Motor		Braking

Only for the cage rotor (or short-circuited rotor), the electromechanical power is

$$P_{elm} = T_e\frac{\omega_r}{p_1} = T_e\frac{\omega_1}{p_1}(1-S) = 3R_r'\left(I_r'^s\right)^2\frac{1-S}{S} \tag{5.84}$$

$$T_e = \frac{3p_1}{\omega_1}\frac{R_r'\left(I_r'^s\right)^2}{S} = \frac{p_1}{\omega_1}P_{em};\quad P_{em} = \frac{3R_r'\left(I_r'^s\right)^2}{S} = \frac{p_{corotor}}{S} \tag{5.85}$$

P_{em} is the electromagnetic power, or the power in the total secondary equivalent resistance. P_{em} is the total active power that passes from the stator to the rotor or vice versa. P_{em} may be either positive (for the motor mode) or negative (for the generator mode) depending on $S \gtrless 0$.

The motor produces torque ($T_e > 0$) in the direction of motion while the generator brakes the rotor ($T_e < 0$ for $\omega_r > 0$). The difference between the generator and the braking modes is that only for the generator $P_{em} < 0$ for $\omega_r > 0$, and thus a part of the kinetic energy from the rotor is returned back to the electric power source. The basic operation modes for positive ideal no-load speed, ω_1, for the cage rotor are shown in Table 5.1.

Let us discuss in detail a few particular symmetric steady-state operation modes.

5.9 IDEAL NO-LOAD OPERATION/LAB 5.1

The ideal no load corresponds to the zero rotor current $\left(I_r'^s = 0\right)$; from Equation 5.81:

$$\underline{V}_r'^s = S_0\underline{V}_{es} \approx -S_0\underline{V}_{s0} \tag{5.86}$$

or

$$S_0 = \frac{\omega_1 - \omega_{r0}}{\omega_1} = -\frac{\underline{V}_r'^s}{\underline{V}_{s0}} \tag{5.87}$$

For $\underline{V}_{r0}'^s$ (rotor voltage in stator coordinates, at stator frequency) about in phase with the stator voltage $\underline{V}_{s0}$, the machine slip at ideal no load is $S_0 < 0$, so $\omega_{r0}/\omega_1 > 1$, which is a supersynchronous operation. For the rotor voltage in stator coordinates, $\underline{V}_{r0}'^s$ in phase opposition with the stator voltage $S_0 > 0$ and thus $\omega_{r0}/\omega_1 < 1$, which is a under-synchronous operation. So the wound rotor (doubly fed) IM can operate as motor or generator for $\omega_r >< \omega_1$, provided the rotor-side PWM converter supplies rotor voltages with an adequate phase angle, at frequency $f_2 = Sf_1$.

For a short-circuited rotor (cage rotor), $S_0 = 0\left(\underline{V}'^{s}_{r0} = 0\right)$ and the ideal no load speed is $n_0 = \omega_1/2\pi p_1 = f_1/p_1$. With the rotor circuit open (electrically: $I'_r = 0$), the equivalent circuit of Figure 5.22 gains a simplified form (Figure 5.23a) with the phasor diagram of Figure 5.23b. The similarities with a transformer at no-load are clearly visible.

Active power absorbed at an ideal load represents the stator copper losses, p_{Co}, plus iron losses p_{iron}:

$$P_0 = 3R_s I_{s0}^2 + 3R_{iron} I_{s0}^2 = p_{Co} + p_{iron} \tag{5.88}$$

When recording measurements under ideal no-load operation, the IM is driven by a synchronous motor with the same number of poles, $2p_1$, to develop exactly the ideal no-load speed, $n_0 = f_1/p_1$. Alternatively, if a variable speed drive is available, the driving motor speed is increased until the stator current, I_{s0}, of the tested IM is minimum (which corresponds to the ideal no load speed). With R_s measured previously (say in d.c. for small machines) and P_0 and I_{s0}, V_{s0} measured via a power analyzer, the iron losses, p_{iron}, may be calculated from Equation 5.88.

Further on,

$$X_{sl} + X_{1m} = \frac{V_{s0}\sin\varphi_0}{I_{s0}}; \quad \sin\varphi_0 = \sqrt{1 - \left(\frac{P_0}{3V_{s0}I_{s0}}\right)^2} \tag{5.89}$$

As for the transformer, the iron losses under load will be only slightly smaller than those under ideal no load, for given stator voltage V_{s0} and frequency f_1. Additional, stray core losses occur on load.

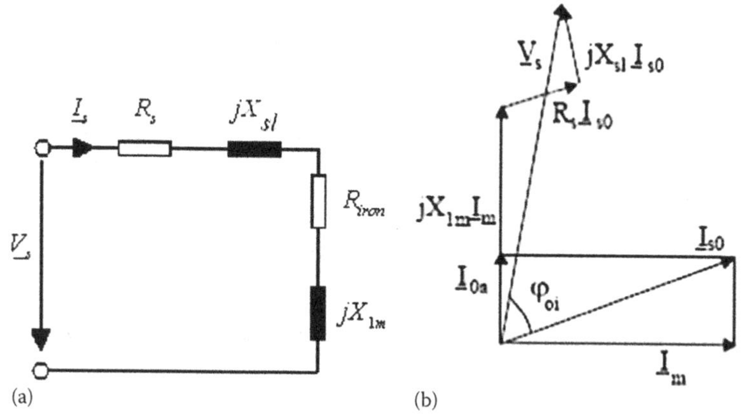

FIGURE 5.23 Ideal no-load operation (cage rotor, $V_r^{\prime s} = 0$): (a) equivalent circuit, (b) phasor diagram, and (c) test rig.

Example 5.2: Ideal No Load of IM

A $2p_1 = 4$ pole cage rotor IM is driven at synchronism at $n_0 = 1800$ rpm, for $f_1 = 60$ Hz, at a rated voltage $V_s = 120$ V (phase RMS) and draws a current $I_{s0} = 4$ A. The power analyzer shows an input power $P_0 = 40$ W; the stator parameters are $R_s = 0.12\ \Omega$ and $X_{sl} = 1\Omega$. Calculate the iron losses p_{iron}, R_{iron}, X_{1m}, and $\cos\varphi_0$.

Solution:

From Equation 5.88, the core losses, p_{iron}, are

$$p_{iron} = P_0 - 3R_s I_{s0}^2 = 40 - 3\times 0.12\times 4^2 = 34.24 \text{ W}$$

Also from Equation 5.88, the core series resistance, R_{iron}, is

$$R_{iron} = \frac{p_{iron}}{3I_{s0}^2} = \frac{34.24}{3\times 4^2} = 0.7133\ \Omega > R_s$$

The power factor, $\cos\varphi_0$, is straightforward from Equation 5.89

$$\cos\varphi_0 = \frac{P_0}{3V_{s0}I_{s0}} = \frac{40}{3\times 120\times 4} = 0.0277;$$

Again, from Equation 5.104

$$X_{1m} = \frac{V_{s0}\sin\varphi_0}{I_{s0}} - X_{sl} \approx \frac{120\times 1}{4} - 1 = 29\ \Omega.$$

5.10 ZERO SPEED OPERATION (S = 1)/LAB 5.2

This time, if we neglect the magnetization current, the equivalent circuit remains with the so-called shortcircuit impedance (Figure 5.24a through c):

$$\underline{Z}_{sc} = R_{sc} + jX_{sc}; \quad R_{sc} = R_s + R'^{s}_{rstart}; X_{sc} = X_{sl} + X'^{s}_{rlstart} \tag{5.90}$$

Both rotor resistance and leakage inductance, R'^{s}_{rstart} and $X'^{s}_{rlstart}$ are in a power grid-connected IM, influenced by skin effect coefficients, $K_R(S\omega_1)$, $K_X(S\omega_1)$, derived in Chapter 2. They are thus different from R'_r and X'_{rl} values for rated (load) conditions at rotor (slip) frequency:

$$f_{2n} = S_n f_{1n} = (0.005 \div 0.05) f_{1n} \tag{5.91}$$

As expected, for the rated voltage, V_{sn}, at zero speed, the starting current, I_{start}, is large:

$$I_{start} = \frac{V_{sn}}{\left|\underline{Z}_{sc}(S=1)\right|} \tag{5.92}$$

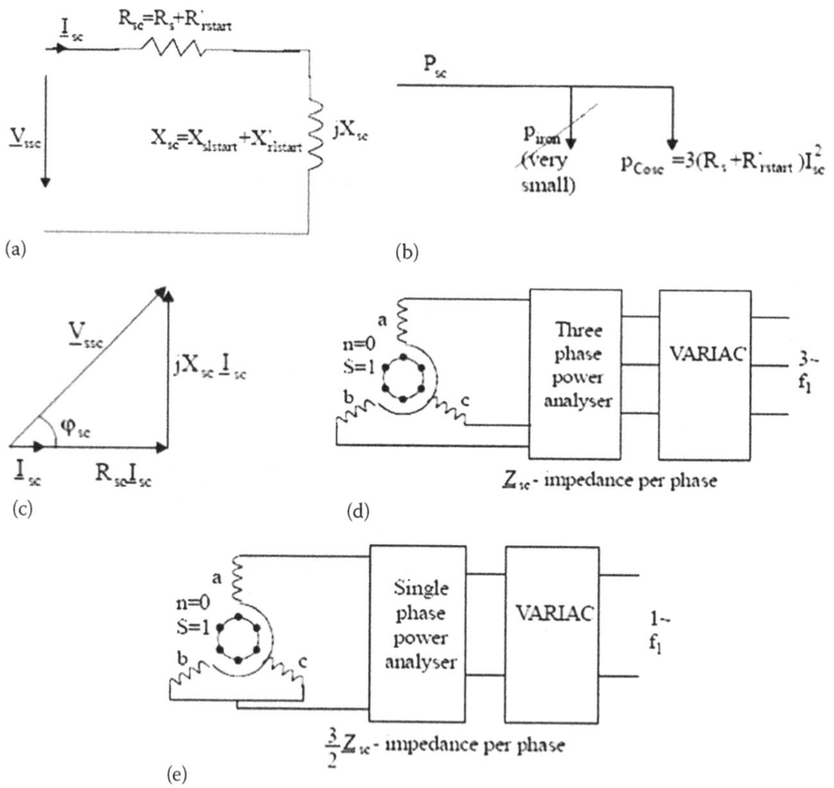

FIGURE 5.24 Zero speed operation ($S = 1$): (a) simplified equivalent circuit, (b) power flow, (c) phasor diagram, (d) three-phase zero-speed testing, and (e) single-phase zero-speed testing.

In general, for well-designed cage rotor IMs to be connected directly at a power grid, $I_{start}/I_{rated} = (4.5–7.5)$.

Testing, however, at zero speed (Figure 5.24d) implies a lower voltage, obtained through a Variac (variable voltage transformer), such that the current does not reach values above the rated current.

And we measure V_{sc}, I_{sc}, and P_{sc} to obtain

$$P_{sc} \approx 3R_{scstart}I_{start}^2; \quad \cos\varphi_{sc} = \frac{P_{sc}}{3V_{sc}I_{sc}} \tag{5.93}$$

$$X_{sc} = \frac{V_{sc}\sin\varphi_{sc}}{I_{sc}}; \quad R_{rstart}'^{s} = R_{scstart} - R_s \tag{5.94}$$

The starting torque is

$$T_{estart} \approx \frac{3R_{rstart}'^{s}I_{sc}^2}{\omega_1} p_1 \tag{5.95}$$

In general, for cage rotor IMs,

$$\frac{T_{\text{estart}}}{T_{\text{erated}}} = (0.7 - 2.5) \tag{5.96}$$

Because the voltage $V_{sc} \ll V_{sn}$, the core losses are neglected at zero-speed testing around the rated current. We may not separate X_{sl} from $X'^{s}_{rlstart}$ in X_{sc}, so, in general, in the industry they adopt the equality condition: $X_{sl} = X'^{s}_{rlstart} = X_{scstart}/2$.

Now if we represent graphically the measured short-circuit current, I_{sc}, versus voltage, we get the results as shown in Figure 5.25.

While the zero-speed test at the rated frequency and rated current is good to estimate the starting current and torque for the rated voltage by squared proportionality, the skin effect makes the use of R_{sc} in calculating copper losses under load much less reliable.

Tests at zero speed with an PWM inverter at low frequency $f_1' = S_n f_1 = f_{2n}$ should be proper for the purpose. To avoid forced shaft stalling (due to starting torque), a single-phase frequency test at zero speed may be performed (Figure 5.24e).

Note. For closed slots on the rotor (to reduce noise and rotor surface additional core losses at the cost of lower breakdown torque and rated power factor), the I_{sc} (V_{sc}) straight line from tests at zero speed (Figure 5.25a) intersects the abscissa at E_s (6–12

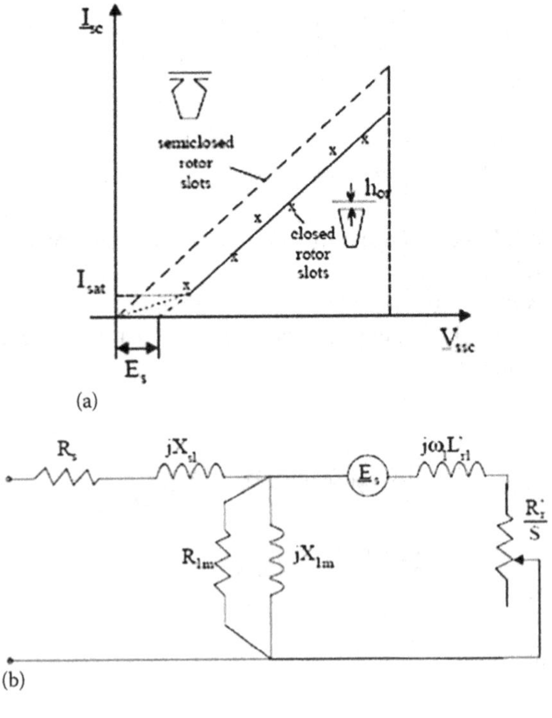

FIGURE 5.25 The case of closed rotor slots: (a) V_{sc} vs. I_{sc} and (b) the equivalent circuit with E_s added for closed rotor slots.

V in 220 V per phase, 50(60) Hz IMs). This additional emf, E_s, is due to the closed rotor slot bridge's early magnetic saturation. E_s is 90%, phase-shifted with respect to the rotor current and is equal to

$$\left(E_s\right)_{\text{closeslots}} \approx \frac{4}{\pi}\pi\sqrt{2}\left(fW_1\right)\Phi_{\text{bridge}}K_{w1};\quad \Phi_{\text{bridge}} = B_{\text{sbridge}}h_{\text{or}}L_e \tag{5.97}$$

$B_{\text{sbridge}} = (2 - 2.2)$ T is the magnetic saturation flux density of the iron core material, $h_{\text{or}} = (0.3 - 1)$ mm is the rotor bridge width, and L_e is the stack length.

Example 5.3: Zero-Speed Operation of IM

The same IM (with semiclosed slots) as that referred to in Example 5.2 is tested at zero speed and rated current (star connection) $I_n = 12$ A, for phase voltage $V_{sc} = 20$ V (rated voltage is 120 V) and power $P_{sc} = 155$ W. The design values of $R_s = 0.12\ \Omega$, $R_r'^s = 0.12\ \Omega$, and $X_{sl} = X_{rl}' = 1\ \Omega$ hold. Calculate R_{scstart}, X_{scstart}, R'^s_{rstart}, X'^s_{rlstart} and determine the skin effect factors, and K_R and K_X, and the starting torque at the rated voltage.

Solution:

From Equation 5.93:

$$R_{\text{scstart}} = \frac{P_{sc}}{3I_n^2} = \frac{155}{3\times 12^2} = 0.358\ \Omega$$

So

$$R'^s_{\text{rstart}} = R_{\text{scstart}} - R_s = 0.358 - 0.12 = 0.238\ \Omega$$

So

$$K_R = \frac{R'^s_{\text{rstart}}}{R_r'^s} = \frac{0.238}{0.12} = 1.983 > 1!$$

Also from Equations 5.93 and 5.94

$$\cos\varphi_{sc} = \frac{P_{sc}}{3V_{sc}I_n} = \frac{155}{3\times 20\times 12} = 0.2153;\quad \sin\varphi_{sc} = 0.976$$

$$X_{\text{scstart}} = \frac{V_{sc}\sin\varphi_{sc}}{I_n} = \frac{20}{12}\times 0.976 = 1.627\ \Omega$$

Now

$$X'^s_{\text{rlstart}} = X_{\text{scstart}} - X_{sl} = 1.627 - 1 = 0.627\ \Omega$$

The skin effect factor on reactance K_X is

$$K_X = \frac{X'^{s}_{rlstart}}{X'_{lr}} = \frac{0.627}{1} = 0.627 < 1$$

The starting torque, T_{estart}, Equation 5.95 is for rated voltage (120 V):

$$T_{estart} \approx \frac{3R'^{s}_{rstart} I^2_{start}}{\omega_1} p_1 = \frac{3 \times 0.238 \times \left(12 \times \frac{120}{20}\right)^2 \times 2}{2\pi \times 60} = 19.646 \text{ Nm}$$

5.11 NO-LOAD MOTOR OPERATION (FREE SHAFT)/LAB 5.3

When no load is applied to the shaft, the IM works in a no-load motor mode. The equivalent circuit (Figure 5.25) may not consider the rotor current zero in principle, because then there would be no torque to cover mechanical losses. However, I'_{r0} is so small that for power balance calculations, in a first instance, it may be neglected.

The test arrangements in Figure 5.26 allow again to measure P_0, I_0, and V_s; the test should be driven to the decreasing voltage until the stator current starts rising (Figure 5.26b):

$$P_0 - 3R_s I_0^2 = p_{iron} + p_{mec} = f\left(\left(\frac{V_s}{V_{sn}}\right)^2\right) \tag{5.98}$$

The intersection of the rather straight line represented by Equation 5.98 with the vertical axis leads to p_{mec} (Figure 5.26b). Then p_{iron} at the rated voltage is segregated simply.

Now that we have p_{mec}, which has been considered constant, while core losses depend on voltage squared ($V \approx \omega_1 W_1 K_{w1} \Phi$), we may approximately write (from Figure 5.26a)

$$\frac{R'_r}{S_{0n}} I'_{r0} \approx V_{sn}; \quad p_{mec} = \frac{3R'_r I'^2_{r0}}{S_{0n}}\left(1 - S_{0n}\right) \approx 3V_{sn} I'_{r0} \tag{5.99}$$

In Equation 5.99, the rotor circuit at no load with a very small slip S_{0n} ($S_{0n} < 10^{-3}$) is considered purely active. From Equation 5.99, we may first calculate I'_{r0}, and then, with R'_r known, S_{0n}!

Example 5.4

The same motor as that referred to in Examples 5.2 and 5.3 is tested as motor on no load at $V_{sn} = 120$ V, 60 Hz, with $I_0 = 4.2$ A; after tests, $P_{0n} = 95$ W, $p_{mec} = 50$ W, and $R_s = 1.2R'_r = 0.12\ \Omega$. Calculate the iron losses and the slip on no load, S_0.

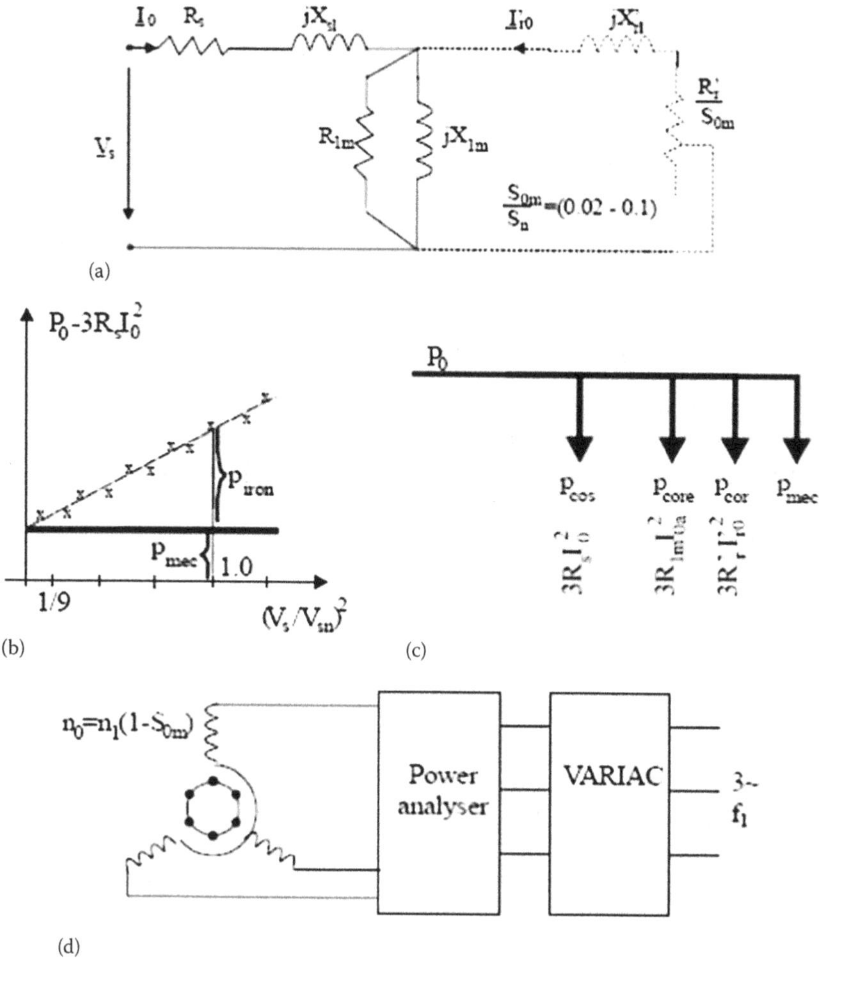

FIGURE 5.26 No-load motor operation: (a) equivalent circuit, (b) no load loss segregation, (c) power balance, and (d) test arrangement.

Solution:

From Equation 5.98 at the rated voltage,

$$p_{iron} = P_{0n} - 3R_s I_0^2 - p_{mec} = 95 - 3\times 0.12\times 4.1^2 - 50$$
$$= 39\ \text{W} \approx 40\ \text{W}\ (\text{at ideal noload})$$

Now from Equation 5.99

$$I'_{r0} \approx \frac{p_{mec}}{3V_{sn}} = \frac{50}{3\times 120} = 0.1388\text{A}$$

and

$$S_{0n} = \frac{R'_r I'_{r0}}{V_{sn}} = \frac{0.1 \times 0.1388}{120} = 1.15 \times 10^{-4}!$$

Now it is clear why, in reality, S_{0n} may be neglected.

Note: A digital scope recording of no-load IM stator current, with star and delta phase connections, would show different shapes with visible time harmonics, mainly due to magnetic saturation of the machine iron cores.

5.12 MOTOR OPERATION ON LOAD (1 > S > 0)/LAB 5.4

On load operation takes place when the IM drives a mechanical load (pump, compressor, drive-train, machine tool, etc.). For motoring, as seen in Table 5.1, $0 < n < n_0 = f_1/p_1$ or $1 > S > 0$. The complete equivalent circuit should be used for the load operation, while the power balance is reiterated in Figure 5.27.

The efficiency, η, is

$$\eta = \frac{\text{Shaft power}}{\text{Shaft power}} = \frac{P_m}{P_{1e}} = \frac{P_m}{P_m + p_{cos} + p_{iron} + p_s + p_{Cor} + p_{mec}} \tag{5.100}$$

The rated speed, n_n, is

$$n_n = \frac{f_1}{p_1}(1 - S_n) \tag{5.101}$$

The rated slip, $S_n = (0.06–0.006)$, with the larger values for low-power (below 200 W) IMs.

The stray load losses refer to additional losses on the stator and rotor surface, due to slotting and magnetic saturation and in the rotor cage, due to stator and rotor mmfs space harmonics (see Ref. [4, Chapter 11]).

The second definition of slip S (the first one is contained in Equation 5.101) from Equation 5.85 is

$$S = \frac{p_{Cor}}{P_{em} - p_s} \approx \frac{p_{Cor}}{P_{em}} \tag{5.102}$$

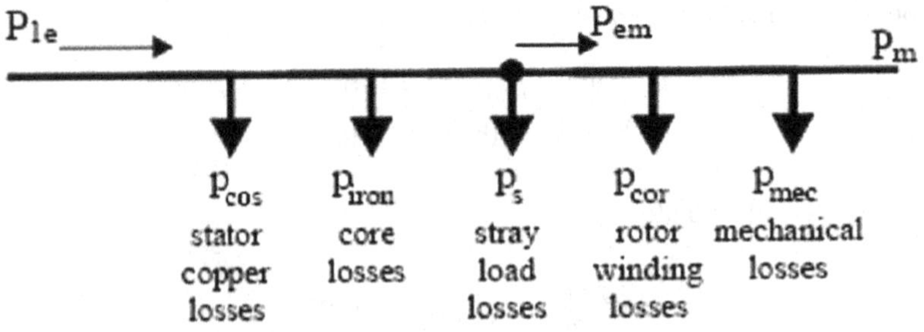

FIGURE 5.27 Power balance for motoring.

So, the larger the slip (the lower the speed n), the larger the rotor winding losses, p_{Cor}, for a given electromagnetic power (or torque).

5.13 GENERATING AT POWER GRID (N > F1/P1,S < 0)/LAB 5.5

As already shown in Equation 5.85 and Table 5.1, when the cage-rotor IM is driven above the no load ideal (synchronous or zero rotor current) speed, $S < 0$ and the electromagnetic torque becomes negative (Figure 5.28).

The driving motor may be a diesel engine, a hydraulic, or a wind turbine; in the lab, it should be a variable speed drive.

To calculate IM performance based on the equivalent circuit, we may calculate the IM equivalent resistance and reactance, R_e and X_e, as the function of slip S (Figure 5.28b).

It is evident from Figure 5.28b that while R_e changes sign to allow for generating active power for $S < 0$, X_e remains positive, and thus the IM always draws reactive power from the power grid. The latter is the main drawback of an IM as a generator. The situation today is mended by introducing a two-stage PWM converter interface whose d.c. link capacitor produces the necessary reactive power to IM and also can deliver a substantial amount to the power grid in a controlled manner.

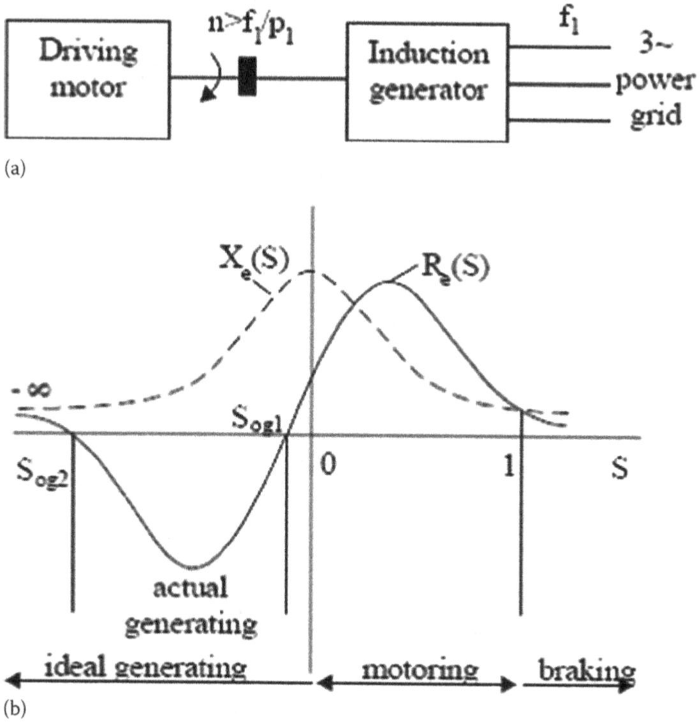

FIGURE 5.28 Induction generator: (a) at power grid and (b) equivalent IM, R_e (s), X_e (s).

5.14 AUTONOMOUS GENERATOR MODE (S < 0)/LAB 5.6

As already demonstrated, the IM operates as a generator for $S < 0$ or at the supersynchronous speed ($n > f_1/p_1$). The frequency is fixed for a power grid connection and the reactive power is drawn from the grid for machine magnetization.

For the autonomous generator mode, still $S < 0$ but the output frequency, f_1, is in relation to the capacitor (or synchronous condensor (Figure 5.29a)), which provides the reactive power, and to machine parameters. The capacitors are delta connected to reduce their capacitance. Let us explore the ideal no-load operation with capacitor excitation (Figure 5.29b).

The d.c. remanent magnetization in the rotor produces on the stator (by motion) an emf, $\underline{E}_{rem}$, at frequency $f_1 = p_1 n_0$ (n_0—rotor speed). This way a.c. currents flow into the machine magnetization reactance and capacitors. They add to the initial $\underline{E}_{rem}$ and thus voltage buildup takes place, until the voltage settles at a certain value, V_{s0}.

Changing the speed will change both the frequency, $f_1' = n_0' p_1$, and the settling voltage, V_{s0}'. Neglecting the stator resistance, R_s, and the stator leakage reactance, X_{sl}, for zero rotor currents (ideal no-load operation), leads to the equivalent circuit in Figure 5.29b.

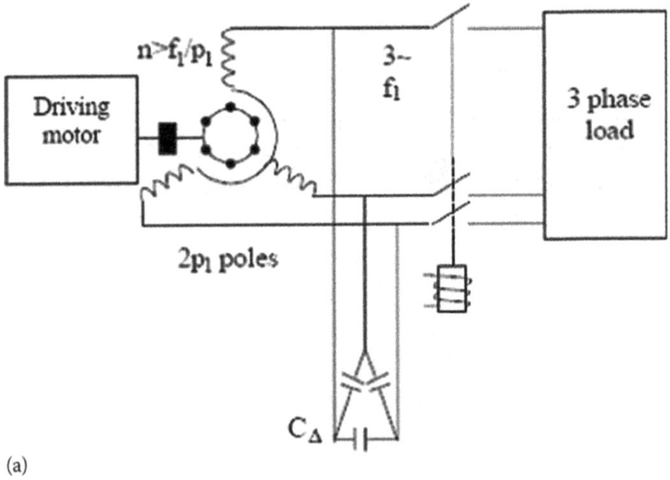

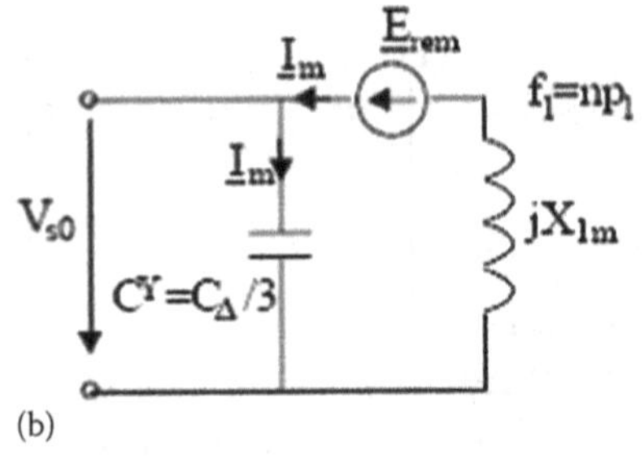

FIGURE 5.29 Autonomous induction generator with cage rotor: (a) with capacitor self-excitation and (b) equivalent circuit with capacitor excitation under ideal no load.

The machine equation for selfexcitation becomes

$$\underline{V}_{s0} = \left| jX_{1m}I_m + E_{rem} \right| = -\frac{j}{\omega_1 C_Y}\underline{I}_m = V_{s0}\left(I_m\right) \tag{5.103}$$

Due to magnetization inductance saturation, X_{1m} depends on I_m and thus $V_{s0}(I_m)$ is a nonlinear function that starts at E_{rem} (Figure 5.30a). Graphically, the intersection of $V_{s0}(I_m)$ with the capacitor straight line (eqn. 5.103) leads to self-excitation voltage V_{s0} (point A) for a given speed, n_0. If the speed is reduced to n_0', the operation point A moves to A' at smaller voltage and smaller frequency $f_1' < f_1$.

E_{rem} is mandatory for the self-excitation to initiate, the "self-excitation process" but magnetic saturation is also necessary to provide a safe intersection of no-load magnetization curve, $V_{s0}(I_m)$, with the capacitor line (Figure 5.30a).

To calculate the performance on load, the complete equivalent circuit (Figure 5.30b), with L_{1m} (I_m) given—magnetic saturation considered—has to be solved, iteratively in general, to obtain the external characteristic $V_s(I_L)$ for resistive–inductive, R_L,L_L, or resistive–capacitive, R_L, C_L, loads (Figure 5.30c), for, say, a constant speed (see Ref. [6]).

It should be noticed that, as expected, slip S increases with load, but also frequency f_1 decreases with load, at a constant speed. Frequency-sensitive loads should be avoided in such a simple scheme. A rather notable voltage drop (regulation) is already visible with a pure resistive load.

Note. A fully variable single capacitor connected to the IM terminals through a PWM voltage-source inverter may provide a constant voltage and, to some extent, constant frequency, for constant speed driving by a prime mover.

5.15 ELECTROMAGNETIC TORQUE AND MOTOR CHARACTERISTICS

By electromagnetic torque, T_e, we understand here the interaction torque between the stator current produced fundamental airgap flux density and the fundamental (sinusoidal) rotor currents.

Its basic expression has already been derived in Equation 5.85, for the singly fed (cage or shortcircuited rotor) IM.

The wound rotor IM may be considered as singly fed only if a passive impedance R_L', L_L', C_L' is lumped into the rotor circuit. Let us consider only an additional rotor resistance R_L' (even a diode rectifier with a single resistance load qualifies). Then Equation 5.85 becomes

$$T_e = \frac{3R_{re}'I_r'^2}{S}; \quad R_{re}' = R_r' + R_L' \tag{5.104}$$

From the equivalent circuit (Figure 5.21), the rotor current, $\underline{I}_r'$ is

$$\underline{I}_r' = \frac{-\underline{I}_s\underline{Z}_{1m}}{\frac{R_{re}'}{S} + jX_{rl}'\underline{Z}_{1m}} \tag{5.105}$$

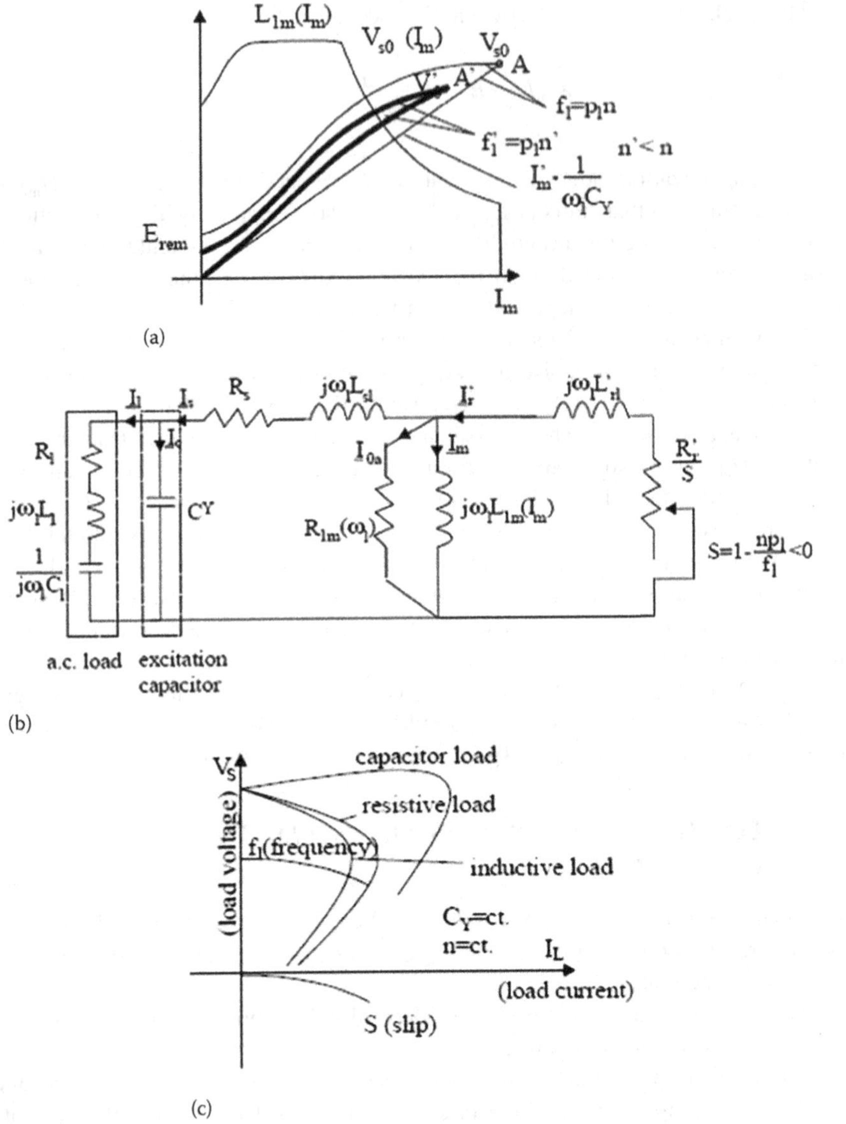

FIGURE 5.30 Autonomous induction generator: (a) self-excitation on no load, (b) complete equivalent circuit with load, and (c) V_L (I_L) curves, slip $S(I_L)$, and frequency $f_1(I_L)$ for constant speed operation.

and with $(Z_{1m} + jX_{sl}/Z_{1m}) \approx (X_{1m} + X_{sl})/X_{1m} = C_1 \approx (1.02\text{–}1.08)$, the stator current is

$$\underline{I}'_r \approx \frac{-\underline{V}_s}{R_s + \dfrac{C_1R'_{re}}{S} + j\left(X_{sl} + C_1X'_{rl}\right)} \tag{5.106}$$

Substituting $\underline{I}_r'$ into Equation 5.104 yields the torque, T_e:

$$T_e = 3V_s^2 \frac{p_1}{\omega_1} \frac{\frac{R_{re}'}{S}}{\left(R_s + C_1 \frac{R_{re}'}{S}\right)^2 + \left(X_{sl} + C_1 X_{rl}'\right)^2} \tag{5.107}$$

The maximum torque is obtained for $\partial T_e/\partial S = 0$ at

$$S_k = \frac{\pm C_1 R_{re}'}{\sqrt{R_s^2 + \left(X_{sl} + C_1 X_{rl}'\right)^2}} \approx \frac{\pm R_{re}'}{\omega_1 L_{sc}} \tag{5.108}$$

$$T_{ek} = \frac{3p_1}{\omega_1} \frac{V_s^2}{2C_1\left[R_s \pm \sqrt{R_s^2 + \left(X_{sl} + C_1 X_{r1}'\right)^2}\right]} \approx 3p_1 \left(\frac{V_s}{\omega_1}\right)^2 \frac{1}{2L_{sc}} \tag{5.109}$$

A few remarks are in order:

- The peak (breakdown) torque is independent of the rotor total resistance, R_{re}', and in general is inversely proportional to the short-circuit inductance, L_{sc}.
- The sign ± in Equations 5.108 and 5.109 refers to motor and generator modes, respectively, as seen in Figure 5.31.
- The neglection of R_s in Equations 5.107 through 5.109 is valid only for $f_1 > 5$ Hz, even for large IMs.
- While the torque expression (Equation 5.109) with C_1 coefficient approximation produces small errors, a reliable I_s expression is:

$$\underline{I}_s \approx \underline{I}_r' + \frac{\underline{V}_s}{R_s + jX_{sl} + \underline{Z}_{1m}} = I_{sa} - jI_{sr} \tag{5.110}$$

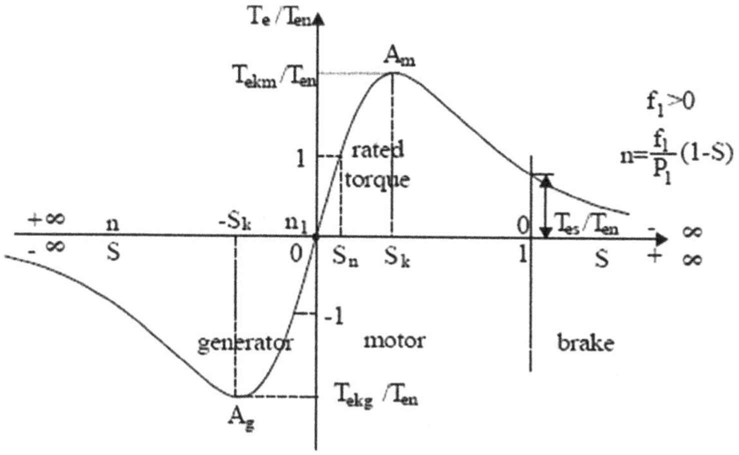

FIGURE 5.31 Torque vs. slip S for constant voltage amplitude V_s and frequency f_1.

With $\underline{I}'_r$ from Equation 5.110, a good precision in the power factor, cos φ_1, calculation is obtained

$$\cos \varphi_1 \approx \frac{I_{sa}}{\sqrt{I_{sa}^2 + I_{sr}^2}} = \frac{|R_e|}{Z_e} \tag{5.111}$$

Typical I_s (S), T_e/T_{en} (S), and cos $\varphi_1(S)$ and, finally, efficiency $\eta(S)$ or versus speed are obtained from the equivalent circuit and constitute the steady-state curves of the IM (Figure 5.32).

For highly variable loads (from 25% to 125% rated load), the design of the IM should provide a large plateau of high efficiency to save energy.

Example 5.5: IM Torque and Performance

An IM with deep bars is characterized by a rated P_n = 25 kW, at V_{sline} = 380 V (star connection), f_1 = 50 Hz, efficiency η_n = 0.92, cos φ_1 = 0.9, p_{mec} = 0.005 P_n, p_s = 0.005 P_n, p_{iron} = 0.015P_n, p_{cosn} = 0.03P_n 2p_1 = 4 poles, starting current (at rated voltage) I_{sc} = 5.2I_n at cos φ_{sc} = 0.4, and the no load current I_{0n} = 0.3I_n. Calculate:

a. Rotor cage rated losses, p_{corn}, electromagnetic power, P_{em}, slip, S_n, speed, n_n, rated current, I_n, rotor resistance, R'_r, for rated load
b. Stator resistance, R_s, and rotor resistance at start, R'_{rstart}
c. Rated and starting electromagnetic torque, T_{en} and T_{estart}
d. Breakdown torque, T_{ek}

Solution:

a. The rotor cage losses, p_{corn}, is the only unknown component of losses so

$$p_{corn} = \frac{P_n}{\eta_n} - (p_{cosn} + p_{iron} + p_s + p_{mec}) - P_n$$

$$= 25{,}000\left[\left(\frac{1}{0.92} - 1\right) - (0.03 + 0.015 + 0.005 + 0.015)\right] = 594 \text{ W}$$

The rated current, I_n, comes directly from the input power, P_n/η_n:

$$I_n = \frac{P_n}{\eta_n \sqrt{3} V_{sline} \cos\varphi_n} = \frac{25{,}000}{0.92 \times \sqrt{3} \times 380 \times 0.9} = 45.928\text{A}$$

The rated slip, S_n, from Equation 5.102 is

$$S_n = \frac{p_{corn}}{\frac{P_n}{\eta_n} - p_{cosn} - p_{iron} - p_s} = \frac{549}{25{,}000\left[\frac{1}{0.92} - 0.05\right]} = 0.02106$$

The rated speed, n_n, is then

$$n_n = \frac{f_1}{p_1}(1 - S_n) = \frac{50}{2}(1 - 0.02106) = 24.4735 \text{ rps} = 1468.4 \text{ rpm}$$

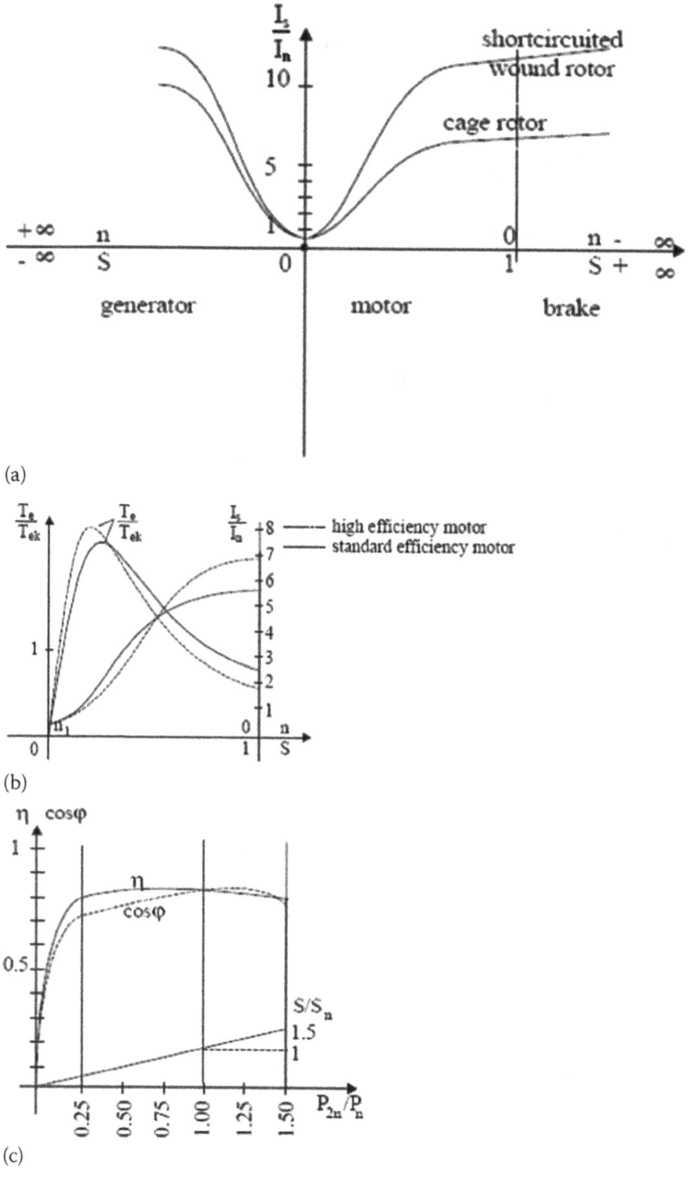

FIGURE 5.32 IM steady-state characteristics: (a) stator current vs. slip (speed), (b) electromagnetic torque vs. slip (speed), and (c) efficiency and power factor vs. slip (speed).

The electromagnetic power, P_{em}, from Equation 5.85 is

$$P_{em} = \frac{p_{corn}}{S_n} = \frac{549}{0.02106} = 26{,}083 \text{ W}$$

(Continued)

Example 5.5: (Continued)

To find the rotor resistance, we first use the rated rotor current, I'_{rn}. By considering the magnetization current as purely reactive and the rotor current (at small slip) as purely active, the latter is simply

$$I'_{rn} = \sqrt{I_n^2 - I_{0n}^2} = 45.928\sqrt{1-0.3^2} = 43.81\ \text{A}$$

So the rotor resistance reduced to the stator, R'_r, is simply

$$R'_r = \frac{p_{corn}}{3I'^2_{rn}} = \frac{594}{3\times 43.81^2} = 0.10316\ \Omega$$

b. The stator resistance, R_s, is straightforward:

$$R_s = \frac{p_{cosn}}{3I_n^2} = \frac{0.03\times 25{,}000}{3\times 45.928^2} = 0.1185\ \Omega$$

The rotor resistance, R'_{rstart}, at start is simply

$$R'_{rstart} = \frac{V_{sline}}{\sqrt{3}}\frac{\cos\varphi_{sc}}{I_{sc}} - R_s = \frac{380\times 0.4}{\sqrt{3}\times 5.2\times 45.928} - 0.1185 = 0.25\ \Omega$$

So the skin-effect resistance factor, K_R, is

$$K_R = \frac{R'_{rstart}}{R'_r} = \frac{0.25}{0.10316} = 2.42$$

This result suggests a strong skin effect (deep-bars rotor cage).

c. The rated electromagnetic torque

$$T_{em} = \frac{P_{em}p_1}{\omega_1} = \frac{26{,}083\times 2}{2\times\pi\times 50} = 166.133\ \text{Nm}$$

The starting torque, T_{estart}, is

$$T_{estart} \approx \frac{3R'_{rstart}I_{sc}^2 p_1}{\omega_1} = \frac{3\times 0.25\times(5.2\times 45.928)^2\times 2}{2\times\pi\times 50} = 272.47\ \text{Nm}$$

d. For the breakdown torque, we need the short-circuit reactance, X_{sc}

$$X_{scstart} = \frac{V_{sline}}{\sqrt{3}}\frac{\sin\varphi_{sc}}{I_{sc}} = \frac{380\times\sqrt{1-0.4^2}}{\sqrt{3}\times 5.2\times 45.928} = 0.843\ \Omega$$

This is an underestimated value for rated torque calculation but not so bad for the breakdown torque, when some notable skin effect is still manifest.

Apparently, from Equation 5.109, the breakdown torque T_{ek} is

$$T_{ek} \approx 3\left(\frac{V_{sline}}{\sqrt{3}}\right)^2 \frac{p_1}{\omega_1}\frac{1}{2X_{sc}} = 3\left(\frac{380}{\sqrt{3}}\right)^2 \frac{2}{2\times\pi\times 50}\frac{1}{2\times 0.843} = 548.54\ \text{Nm}$$

Now, the $T_{ek}/T_{em} = 548.54/166.133 = 3.3$, a value which is larger than usual, a signal that the problem data are not entirely coherent with an industrial motor with a deep-bar cage rotor.

5.16 DEEP-BAR AND DUAL-CAGE ROTORS

For heavy and frequent starting torque application IMs, connected directly to the local power grid, the skin effect is used to reduce the starting current down to around $5 \times I_n$ and increase the starting torque above $2 \times T_{en}$; deep-bar and dual-cage rotors are used for this scope (Figure 5.33a and b).

In both cases, the equivalent rotor resistance, R_r', and leakage reactance, X_{rl}', depend on the slip (on rotor frequency, in fact):

$$\begin{aligned} R_r'(\omega_2) &= K_R(\omega_2)R_{r0}'; \quad S = \frac{\omega_2}{\omega_1} \\ R_{rl}'(\omega_2) &= K_X(\omega_2)X_{rl0}' \end{aligned} \tag{5.112}$$

As derived in Chapter 2,

$$K_R = \xi\frac{\sinh 2\xi + \sin 2\xi}{\cosh 2\xi - \cos 2\xi}; \quad K_X = \frac{3}{2\xi}\frac{\sinh 2\xi - \sin 2\xi}{\cosh 2\xi - \cos 2\xi} \tag{5.113}$$

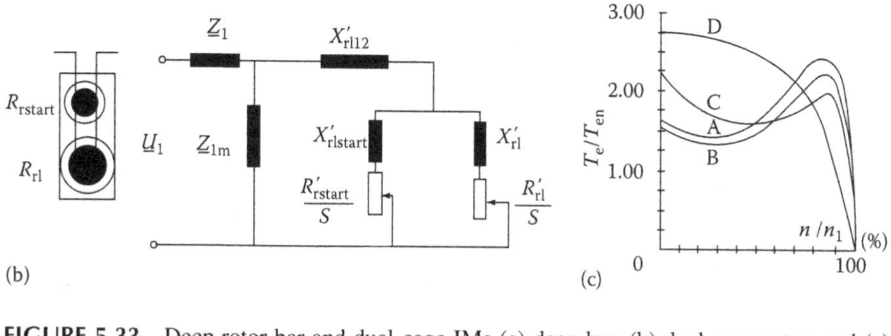

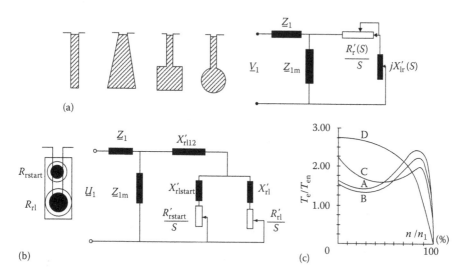

FIGURE 5.33 Deep rotor bar and dual-cage IMs (a) deep-bar, (b) dual-cage rotor, and (c) torque–speed curve types (NEMA standards).

$$\xi = h_s\sqrt{\frac{\omega_2\mu_0\sigma_{AC}}{2}} \tag{5.114}$$

To calculate performance, I_s (S), T_e (S), $\eta(S)$, and cos $\varphi(S)$, we can still make use of the equivalent circuit (Figure 5.33a) but with variable parameters.

For the dual-cage rotor, the upper (-starting) cage may be made of brass (of higher resistivity), while the inner (running) cage is made of aluminum. In such a case, separate end rings may be necessary due to different thermal expansion rates.

At start and low speeds, the starting cage works, as the field penetration depth is small (ω_2 is large), while at high speeds (low slips), the running cage is prevalent.

The constant parameter dual-cage concept may also be used to simulate the deep-bar effect.

The NEMA standards suggest four main designs, and while A and B refer to low skin effect, C and D refer to the deep-bar and dual-cage rotors used for heavy and frequent starts (Figure 5.33c).

5.17 PARASITIC (SPACE HARMONICS) TORQUES

As already mentioned in Section 5.3 on a.c. windings, their mmf and airgap magnetic field (flux density) show space harmonics in the order of $\nu = km \pm 1$, $m = 3$ and $k \geq 1$ with the pole pitch $\tau_\nu = \tau/\nu$, and their synchronous speed with respect to stator $n_{1\nu} = n_1/\nu$. Among these harmonics, the first-slot harmonics $\nu_c = 2qm \pm 1$ ($k = 2q$) are very important as their distribution factor is $K_q\nu_c = K_{q1}$. The open slots tend to magnify these mmf harmonic effects. The currents induced in the rotor by these harmonic fields produce their own fields, full of space harmonics, of ν' orders.

If for the wound rotor, with three phases in the rotor and stator, both stator and rotor windings mmf have their own harmonics orders, for the cage rotor the relationship between a stator harmonic, ν, and a rotor harmonic, ν', is

$$p_1\nu - p_1\nu' = K_2N_r \tag{5.115}$$

where N_r is the number of rotor slots.

The slip for the ν mmf space harmonic in the stator S_ν is

$$S_\nu = \frac{n_{1\nu} - n}{n_{1\nu}} = \frac{\frac{n_1}{\nu} - n_1}{\frac{n_1}{\nu}} = 1 - \nu(1 - S) \tag{5.116}$$

Now, the ν' rotor mmf harmonic speed, $n_{2\nu,\nu'}$, with respect to the rotor is

$$n_{2\nu,\nu'} = \frac{f_{2\nu}}{\nu' p_1} = \frac{f_1 S_\nu}{\nu' p_1} = \frac{n_1}{\nu'}\left(1 - \nu(1 - S)\right) \tag{5.117}$$

But, with respect to the stator, the ν' rotor (produced by ν stator harmonic) moves at speed $n_{1\nu,\nu'}$:

$$n_{1\nu,\nu'} = n_{2\nu,\nu'} + n = \frac{n_1}{\nu'}\left(1+\left(\nu'-\nu\right)\left(1-S\right)\right) \quad (5.118)$$

or with Equation 5.115:

$$n_{1\nu,\nu'} = \frac{n_1}{\nu'}\left[1+\frac{K_2 N_r}{p_1}\left(1-S\right)\right] \quad (5.119)$$

So every stator mmf space harmonic, ν, produces an infinity of harmonics ν' in the cage rotor mmf whose traveling speed with respect to the stator is $n_{1\nu,\nu'}$. These harmonics act upon an equivalent circuit of an IM with parameters corresponding to the same frequency, but in most cases, the magnetization branch may be neglected.

The main effects of mmf space harmonics are parasitic torques and uncompensated radial forces that produce noise and vibration.

- The parasitic torques manifest themselves between stator and rotor mmf harmonics of the same order (according to the frequency theorem (Chapter 3)).
- *Asynchronous parasitic torques* are produced by the stator harmonic, ν, in the stator and one in the rotor produced by the former ($\nu = \nu'$).

  With their synchronism $S_\nu = 0$ for $\nu = 2km - 1$ (5, 11, 17), they are inverse harmonics (see Section 5.3 on a.c. windings), and for $\nu = 2km + 1$ (7, 13, 19,…), they are direct:

$$\begin{aligned} S_5 &= 0 = 1-\left(-5\right)\left(1-S_{05}\right); \quad S_{05} = \frac{6}{5} \\ S_7 &= 0 = 1-\left(+7\right)\left(1-S_{07}\right); \quad S_{07} = \frac{6}{7} \end{aligned} \quad (5.120)$$

On the torque/speed (slip) curve, the fifth and seventh asynchronous torques tend to be the most visible (Figure 5.34b). Coil chording is used to attenuate the fifth harmonic.

$$K_{y5} = \sin\frac{\pi}{2}\frac{5y}{\tau} = 0; \quad \frac{y}{\tau} = \frac{2K_1}{5} \approx \frac{5}{6} \quad (5.121)$$

To attenuate the first-slot harmonic $\nu_c = 2qm \pm 1 = (N_s/p_1) \pm 1$, slot skewing is used

$$K_{c\nu} = \frac{\sin\frac{\pi}{2}\frac{c}{\tau}\nu}{\frac{\pi}{2}\frac{c}{\tau}\nu} = 0; \quad \frac{c}{\tau} = \frac{2K_2}{\frac{N_{c1}}{p_1} \pm 1} \quad (5.122)$$

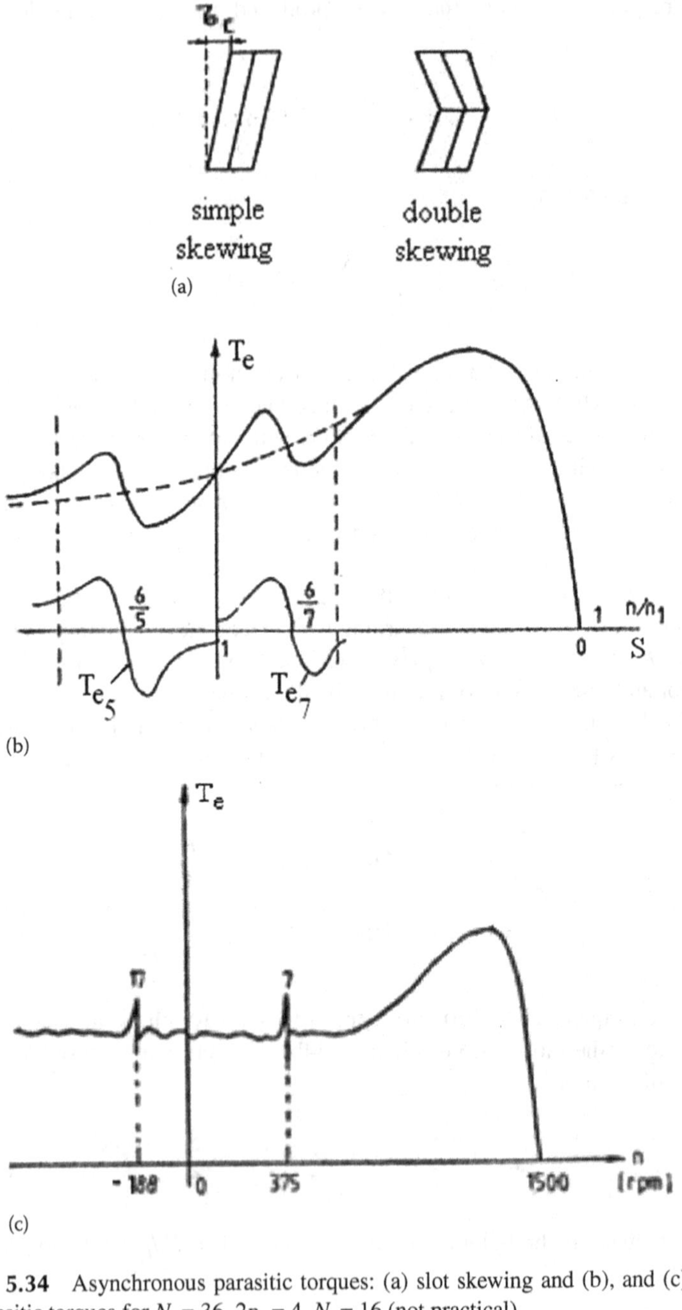

FIGURE 5.34 Asynchronous parasitic torques: (a) slot skewing and (b), and (c) synchronous parasitic torques for $N_s = 36$, $2p_1 = 4$, $N_r = 16$ (not practical).

Skewing by one to two slot pitches is typical.

- *Synchronous parasitic torques* occur when the two (stator and rotor) harmonics, ν_1 and ν', have different origins (Equation 5.118).

$$n_{1\nu_1} = \frac{n_1}{\nu_1} = n_{1\nu,\nu'} = \frac{n_1}{\nu'}\left[1 + \frac{K_2 N_r}{p_1}(1-S)\right] \tag{5.123}$$

ν' and ν_1 should be equal to each other but one may be positive and the other positive or negative:

$$\nu_1 = \nu'; \quad S = 1 \tag{5.124}$$

$$\nu_1 = -\nu'; \quad S = 1 + \frac{2p_1}{K_2 N_r} \tag{5.125}$$

So synchronous parasitic torques occur at zero speed (S = 1) or close to zero speed (Equation 5.125). They appear to us as oscillatory at respective slips (Equations 5.124 and 5.125) on the T_e (S) curve (Figure 5.34c); and, if at standstill (S = 1), they may lead to unsafe starting.

By choosing right combinations between the stator and the rotor slot numbers, N_s, N_r, and pole pairs p_1, the main synchronous parasitic torques may be suppressed. In general,

$$\begin{aligned} &N_s \neq N_r; \quad 2N_s \neq N_r, N_r \pm 2p_1 \\ &N_r \neq \frac{N_s}{2} \pm p_1 \end{aligned} \tag{5.126}$$

For more on parasitic torques in IMs, see Ref. [4, Chapter 10] and Ref. [7].

5.18 STARTING METHODS

The starting methods refer to the machine connected to the power grid without a frequency changer (PWM converter). It is, in fact, a transient process, both in terms of electrical variables (flux linkages, currents) and mechanical variables (torque, speed).

These transients will be investigated in the companion book. Here, the main starting methods and their current and torque versus speed characteristics are given.

For the cage rotor IM, these are

- Direct starting to three-phase a.c. power grid
- Reduced stator voltage (soft starter, by star/delta connection switch or autotransformer)

For the wound rotor IM

- Controlled rotor additional resistance

5.18.1 Direct Starting (Cage Rotor)

For simplification, let us consider the IM with a large inertia at shaft. Thus, the starting will be slow, and the machine will experience only mechanical transients. Typical steady-state characteristics versus speed are shown in Figure 5.35.

What interests us most, for frequent starts (say, for compressor loads), is the energy for a start from zero speed to almost ideal no-load speed $n_1 = f_1/p_1$, under no load at shaft. However, the investigation may be solved numerically under load if $T_{load}(\omega_r)$ is known:

$$\frac{J}{p_1}\frac{d\omega_r}{dt} = T_e(\omega_r) - T_{load}; \quad \omega_r = \omega_1(1-S) \tag{5.127}$$

We may define a so-called electromechanical time constant:

$$\tau_{em} = \frac{\omega_1}{T_{ek}}\frac{J}{p_1} \tag{5.128}$$

τ_{em} is a few tens of milliseconds for small machines and seconds for large machines.

Now, for no load $T_{load} = 0$, and thus Equation 5.127 may be integrated from zero to ideal starting time, t_p, that corresponds roughly to synchronous speed n_1, to obtain rotor winding losses:

$$W_{cor} = \int_0^{t_p} P_{em} S\, dt = \int_0^{t_p} \frac{J}{p_1}\frac{d\omega_r}{dt}\frac{\omega_1}{p_1} S\, dt = -\frac{J}{p_1}\int_0^1 \omega_1^2 S\, dS = \frac{J}{2}\left(\frac{\omega_1}{p_1}\right)^2 \tag{5.129}$$

So the rotor energy winding losses for direct starting are equal to the kinetic energy of the rotor. Now, the stator copper losses W_{cos} is

$$W_{cos} \approx W_{cor}\frac{R_s}{R_r'} \tag{5.130}$$

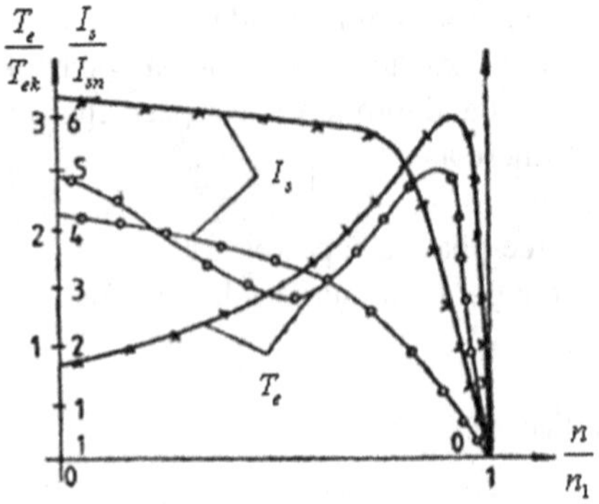

FIGURE 5.35 Torque and current vs. speed n/n_1. x—single- cage rotor; 0—dual-cage rotor.

The total winding energy losses for no-load starting is

$$W_{co} = W_{cos} + W_{cor} \approx \frac{J}{2}\left(\frac{\omega_1}{p_1}\right)^2 \left(1 + \frac{R_s}{R_r'}\right) \tag{5.131}$$

Larger rotor resistance (deep-bar-cage) rotors lead apparently to lower energy losses for frequent starts, as expected.

5.18.2 Reduced Stator Voltage

The starting current (Equation 5.92) at zero speed is (5–8) I_n, where I_n is the rated current. In many situations, the weak power grid or the application requires reduced starting current for light (low load at low speed) starts.

For such cases, there are three main devices: the soft (thyristor) starter (Figure 5.36a), the star–delta switch (Figure 5.36b), and the autotransformer (Figure 5.36c). Soft starters may reduce the starting current to (2.5–3) I_n and, by special control, provide additional torque below 33% of rated speed during starting; they are commercial up to about 1.5 MVA/unit.

The star–delta switch relies on phase voltage increase by $\sqrt{3}$ times after a settled time after motor start initialization. Unfortunately, the machine is designed for delta connection, so star connection means voltage and current reduction by $\sqrt{3}$ and thus a torque reduction by three times occurs, because the stator current is proportional to the phase voltage and the torque with the stator voltage squared:

$$V_Y^{ph} = \frac{V_\Delta^{ph}}{\sqrt{3}};\quad I_Y^{ph} = \frac{I_\Delta^{ph}}{\sqrt{3}};\quad T_{eY} = \frac{T_{e\Delta}}{3} \tag{5.132}$$

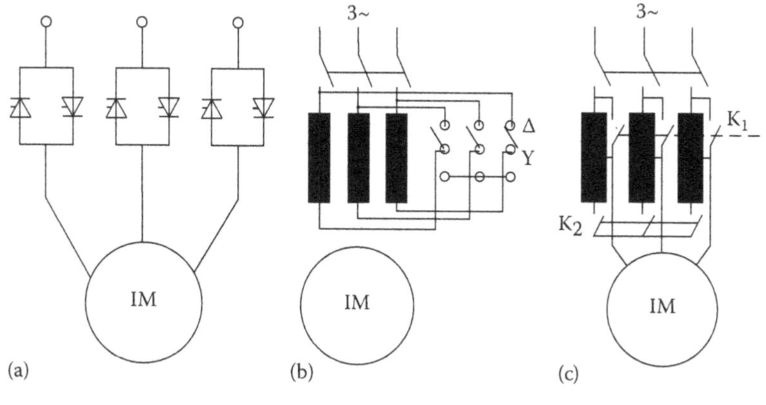

FIGURE 5.36 Reduced voltage starting: (a) with soft starter, (b) with star–delta switch, and (c) with autotransformer.

This explains why only light starts are feasible this way. A similar situation is obtained with the autotransformer that produces, with K_2 closed and K_1 open, low voltage and then, with K_2 open and K_1 closed, full voltage.

5.18.3 Additional Rotor Resistance Starting

Wound rotor IMs are applied at large power loads (above 100 kW) when limited (10%–20%) speed control range (interval) is required.

To cut costs, a diode-rectifier variable resistance is used (Figure 5.37). The static switch K_1 (or the d.c.–d.c. converter) controls the on–off process in the resistance current, eventually with a stator current close-loop regulator for limiting the stator current at the desired value.

An additional speed regulator may be used to produce the reference current and thus limited speed range close-loop control with the same device is obtained (Figure 5.37).

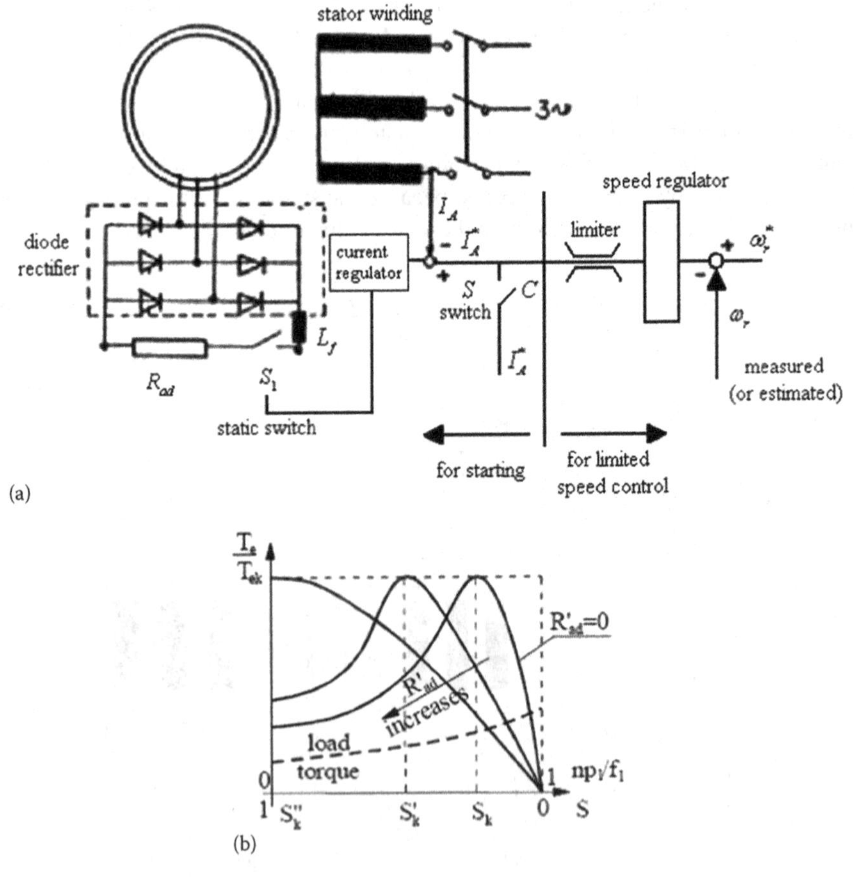

FIGURE 5.37 Additional rotor resistance starting of wound rotor IMs: (a) wound rotor IM starting with additional controlled resistance and (b) torque–speed characteristics (for limited speed control).

As the breakdown torque—independent of rotor total resistance (Figure 5.37)—moves toward zero speed, the method is characterized by large torque at start, and thus heavy starts are approachable, provided energy consumption is not a problem in the application due to the small number of starts (only a few per hour).

5.19 SPEED CONTROL METHODS

For now, we are concerned with cage rotor IMs for which the speed n is

$$n = \frac{f_1}{p_1}(1 - S) \tag{5.133}$$

The speed control methods are then related to the terms in Equation 5.133:

- Changing the slip S for a given torque: by reducing voltage for limited slip (speed) control range (10%–20%) (see Section 5.18) (Figure 5.38a)
- Pole count (pair) p_1 changing: by using two different stator windings or a pole changing winding (with $p_1/p_1' = 1/2$ in general) for constant power or constant torque applications (Figure 5.38b)
- Stator frequency, f_1, and voltage, V_s, control (Figure 5.38c): to control the flux linkage level within a reasonable (or desired) magnetic saturation of motor cores

The characteristics in Figure 5.38 are drawn by analysis of torque expression in Section 5.15. While reduced voltage speed control is applicable only to light starts and provides for limited speed control, pole changing (2:1) and frequency (and voltage) control provide for heavy starts and for wide speed control range (up to 1000/1), and good energy conversion. Frequency and voltage coordinated control with tight flux control—called vector control or direct torque and flux control—to produce fast torque response, with PWM converter supplies for IMs, are now a worldwide industry and will be treated in the companion book as a mere brief introduction in *Electric Drives*, a separate topic in itself.

5.19.1 Wound Rotor IM Speed Control

Wound rotor IM speed control may be accomplished either through a diode rectifier and d.c.–d.c. converter and a fixed resistor in the rotor, for starting and limited speed control range (10%–15%), to limit the total energy loss, or variable frequency ω_2 and voltage V_r' in the rotor via a bidirectional two-stage PWM converter (Figure 5.39). The rotor flux linkage in the machine is controlled together with frequency $\omega_2 = \omega_1 - \omega_r$ (ω_1 = constant, ω_r = variable). As already shown, this is a limited speed range control method (± 30%, $|S_{max}| < 0.3$), for which the rotor-connected converter rating is about $|S_{max}|P_n$ (P_n—stator rated power). At maximum (oversynchronous) speed, the total machine power is $P_n + |S_{max}|P_n$ for rated torque. This method is used for variable speed wind generators up to 5 MW, or by hydrogenerator systems (including pump storage up to 400 MW) to cut the equipment costs but retain good performance.

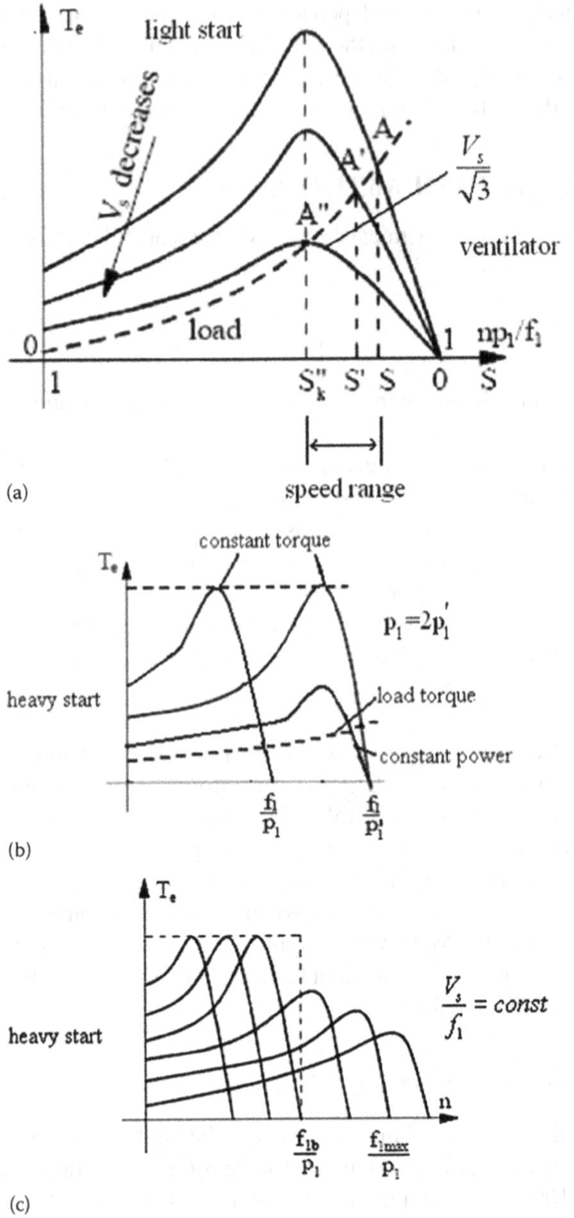

FIGURE 5.38 Speed control methods: (a) with soft starter, (b) with changing pole, and (c) variable frequency and voltage.

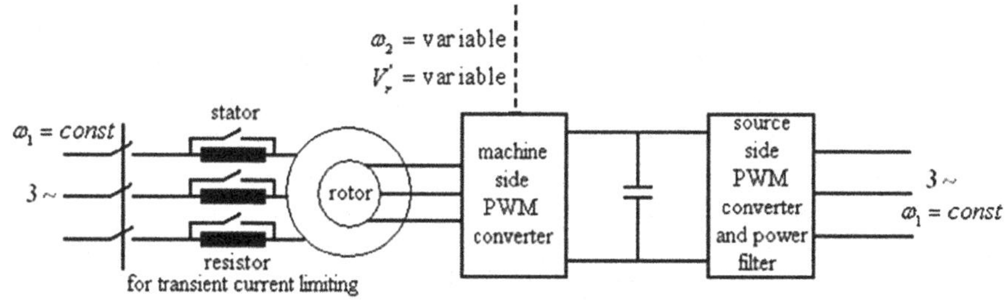

FIGURE 5.39 Variable speed wound rotor IM with variable frequency and voltage in the rotor for under and supersynchronous motor–generator operation modes.

More on the doubly fed induction generator (DFIG) can be found in Ref. [5, vol. 2, 3rd ed, Chapter 3].

Example 5.6: V/f Speed Control

An induction motor with cage rotor has the following design data: $P_n = 5.5$ kW, $V_{nl} = 440V$ (stator), $f_{1b} = 60$ Hz, $\eta_n = 0.92$, $\cos \varphi_n = 0.9$, $2p_1 = 4$, $I_{start}/I_n = 6/1$, $I_{0n}/I_n = 0.33$, $p_{iron} = p_{mec} = p_s = 0.015P_n$, $R_s = 1.2R_r'$, and $X_{sl} = X_{rl}'$, and skin effect is negligible.

Calculate rated stator current, rated rotor current, R_s, R_r', X_{1m}, X_{sc} and critical slip S_k, and breakdown torque at base frequency f_{1b} and full voltage and at $f_{1max} = 2 \cdot f_{1b} = 120$ Hz. To preserve breakdown torque at start and at all frequencies, at $f_{min} = S_k f_{1b}$, determine the required f_{min} and stator voltage and V/f dependence $V_s = V_0 + k{\cdot}f$.

Solution:

The problem in its first part is similar to the previous numerical example (Example 5.5). Efficiency is

$$\eta_n = \frac{P_n}{\sqrt{3}V_{nl}I_n \cos \varphi_n} = \frac{5500}{\sqrt{3} \times 440 \times I_n \times 0.9} = 0.92$$

The rated current, I_n, is

$$I_n = 8.726 \text{ A}$$

The rotor-rated current, I_{rn}', is

$$I_{rn}' = \sqrt{I_n^2 - I_{0n}^2} = 8.726\sqrt{1 - 0.33^2} = 8.226 \text{ A}$$

The copper stator and rotor losses, $p_{Cos} + p_{Cor}$, are

$$p_{cos} + p_{Cor} = \frac{P_n}{\eta_n} - P_n - p_{mec} - p_{iron} - p_{ps} = 5500\left(\frac{1}{0.92} - 1 - 3 \times 0.015\right)$$
$$= 230.76 \text{ W}$$

But also

$$p_{\cos} + p_{\text{Cor}} = 3I_n^2\left[1.2R_r' + R_r'\left(\frac{I'_{rn}}{I_n}\right)^2\right] = 3\times R_r'\times 159$$

So

$$R_r' = 0.4836\ \Omega;\quad R_s = 1.2\times 0.4836 = 0.588\ \Omega$$

The short-circuit reactance is calculated from the starting impedance:

$$X_{sc} = \sqrt{\left(\frac{V_{nl}}{6\sqrt{3}I_n}\right)^2 - \left(R_s + R_r'\right)^2} = \sqrt{\left(\frac{440}{\sqrt{3}\times 6\times 8.726}\right)^2 - (0.4836+0.588)^2}$$
$$= 4.72\ \Omega$$

The critical slip (Equation 5.108) is

$$\left(S_k\right)_{60\text{Hz}} \approx \frac{R_r'}{\sqrt{R_s^2 + X_{sc}^2}} = \frac{0.4836}{\sqrt{0.588^2 + 4.72^2}} = 0.1067$$

and the breakdown torque is

$$\left(T_{ek}\right)_{60\text{ Hz}} \approx \frac{3}{2}p_1\frac{\left(V_{nl}/\sqrt{3}\right)^2}{2\pi f_{1b}}\frac{1}{R_s + \sqrt{R_s^2 + X_{sc}^2}} = 100.345\ \text{Nm}$$

For $f_{1\max} = 120$ Hz,

$$\left(S_k\right)_{120\text{ Hz}} = \frac{0.4836}{\sqrt{0.588^2 + (4.72\times 2)^2}} = 0.0511!$$

Though the critical slip decreases twice, the corresponding frequency, $\omega_2 = Sf_1$ remains the same. The breakdown torque is now

$$\left(T_{ek}\right)_{120\text{ Hz}} = \frac{3}{2}\times 2\frac{\left(440/\sqrt{3}\right)^2}{2\pi 120}\frac{1}{0.588 + \sqrt{0.588^2 + (4.72\times 2)^2}} = 25.55\ \text{Nm}$$

The critical rotor frequency

$$f_{2k} = f_{1b}\left(S_k\right)_{f_{1b}} = 60\times 0.1067 = 6.402\ \text{Hz}$$

Now, the machine has to develop a breakdown torque of 100.345 Nm at zero speed ($S = 1$) and $f_1' = f_2 = 6.402$ Hz, and we need to calculate the required stator voltage to do so

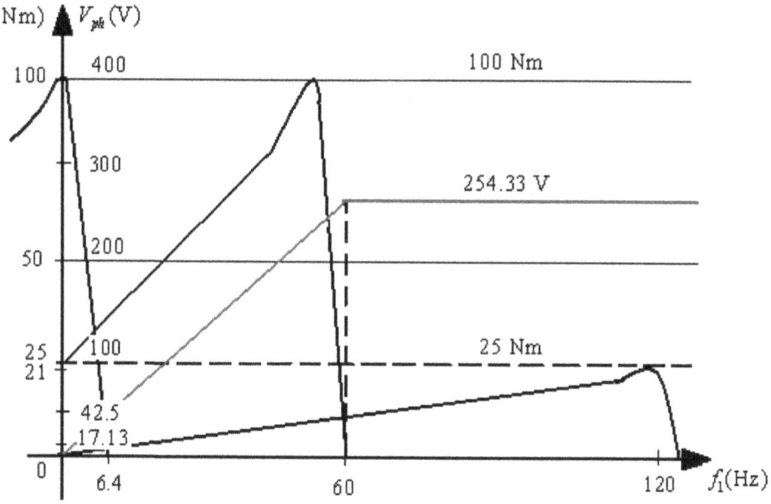

FIGURE 5.40 $V_s = V_0 + kf_1$ and torque–speed curves for V/f control for constant breakdown torque from 60 Hz to zero speed (6.402 Hz).

$$(T_{ek}) = 100.345 = \frac{3}{2} p_1 \frac{(V'_{ph})^2}{2\pi f'_1} \frac{1}{R_s + \sqrt{R_s^2 + (X_{sc} S_k)^2}}$$

$$V'_{ph} = 42.448 \text{ V}$$

To find an appropriate formula for $V_s(f_1) = V_0 + kf$, we notice that at 60 Hz $V_s = V_{nl}/\sqrt{3} = 440/\sqrt{3} = 254.33$ V. So

$$42.448 = V_0 + k \times 6.402$$
$$254.33 = V_0 + k \times 60$$

Consequently

$$k = \frac{254.33 - 42.448}{60 - 6.402} = 3.9545$$

and

$$V_0 = 17.13 \text{ V}$$

The torque–speed curves for 120, 60, and 6.402 Hz at 42.448 V and 254.33 V, and (phase voltage RMS), respectively, are shown in Figure 5.40.

5.20 UNBALANCED SUPPLY VOLTAGES

In real, local, power grids, the three-phase line voltages are not purely balanced (equal amplitudes, 120° phase shifts). For a three-phase power supply, we may decompose them in forward (+) and backward (−) components

$$
\begin{aligned}
&\underline{V}_{a+} = \frac{1}{3}\left(\underline{V}_a + a\underline{V}_b + a^2\underline{V}_c\right); \quad a = e^{j\frac{2\pi}{3}} \\
&\underline{V}_{a-} = \frac{1}{3}\left(\underline{V}_a + a^2\underline{V}_b + a\underline{V}_c\right) \\
&\underline{V}_{b+} = a^2\underline{V}_{a+}; \quad \underline{V}_{c+} = a\underline{V}_{a+} \\
&\underline{V}_{b-} = a\underline{V}_{a-}; \quad \underline{V}_{c-} = a^2\underline{V}_{a-}
\end{aligned}
\tag{5.134}
$$

Now, the slip for the direct (⊕) component is $S_+ = S$, but for the reverse (⊖) component, $S-$ is

$$
S_- = \frac{-\frac{f_1}{p_1} - n}{-\frac{f_1}{p_1}} = 2 - S \tag{5.135}
$$

So, if we neglect the skin effect (the rotor-slip frequency is large for the ⊖ component: $f_{2-} = S_- f_1$) and the magnetic saturation (which may be considered constant or only dependent on the ⊕ component), we do have two different fictitious machines with their $\underline{V}_+$ and $\underline{V}_-$ voltages and S and $2 - S$ slip values.

The torque is composed of two components, and making use of torque/electromagnetic power (copper loss, slip) definitions, it becomes

$$
T_e = T_{e+} + T_{e-} = \frac{3R_r'\left(I_{r+}'\right)^2}{S}\frac{p_1}{\omega_1} + \frac{3R_r'\left(I_{r-}'\right)^2}{2-S}\frac{p_1}{-\omega_1} \tag{5.136}
$$

The phase voltage unbalance index (in percent) can be defined as

$$
V_{\text{unbalance}} = \frac{\Delta V_{\max}}{V_{\text{av}}}; \quad \Delta V_{\max} = V_{\max} - V_{\min}; \quad V_{\text{av}} = \frac{V_a + V_b + V_c}{3} \tag{5.137}
$$

where $V_{\max}$ and $V_{\min}$ are the maximum and minimum of the three voltages, respectively, or

$$
V_{\text{unbalance}}(\%) = \frac{V_{a-}}{V_{a+}}100 \tag{5.138}
$$

The efficiency is

$$
\eta = \frac{T_e\omega_1(1-S)}{3p_1 \operatorname{Re}\left[\underline{V}_{a+}\underline{I}_{a+}^* + \underline{V}_{a-}\underline{I}_{a-}^*\right]} \tag{5.139}
$$

A small voltage unbalance leads to a notable phase current unbalance. The ⊖ component torque is in general negative and small but it contributes to more losses and lower efficiency, for a given torque.

NEMA standards recommend an IM derating with a voltage unbalance as in Equation 5.137 to 97.5% at $\Delta V_{\text{unbalance}} = 2\%$ and to 75% at $\Delta V_{\text{unbalance}} = 5\%$ voltage!

5.21 ONE STATOR PHASE OPEN BY EXAMPLE/LAB 5.7

Let us consider the case when phase A is open (Figure 5.41): when $I_A = 0$; $I_B + I_C = 0$ (star connection).

For $V_{nl} = 220\text{V}, f_1 = 60$ Hz, $2p_1 = 4$, $R_s = R_r' = 1\Omega$, $X_{sl} = X_{r1}' = 2.5\ \Omega$, $X_{1m} = 75\ \Omega$, $R_{1m} = \infty$ (no core losses), and $S = 0.03$. Calculate stator current, torque, and power factor before and after opening phase A.

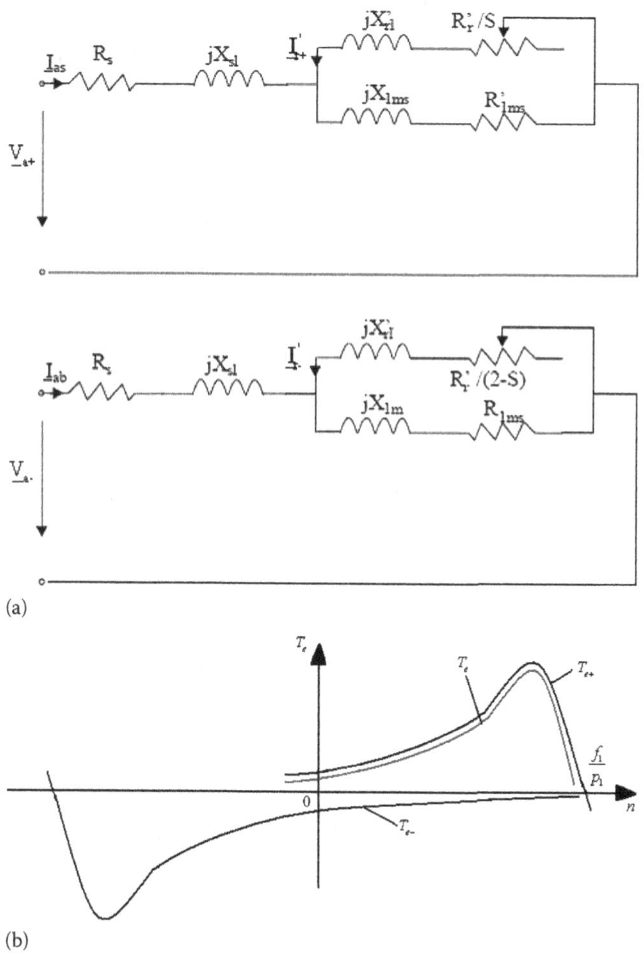

FIGURE 5.41 One stator phase is open (a) equivalent circuit and (b) torque–speed curve.

Solution:

First, the ⊕ and ⊖ stator currents are calculated

$$\underline{I}_{A+} = \frac{1}{3}\left(\underline{I}_A + a\underline{I}_B + a^2\underline{I}_C\right) = \underline{I}_B \frac{a - a^2}{3} = j\frac{\underline{I}_B}{\sqrt{3}}$$
$$\underline{I}_{A-} = \frac{1}{3}\left(\underline{I}_A + a^2\underline{I}_B + a\underline{I}_C\right) = -\underline{I}_{A+} \tag{5.140}$$

So, in fact, the two equivalent impedances, Z_+ (S) and $Z_-(2 - S)$, are related to their phase voltage components (Figure 5.42a)

$$\underline{V}_{A+} = \underline{Z}_+\underline{I}_{A+}; \quad \underline{V}_{A-} = \underline{Z}_-\underline{I}_{A-} \tag{5.141}$$

with $V_{B+} = a^2V_{A+}$ and $V_{B-} = a\underline{V}_{A-}$.

The line voltage, $V_B - V_C$, which is known, has the expression

$$\underline{V}_B - \underline{V}_C = \underline{V}_{B+} + \underline{V}_{B-} - \underline{V}_{C+} - \underline{V}_{C-} = \left(a^2 - a\right)\underline{I}_{A+}\left(\underline{Z}_+ + \underline{Z}_-\right)$$
$$\underline{V}_B - \underline{V}_C = \underline{I}_B\left(\underline{Z}_+ + \underline{Z}_-\right) \tag{5.142}$$

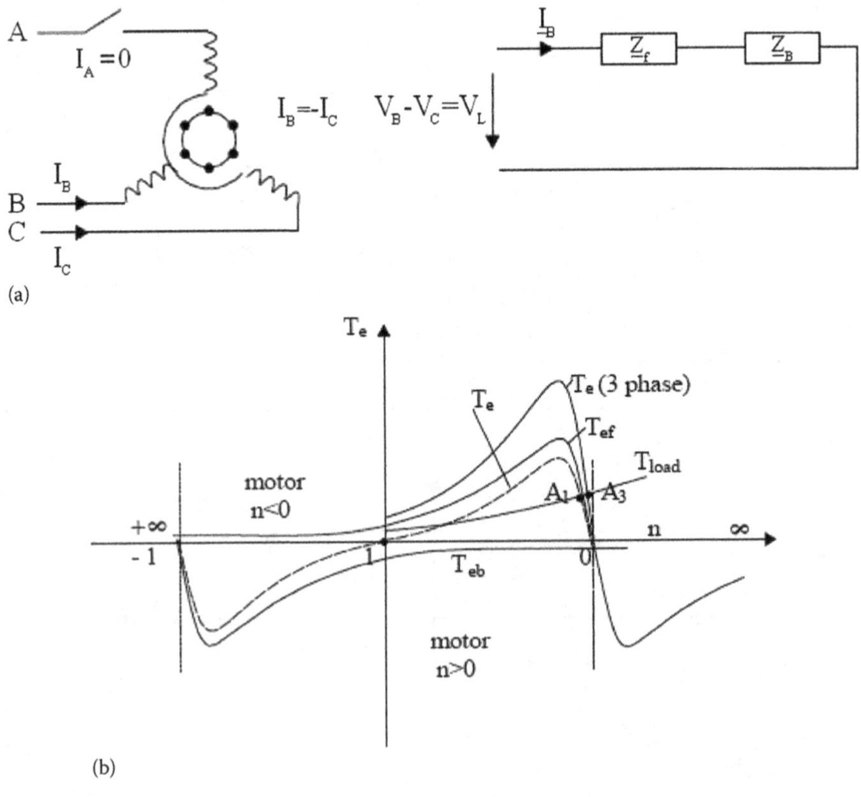

FIGURE 5.42 Unbalanced voltage IM ⊕ and ⊖ component equivalent circuits of IM.

Equation 5.136 is reflected in Figure 5.41a.

For zero speed ($S = 1$): $(Z_+)_{S=1} = (Z_-)_{S=1} = Z_{sc}$ and thus

$$\underline{I}_{sc1} = \left(\underline{I}_B\right)_{S=1} = \frac{\underline{V}_{nl}}{2\underline{Z}_{sc}} = \frac{\sqrt{3}}{2}\underline{I}_{sc3} \tag{5.143}$$

Consequently, at standstill ($S = 1$), because $Z_+ = Z_-$, the direct and inverse torque components are equal, so the resultant torque is zero, and the phase B (C) current is a bit smaller than for the three-phase operation.

From Figure 5.42 and Example 5.5, before opening phase A

$$\underline{V}_A = \underline{Z}_+(S)\underline{I}_A; \quad c_1 \approx \frac{X_{sl} + X_{1m}}{X_{1m}} = \frac{75 + 2.5}{75} = 1.033$$

$$I'_{rn} \approx \frac{V_{nl}/\sqrt{3}}{\sqrt{\left(R_s + c_1\dfrac{R'_r}{S}\right)^2 + \left(X_{sl} + c_1 X^{\#}_{lr}\right)^2}} = 3.55\text{ A}$$

The no load current is

$$I_{0n} \approx \frac{V_{nl}/\sqrt{3}}{\left|R_s + j\left(X_{1m} + X_{sl}\right)\right|} = \frac{220}{\sqrt{3}\sqrt{1^2 + 77.5^2}} = 1.64\text{ A}$$

So the stator current, I_{sn}, is

$$I_{sn} \approx \sqrt{I'^2_{rn} + I^2_{0n}} = \sqrt{3.55^2 + 1.64^2} = 3.91\text{ A}$$

$$\cos\varphi_n \approx \frac{I'_{rn}}{I_n} = \frac{3.55}{3.91} = 0.9078$$

The torque for symmetric steady state at $S_n = 0.03$ is

$$T_{e3} = \frac{3R'_r I'^2_{rn}}{S_n}\frac{p_1}{\omega_1} = \frac{3 \times 1 \times 3.55^2}{0.03}\frac{2}{2\pi \times 60} = 6.69\text{ Nm}$$

For the same slip $S = 0.03$ but with phase A open, based on Equation 5.142, we can calculate I_B and then I_{A+} and I_{A-}:

$$\underline{I}_B = \frac{\underline{V}_{BC}}{\underline{Z}_+ + \underline{Z}_-} = \frac{\underline{V}_{BC}}{\left[2\left(R_s + jX_{sl}\right) + \dfrac{R'_r}{2 - S} + jX'_{rl} + j\dfrac{X_{1m}R'_r/S}{\dfrac{R'_r}{S} + j\left(X_{1m} + X_{rl}\right)}\right]}$$

$$= \frac{220}{\left[2(1 + j2.5) + \dfrac{1}{2 - 0.03} + j2.5 + \dfrac{j75/0.03}{1/0.03 + j(75 + 2.5)}\right]} = 4.705 - j2.752$$

So

$$\cos\varphi_1 = \frac{4.706}{\sqrt{4.706^2 + 2.752^2}} = \frac{4.706}{5.45} = 0.8633$$

The current $I_B = 5.45$ A is notably larger than for three-phase symmetric (balanced) voltages ($I_n = 3.91$ A). Now, for the calculation of torque components, we need I'_{r+} and I'_{r-} (Figure 5.42)

$$I'_{r+} = I_{A+}\left|\frac{jX_{1m}}{j(X_{1m} + X'_{rl}) + R'_r/S}\right| = \frac{5.45}{\sqrt{3}}\left|\frac{j75}{j(75+2.5)+1/0.03}\right| = 2.8\text{ A}$$

$$I_{r-} = I_{A-} = \frac{I_B}{\sqrt{3}} = \frac{5.45}{\sqrt{3}} = 3.15\text{A}$$

The torque (Equation 5.136) is

$$T_e = T_{e+} + T_{e-} = \frac{3p_1}{\omega_1}R'_r\left[\frac{(I'_{r+})^2}{S} - \frac{(I'_{r-})^2}{2-S}\right] = \frac{3\times 2\times 1}{2\pi\times 60}\left(\frac{2.8^2}{0.03} - \frac{3.15^2}{2-0.03}\right)$$

$$= 4.16 - 0.08 = 4.08\text{ Nm}$$

So, the torque with one phase open is reduced for $S = 0.03$ from 6.69 Nm to 4.08 Nm. The power factor is a bit reduced from 0.9078 to 0.8633. The copper losses for three-phase and two-phase operations are

$$P_{Co3} = 3\left(R_s I_n^2 + R'_r I'^2_{rn}\right) = 3\left(1\times 3.91^2 + 1\times 3.55^2\right) = 83.67\text{ W}$$

$$P_{Co1} \approx V_{BC} I_B \cos\varphi_1 - T_e\frac{\omega_1}{p_1}$$

$$\approx 220\times 5.45\times 0.8633 - 4.08\times\frac{2\pi\times 60}{2}$$

$$\approx 1035.1 - 748.68 = 286.42\text{ W}$$

The copper losses are much higher for one phase open, and thus the motor gets overheated if it is not disconnected in due time.

This example illustrates implicitly the case of a single-phase winding induction motor which may not start as such but, if on speed, it may operate acceptably in terms of power factor and torque, albeit at higher copper losses.

5.22 ONE ROTOR PHASE OPEN

It is not seldom that some rotor bars or end rings break, and thus the rotor winding becomes asymmetric. An exact treatment of one or more broken bars needs bar-by-bar circuit simulation circuits (see Ref. [4, Chapter 13]), but for a three-phase wound rotor, the case of one phase open is easy to handle (Figure 5.43a and b).

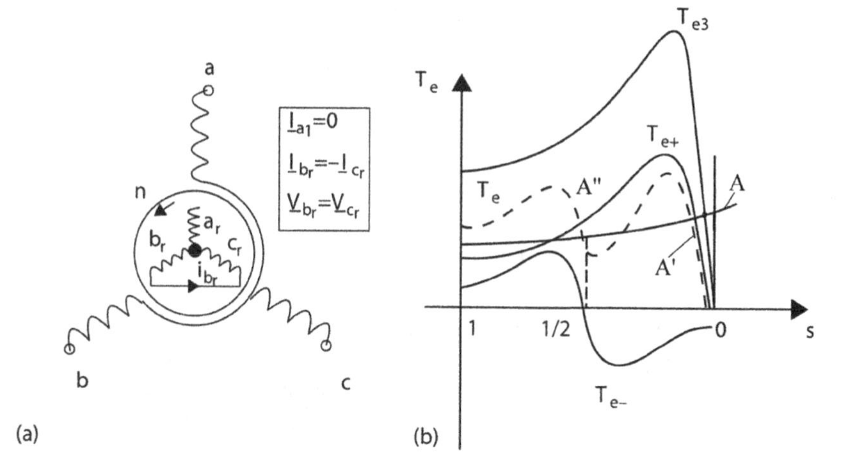

FIGURE 5.43 One rotor phase open: (a) one rotor phase open and (b) torque/speed components and operation points A', A'', and A'''.

In effect now, the rotor current, $I'_{br'}$, is decomposed into ⊕ and ⊖ components:

$$I'_{ar+} = -I'_{ar-} = +\frac{j}{\sqrt{3}} I'_{br}$$
$$\underline{V}'_{ar+} = \underline{V}'_{ar-} = \frac{1}{3}\left(V'_{ar} - V'_{br}\right) \tag{5.144}$$

because $\underline{I}'_{br} = -\underline{I}'_{cr}\ V'_{br} = V'_{cr}$ (Figure 5.43a).

For this situation, ⊕ and ⊖ equations "spring" from the rotor whose ⊖ mmf component leads to an n' speed with respect to the stator, which is considered short-circuited for the ⊖ component (infinite power source):

$$I'_{r+}R'_r - \underline{V}'_{r+} = -jS\omega_1\underline{\Psi}'_{r+};\qquad \underline{\Psi}'_{r+} = L'_r\underline{I}'_{r+} + L_{1m}\underline{I}_{s+}$$
$$I_{s+}R_s - \underline{V}_s = -j\omega_1\underline{\Psi}_{s+};\qquad \underline{\Psi}_{s+} = L_s\underline{I}_{s+} + L_{1m}\underline{I}_{r+} \tag{5.145}$$

$$n' = n - S\frac{f_1}{p_1} = \frac{f_1}{p_1}\left(1-2S\right) = \frac{f'_1}{p_1};\quad f'_1 = f_1\left(1-2S\right) \tag{5.146}$$

and

$$I'_{r-}R'_r - \underline{V}'_{r-} = -jS\omega_1\underline{\Psi}'_{r-};\qquad \underline{\Psi}'_{r-} = L'_r\underline{I}'_{r-} + L_{1m}\underline{I}_{s-}$$
$$I_{s-}R_s = -j\left(1-2S\right)\omega_1\underline{\Psi}_{s-};\qquad \underline{\Psi}_{s-} = L_s\underline{I}_{s-} + L_{1m}\underline{I}'_{r-} \tag{5.147}$$

The unknowns of Equations 5.145 through 5.147 are $I'_{r+} = -I'_{r-}$, $V'_{r+} = V'_{r-}$, I'_{s+}, I'_{s-} for given slip and motor parameters. The relationship between rotor and stator ⊖ components (from Equation 5.147) is

$$\underline{I}_{s-} = -\frac{j\omega_1(1-2S)L_{1m}\underline{I}'_{r-}}{R_s + j\omega_1(1-2S)L_s} \tag{5.148}$$

$$L_s = L_{1m} + L_{sl}; L'_r = L'_{r1} + L_{1m}.$$

The torque is composed of two components:

$$T_e = 3p_1 L_{1m}\text{Imag}\left[\underline{I}_{s+}\underline{I}'^*_{r+} - \underline{I}_{s-}\underline{I}'^*_{r-}\right] \tag{5.149}$$

It is clear from Equation 5.148 that the stator ⊖ current $I_{s-} = 0$ (its synchronism) for $S = 1/2$. The ⊖ torque component is positive for $S > 1/2$ and negative for $S < 1/2$, and it is zero at synchronism ($S = 0$). The ⊖ torque component is also called single-phase monoaxial (Georges') torque (Figure 5.43b). If the load torque at low speeds is large, the machine may accelerate only to point A'' (around 50% rated speed) due to one rotor phase open—instead of accelerating to A'. Limit A corresponds to the symmetric rotor operation.

5.23 CAPACITOR SPLIT-PHASE INDUCTION MOTORS/LAB 5.8

Induction motors connected directly to a single-phase a.c. power grid are called split-phase motors. Such a motor has a main winding always on, during starting and operation, and an auxiliary winding (displaced by 90° (electrical), in general) for starting and also for running, sometimes.

The auxiliary phase has a series-connected resistance or a capacitor (C_{start}) or two (a running capacitor $C_n < C_{start}$), to produce at a desired slip ($S = 1$ and $S = S_n$) an almost traveling resultant mmf (Figure 5.44a and b). The two windings have different numbers of turns, $W_m \neq W_a$ ($a = W_a/W_m$), for unidirectional motion and $W_m = W_a$ for bidirectional motion, when the capacitor is switched from one phase to the other.

It may be demonstrated that if the amplitudes of the main and the auxiliary windings mmfs, F_{1m} and F_{1a}, are not equal to each other or their time phase angle of currents I_m and I_a is not 90°, the total mmf may be decomposed into a positive (+) mmf, F_{1+}, and an inverse (−) mmf, F_{1-}.

For steady state, we can use the symmetrical components theory (Figure 5.45a):

$$\underline{F}_m = \underline{F}_{m+} + \underline{F}_{m-}; \qquad \underline{F}_a = \underline{F}_{a+} + \underline{F}_{a-}$$
$$\underline{A}_{m+} = \frac{1}{2}(\underline{A}_m - j\underline{A}_a); \qquad \underline{A}_{m-} = \underline{A}^*_{m+} \tag{5.150}$$

The machine acts as two separate machines for the two components:

$$\underline{V}_{m+} = \underline{Z}_{m+}\underline{I}_{m+}; \qquad \underline{V}_{m-} = \underline{Z}_{m-}\underline{I}_{m-}$$
$$\underline{V}_{a+} = \underline{Z}_{a+}\underline{I}_{a+}; \qquad \underline{V}_{a-} = \underline{Z}_{a-}\underline{I}_{a-} \tag{5.151}$$
$$\underline{V}_m = \underline{V}_{m+} + \underline{V}_{m-}; \qquad \underline{V}_a = \underline{V}_{a+} + \underline{V}_{a-}$$

The ⊕ and ⊖ impedances of the machine correspond to S and $2 - S$ slip values (as shown in Figure 5.42). $Z_{m\pm}$ and $Z_{a\pm}$ have the rotor reduced to the main and auxiliary windings, respectively (Figure 5.45b).

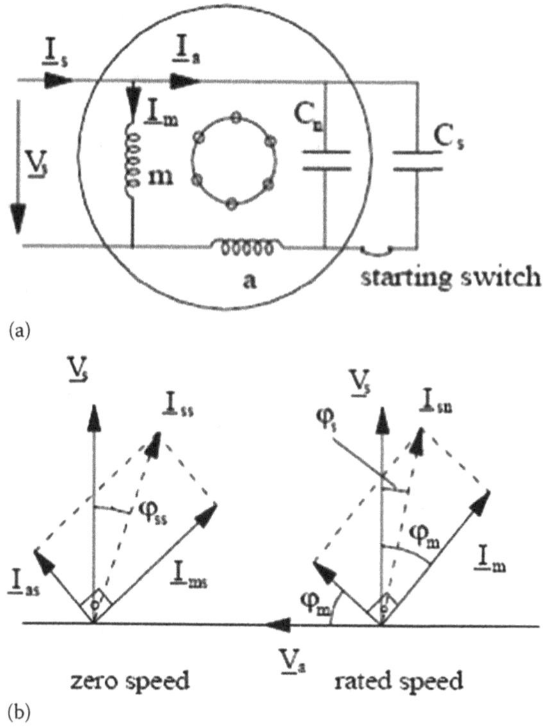

FIGURE 5.44 The dual capacitor IM: (a) equivalent scheme and (b) phasor diagram for zero and rated speeds.

The relationship between V_m and V_a voltages and the source voltage V_s is

$$\underline{V}_{sn} = \underline{V}_s; \underline{V}_a = \underline{V}_s - \left(\underline{I}_{a+} + \underline{I}_{a-}\right)\underline{Z}_a \tag{5.152}$$

where Z_a is the auxiliary resistance or capacitance in series with the auxiliary winding.

Solving Equations 5.151 and 5.152 with Z_{m+} and Z_{a+} from Figure 5.45b, the machine currents I_{a+}, I_{m+}, and I_{rm+} can be found (see Figure 5.45a).

So the torque T_e is

$$T_e = T_{e+} + T_{e-} = \frac{2p_1}{\omega_1} R_{rm} \left[\frac{I_{rm+}^2}{S} - \frac{I_{rm-}^2}{2-S}\right] \tag{5.153}$$

When $Z_a = \infty$ (the auxiliary phase is open), $I_{rm+} = I_{rm-}$ and thus at zero speed ($S = 1$), the torque is zero (as demonstrated in Section 5.21).

A typical T_e (ω_r) characteristic is shown in Figure 5.45c for a split-phase capacitor induction motor. The positive sequence torque, T_{e+}, has the synchronism at ($+\omega_1/p_1$) while the negative sequence has it at ($-\omega_1/p_1$). The negative sequence torque, T_{e-} is not large, but, due to its large slip ($2 - S$), it produces large rotor winding losses.

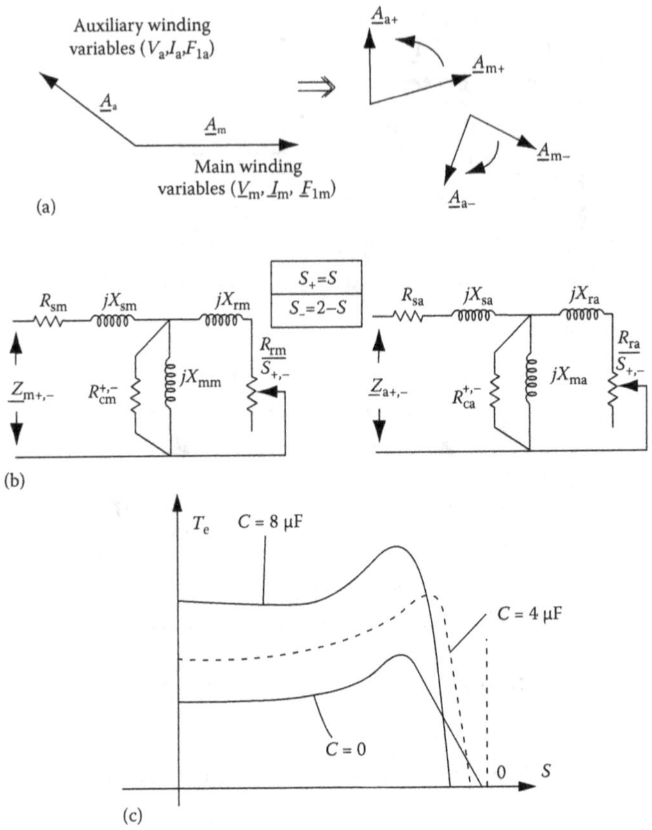

FIGURE 5.45 The symmetrical components model: (a) the +/− model components, (b) equivalent impedances Z_{m+} and Z_{a+}, and (c) typical T_e (ω_r) curve.

Full symmetrization of the two windings is thus required for reducing winding losses:

$$\underline{I}_{rm-} = 0 \tag{5.154}$$

This leads to the required turns ratio a and the required capacitor C for a given slip. Now, a, once chosen, say for symmetrization at $S = 1$, may not be changed for symmetrization at a rated slip S_n, but a smaller capacitor, C_n, should be chosen.

For the case of the main and the auxiliary windings using the same quantity of copper ($R_{sm} = R_{sa}/a^2$, $X_{sm} = X_{sa}/a^2$), the symmetrization conditions are (Ref. [4, Chapter 24])

$$\begin{aligned} a &= X_{m+}/R_{m+} = \tan\varphi_+ \\ X_C &= 1/\omega_1 C = Z_+ a\sqrt{a^2+1} \end{aligned} \tag{5.155}$$

The $T_e(\omega_r)$ for two different capacitors (and for open auxiliary phase: $C = 0$) are shown in Figure 5.45 only to illustrate that a very large capacitor is not adequate for running conditions. Designing split-phase permanent capacitor motors is thus a hard-to-bargain compromise between good starting and running performance, even with two capacitors $C_{start} \gg C_{run}$.

As split-phase capacitor motors with efficiency above 85% at 100 W are used predominantly in home appliances such as refrigerator compressors, a numerical example is solved here to grasp a feeling of magnitudes.

Example 5.7: Capacitor-Run Split-Phase IM

Let us consider a capacitor run (C = 4 μF), split-phase IM fed at 230 V and 50 Hz with n_n = 940 rpm ($2p_1$ = 6). The main and auxiliary winding parameters are a = 1.73, R_{sm} = 34 Ω, X_{sm} = 35.9 Ω, $R_{sa} = R_{sm}/a^2$, $X_{sa} = X_{sm}/a^2$, X_{rm} = 29.32 Ω, R_{rm} = 23.25 Ω, and X_{mm} = 249 Ω.

Calculate the supply current, torque, power factor, input power, and efficiency at S = 0.06 (core and mechanical losses are neglected).

Solution:

From Equations 5.151 and 5.152, we may derive [4, pp. 84]

$$\underline{I}_+ = \frac{\underline{V}_s}{2}\frac{(1-j/a)\left(\underline{Z}_- + 2\underline{Z}_a^m\right)}{\underline{Z}_+\underline{Z}_- + \underline{Z}_a^m\left(\underline{Z}_+ + \underline{Z}_-\right)}$$
$$\underline{I}_- = \frac{\underline{V}_s}{2}\frac{(1+j/a)\left(\underline{Z}_+ + 2\underline{Z}_a^m\right)}{\underline{Z}_+\underline{Z}_- + \underline{Z}_a^m\left(\underline{Z}_+ + \underline{Z}_-\right)} \quad (5.156)$$

$$T_e = \frac{2p_1}{\omega_1}\left[I_{m+}^2\,\mathrm{Re}\left(\underline{Z}_+\right) - I_{m-}^2\,\mathrm{Re}\left(\underline{Z}_-\right) - R_{sm}\left(I_{m+}^2 + I_{m-}^2\right)\right] \quad (5.157)$$

$$\underline{Z}_a^m = -\frac{j}{2\omega C a^2} = -\frac{j}{2\pi 50\times 4\times 10^{-6}\times 1.73^2} = -j133.4\Omega$$

$$\left(\underline{Z}_+\right)_{S=0.06} = R_{sm} + jX_{sm} + \frac{jX_{mm}\left(R_{rm}/S + jX_{rm}\right)}{R_{rm}/S + j\left(X_{mm} + X_{rm}\right)} = 139.4 + j197.75$$

$$\left(\underline{Z}_-\right)_{S=0.06} = R_{sm} + jX_{sm} + \frac{jX_{mm}\left(R_{rm}/(2-S) + jX_{rm}\right)}{R_{rm}/(2-S) + j\left(X_{mm} + X_{rm}\right)} = 46.6 + j50.25$$

So from Equations 5.156 and 5.157

$$\underline{I}_{m+} = 0.525 - j0.794$$
$$\underline{I}_{m-} = 0.1016 + j0.0134$$
$$T_{e+} = \frac{2\times 3}{2\pi 50}0.9518^2\left(139.4 - 34\right) = 1.8245\ \mathrm{Nm}$$
$$T_{e-} = -\frac{2\times 3}{2\pi 50}0.1025^2\left(46.6 - 34\right) = -2.53\times 10^{-3}\ \mathrm{Nm}$$
$$T_e = T_{e+} + T_{e-} = 1.822\ \mathrm{Nm}$$

The main and the auxiliary winding currents, I_m and I_a, are (Equation 5.151)

$$\underline{I}_m = \underline{I}_{m+} + \underline{I}_{m-} = 0.525 - j0.794 + 0.1016 + j0.0134 \approx 0.62 - j0.78$$

$$\underline{I}_a = j\frac{\underline{I}_{m+} + \underline{I}_{m-}}{a} = 0.466 + j0.274$$

The total stator current, I_s, is

$$\underline{I}_s = \underline{I}_a + \underline{I}_m \approx 1.092 - j0.506$$

So the motor power factor, cos φ_s, is

$$\cos\varphi_s = \frac{\mathrm{Re}(\underline{I}_s)}{I_s} = \frac{1.092}{1.204} \approx 0.90!!$$

The input active power, P_{1e}, is

$$P_{1e} = V_s \mathrm{Re}(\underline{I}_s) = 230 \times 1.092 = 251.16\ \mathrm{W}$$

Now the mechanical power, P_{out}, is

$$P_{out} = T_e \frac{\omega_1}{p_1}(1 - S) = 1.822 \times \frac{2\pi 50(1 - 0.06)}{3} = 179.26\ \mathrm{W}$$

So the efficiency, $\eta_{S=0.06}$, is

$$\eta_{S=0.06} = \frac{P_{out}}{P_{1e}} = \frac{179.26}{251.16} = 0.7137$$

Notes:

- The reverse ⊖ torque component is small.
- The power factor is good (due to the capacitor's presence).
- The efficiency is not very high, partly due to $2p_1 = 6$ ($2p_1 = 2$ would lead to better efficiency, in general).
- The phase shift between I_m and I_a is about 30.45 – (–51.52) ≈ 82°, not far away from the ideal 90°.
- The ratio of *auxiliary* and *main* mmf amplitudes is $I_a W_a / I_m W_m = a I_a / I_m = 1.73 \times 0.54/0.996 = 0.9308$. The ratio of mmf amplitudes is not far away from unity. So the situation is not far away from the symmetrization conditions for $S = 0.06$.
- The capacitor split-phase stator may also be built with a cage PM (or reluctance) rotor as a line-start premium efficiency synchronous motor (to be dealt with in Chapter 6).

5.24 LINEAR INDUCTION MOTORS

By the imaginary process of unrolling a cage-rotor conventional three-phase induction motor and spreading the cage (now turned ladder), a short-primary long-secondary single-sided linear induction motor is obtained (Figure 5.46).

In applications such as urban and interurban people movers or in industrial short-distance transport, the secondary ladder secondary on ground may be replaced, for cost reasons, by an aluminum sheet over one to four pieces of solid back iron (Figure 5.46).

With the solid back iron on ground, there will be skin effect and eddy currents in it from the primary traveling field. So the solid iron contributes to thrust but also leads to an increase in the magnetization current and losses.

The mechanical airgap g = 1–15 mm, 1 mm for short distance (say, a clean room) transport and 8–15 mm for urban and interurban people movers.

The three-phase windings stem from those of rotary machines by cutting and unrolling to obtain

- Single layer $2p_1 = 2,4,6,\ldots$(even number) windings (Figure 5.47a)
- Double-layer full-pitch (or chorded-coil) windings with $2p_1 + 1$ poles, with half-wound end poles, to reduce the additional (Gauss law caused) back iron a.c. (fix) flux density to almost zero in the $2p_1 - 1$ central poles (Figure 5.47b).

5.24.1 End and Edge Effects in LIMs

When the number of poles is small, say $2p_1 = 2,4$, the single-layer winding (Figure 5.47a) appears more adequate as it allows full utilization of all primary cores, but phase B is in a different position with respect to phases A and C relative to the limited core length along the direction of motion.

A so-called **static end effect** occurs in such cases, which is characterized by unbalanced phase currents. A two-phase winding would avoid this inconveniency.

For $2p_1 > 4$, in general, $2p_1 + 1$ double-layer windings with half-filled end poles (Figure 5.47b) are to be used as the static end effect is thus diminished.

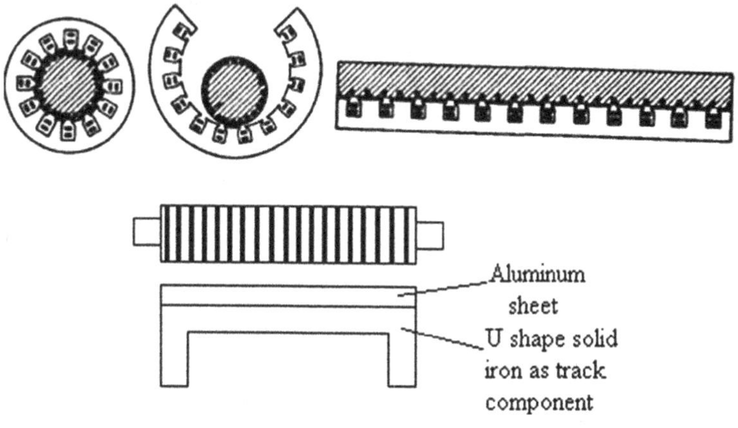

FIGURE 5.46 Obtaining a linear induction motor with flat geometry.

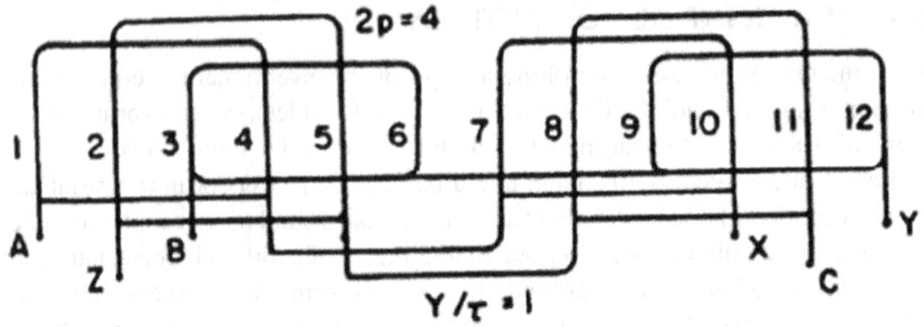

(a)

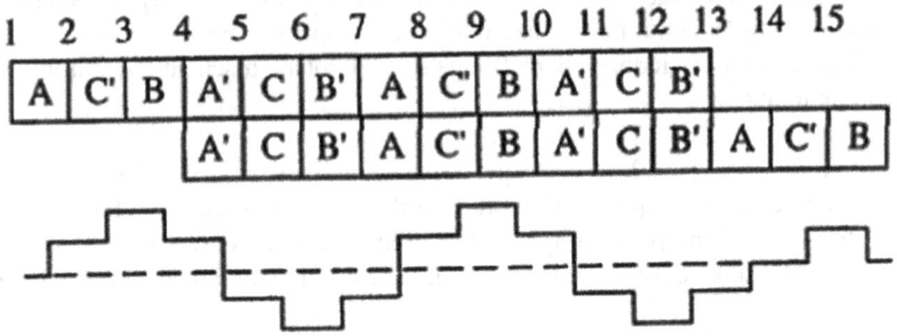

(b)

FIGURE 5.47 Linear induction motor windings: (a) single-layer, full-pitch windings (N_1 = 12 slots, $2p_1 = 4$, $q = 1$) and (b) double-layer, full-pitch windings (N_1 = 15 slots, $2p_1 + 1 = 5$, $q = 1$).

Because the aluminum sheet is a continuum, the second current density has a longitudinal component J_x (Figure 5.48) besides the useful, transverse, component J_y. On an overall basis, this effect produces an increase, K_T, in the equivalent resistance of the secondary, when reduced to the primary. This effect accounts also for the current density lines returning outside the active zone (as in end rings) and is called the transverse edge effect.

Approximately [8]

$$K_T = \frac{1}{1 - \lambda \dfrac{\tanh \frac{\pi}{\tau} a_e}{\frac{\pi}{\tau} a_e}} > 1; \quad a_e = a + g_e$$
$$\lambda = \frac{1}{1 + \tanh \frac{\pi}{\tau} a_e \tanh \frac{\pi}{\tau}(c - a_e)} \tag{5.158}$$

In general, the aluminum overhang is $c - a_e \le \tau/\pi$; τ is the pole pitch of the primary winding.

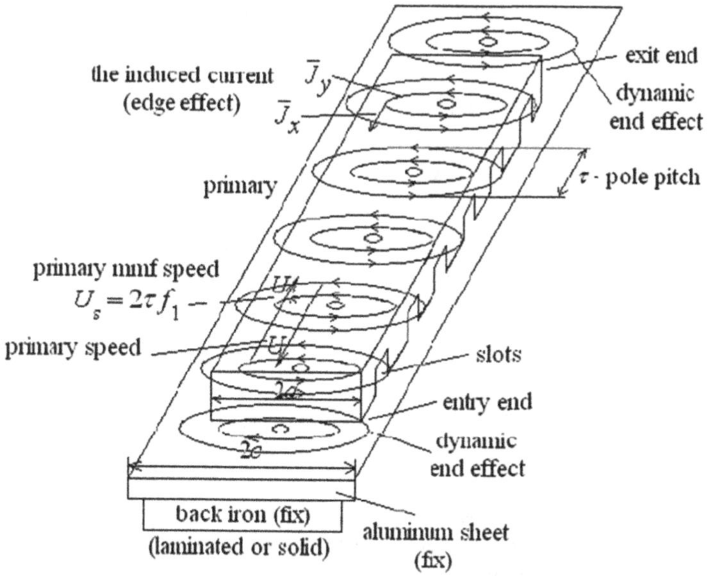

FIGURE 5.48 LIM with secondary current density paths reflecting both edge effect and longitudinal (dynamic) end effect.

The coefficient $K_T > 1$ may be lumped into the aluminum conductivity $\sigma_{Ale} = d_{Ale}\sigma_{Al}/K_T$ and then we may suppose that the secondary current density has only the transverse component J_y, which is thrust producing.

Solving the airgap flux density distribution, B_g, with pure traveling mmf and only axial x variation, and accounting for Gauss flux law [9], the Poisson equation yields

$$\frac{\partial B_g}{\partial x^2} - \frac{\mu_0 \sigma_{Ale} U}{g_e}\frac{\partial B_g}{\partial x} - \frac{\mu_0 \sigma_{Ale}}{g_e}\frac{\partial B_g}{\partial t} = \frac{\mu_0}{g_e}\frac{\partial A_s}{\partial x} \tag{5.159}$$

$$A_s(x,t) = A_m e^{j\left(\omega_1 t - \frac{\pi}{\tau}x\right)} \quad \text{stator current sheet} \tag{5.160}$$

with

$$A_m = \frac{3W_1 k_{W1} I\sqrt{2}}{p_1 \tau}; \quad U \quad \text{speed (m/s)} \tag{5.161}$$

In the absence of a dynamic end effect, the airgap flux density is also a traveling wave:

$$B_{gc}(x,t) = B_g e^{j\left(\omega_1 t - \frac{\pi}{\tau}x\right)} \tag{5.162}$$

In this case, Equation 5.159 yields

$$B_{gc} = \frac{\mu_0 F_{1m}}{g_e\left(1+jSG_e\right)}; \quad F_{1m} = A_{1m}\frac{\tau}{\pi} \tag{5.163}$$

where S is the slip, as for rotary IMs.

$$S = \frac{U_s - U}{U_s}; \quad U_s = \tau\frac{\omega}{\pi} = 2\tau f_1 \tag{5.164}$$

and

$$G_e = \frac{2f_1\mu_0\sigma_{Ale}\tau^2}{\pi g_e} = \frac{X_{1m}}{R_r'}; \quad g_e = g + d_{Ale} \tag{5.165}$$

G_e is called the equivalent goodness factor.

X_{1m} and R_r' are magnetization and secondary resistances reduced to the primary (in the absence of the end effect).

d_{Ale} is the equivalent aluminum plate thickness, which may also include the back iron contribution with its conductivity and field penetration depth.

The larger the goodness factor, G_e, the better the conventional performance. The concept may also be extended to cage rotor IMs, which has not been done so far! However, the larger the goodness factor and the lower the number of machine poles, $2p_1 + 1$, the larger the damaging consequence of dynamic end effects in producing additional secondary losses, lower thrust, and lower power factor.

Considering only the active (primary) length, a simplified solution for the airgap flux density, B_g (Equation 5.159) is

$$\underline{B}_g(x,t) = \underbrace{\underline{A}e^{\gamma_1 x}}_{\text{Entry end effect wave}} + \underbrace{\underline{B}e^{\gamma_2(x-L_p)}}_{\text{Entry end effect wave}} + \underbrace{\underline{B}_{gc}e^{-j\frac{\tau}{\pi}x}}_{\text{Conventional traveling field}} \tag{5.166}$$

where L_p is the machine primary length.

With

$$\gamma_{1,2} = \pm\frac{a_1}{2}\left(\sqrt{\frac{b_1+1}{2}} \pm 1 + j\sqrt{\frac{b_1-1}{2}}\right) = \gamma_{1,2r} \pm j\gamma_i$$
$$a_1 = \frac{\pi}{\tau}G_e(1-s); \quad b_1 = \sqrt{1+\frac{16}{G_e^2(1-s)^4}} \tag{5.167}$$

Coefficients A and B are obtained from boundary conditions at primary entry and exit ends. The thrust is simply $\bar{j} \times \bar{B} l$ along the primary length:

$$F_x \approx -a_e d_{\text{Ale}} \operatorname{Re}\left[\int_0^{L_p} \underline{A}^*(x) B_g(x) \mathrm{d}x\right] = F_{xc} + F_{\text{xend}} \tag{5.168}$$

The end effect thrust, F_{xend}, is additional to the conventional one, F_{xc}, and both are given by

$$F_{\text{xend}} = \frac{a_e \mu_0 \tau}{g_e} A_m^2 \operatorname{Re}\left[-\frac{j(\gamma_1 \tau)\left(e^{\gamma_2 \tau - j\pi} - 1\right)}{\left(\underline{\gamma_2 \tau} - \underline{\gamma_1 \tau}\right)\left(\dfrac{\underline{\gamma_2 \tau}}{\pi} - j\right)}\right] \tag{5.169}$$

$$F_{xc} = 2 a_e p_1 \frac{\tau^2 A_m^2}{g_e \pi} \frac{\mu_0 S G_e}{1 + S^2 G_e^2} \tag{5.170}$$

We may extract in the same way an end effect reactive power, Q_{end}, and the secondary aluminum losses, P_{Alend}. To simplify the optimization design, in the absence of reliable end effect compensation schemes despite more than 40 years of efforts [9], an optimum goodness factor G_e was proposed [9] such that the dynamic end effect thrust is zero at zero slip ($S = 0$) (Figure 5.49). G_e depends only on the number of pole pairs, p_1.

Example: For $2p_1 = 12$, at $S = 0.07$ and $U = 110$ m/s, efficiency $\eta_2 = 0.89$ and cos $\varphi_2 = 0.82$ (these are secondary efficiency and power factor: primary losses and primary leakage reactive power were not considered). This is still very good performance at high speeds.

To simplify the design of LIM with end effect (which is notable even for urban people movers with LIM ($U_{\max} < 30$ m/s)), dynamic end effect correction terms depending on SG_e and p_1 may be introduced in the rotary IM equivalent circuit to streamline the LIM design [10].

We conclude the presentation of LIM theory, urging the interested reader to follow the abundant pertinent literature [8–14].

5.25 REGENERATIVE AND VIRTUAL LOAD TESTING OF IMS/LAB 5.7

The availability of PWM bidirectional power flow two or single-stage converters (frequency changers) allows for regenerative and virtual load testing for performance and for temperature (endurance) tests (Figure 5.50a).

The load (drive) IM may act as a motor or as a generator with the tested machine as either a generator ($S < 0$) or a motor ($S > 0$). With the load (drive) system yielding electromagnetic torque and speed estimations and with the power analyzer delivering the input power, cos φ, current of the tested IM, the latter may be tested as a motor and a generator up to 150% rated torque (thrust), if needed by the application.

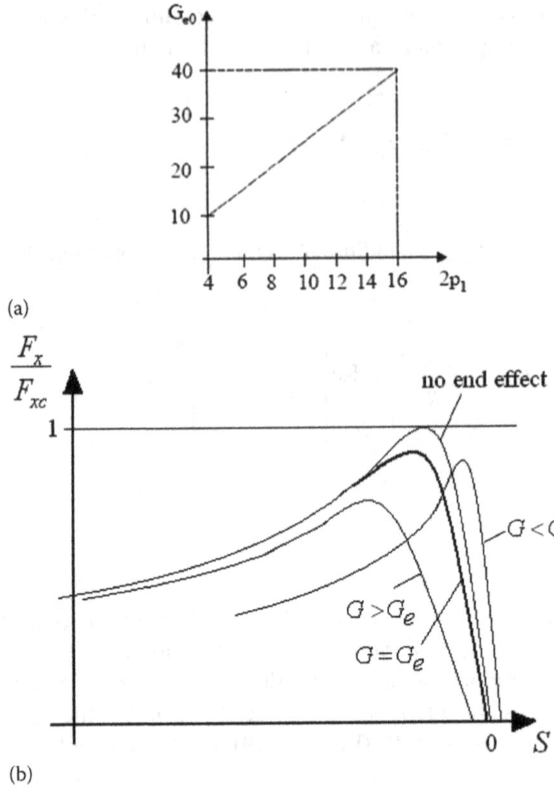

FIGURE 5.49 Dynamic end effect in LIMs: (a) the optimum goodness factor vs. pole number $2p_1$ and (b) thrust/slip with end effect.

For virtual loading (say for vertical shaft IMs where coupling a driver load is not easy), the tested IM itself (with a free shaft) is "driven" by the bidirectional PWM converter, and the speed reference is oscillated with an amplitude and frequency that produces the desired RMS stator current. If we recover the motor input active power (positive or negative), its average after a few cycles represents the machine energy loss, W_{loss}. But if we integrate and average only positive power, we get the input energy for motoring, W_{motor}.

So, the efficiency η_m is

$$\eta_m = \frac{W_{motor} - W_{loss}}{W_{motor}} \tag{5.171}$$

This test is called artificial loading by mechanically forcing the IM to switch from motor to generator by notable speed oscillation. It is also possible for PWM to have two frequencies f_1 and $f_1' = (0.8 - 0.9) f_1$, and thus the machine will have speed n about constant $f_1'/p_1 < n < f_1/p_1$, but the machine will be electrically forced to switch from a motor to a generator. Thus, the frequency mixing method introduced in 1929, with a transformer and an a.c. generator as power sources at f_1 and f_1', nowadays can be done elegantly with power electronics.

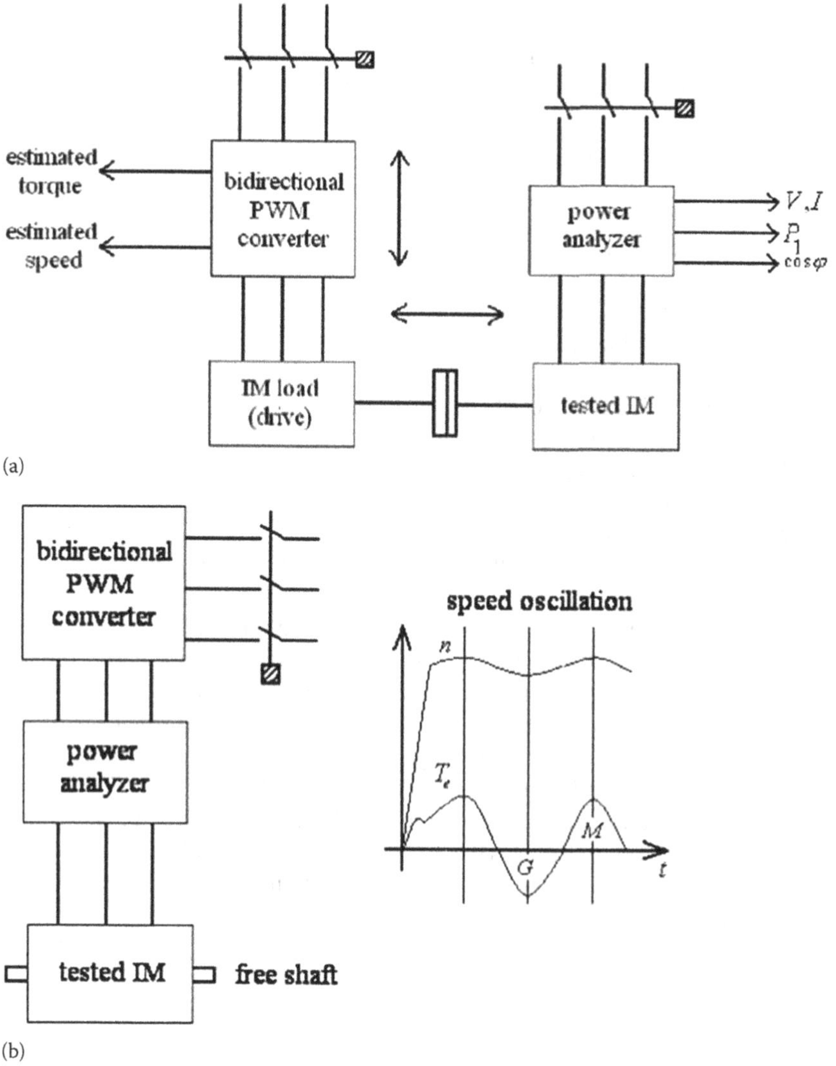

FIGURE 5.50 IM testing: (a) IM regenerative braking testing and (b) virtual load testing.

For more on IM testing, see Ref. [4, Chapter 22], ANSI-IEEE 112, IEC Publication 34, Part 2, IEC Publication 37, ANSI-NEMA Publication MG1, standards IEEE Standard 114 and 839/1986 for single-phase IMs.

5.26 PRELIMINARY ELECTROMAGNETIC IM DESIGN BY EXAMPLE

It is needed to design (size) a cage rotor conduction motor at $P_n = 5$ kW, $V_{nl} = 380$ V (star connection), and $f_1 = 50$ Hz with $2p_1 = 4$ poles and an efficiency around (above) 85%. The breakdown torque is $T_{ek}/T_{en} \geq 2.25$ and the starting torque is $T_{es}/T_{en} \geq 1.2$. The starting current is $I_{stator}/I_n < 6.5$ and cos $\varphi_n \geq 0.85$.

Solution:

The above data are called "main specifications." A realistic analytical model of the IM is needed. Such a model will then require a set of variables to size the machine to suit the specifications.

Let us mention here a possible set of variables:

- Stator outer diameter: D_{out}
- Stator inner diameter: D_{is}
- Airgap: g
- Shaft diameter: D_{shaft}
- Stator core axial length: L_e
- Stator slots height: h_{ss}
- Stator slot width/slot pitch: b_{ss}/τ_{ss}
- Rotor slot height: h_{sr}
- Rotor slot width/slot pitch: b_{sr}/τ_{sr}
- Number of stator slots: N_s
- Number of rotor slots: N_r
- Number of turns per phase W_s (and the number of current paths a)

Earlier in this chapter, we have introduced (derived) analytical expressions of machine parameters, R_s, R_r, L_{sl}, L_{rl}, L_{1m}, and R_{iron}, that depend on the above variables and have entered the expressions for the above design specifications.

As the relationships are nonlinear we cannot derive directly a single set of variables above—which define the machine geometry for the potential manufacturer—from the specifications and from analytical expressions.

But if we add some additional data from past design experience, we can produce a preliminary initial set of variables that can lead to a rather complete machine sizing. This initial design serves as a good basis for design optimization.

The main preliminary design issues are

- Magnetic circuit
- Electric circuit
- Parameters
- Starting torque and current
- Magnetizing reactance X_m
- No-load current
- Rated current
- Efficiency and power factor

5.26.1 Magnetic Circuit

As the basic design constant, we choose the tangential specific force, f_t, at rated torque. This specific force, f_t, increases with rotor diameter and has values in the interval:

$$f_t = 0.3 - 3\ \text{N/cm}^2 \quad \text{for} \quad D_{is} = (0.05 - 0.2)\ \text{m} \tag{5.172}$$

Also, the ratio between the stator core length and the stator bore diameter, L_e/τ_{is}, is

$$\frac{2p_1L_e}{\pi D_{is}} = \frac{L_e}{\tau} = 0.5-2 \tag{5.173}$$

A longer core means a relatively shorter coil-end-connection length and thus lower stator winding losses.

The fundamental amplitude of airgap flux density, about the same at no load and at load, is

$$B_{g1} = (0.4-0.8)\text{T} \tag{5.174}$$

Smaller values are characteristic of small power (below 0.5 kW) and large frequency (speed) IMs.

From industrial experience and design optimization results,

$$\begin{aligned} k_D &= \frac{D_{is}}{D_{out}} \approx 0.5-0.6;\ 2p_1 = 2 \\ k_D &= \frac{D_{is}}{D_{out}} \approx 0.6-0.67;\ 2p_1 = 4 \\ k_D &= \frac{D_{is}}{D_{out}} \approx 0.67-0.72;\ 2p_1 = 6 \\ k_D &= \frac{D_{is}}{D_{out}} \approx 0.7-0.75;\ 2p_1 \geq 6 \end{aligned} \tag{5.175}$$

The airgap, g, is mechanically limited at the lower end, and by the necessity to limit additional core and cage losses due to space harmonics caused by the windings in slots and by slot opening: $g = (0.3–3)$ mm, with the larger values suited for the MW power range. Too large airgap means low power factor.

The design (rated) current densities depend on the duty cycle, type of cooling, machine size, and the desired efficiency.

For the case in discussion, $f_{bt} = 1.7\ \text{N/cm}^2$, $L_e/\tau = 0.820$, $B_{g1} = 0.75$ T, $D_{is}/D_{out} = 0.623$, and $g = 0.4$ mm.

Though the rated slip is not yet known, we may choose an initial value $S_n = 0.025$, as the speed reduction with load is small with IMs.

So the rated torque, T_{en}, is

$$T_{en} \approx \frac{P_n}{2\pi \frac{f_1}{p_1}(1-S_n)} = \frac{5000}{2\pi \frac{50}{2}\times(1-0.025)} = 32.66\ \text{Nm} \tag{5.176}$$

But

$$T_{en} \approx f_{tn}\frac{D_{is}}{2}\left(\frac{L_e}{\tau}\frac{\pi}{2p_1}\right)\pi D_{is}^2 = 1.7\times10^4\times D_{is}^3\times\frac{\pi}{2}\times0.820\times\frac{3.14}{2\times2} \tag{5.177}$$

$$D_{is} = 0.12345\text{m}$$

So the stator stack length, L_e, is

$$L_e = \frac{L_e}{\tau}\frac{\pi D_{is}}{2p_1} = 0.828 \times \pi \times \frac{0.12345}{2 \times 2} \approx 0.08 \text{ m} \tag{5.178}$$

The stator diameter, D_{out}, is thus

$$\left(D_{out}\right)_{2p_1=4} = \frac{D_{is}}{k_D} = \frac{0.12345}{0.623} = 0.198 \text{ m} \tag{5.179}$$

The shaft diameter is chosen in relation to the breakdown torque, D_{shaft} = 30 mm.

We consider that magnetic saturation does not flatten the sinusoidal airgap flux density, and the slot opening will be accounted for only by the Carter coefficient, K_c.

Thus, it is rather handy to calculate the stator back iron (yoke) height, h_{ys} (B_{ys} = 1.5 T)

$$h_{ys} = \frac{\frac{\Phi_p}{2}}{B_{ys}L_e} = \frac{\frac{B_{g1}\tau}{\pi}}{B_{ys}} = \frac{\frac{0.75 \times \pi \times 0.12345}{\pi \times 2 \times 2}}{1.5} = 15.428 \times 10^{-3} \text{ m} \tag{5.180}$$

For the rotor yoke, h_{yr} (B_{yr} = 1.6 T) is

$$h_{yr} = \frac{\frac{\Phi_p}{2}}{B_{yr}L_e} = \frac{\frac{B_{g1}\tau}{\pi}}{B_{yr}} = \frac{\frac{0.75 \times 0.0969}{\pi}}{1.6} = 14.5 \times 10^{-3} \text{ m} \tag{5.181}$$

$$\tau = \frac{\pi D_{is}}{2p_1} = \pi \times \frac{0.12345}{2 \cdot 2} = 0.0969 \text{ m} \tag{5.182}$$

As the outer stator and shaft diameters, D_{out} and D_{shaft}, are now known, the total radial height of stator and rotor slots, h_{ss} and h_{sr}, may be calculated as

$$h_{ss} = \frac{D_{out} - D_{is}}{2} - h_{ys} = \frac{(0.198 - 0.1234)}{2} - 0.0154 \approx 21.6 \text{ mm} \tag{5.183}$$

$$h_{sr} = \frac{D_{is} - D_{shaft}}{2} - h_{yr} - g = \frac{(0.1234 - 0.03)}{2} - 0.0145 - 0.0004 \approx 30.1 \text{ mm} \tag{5.184}$$

It is now time to choose the number of stator slots per pole and phase $q = 3$ (the pole pitch is $\tau = 0.09687$ m) and thus with $2p_1 = 4$, $N_s = 2p_1 q m_1 = 2 \times 2 \times 3 \times 3 = 36$ slots. The rotor slot number $N_r = 30$ (there are tables of suitable N_s, N_r, $2p_1$, combinations to minimize parasitic torque effects [4, Chapter 15]). The tooth flux density in the stator and rotor are again chosen, at $B_{ts,r}$ = 1.5 T and thus the tooth/slot pitch ratio is

$$\frac{b_{ts}}{\tau_{ss}} = \frac{b_{tr}}{\tau_{sr}} = \frac{B_{g1}}{B_{ts,r}} = \frac{0.75}{1.5} = 0.5 \tag{5.185}$$

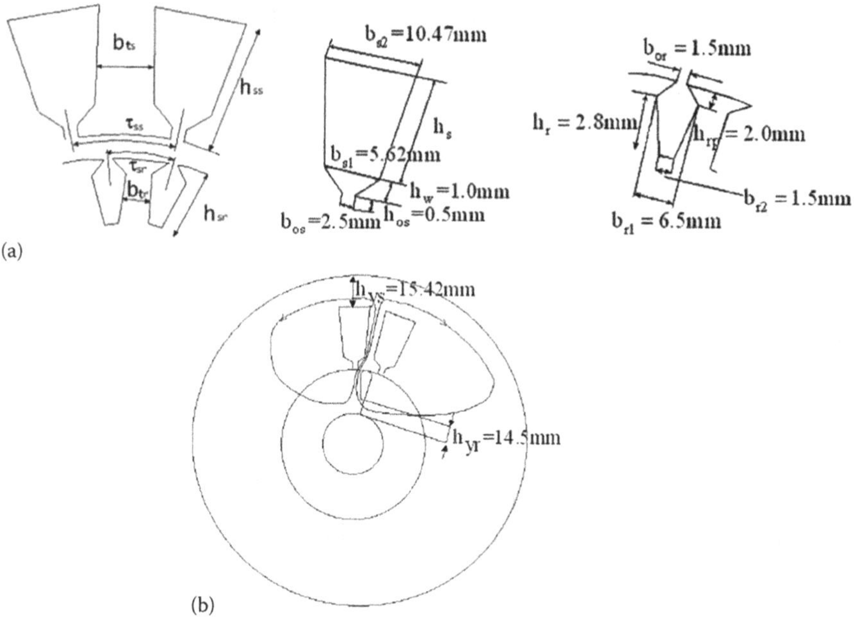

FIGURE 5.51 The magnetic circuit: (a) typical stator and rotor slots and calculated geometry (b) cross-section.

Now, we may completely size the stator if in Figure 5.51 we adopt $h_{0s} = 0.5$ mm, $b_{0s} = 2.5$ mm, $h_w = 1$ mm, $h_{rp} = 2$ mm, and $b_{0r} = 1.5$ mm. The active slot top and bottom widths are

$$
\begin{aligned}
b_{s1} &= \frac{\pi\left(D_{is} + 2\left(h_{0s} + h_w\right)\right)}{N_s} - b_{ts} \\
&= \frac{\pi \times \left(123.4 + 2 \times (0.5 + 1)\right)}{36} - 0.5387 \times \frac{96.87}{9} = 5.62 \text{ mm} \\
b_{s2} &= \frac{\pi\left(D_{is} + 2h_{ss} + h_{ys}\right)}{N_s} - b_{ts} \\
&= \frac{\pi \times (123.4 + 2 \times 21.6)}{36} - 5.4 = 10.47 \text{ mm} \\
b_{r1} &= \frac{\pi\left(D_{is} - 2g - 2h_{rp}\right)}{N_r} - b_{tr} = \frac{\pi \times (123.4 - 0.8 - 4)}{30} - 6.48 = 6.5 \text{ mm} \\
b_{r2} &= \frac{\pi\left(D_{ir} - 2g - 2h_{sr}\right)}{N_r} - b_{tr} = \frac{\pi \times (123.4 - 0.8 - 60)}{30} - 6.48 = 1.5 \text{ mm!!}
\end{aligned}
\tag{5.186}
$$

The slot active areas (filled with coils) in the stator and rotor are

$$A_{\text{ssa}} = \left(h_{\text{ss}} - h_{\text{os}} - h_{\text{w}}\right)\frac{\left(b_{\text{s1}} + b_{\text{s2}}\right)}{2} = \frac{\left(21.5 - 1.5\right)\times\left(5.62 + 10.47\right)}{2} \tag{5.187}$$
$$= 160.90\ \text{mm}^2$$

$$A_{\text{sra}} = h_{\text{rp}}\frac{\left(b_{\text{or}} + b_{\text{r1}}\right)}{2} + \left(h_{\text{sr}} - h_{\text{rp}}\right)\frac{\left(b_{\text{r1}} + b_{\text{r2}}\right)}{2} = 2\times\frac{\left(1.5 + 6.5\right)}{2} + 28\times\frac{8}{2} \tag{5.188}$$
$$= 120\ \text{mm}^2$$

The ratio of stator and rotor slot areas is

$$\frac{A_{\text{ss}}}{A_{\text{sr}}} = \frac{N_{\text{s}}A_{\text{ssa}}}{\left(N_{\text{r}}A_{\text{sra}}\right)} = \frac{36\times 160.90}{\left(30\times 120\right)} = 1.609 > 1 \tag{5.189}$$

As expected, $A_{ss}/A_{sr} > 1$ as less slot space is available in the interior rotor, to avoid very heavy magnetic saturation: even values of 2/1 or a bit more are feasible.

5.26.2 Electric Circuit

It is known from industrial practice that the emf (airgap flux voltage) in the stator, V_{en}, is about 0.93–0.98 of the supply rated voltage, V_n:

$$\frac{V_{\text{en}}}{V_n} \approx \left(0.93 - 0.98\right) \tag{5.190}$$

The smaller values are valid for sub-kW IM and for larger number of poles $2p_1$ (when the stator leakage reactance in p.u. is increasing). But V_{en} is, from Equation 5.34,

$$V_{\text{en}} = \pi\sqrt{2}f_1 W_{\text{s}} k_{\text{w1s}}\frac{2}{\pi}B_{\text{g1}}\tau L_{\text{e}} \tag{5.191}$$

With Equation 5.9, the stator winding factor, k_{w1s}, is

$$k_{\text{w1s}} = \frac{\sin\dfrac{\pi}{6}}{q\sin\dfrac{\pi}{6}}\sin\frac{y}{\tau}\frac{\pi}{2} \tag{5.192}$$

With $q = 3$ and y/τ (coil span/pole pitch) = 8/9, $k_{w1} = 0.925$. The only unknown in Equation 5.191 is the number of turns per phase, W_s:

$$W_{\text{s}} = \frac{\dfrac{380}{\sqrt{3}}\times 0.93}{\pi\sqrt{2}\times 50\times 0.925\times\dfrac{2}{\pi}\times 0.75\times 0.0968\times 0.0825} = 261\ \text{turns/phase}$$

Now, the number of turns/coil W_c for a two layer winding is

$$W_c = \text{Integer}\left(\frac{W_s}{2p_1q}\right) = \text{integer}\left(\frac{261}{2\times2\times3}\right) = 21 \text{ turns/coil}$$

So the number of turns per phase is $W_s = 252$ turns, indeed. At this point in design, we may not calculate the rated current, I_n, unless we assume certain values for rated efficiency and power factor (in traditional design methods η_n and $\cos\varphi_n$ are assumed initial values, based on experience). Here, we first proceed with machine parameters and then calculate I_n, η_n, and $\cos\varphi_n$.

5.26.3 Parameters

Knowing the number of turns per phase and the stator slot area, we may calculate the stator resistance:

$$R_s = \rho_{Co} l_{cs}\frac{W_s}{A_{Co}} = \frac{2.1\times10^{-8}\times0.4353\times252}{1.408\times10^{-6}} = 1.63682\ \Omega$$
$$A_{Co} = \frac{A_{ss}k_{fill}}{2W_c} = \frac{160.9\times0.45}{2\times21} = 1.408\ \text{mm}^2 \tag{5.193}$$

The coil turn length, l_{cs}, is

$$l_{cs} \approx 2L_e + 2l_{ec} = 2L_e + \pi y = 2\times(0.0825+0.13518) = 0.4353\ \text{m} \tag{5.194}$$

The rotor aluminum bar resistance, R_b, is

$$R_b = \rho_{Al}\frac{L_e+0.01}{A_{sra}} = \frac{3.5\times10^{-8}\times0.0925}{120\times10^{-6}} = 0.27\times10^{-4}\ \Omega \tag{5.195}$$

The end ring-to-bar current ratio (Equation 5.24) is

$$\frac{I_r}{I_b} = \frac{1}{2\sin\dfrac{\alpha_{esr}}{2}} = \frac{1}{2\sin\dfrac{2\times\pi\times2}{2\times30}} = 2.38 \tag{5.196}$$

So the end ring area, A_{ring}, is

$$A_{ring} \approx A_{sra}\frac{I_r}{I_b} = 120\times2.38 = 286.62\ \text{mm}^2 \tag{5.197}$$

The rotor bar/ring relationships are illustrated in Figure 5.52.

The end ring cross-sectional dimensions, a and b, are $a = 10$ mm and $b = 28.6$ mm. The length of the end ring segment, corresponding to a bar, is

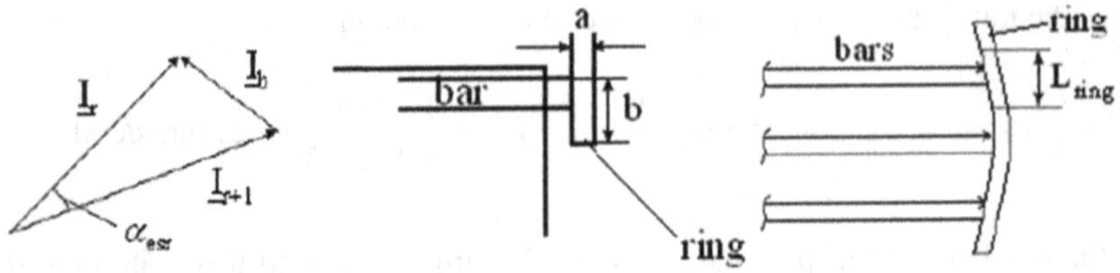

FIGURE 5.52 The rotor end ring sizing.

$$L_{ring} \approx \frac{\pi\left(D_{is} - b - 2g\right)}{N_r} = \frac{\pi \times \left(123.4 - 28.6 - 0.8\right)}{30} \approx 9.92 \text{ mm} \tag{5.198}$$

So the end ring segment resistance, R_r, is

$$R_r = \rho_{Al}\frac{L_{ring}}{A_{ring}} = \frac{3.5 \times 10^{-8} \times 0.00992}{286 \times 10^{-6}} = 0.1211 \times 10^{-5}\ \Omega \tag{5.199}$$

The equivalent bar resistance, R_{be} (Equation 5.26), is

$$R_{be} = R_b + \frac{R_r}{2\sin^2\dfrac{\alpha_{esr}}{2}} = 0.27 \times 10^{-4} + \frac{0.1211 \times 10^{-5}}{2\sin^2\dfrac{\pi}{12}} \approx 0.4083 \times 10^{-4}\ \Omega \tag{5.200}$$

The reduction factor to the stator is (Equation 5.51)

$$R_r' = R_{be}\frac{N_r}{3k_i^2}; \quad k_i = \frac{N_r k_{skew}}{6W_s k_{w1s}} \tag{5.201}$$

With 1 rotor slot pitch ($c = \tau_{ss}$) skewing (Equation 5.48),

$$k_{skew} = \sin\frac{\dfrac{\pi}{2}\dfrac{c}{\tau}}{\dfrac{\pi}{2}\dfrac{c}{\tau}} \tag{5.202}$$

So

$$k_i = \frac{30 \cdot 0.95}{6 \cdot 252 \cdot 0.925} = 0.0203 \tag{5.203}$$

and

$$R_r' = 0.4083 \times 10^{-4}\frac{30}{3 \times 0.0203^2} = 0.99\ \Omega < R_s = 1.636\ \Omega$$

Note. The cage slot fill factor is about $k_{fillc} = 1$ and the ring segment length/bare is small and thus rotor cage resistance $R_r' < R_s$, in general.

5.26.3.1 Leakage reactances

According to Equation 5.54, the stator leakage reactance (see Section 5.4.2) is

$$X_{sl} = 2\mu_0\omega_1 \frac{W_1^2 L_e}{p_1 q} \Sigma\lambda$$
$$\Sigma\lambda = \lambda_{sls} + \lambda_{slc} + \lambda_{sld} + \lambda_{slz} \tag{5.204}$$

We will consider here the differential leakage lumped into the zig-zag leakage λ_z. The slot permeance coefficient is

$$\lambda_{sls} = \frac{hs}{3(b_{s1} + b_{s2})} + \frac{h_{0s}}{b_{0s}} + \frac{2h_w}{b_{0s} + b_{s2}} = 1.347 \tag{5.205}$$

The zig-zag permeance coefficient is

$$\lambda_{slz} = \frac{5g/b_{0s}}{5 + 4g/b_{0s}} = 0.1418 \tag{5.206}$$

The end-connection permeance coefficient is

$$\lambda_{sec} = 0.34 \frac{q}{L_e}(l_{ac} - 0.64y) = 0.99 \tag{5.207}$$

So the total stator leakage reactance X_{sl} is (Equation 5.204)

$$X_{sl} = 2 \times 1.256 \times 10^{-6} \times 2\pi \times 50 \times 252^2 \frac{0.0825}{2 \times 3}$$
$$\times (1.347 + 0.1418 + 0.99) \approx 1.707\ \Omega \tag{5.208}$$

The rotor cage leakage reactance is calculated in the same way (see again Section 15.4.2)

$$\lambda_{rls} = \frac{2h_{rp}}{3(b_{r1} + b_{r2})} + \frac{2h_r}{3(b_{r1} + b_{r2})} = 2.506 \tag{5.209}$$

For the end ring,

$$\lambda_{ring} = \frac{L_{ring}}{L_e} \ln\left(\frac{L_{ring} N_r}{2\sqrt{\frac{ab}{\pi}}}\right) = 0.331 \tag{5.210}$$

So

$$X_{rl} = X_{be}\frac{N_r}{3k_i^2}$$
$$X_{be} = \mu_0\omega_1 L_e\left(\lambda_{rls} + \lambda_{ring}\right) = 0.922\times10^{-4}\ \Omega \quad (5.211)$$
$$X'_{re} \approx 2.21\ \Omega$$

Note. The skewing leakage inductance was neglected. Also, the main reactance, X_{1m}, should be reduced by k_{skew}

5.26.4 Starting Current and Torque

Due to the skin effects, the rotor parameters, R'_r and X'_{rl}', have to be corrected when calculating starting current and torque.

Let us consider the rotor slot as rectangular.

The skin effect parameter, ξ (Equation 5.104), is

$$(\xi)_{start} = h_{rs}\sqrt{\frac{\omega_1\mu_0\sigma_{se}}{2}} = 2.25 \quad (5.212)$$

Approximately from Equation 5.209

$$\phi(\xi) \approx (\xi)_{start} = 2.25$$
$$\psi(\xi) \approx \frac{3}{2\xi} = 0.66 \quad (5.213)$$

Because the actual rotor slot is thinner toward the bottom, we adopt $\varphi(\xi) = 1.5$, $\psi(\xi) = 0.8$ (smaller skin effect).

The skin effect in the end ring is, in general, smaller because it is placed mostly in air, but we will still consider it (conservatively) to be the same as is in the slot zone.

So

$$R'_{rstart} = \phi(\xi)R'_r = 1.5\times0.99 = 1.485\ \Omega$$
$$X'_{rlstart} = \psi(\xi)X'_{re} = 0.8\times2.215 = 1.772\ \Omega \quad (5.214)$$

So the starting current, I_{start}, is simply (Equation 5.56)

$$I_{start} = \frac{V_{nl}/\sqrt{3}}{\sqrt{\left(R_s + R'_{rstart}\right)^2 + \left(X_{sl} + X'_{rlstart}\right)^2}} \approx 47.12\text{A} \quad (5.215)$$

The starting torque is obtained applying Equation 5.95 for $S = 1$:

$$T_{\text{estart}} \approx \frac{3p_1}{\omega_1} I_{\text{start}}^2 R'_{\text{rstart}} \approx 63\ \text{Nm} \tag{5.216}$$

5.26.5 Breakdown Slip and Torque

S_k, T_{ek} (Equations 5.108 and 5.109):

$$S_k = \frac{R'_r}{\sqrt{R_s^2 + (X_{sl} + X_{rl})^2}} = \frac{0.99}{\sqrt{1.636^2 + (1.70 + 2.215)^2}} = 0.2338 \tag{5.217}$$

$$T_{ek} \approx \frac{3p_1}{2} \frac{\left(\frac{V_{nl}}{\sqrt{3}}\right)^2}{\omega_1} \cdot \frac{1}{X_{sl} + X'_{rl}} = \frac{3\times 2}{2} \times \frac{220^2}{2\pi \times 50} \times \frac{1}{(1.70 + 2.215)} = 118\ \text{Nm} \tag{5.218}$$

Note: The rated torque, still to be calculated, will be about 32 Nm, so the starting and breakdown torques are large (at the price of lower efficiency, perhaps).

5.26.6 Magnetization Reactance, X_M, and Core Losses, p_{IRON}

The magnetization reactance may be calculated after we calculate the iron mmf contribution coefficient, K_s, to magnetic saturation

$$K_s = \frac{2H_{ts}h_s + H_{ys}l_{ys} + H_{yr}l_{yr} + 2H_{tr}l_r}{2\frac{B_{g1}}{\mu_0} g K_c} \tag{5.219}$$

The Carter coefficient, K_c (which accounts for both stator and rotor slotting), is

$$K_c = K_{cs} K_{cr}$$

$$K_c = \frac{\tau_{ss}}{\tau_{ss} - \gamma_s g} \frac{\tau_{sr}}{\tau_{sr} - \gamma_r g}; \quad \gamma_{s,r} = \frac{(b_{0s,r}/g)^2}{5 + b_{0s,r}/g} \tag{5.220}$$

$$\tau_{ss} = \frac{\pi D_{is}}{N_s} = \frac{\pi \cdot 0.1234}{36} = 0.0107\ \text{m}; \quad b_{0s} = 2.5\ \text{mm} \tag{5.221}$$

$$\tau_{sr} = \frac{\pi \cdot (D_{is} - 2g)}{N_s} = \frac{\pi(0.1234 - 2\times 0.0004)}{30} = 0.01283\ \text{m}; \quad b_{0r} = 1.5\ \text{mm} \tag{5.222}$$

Finally, $K_c = 1.497$

The stator wire diameter is

$$d_{co} = \sqrt{\frac{4}{\pi} \cdot A_{co}} = \sqrt{\frac{4}{\pi} \cdot 1.408} = 1.34 \text{ mm} \tag{5.223}$$

As the stator slot opening is $b_0 = 2.5$ mm, there is enough room to insert the turns one by one in the slot.

Now with $B_{ts} = B_{tr} = 1.5$ T $= B_{ys} = B_{yr}$ from the $B(H)$ curve of the silicon steel (Chapter 4), $H_{ts} = H_{tr} = 500$ A/m.

The average field path lengths in the stator and rotor yokes based on conservative industrial design experience is

$$l_{ys} = \frac{\pi \cdot (D_{out} - h_{ys})}{2p} = \frac{\pi \cdot (0.198 - 0.0154)}{4} = 0.143 \text{ m} \tag{5.224}$$

$$l_{yr} = \frac{\pi \cdot (D_{shaft} - h_{yr})}{2p_1} = \frac{\pi \cdot (0.03 + 0.0154)}{4} = 0.03564 \text{ m} \tag{5.225}$$

So finally K_s (Equation 5.219) is

$$K_s = \frac{2 \times 500 \times 0.0215 + 500 \times 0.9433 + 500 \times 0.03564 + 2 \times 500 \times 0.028}{2 \times \frac{0.75}{1.256 \times 10^{-6}} \times 0.4 \times 10^{-3} \times 1.497}$$
$$= 0.194$$

The magnetic saturation coefficient, K_s, is rather small (values from 0.3 through 0.5 are common for $2p_1 = 4$) but the Carter coefficient is rather large, so an increase in airgap or a reduction of b_{0s}, say, to 2 mm is necessary to keep not only the magnetization current (and power factor) but the additional core (stray) losses within practical limits.

The magnetization reactance (Equation 5.41) is

$$X_{1m} = \frac{6\mu_0\omega_1}{\pi^2}\left(W_s k_{w1s}\right)^2 \frac{\tau L_e}{p_1 g K_c (1 + K_s)} = 65.77\ \Omega \tag{5.226}$$

The specific core losses at 1.5 T, 50 Hz are considered here:

$$\left(p_{iron}\right)_{1.5T,50Hz} = 4.2 \text{ W/kg} \tag{5.227}$$

As 1.5 T was considered in both parts of stator iron core (yoke and teeth), we only need to calculate the total iron weight. Stator yoke weight, G_{ys}, is

$$G_{ys} \approx \pi\left(D_{int} - h_{ys}\right) \times h_{ys} \times L_e \gamma_{iron} = 5.536 \text{ kg} \tag{5.228}$$

The stator teeth weight is

$$G_{ts} = h_{ss} \times L_e \times b_{ts} \times N_s \times \gamma_{iron} = 2.606 \text{ kg} \tag{5.229}$$

The rotor iron losses are negligible because the rotor (slip) frequency $f_2 = S_n f_1 <$ (1.5–2) Hz in our case. So the total core loss, p_{iron}, is

$$p_{iron} = \left(p_{iron}\right)_{1.5T,50\text{ Hz}} \times G_{ts} + \left(p_{iron}\right)_{1.5\text{ T},50\text{Hz}} G_{ys} = 34.1964 \text{ W} \tag{5.230}$$

To account for additional even core losses, we may double the value of fundamental core losses above:

$$p_{iront} = 2p_{iron} = 68.4 \text{ W}$$

5.26.7 No-Load and Rated Currents, I_0 and I_N

Based on

$$\frac{V_{nl}}{\sqrt{3}} = I_0\sqrt{\left(R_s + R_{iron}\right)^2 + \left(X_{sl} + X_{1m}\right)^2} \tag{5.231}$$

and

$$3R_{iron}I_0^2 = p_{iront} \tag{5.232}$$

we may iteratively calculate both the unknown I_0 and R_{iron}. But, neglecting R_{iron} in Equation 5.231 does not produce unacceptable errors:

$$I_{0i} \approx \frac{\left(V_{nl}/\sqrt{3}\right)}{\sqrt{R_s^2 + \left(X_{sl} + X_{1m}\right)^2}} = 3.26 \text{ A} \tag{5.233}$$

The iron loss resistance, R_{iron}, is

$$R_{iron} = \frac{p_{iront}}{3I_0^2} = 2.158\ \Omega \tag{5.234}$$

I_{0i} is the ideal no-load current which is close to the motor no-load current I_0 because the iron losses plus mechanical losses do not add much to I_{0i} and the flux density in the airgap is about the same for ideal no-load and motor load modes.

The rated current I_n contains the rated slip which is still unknown:

$$I_r' = \frac{\left(V_{nl}/\sqrt{3}\right)}{\sqrt{\left(R_s + C_1\dfrac{R_r'}{S_n}\right)^2 + \left(X_{sl} + C_1X_{r1}'\right)^2}};\quad C = 1 + \frac{1.7}{65.7} = 1.0258 \tag{5.235}$$

$$I_n \approx \sqrt{I_r'^2 + I_{0i}^2} \tag{5.236}$$

To avoid tedious iteration calculations, we vary S_n from 0.01 to 0.06 and calculate I_r' from Equation 5.235 and I_n from Equation 5.236, and then

$$T_{\text{en}} = \frac{3R_r'\left(I_{\text{rn}}'\right)^2}{S_n}\frac{p_1}{\omega_1} \tag{5.237}$$

For nominal power and, say (with $p_{\text{mecn}} = 0.01\ P_n$),

$$p_{\text{mecn}} + P_n = T_{\text{en}}\frac{2\pi f_1\left(1-S\right)}{p_1} = 5050\ \text{W} \tag{5.238}$$

We vary slip until Equation 5.238 is satisfied (with I_n from Equations 5.235 through 5.237). Let us consider $S_n = 0.0495$ and obtain $I_n = 9.88$ A, $I_{rn}' = 9.4$ A, and $T_{\text{em}} = 33.8$ Nm. The developed shaft power is

$$P_{\text{shaft}} = T_{\text{en}}\frac{2\pi f_1\left(1-S_n\right)}{p_1} - p_{\text{mec}} = 5000\ \text{W} \tag{5.239}$$

5.26.8 Efficiency and Power Factor

The efficiency, η_n, is simply

$$\eta_n \approx \frac{P_n}{T_e\frac{2\pi f_1}{p_1} + p_{\text{iront}} + 3R_s I_n^2} = 0.855 \tag{5.240}$$

The power factor, $\cos\varphi_n$, is

$$\cos\varphi_n = \frac{P_n}{\eta_n 3\left(V_{nl}/\sqrt{3}\right)I_n} = 0.9 \tag{5.241}$$

5.26.9 Final Remarks

- The stator resistance should be reduced by 20%–30% by deepening the stator slots with mild additional magnetic saturation of the stator yoke. The efficiency may be increased by a few percent.
- The rather large $T_{\text{estart}}/T_{\text{en}} = 63/33.8$ and $T_{\text{ek}}/T_{\text{en}} = 118/33.8$ are responsible to a great existent for the lower efficiency. In essence, the machine is a bit too small (in volume) for the tasks. For design optimization, the design should go back

and change the specific force (rotor shear stress) f_{tn} (in N/cm^2) and the stack length (pole pitch) and redo the whole design. As shown in the companion book, design optimization with various methods may improve performance notably, observing the important constraints (more on IM design in Refs. [15–19]).

5.27 DUAL STATOR WINDINGS INDUCTION GENERATORS (DWIG)

Applications such as aircraft and ships power systems use traditionally d.c. excited synchronous generators that deliver constant voltage at variable frequency for variable speed in a certain range.

Loads that are frequency insensitive (resistances) may be supplied this way and there is plenty of them. Alternatively, constant voltage and frequency loads have to be dealt with separately or the wound rotor induction generators with bidirectional PWM a.c.-d.c.-a.c. partial ratings converters (DFIG) are proposed for the scope.

Finally, the most recent trend in vehicular power systems is the d.c. power bus, with the prime movers and the generators operating at variable speed.

For both situations, cage (symmetric or nested type) rotor induction generators have been proposed as they are brushless (in contrast to DFIG) and they use partially rated PWM converter connected to one of the winding (for a.c. output) and a diode rectifier output (for d.c. output) [20,21].

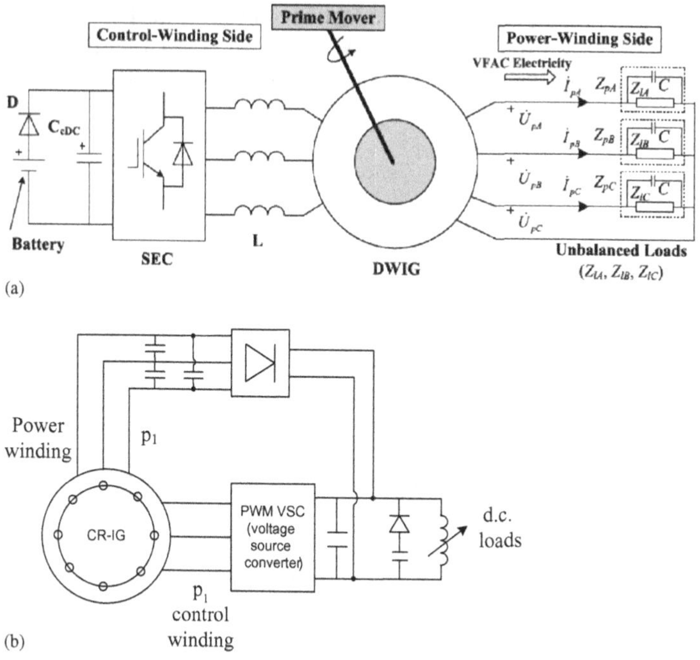

FIGURE 5.53 Dual stator winding cage rotor IG: (a) with a.c. variable frequency f_1 and constant – controlled a.c. output voltage; (b) with d.c. constant-controlled voltage output.

Basic such schemes for the cage rotor case are presented in Figure 5.53a,b.

In Figure 5.53a the partially rated VSC (voltage source converter) feeds the control winding that excites the IG power winding. Both the control and power winding have the same number of pole pairs. The same is true for the scheme in Figure 5.53b, where, in addition, the "power winding" now intervenes only above a certain speed when the diodes in the rectifier open, adding more d.c. power. Such a scheme (Figure 5.53b) seems suitable when the required output power increases with speed (wind energy, for example). In both schemes the stator frequency, the same for the dual windings, varies with speed to keep the slip frequency in the symmetric rotor cage small and thus retain a good efficiency with variable speed.

A.c. unbalanced loads may be handled (Figure 5.53a). The steady state of such a machine, with, say, unbalanced a.c. load is handled by the positive, negative, and zero sequence standard equivalent circuits (Figure 5.54) [21].

The presence of the ±0 sequence equivalent circuits is visible in Figure 5.54. The rotor cage suppresses notably the inverse (–) sequence of currents (at the "price" of additional cage losses) and the inverse (–) voltage is small. Finally, the zero-sequence currents flow in the stator power winding resistances and leakage reactances with small (zero) voltages.

The battery in the d.c. bus in Figure 5.54 a,b serves only to ignite the selfexcitation process (if the generator operates in standalone mode), but a full voltage battery could be used and, the excess energy when available could be absorbed by recharging the battery [20].

The steady-state operation of the d.c. output DSWIG (Figure 5.54b) may be investigated also by using the standard equivalent circuit of IMs [22].

The nested—cage—rotor brushless DWIG, known as BDFIG [23–25] has been also proposed as a constant frequency f1 and voltage a.c. generator for variable speed (Figure 5.55).

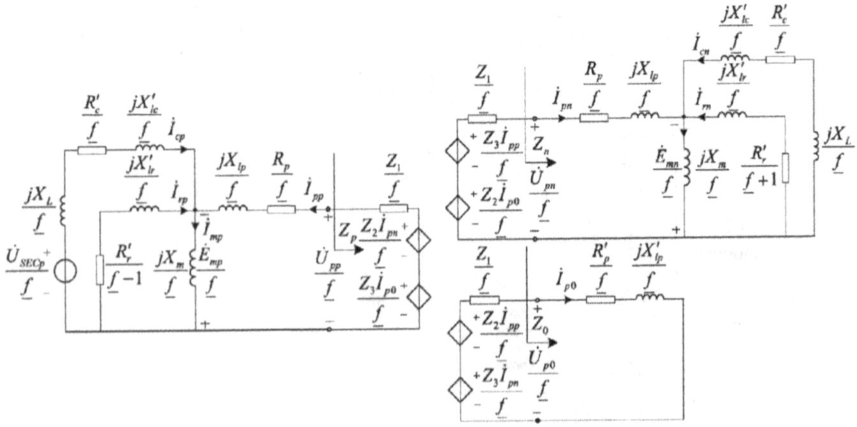

FIGURE 5.54 Equivalent circuit of a.c. DSWIG with unbalanced load; $f_s = f_1/(p_1 n)$ (n – rotor speed in Hz, p_1 – pole pairs of stator windings, n – symmetric cage rotor speed in rps, f_1 – stator windings fundamental frequency in Hz).

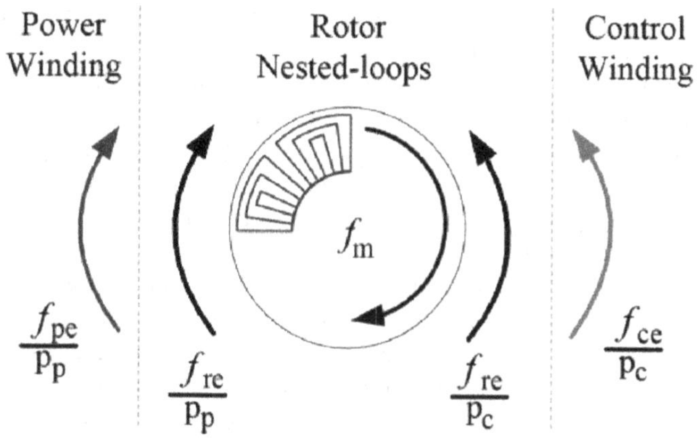

FIGURE 5.55 Nested – cage – rotor DWIG: the main (f_{pe}/p_p), control (f_{ce}/p_c) windings and nested–cage (f_{re}/p_p and f_{re}/p_c) magnetic fields.

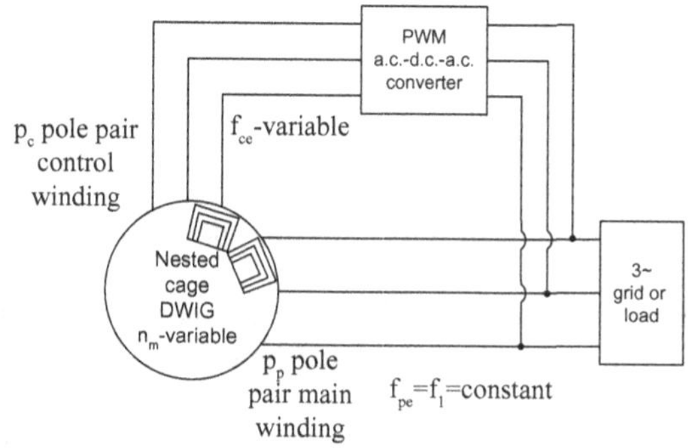

FIGURE 5.56 Nested cage BDFIG.

The two windings have now different numbers of pole pairs p_p and p_c while the unique rotor currents frequency f_{rc} (for a nested cage with $p_r = p_p + p_c$ poles (nests)) is:

$$f_{re} = f_{pe} - p_p f_m = f_{ce} + p_c f_m;\ f_m = n_m \tag{5.242}$$

The two windings are coupled synchronously at a speed $f_m = n_m$:

$$n_m = \left(f_{pe} - f_{ce}\right)/p_r;\ \text{again } p_r = p_p + p_c \tag{5.243}$$

It should be noted that the magnetic fields of the two windings rotate in opposite directions. In general (so far) $f_{pe} = f_1$ = const and the control winding is fed through a bidirectional PWM a.c.-d.c.-a.c. partial rating static PWM converter (Figure 5.56).

The closed loop control of PWM bidirectional converters modifies the voltage and frequency f_{ce} of control winding "in tact" with the rotor speed n_m, with and without an encoder [25].

Note. It is also possible to control the nested-cage-rotor BDFIG with both frequencies, f_{ce} and f_1 as variable such that to limit slip frequency and thus control optimally a stand-alone operation mode with a.c. output or with diode-rectified d.c output as with the symmetric—cage—rotor DSWIG (Figure 5.53b).

The advantages of nested-cage over symmetric-cage rotor is mainly due to different pole pairs p_p+p_c of the two windings, which leads to operation at lower speed for given frequency f_1, $p_1 = p_p$ (5.243). A kind of "torque magnification" occurs as in "flux-modulation" machines.

However, both windings have an additional leakage inductance due to the very principle (the nested cage of $p_r = p_p + p_c$ nests couples windings of different pole pairs) which implicitly increases the absorbed reactive power required to magnetize the machine; so, the torque density is reduced with respect to a regular cage IM.

The steady state investigation of nested-cage-rotor BDFIG may be done by using properly the equivalent circuit of IM concept [22], while the transients and control investigations are based on the orthogonal (dq) model concept [25].

Final Note. This paragraph should be considered a mere introduction to brushless DFIGs, which are still in the labs, but with fair chances to reach markets soon as vehicular (stand alone) a.c. or d.c. output or as grid-connected variable speed generators.

5.28 SUMMARY

- IMs are a.c. stator and a.c. rotor traveling magnetic field machines with $2p_1$ poles.
- They have both the stator and the rotor cores made of nonoriented grain silicon steel sheets, with stamped uniform (for three phases) slots, around the periphery toward the airgap.
- The IM airgap should be small enough to reduce the magnetization current, I_{0s}, but large enough to reduce supplementary (additional) load losses, p_s, due to mmf and slot opening magnetic field space harmonics. In general, $g = 0.3 - 3$ mm for powers from 100 W to 30 MW, but $g = (1–15)$ mm for linear IMs.
- The stator slots are filled, in general, with three-phase single- or double-layer a.c. windings with traveling mmf and thus a traveling airgap field occurs. The IM windings have the number of slots/pole/phase $q \geq 2$, integer or even fractional.
- The rotor slots are filled either with aluminum uninsulated die-cast (or copper) bars and end rings (the cage rotor) or with a three-phase a.c. winding with the same number of poles, $2p_1$, as in the stator. The cage rotor adapts to any number of stator poles and thus pole changing stator windings may be used with cage rotor, to change the ideal no-load or synchronous speed $n_1 = f_1/p_1$: the speed for zero rotor current and torque.
- The a.c. winding mmf shows space harmonics ν ($\nu > 1$) above the fundamental due to winding placement in slots. The mmf harmonics, $\nu = -(5, 7, 11, 17, \ldots)$, are inverse rotating at n_1/ν ($\nu < 0$) but they are forward rotating for $\nu = +(7, 13, 19, \ldots)$. These harmonics produce their own rotor cage currents and asynchronous parasitic torques that may hamper starting under heavy loads. Chording

the coils in the winding to $y/\tau \approx 0.8$ leads to harmonic reduction to acceptable levels in industry. The first slot harmonics $\nu_c = 2kqm \pm 1$ also produce parasitic torques, asynchronous and synchronous ones, that are reduced by slot skewing and, respectively, by choosing proper stator and rotor slot count combinations ($N_s \neq N_r$).

- Rotor three-phase or cage windings may be mathematically reduced to the stator as done for transformers. This way the mutual stator/rotor phase inductance amplitudes are equal to the stator phase self main flux inductance.
- The main and leakage magnetic fields in the IM are "translated" into main L_{1m} and leakage inductances L_{sl}, L_{rl}; in general, l_{1m} $(p.u.) = L_{1m}\omega_n I_n/V_n = (1.2–4)$, $l_{sl}(p.u.) \approx l_{rl}(p.u.) = 0.03–0.1$; smaller l_{1m} for linear induction motors with large airgap
- Once the mutual self and leakage inductances and resistances of stator/rotor are defined (calculated), the machine may be considered as an "ensemble" of coupled electric circuits with only stator/rotor mutual inductances to be dependent on rotor electrical position $\Theta_{er} = p_1\Theta_r$: Θ_r—mechanical rotor position angle.
- After changing the rotor variables to stator coordinates (by counterrotation along Θ_{er}), the equations of an IM become independent of the rotor position for steady state and phasors may be applied. The obtained equations, allowing for apparent phase segregation (star connection) resemble those of the transformer, but to the cage rotor an additional resistance $R_r'\left(1-S\right)/S$ is added, which corresponds to the mechanical power developed by the actual IM.
- S is the slip; $S = (n_1 - n)/n_1$; for $S = 0$ (and cage rotor) the synchronous operation (zero torque) is obtained. For motoring $1 > S > 0$ ($n = 0$ to n_1), and for generating $S < 0$ ($n > n_1$ for $n_1 > 0$).
- IM ideal no-load motor and stalled rotor operation modes are used to segregate the losses and assess equivalent circuit parameters.
- Self-excited induction generator operation with capacitors is dependent on magnetic saturation, speed, and machine parameters. Voltage regulation is substantial, so a controlled capacitor would be needed for voltage (or frequency)-sensitive loads even at a constant speed.
- The electromagnetic torque has a breakdown value for motoring ($S = S_k$) and for generating ($S = -S_k$), which is independent of rotor resistance R_r', but S_k is proportional to R_r'.
- Speed control (variation) for a given torque may be done by pole-changing windings or frequency f_1 control through full power PWM converters. Only for starting, the reduced voltage method is feasible. Rotor resistance increases, feasible with wound rotors, produce large torque at all speeds but for very large losses; so it is used for limited speed range control. Supplying the wound rotor from a variable frequency f_2 PWM converter ($f_2 = f_1 - np_1$) allows for limited speed control for lower PWM converter rating (and costs).
- Deep bars or a dual cage in the rotor allows for heavy starts at reduced starting currents (down to 5 I_n); I_n is the stator rated current.
- Unbalanced supply voltages lead to negative sequence current components I_{s-}, besides the direct sequence I_{s+}, and they contribute to larger losses and lower

torque and power factor in the machine. Machine derating is recommended in the NEMA standards in relation to voltage unbalance index.

- The one stator phase open—which is equivalent to a single-phase supply—leads to zero starting torque and to larger loss and smaller speed for given torque and nonzero speed.
- Asymmetric rotor circuits (broken bars or one rotor phase open) lead to additional reverse sequence stator currents at frequency $f_1' = f_1(1-2S)$ which, in small machines (with large R_s (in p.u.)), produce Georges' effect (torque), that is zero at $S = 1/2$, and thus the motor, during starting on load, may be "arrested" around 50% of rated speed. This phenomenon is typical to asynchronous starting of synchronous motor when the d.c. excitation circuit is short-circuited (see Chapter 6). Broken bars have to be diagnosed early and due measures of rotor replacement be taken as soon as feasible.
- For low-power and residential applications (refrigerator compressors, heater's circulating pumps), single-phase a.c. source fed capacitor split-phase IMs are used. They have a main and a starting (or permanent) orthogonally placed auxiliary winding. The mmf of such a two-phase winding has a positive ⊕ and a negative ⊖ sequence. Destroying the negative sequence stator current component (symmetric operation) is met at a single slip S for

$$a = \frac{X_{m+}}{R_{m+}} = \tan\varphi_+; \quad a = \frac{W_a}{W_m}; \quad X_C = \frac{1}{\omega_1 C} = Z_+ a\sqrt{a^2+1}$$

where

W_a is the auxiliary winding turns,
W_m is the main winding turns,
C is the capacitor in series with the auxiliary winding, and
X_{m+}, R_{m+} are the total ⊕ sequence impedance components as seen in the main winding.

- To design a good capacitor IM, a hard-to-get compromise between starting and running performance and motor total costs has to be worked out.
- Linear induction motors may be obtained by cutting and unrolling a rotary IM.
- LIMs have a longitudinally (along motion direction) open magnetic circuit. Consequently, the traveling mmf of the primary at linear speed $U_s = 2\tau f_1$ induces at primary core entry and exit, at small slip values $S = 1 - U/U_s$, for a small number of poles $2p_1$ ($2p_1 + 1$), end effect secondary currents which reduce the thrust and efficiency and decrease the power factor. These dynamic end effects may be reduced to reasonable proportions if the LIM is designed at the optimum goodness factor, $G_e(2p_1) = X_m/R_r'$ (Figure 5.49), where X_m is the magnetizing reactance and R_r' is the secondary resistance. The goodness factor concept has not been yet used in rotary IMs!
- The thrust of LIM suits well urban and suburban transportation or industrial short travel transport with wheels or with controlled magnetic suspension (MAGLEV).

- Induction motor full-load testing shows low-energy consumption by using bidirectional PWM converter motor loads/drivers for regenerating braking or for virtual load (of the free-shaft IM). For a full account of IM industrial testing, see IEEE-112B Standard, Ref. [4, Chapter 22] and Ref. [15, Chapter 7].
- PWM converters provide variable frequency and voltage that recently transformed the IM from the workhorse into the racehorse of industry.

5.29 PROPOSED PROBLEMS

5.1 Build a two-layer $2p_1 = 2$, $q = 15$ turbogenerator three-phase a.c. winding with chorded coils ($y/\tau = 36/45$), and calculate its distribution and chording factors $k_{q\nu}$ and $k_{y\nu}$ for $\nu = (-5), (+7), (-11), (+13), (-17), (+19), 6q \pm 1$.
Hint: See Section 5.3.1, Equation 5.20.

5.2 Build a two-phase single-layer winding for a capacitor-split IM for $2p_1 = 2$, $N_s = 24$ with 16 slots for the main winding and 8 slots for the auxiliary winding.
Hint: See Figure 5.14 for clues.

5.3 For the winding in Example 5.1, calculate the ratio between the differential leakage, L_{sld}, and the main (fundamental) inductance, L_{1m}, and discuss the result.
Hint: See Equations 5.41 and 5.43.

5.4 A three-phase cage rotor induction motor of $P_n = 1.5$ kW, $V_{nl} = 220$ V (star connection), $f_1 = 60$ Hz, $2p_1 = 4$ has a rated efficiency $\eta_n = 0.85$ and a power factor $\cos\varphi_n = 0.85$. The iron, stray, and mechanical losses are $p_{iron} = p_s = p_{mec} = 1.0\%$ of P_n and $p_{cosn} = 1.2p_{corn}$. Calculate:

a. The rated phase current I_n.
b. The no-load current $I_0 = I_n \sin\varphi_n$, and the rated rotor current $I'_m \approx \sqrt{I_n^2 - I_0^2}$.
c. The total losses, $\sum p_n$, and the stator and rotor winding losses, p_{cosn}, p_{corn}, R_s, R_r, and R_{sc}.
d. The electromagnetic power $P_{elm} = P_1 - p_{cosn} - p_{iron} - p_s$ and the rated slip S_n.
e. The rated electromagnetic torque, T_{en}, and the shaft torque, T_{shaft}
f. If the peak torque is $T_{ek} = 2.2T_{en}$, calculate approximately the short-circuit reactance, $X'_{sc'}$, of the machine and the critical slip, S_k ($C_1 = 1.05$).
g. If there is no skin effect, calculate the starting current, I_{start}, and torque, T_{estart}.
h. Calculate the stator current, I_s, the delivered electric power (as generator), P_1, and the absorbed reactive power, Q_s, for $S = -0.03$ and determine the rotor speed (in rps) required for this, at power grid.

Hints: Check Example 5.5 and Equations 5.108 and 5.109.

5.5 For a cage rotor three-phase IM with $N_s = 36$ slots, $N_r = 16$ slots and $2p_1 = 4$, $f_1 = 60$ Hz, calculate the speeds (in rps) where teeth harmonics ($2q_{1,2}m \pm 1$) and mmf space harmonics ($km \pm 1$) produce synchronous parasitic torques.
Hints: Check Equations 5.115 through 5.119.

5.6 A three-phase IM with $V_{nl} = 220$ V (star connection), $f_1 = 60$ Hz, $2p_1 = 4$, $R_s = R_r' = 1\ \Omega$, $X_{sl} = X_{rl}' = 2.5\ \Omega$, $X_m = 75\ \Omega$, $R_{iron} = \infty$ (no iron losses) remains in two-phases (B and C) because phase A gets disconnected while working at slip $S_n = 0.04$.
Calculate:

a. The stator current, I_s, the magnetization current, I_0, the rotor current, I_r' at $S = 0.03$ (from the equivalent circuit) with three-phase balanced steady state. Also electromagnetic torque is needed, and so is the direct impedance, Z_+.
b. With phase A open, calculate for the same slip $S = 0.03$, the inverse impedance, Z_-, the stator current, and electromagnetic torque.
c. The starting ($S = 1$) current for three phases fed and for the case of phase A open. Will the motor start with phase A open?

Hints: See Section 5.21.

5.7 Resolve Example 5.6, but for $S = 0$ ($n = n_1 = f_1/p_1$). Discuss the results.

5.8 A flat three-phase linear induction motor with an aluminum sheet on an ideal laminated core has the following data:

- Stack length: $2a = 0.20$ m
- Mechanical airgap: $g = 0.01$ m
- Aluminum plate thickness: $d_{AL} = 6$ mm
- Pole pitch: $\tau = 0.25$ m
- Number of poles: $2p_1 = 8$ (single-layer winding)
- The aluminum plate width: $2e = 0.36$ m
- Rated frequency: $f_n = 50$ Hz
- $\rho_{Al} = 3.5 \times 10^{-8}\ \Omega$m

Calculate:

a. The ideal synchronous speed $U_s = 2\tau f_n$.
b. The goodness factor G_e, after calculating the edge factor k_T.
c. The primary mmf amplitude, F_{1m}, for which the conventional primary airgap flux density, B_{gc}, is 0.4 T at $S = 0.1$.
d. The current loading amplitude A_m, the end effect coefficients γ_1 and γ_2.
e. The end effect thrust, F_{xend}, the conventional thrust, F_{xc} and the total thrust for $S = 0.1, 0.05, 0.01$ for A_m of point d above. Discuss the results.

Hints: Check Section 5.24, Equations 5.158 through 5.170.

REFERENCES

1. Alger, P.L., *Induction Machines*, 2nd edition, Gordon & Breach, New York, 1970 and new edition 1999.
2. Cochran, P.L., *Polyphase Induction Motors*, Marcel Dekker, New York, 1989.
3. Stepina, K., *Single Phase Induction Motors*, Springer Verlag, Berlin, Germany, 1981 (in German).
4. Boldea, I. and Nasar, S.A., *Induction Machine Handbook*, vol. 1 and 2, 3rd edition, CRC Press, Boca Raton, FL, 2020.

5. Yamamura, S., *Spiral Vector Theory of AC Circuits and Machines*, Clarendon Press, Oxford, U.K., 1992.
6. Boldea, I., *Electric Generator Handbook*, Vol. 2, Variable Speed Generators, CRC Press, Boca Raton FL; Taylor & Francis Group, New York, 2005.
7. Heller, B. and Hamata, V., *Harmonics Effects in Induction Motors*, Elsevier, Amsterdam, the Netherlands, 1977.
8. Boldea, I. and Nasar, S.A., *Linear Motion Electromagnetic Systems*, John Wiley & Sons, New York, 1985, Chapter 6.
9. Boldea, I. and Nasar, S.A., *Linear Motion Electric Machines*, John Wiley & Sons, New York, 1976.
10. Cabral, C.M., *Analysis of LIM Longitudinal End Effects*, Record of LDIA, Birmingham, U.K., 2003, pp. 291–294.
11. Yamamura, S., *The Theory of Linear Induction Motors*, John Wiley & Sons, New York, 1972.
12. Gieras, J., *Linear Induction Drives*, Clarendon Press, Oxford, U.K., 1992.
13. Fuji, N., Hoshi, T., and Tanabe, Y., *Characteristics of Two Types of End Effect Compensators for LIM*, Record of LDIA, Birmingham, U.K., 2003, pp. 73–76.
14. Boldea, I., *Linear Electric Machines Drives and MAGLEVs Handbook*, CRC Press, Taylor & Francis Group, New York, 2013.
15. Tolyiat, H. and Kliman, G. (eds), *Handbook of Electric Motors*, Marcel Dekker Inc., New York, 2004, Chapter 7.
16. Levi, E., *Polyphase Motors: A Direct Approach to Their Design*, John Wiley & Sons, New York, 1985.
17. Hamdi, E.S., *Design of Small Electrical Machines*, John Wiley & Sons, New York, 1994, Chapter 5.
18. Vogt, K., *Design of Rotary Electric Machines*, VEB Verlag Technik, Berlin, Germany, 1983 (in German).
19. J. Pyrhönen, T. Jokinen and V. Hrabovcova, *Design of Rotating Electrical Machines*, John Wiley & Sons, UK, Ltd, 2008.
20. O. Ojo, I. E. Davidson, "PWM-VSI inverter – assisted stand-alone dual stator winding induction generator", *IEEE Trans.*, Vol. IA-36, No. 6, 2000, pp. 1604–1611.
21. H. Xu, F. Bu, W. Huang, Y. Hu, H. Liu, "Control and performance of 5 – phase dual stator winding induction generator d.c. generation system", *IEEE Trans*, Vol. IE-64, No. 7, 2017, pp. 5276–5285.
22. P. C. Roberts, R. A. McMahon, P. J. Tavner, J. M. Maciejowski, T. J. Flack, "Equivalent circuit for the brushless doubly fed machine (BDFM) including parameter estimation and experimental verification", *IEE Proc.*, Vol. EPA-152, No.4, 2005, pp. 933–942.
23. D. Zhou, R. Spee, A. K. Wallace "Laboratory control implementation for doubly-fed machines", *Proc. Int. Conf. Ind. Electronics Control Instrumentation*, 1993, pp. 1181–1186.
24. S. Williamson, A. C. Ferreira, A. K. Wallace, "Generalized theory of the brushless doubly – fed machine" *I. Analysis, IEE Proc.*, Vol. EPA-144, No.2, 1997, pp. 111–122.
25. U. Shipurkar, T. D. Strous, H. Polinder, J. A. Ferreira, A. Veltman, "Achieving sensorless control of the brushless doubly fed induction machine, ", *IEEE Trans*, Vol. EC-32, No. 4, 2017, pp. 1611–1619.
26. F. Bu, S. Zhuang, W. Huang, N. Su, Y. Hu, "Asymmetrical operation analysis for dual stator-winding induction generator variable frequency a.c. generating system with unbalanced loads", *IEEE Trans*, Vol. IE-64, No. 1, 2017, pp. 52–59.
27. L. Tutelea, I. Boldea, N. Muntean, S. Deaconu, "Modeling and performance of novel scheme dual winding cage rotor variable IG with d.c. link power delivery", Record of IEEE-ECCE, 2014.

6 Synchronous Machines
Steady State

6.1 INTRODUCTION: APPLICATIONS AND TOPOLOGIES

As shown in Chapter 3, synchronous machines (SMs) are characterized by a.c. multiphase winding currents in the stator, and d.c. (or PM) excitation or magnetically salient (reluctance) passive rotors. These are basically traveling field machines in which both stator and rotor fields are rotating "synchronously" at the rotor electric speed, ω_r:

$$\omega_1 = \omega_r = 2\pi p_1 n_1; \quad n_1 = \frac{f_1}{p_1} \tag{6.1}$$

where

ω_1 is the stator frequency and

p_1 denotes pole pairs.

There are also unipolar rotor-position-triggered or ramped-frequency stator-current pulse multiphase machines with reluctance (and eventually with PMs also) rotors that have a stepping stator field whose average speed is equal to the rotor speed. These are called stepper machines when the current pulses are initiated at a ramping reference frequency independent of the rotor position, or are called switched reluctance machines (SRMs) when the current pulses in each phase (one or two at a time) are rotor-position-triggered.

SMs require stator frequency control for controlling speed, which is carried out by frequency changers (pulse width–modulated (PWM) inverter/rectifier).

SMs have widespread applications. Some examples are as follows:

Power systems electric power plants use synchronous generators (SGs) with salient rotor poles (hydrogenerators, $2p_1 > 4$, Figure 6.1a) or nonsalient rotor poles (turbogenerators, $2p_1 = 2,4$, Figure 6.1b) with d.c. rotor excitation, to power up to 1000 MVA (hydrogenerators) and 1700 MVA (turbogenerators) [1]. Power electronics is used only to supply, through brushes and copper slip rings, energy to the excitation heteropolar excitation placed on the d.c.-fed rotor.

Automobile alternators with single ring-shaped coil heteropolar d.c. excitation claw-pole rotor and with a.c. stator and diode rectifier d.c. output to the on-board battery (Figure 6.2).

DOI: 10.1201/9781003214519-6

FIGURE 6.1 Large SGs for power systems. (a) Rotor cross section of a hydrogenerator ($S_n \leq 770$ [MW], $f = 50(60)$ [Hz], $V_{nl} \leq 24$ [kV]). (b) Rotor cross section of a turbogenerator ($S_n \leq 1500$ [MW], $f = 50(60)$ [Hz], $V_{nl} \leq 28$ [kV]).

Permanent magnet synchronous generators (*PMSGs*), which have been introduced recently for full bidirectional converter control to the power grid for wind energy conversion (up to 3 MVA at 16 rpm, transmission-less driving, Figure 6.3) and up to 8 MW with transmission at 490 rpm.

Automotive permanent magnet synchronous small power actuators and starters/alternators in hybrid electric vehicles (HEVs) (Figure 6.4) [2–11].

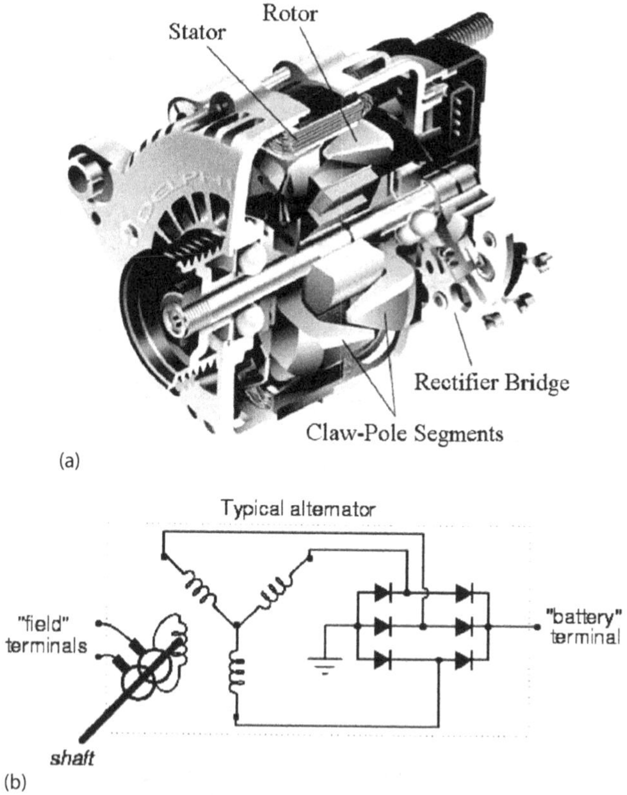

FIGURE 6.2 Automotive alternator with diode rectifier d.c. output with battery backup ($P_n = 0.5 - 2.5$ [kW], $V_{dc} = 14,28,42$ [V]).

Hard disks or other information gadget drives (Figure 6.5).

Single-phase micro-PMSMs (permanent magnet synchronous motors) with parking PM and PWM inverter control for self-starting micropower motors (for disk drives, mobile-phone ringers, or automotive climate-control air blowers), Figures 6.5 and 6.6.

Special configuration PM-assisted reluctance rotor SMs (PM-RSMs), which have been introduced recently as automotive starters for achieving large torque density (Figure 6.7a) or as transverse flux (TF) and flux reversal (FR) PMSMs for large-torque low-speed motors/generators with lower copper losses (Figure 6.7b and c) [12].

Linear SMs, as counterparts of rotary ones, which have been proposed for people movers (up to 400–550 km/h), for industrial (limited) excursion applications, and for linear oscillatory motion (Figure 6.8a through f) as in refrigerator compressors or mobile-phone ringers.

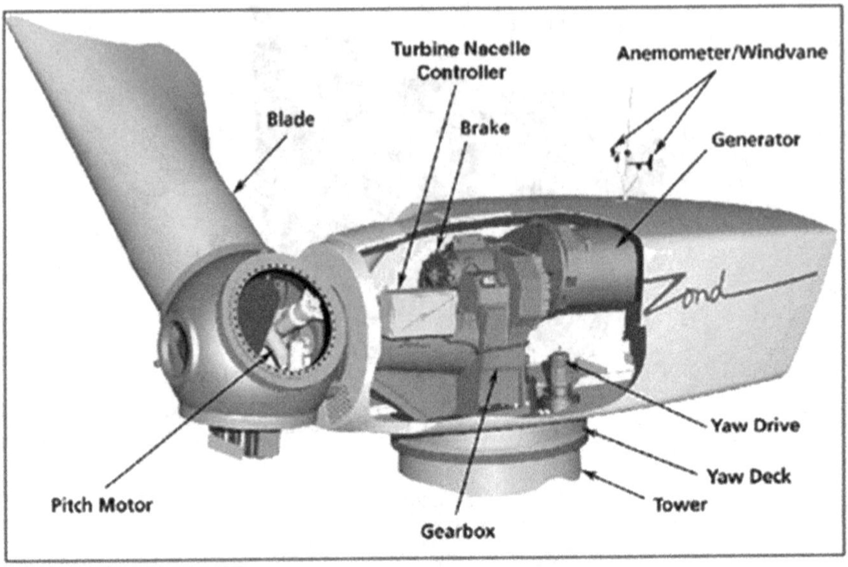

FIGURE 6.3 PMSG for renewable energy (wind generator)—3 MW at 16 rpm.

The applications and configurations presented so far illustrate the unusually large spectrum of power, speed, and topological diversity of synchronous machines.

The stator magnetic core that houses the a.c. one- or three-phase windings is made of silicon sheets of a thickness of 0.5 mm up to 150 Hz fundamental frequency, and 0.1 mm (from 500 Hz to 3 kHz fundamental frequency, which are provided with uniform slots which are open (Figure 6.9a), semiopen (Figure 6.9b), or semiclosed (Figure 6.9c).

In the semiopen slot, each of the two preformed multistrand cables (bars) of the coil is inserted one by one in the slot.

The rotor core of the SMs is in general laminated for interior PM and reluctance rotor configurations, and so are the pole shoes of the salient pole rotor SGs. Solid soft iron rotors are used in large turbogenerators ($2p_1 = 2,4$) and in the rotor yoke of surface PM rotors.

To visualize the construction elements, we show in Figure 6.10 a cross section into a PMSM that identifies the additional elements, such as the shaft and the frame.

6.2 STATOR (ARMATURE) WINDINGS FOR SMS

There are two main types of armature (full power) a.c. windings:

- Distributed a.c. windings—(described in Chapter 5), used for all power plant generators, autonomous medium-power generators, large SMs, and many PMSMs with sinusoidal current control.

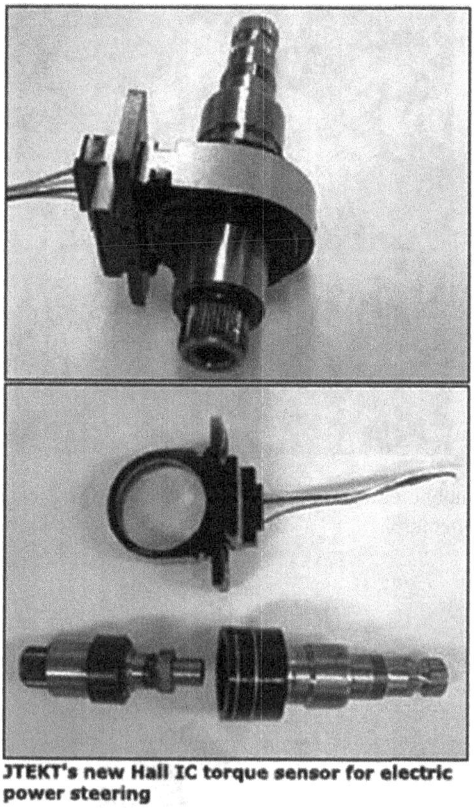

FIGURE 6.4 Automotive electric power steering assist PMSM; and V_{dc} = 12 V, T_{en} = up to 1 Nm, n = 2000 rpm, n_{max} = 4000 rpm, and T_{e_nmax} = 0.3 Nm with PWM inverter control for variable speeds.

FIGURE 6.5 Axial airgap PM synchronous (d.c. brushless) hard disk motor and drive (embedded power electronics control).

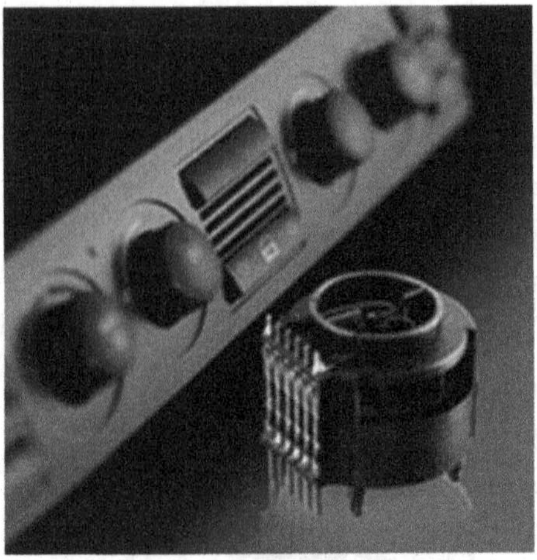

FIGURE 6.6 Automobile climate hot-air-blow micro-PMSM: single phase, with parking PM and PWM inverter control.

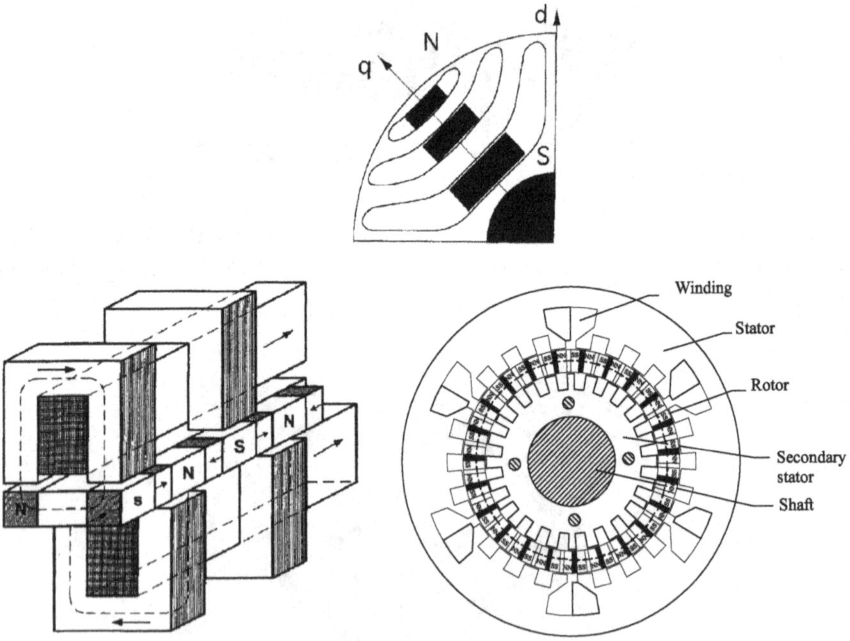

FIGURE 6.7 Special configuration PMSMs: (a) PM-RSM, (b) TF, and (c) flux reversal (FRM) machines.

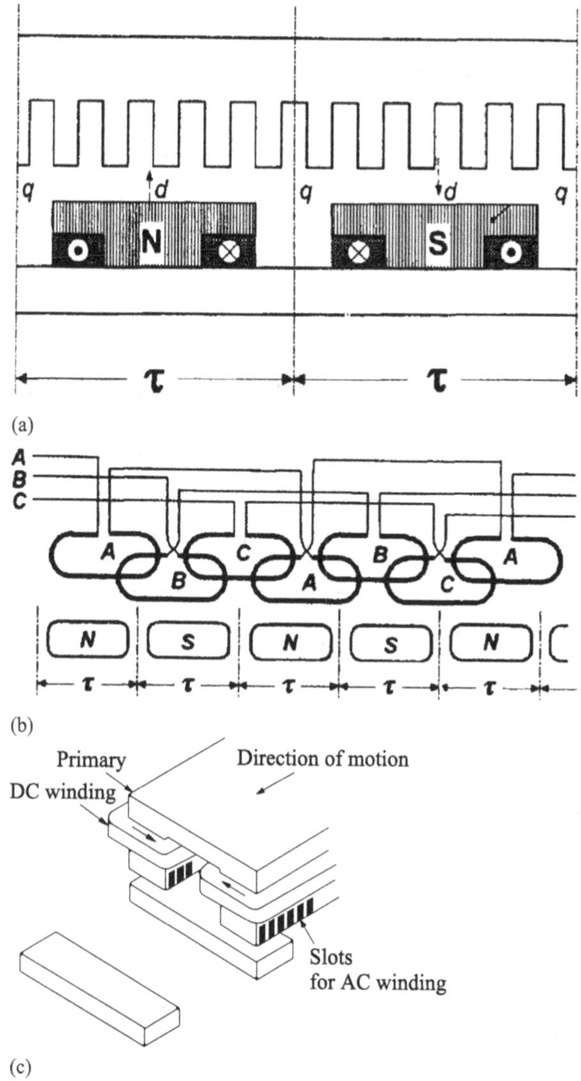

FIGURE 6.8 (a) Linear SMs with d.c. excitation on board and three-phase a.c.-controlled winding on ground (Transrapid Maglev, in Germany). (b) Linear SMs with superconducting d.c. excitation on board and three-phase a.c.-controlled winding on ground (LSM-Japanese Maglev, in Japan). (c) Linear homopolar SM with d.c. and a.c. three-phase-controlled windings on board (Magnibus Maglev, in Romania). (d) Linear flat PMSM for industrial use. (e) Linear flat switched reluctance motor (SRM). (f) Linear tubular oscillatory, PM single-phase SM. (From Boldea, I. and Nasar, S.A., *Linear Electric Actuators and Generators*, Cambridge University Press, Cambridge, U.K., 1997; Boldea, I. and Nasar, S.A., *Linear Motion Electromagnetic Devices*, CRC Press, Taylor & Francis Group, New York, 2001.)

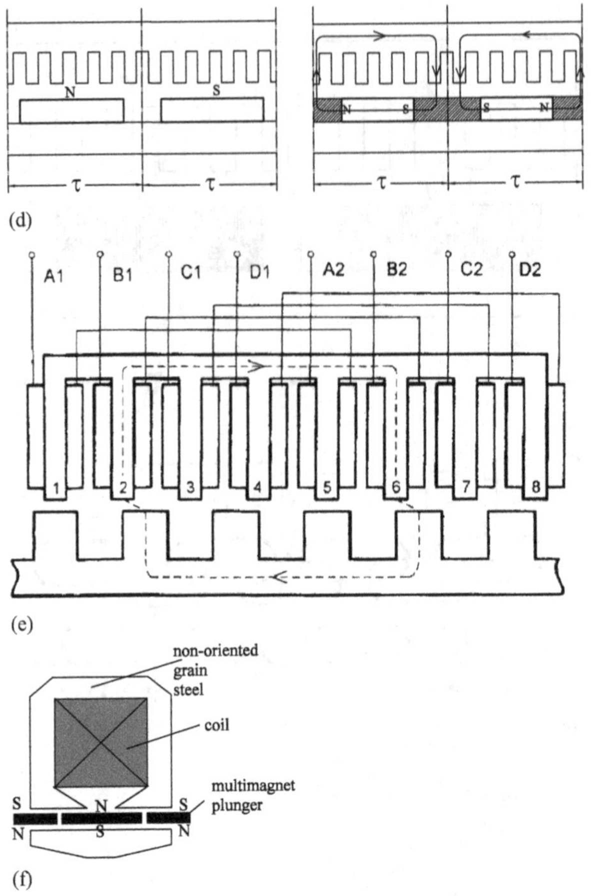

FIGURE 6.8 (Continued)

- Nonoverlapping (tooth-wound or circular) coil windings—used in PMSMs (from hard disk to industrial servodrives and large-torque, low-speed applications) and, for dual saliency, in stepper and SRMs, with unipolar current pulses; circular coils embrace all poles on the periphery in the TF-PMSM (Figure 6.7b) to produce "torque magnification."

6.2.1 Nonoverlapping (Concentrated) Coil SM Armature Windings

In PMSMs, to reduce the end-coil length (and losses), and to reduce the torque at zero current, tooth-wound coil windings with $N_s \neq 2p_1$ but $N_r = 2p_1 + 2k$ (Figure 6.11a and b) are used. For all these PMSMs, $q < 0.5$. For the case in Figure 6.11b and c, $N_s = 6$ slots and $2p_1 = 4$ poles ($k = 1$). There is one coil per phase for the single-layer winding and two coils per phase for the two-layer winding in Figure 6.11a, b.

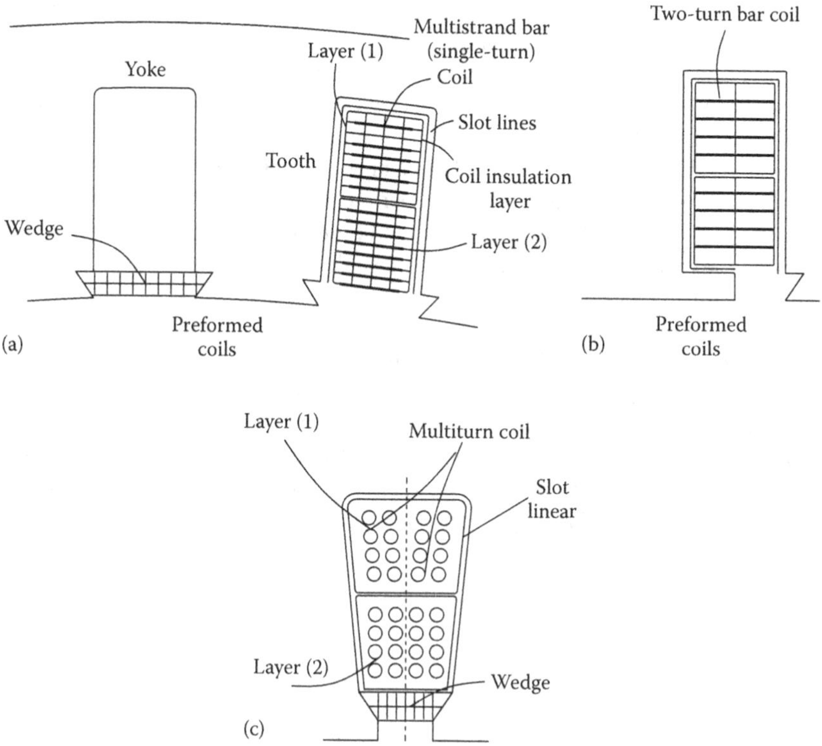

FIGURE 6.9 Typical slot shapes for armature (stator) windings: (a) open slots (for large-power SGs), (b) semiopen slots (for two-turn bar coils), and (c) semiclosed slots for low-torque SMs.

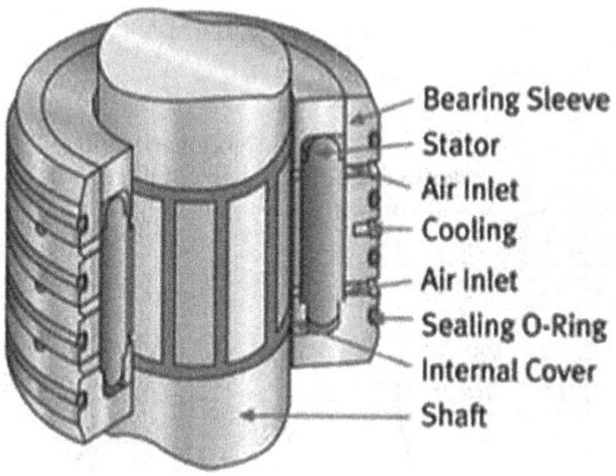

FIGURE 6.10 Open cross section in a PMSM.

There are many other combinations of N_s and $2p_1$ such as 3/2, 3/4, 6/4, 6/8, 9/8, 9/10, 9/12, 12/10, 12/14, 24/16, 24/22, …, 36/42, and so on.

It has been demonstrated that the number of periods of the torque at zero current (cogging torque) is a multiple of the LCM (lowest common multiplier) of N_s and $2p_1$; the larger the LCM the smaller the cogging torque.

These machines operate as $2p_1$ pole machines. So the stator mmfs with $N_s \neq 2p_1$ have rather large winding factors for $2p_1$ periods (Tables 6.1 and 6.2), but they also have sub- and super-harmonics, which may be included in the differential leakage inductance (explained in Chapter 5). It is feasible to produce with such windings a sinusoidal emf (by motion) in the stator phases.

Basically, similar tooth coil windings are used for salient poles passive rotors (N_s, $2p_1$; $N_s = 2p_1 + 2k$; $k = \pm 1, \pm 2$) of SRMs (Figure 6.12).

Concentrated (tooth-wound) coil windings exhibit through their 3 (multi) phase mmfs numerous space harmonics which imply a large differential leakage (airgap leakage) inductance $L_{m\upsilon}$. This means that such machines have larger synchronous inductances (reactances) in p.u. than windings with distributed windings. The ratio between such differential leakage inductance and that for the fundamental ($\upsilon = 1$ for $q \geq 1$, $\upsilon = p'$ for $q < 1$) are:

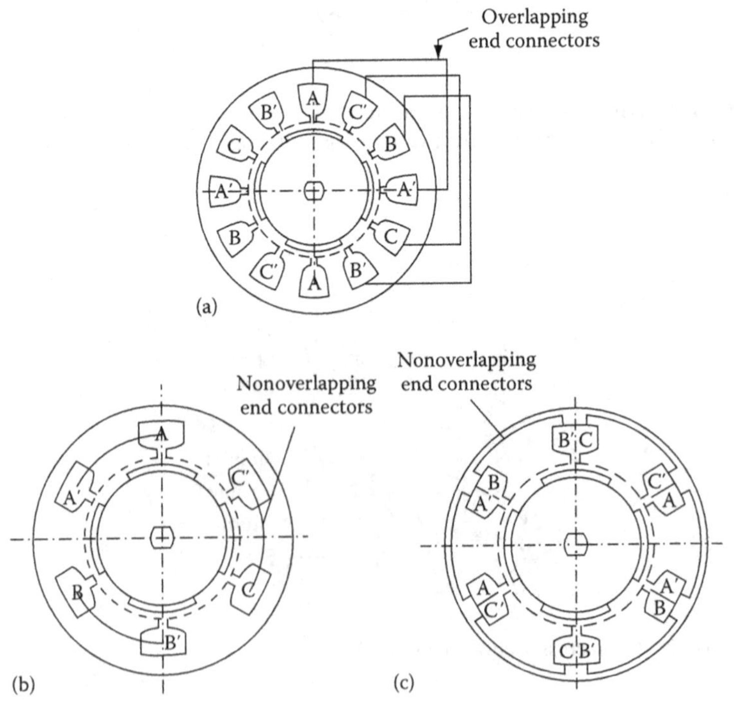

FIGURE 6.11 Four-pole PMSM windings: (a) distributed with $q = 1$ ($N_s = 12$), $2p_1 = 4$; (b) tooth-wound, single-layer ($N_s = 6$), $2p_1 = 4$; and (c) tooth-wound dual-layer ($N_s = 6$), $2p_1 = 4$.

TABLE 6.1
Winding Factors of Concentrated Windings

	2p-Poles															
N_s	2		4		6		8		10		12		14		16	
slots	*	**	*	**	*	**	*	**	*	**	*	**	*	**	*	**
3	*	0.86	*	*	*	*	*	*	*	*	*	*	*	*	*	*
6	*	*	0.866	0.866	*	*	0.866	0.866	0.5	0.5	*	*	*	*	*	*
9	*	*	0.736	0.617	0.667	0.866	0.960	0.945	0.96	0.945	0.667	0.764	0.218	0.473	0.177	0.175
12	*	*	*	*	*	*	0.866	0.866	0.966	0.933	*	*	0.966	0.933	0.866	0.866
15	*	*	*	*	0.247	0.481	0.383	0.621	0.866	0.866	0.808	0.906	0.957	0.951	0.957	0.951
18	*	*	*	*	*	*	0.473	0.543	0.676	0.647	0.866	0.866	0.844	0.902	0.960	0.931
21	*	*	*	*	*	*	0.248	0.468	0.397	0.565	0.622	0.521	0.866	0.866	0.793	0.851
24	*	*	*	*	*	*	*	*	0.930	0.463	*	*	0.561	0.76	0.866	0.866

Source: Boldea, I., *Variable Speed Generators*, Chapter 10, CRC Press, Taylor & Francis Group, New York, 2005.
* One layer.
** Two layers.

TABLE 6.2
LCM of N_s and $2p_1$

N_s	$2p_1$ = 2	4	6	8	10	12	14	16
3	6	12	*	*	*	*	*	*
6	*	12	*	24	30	*	*	*
9	*	**36**	18	72	90	36	126	144
12	*	*	*	24	60	*	84	48
15	*	*	**30**	120	30	60	**210**	240
18	*	*	*	72	90	36	126	144
21	*	*	*	**168**	**210**	**84**	42	**336**
24	*	*	*	*	120	*	168	48

* One layer.

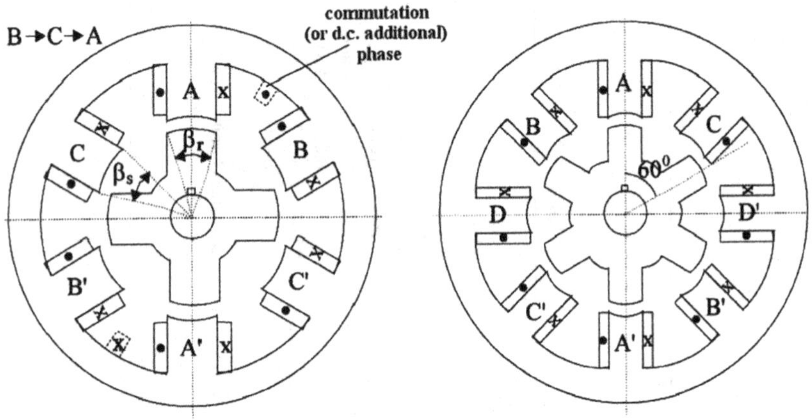

FIGURE 6.12 SRMs: (a) $N_s = 6$, $2p_1 = 4$; 3 phases; and (b) $N_s = 8$, $2p_1 = 6$; 4 phases.

$$\frac{L_{m\nu}}{L_{m1}} = \sum_{\substack{-\infty \\ \nu_1 \neq 1}}^{+\infty} \frac{k_{w\nu}}{k_{w1}} \quad \text{for } q \geq 1 \tag{6.2}$$

$$\frac{L_{m\nu}}{L_{mp_1^*}} = \sum_{\substack{\nu=1 \\ \nu_1 \neq p'}}^{+\infty} \frac{p_1 k_{w\nu}}{k_{wp_1^*}} \quad \text{for } q < 1 \tag{6.3}$$

The total "magnetization" inductance L_{mt} would be:

$$L_{mt} = L_{m1} + L_{m\upsilon}, \text{ for } q \geq 1; \; L_{mt} = L_{mp'} + L_{m\upsilon}, \text{ for } q < 1$$

For same stator slots N_s and 2p rotor pole combinations in concentrated coil windings L_{mt} is 1.5-2 or more than $L_{mp'}$ (say for $N_s = 12$, $2p = 14$) [17] which means:

- Rich space mmf harmonics content with larger additional stator core and rotor losses
- Lower shortcircuit current in generator mode but also larger voltage regulation
- Lower power factor in motoring
- Wider constant power speed range (CPSR) in motoring when fed from a given d.c. link voltage source inverter.

Stepper motors use large numbers of N_s and $2p_1$, and, thus, are good for refined step-by-step motion or slewing; based on reluctance torque (see Chapter 3), they use single-polarity current pulses in open-circuit sequences of phases. In contrast, SRMs have their phase current pulses triggered by the rotor position to preserve synchronism, with rotor motion, phase by phase.

For single-phase configurations, the numbers of stator slots, N_s, and rotor poles are the same (both for PMSMs and SRMs) Figure 6.13. Both they need a self-starting positioning of the rotor by a parking PM or a stepped airgap. The cogging torque is large with $N_s = 2p_1$, but the rotor poles shape may be used to reduce total double-frequency torque pulsations, typical to a.c. single-phase machines.

The same numbers of stator salient poles, N_s, and rotor poles, $2p_1$, ($N_s = 2p_1$) are typical for the so-called TF–PMSM made of 2(3) or more single-phase units placed axially on the shaft with a $2\tau/3$ phase shift for the three-phase case (Figure 6.14). TF–PMSMs enjoy the merit of acquiring the highest torque per watt of copper losses for given PM weight. The PM flux fringing (leakage) is one of their main demerits, besides difficulty in manufacturing and lower power factor.

All these tooth-wound machines, or even TF–PMs [12] or reluctance SMs do not have a cage on the rotor, so they cannot be connected directly to the a.c. power grid. They are fully dependent on full power electronics variable-frequency control. Even

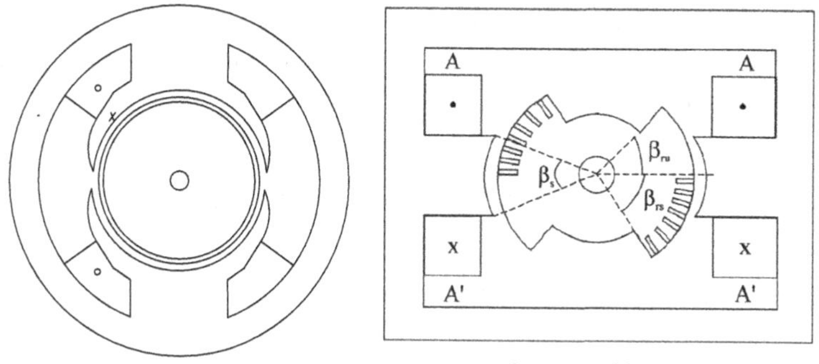

(a) (b) $\beta_{rs} \sim \beta_{ru} \sim 180 - \beta_{ru} - \beta_{rs}$

FIGURE 6.13 Two-pole, self-starting two-slot single-phase tooth-wound, machines: (a) PMSM and (b) SRM.

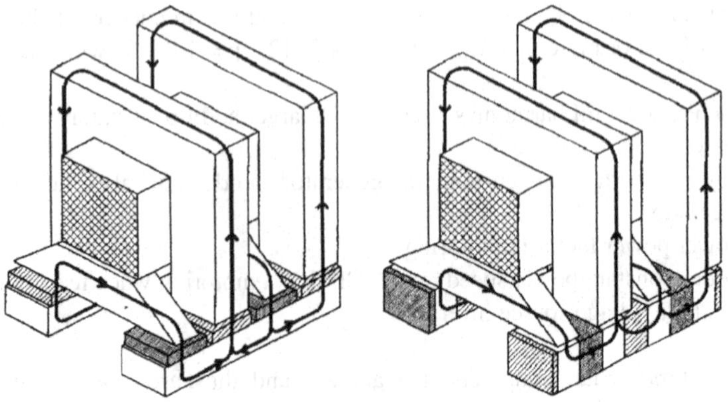

FIGURE 6.14 TF–PMSMs: (a) with surface PM rotors and (b) with interior PM rotors (flux concentration).

if they had a cage rotor, because $N_S \neq 2p_1$ ($N_S = 2p_1 + 2k; k = \pm 1, \pm 2$), the stator mmf has a very rich space harmonics content that would lead to large torque pulsations and additional (eddy current) losses in the rotor core and in the PMs themselves. It goes without saying that axial airgap disk-shaped rotor configurations are also feasible. Also inner or outer rotor configurations have been introduced.

We now return to the SM's distributed armature a.c. windings—with d.c. or PM rotor excitation.

6.3 SM ROTORS: AIRGAP FLUX DENSITY DISTRIBUTION AND EMF

For radial airgap (cylindrical rotor/stator) SMS, there are four basic types of active rotors (Figure 6.15a through d) and variable reluctance (passive) rotors (Figure 6.15e and f).

We will now refer only to active rotors and to armature mmfs.

We already showed in Chapter 5 that a three-phase armature winding produces a traveling mmf/pole distribution:

$$F_1(x_1,t) = F_{1m}\cos\left(\omega_1 t - \frac{\pi}{\tau}x_s - \delta_i\right);\quad F_{1m} = \frac{3\left(W_s k_{\omega_{1s}}\right)I\sqrt{2}}{\pi p_1} \tag{6.4}$$

This wave travels at peripheral (linear) speed:

$$U_s = \frac{\tau}{\pi}\omega_1 = 2\tau f_1 \tag{6.5}$$

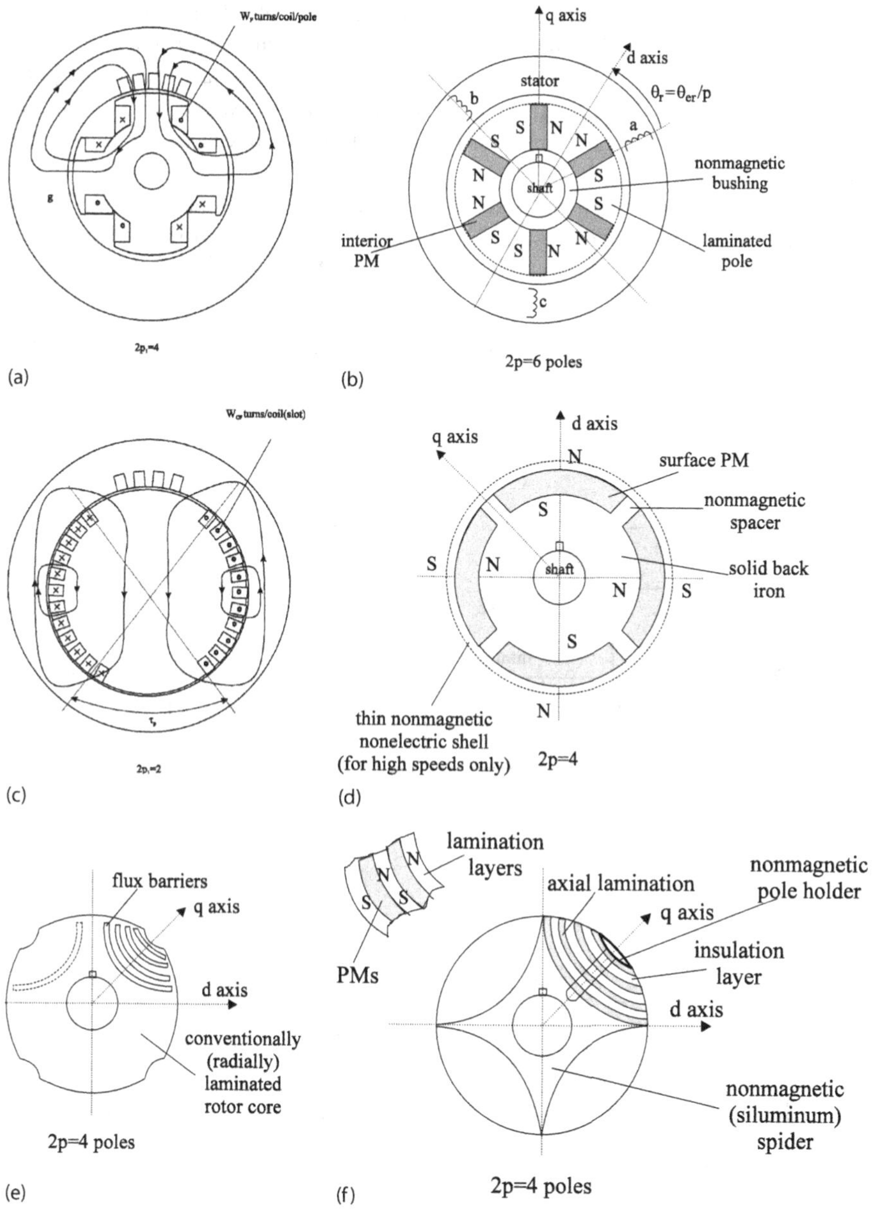

FIGURE 6.15 (a,b) Salient and (c,d) nonsalient poles. (a,c) DC-excited and (b,d) PM rotors. (e) Flux barrier and (f) axially laminated rotors.

which corresponds to the already defined synchronous speed, $n_1 = f_1/p_1$. Now, as demonstrated in Chapter 3, to have a rippless torque, the rotor-produced mmf should travel at the same speed along the stator periphery. If the rotor has a d.c. heteropolar winding (Figure 6.15a) or PMs (Figure 6.15b), their mmf is fixed to the rotor. So, for $\omega_1 = \omega_r = 2\pi p_1 n_1$, the same number of poles has to be produced by the rotor's d.c. excitation or PMs.

With rotor salient poles, the rotor mmf distribution is single stepped (Figure 6.16c) and multiple stepped (Figure 6.16c) for nonsalient poles. So the airgap maximum flux density, B_{gFm}, occurs above the rotor pole shoe:

$$B_{gFm} = \frac{\mu_0 W_F I_F}{K_c g\left(1+K_{s0}\right)}; \quad \text{for } |x| < \tau_p/2; \quad \frac{\tau_p}{\tau} = 0.65 - 0.75 \tag{6.6}$$

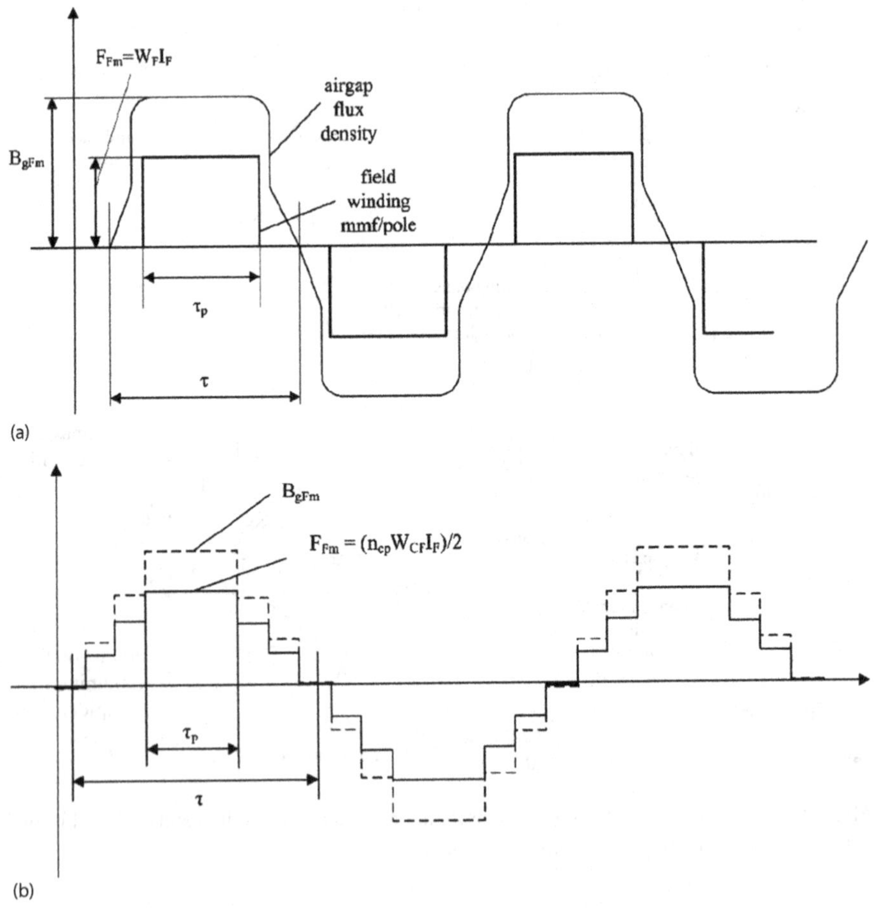

FIGURE 6.16 Excitation mmf and airgap flux density for (a) salient poles and (b) nonsalient poles.

and is zero otherwise for salient poles (Figure 6.16a), and

$$B_{\text{gFm}} = \frac{\mu_0 \left(n_{\text{cp}}/2\right) W_{\text{cF}} I_{\text{F}}}{K_{\text{cg}}\left(1+K_{s0}\right)}; \quad \text{for} \mid x \mid < \tau_p/2; \quad \frac{\tau_p}{\tau} = 0.30-0.4 \tag{6.7}$$

and is stepwise decreasing otherwise (Figure 6.16b).

We may decompose this distribution into a fundamental and harmonics:

$$B_{\text{gF}\nu}\left(x_{\text{r}}\right) = K_{\text{F}\nu} B_{\text{gFm}} \cos \nu \frac{\pi}{\tau} x_{\text{r}}; \quad \nu = 1,3,5 \tag{6.8}$$

with

$$K_{\text{F}_\nu} = \frac{4}{\nu\pi} \sin \nu \frac{\tau_p}{\tau} \frac{\pi}{2} \tag{6.9}$$

for salient rotor poles and

$$K_{\text{F}_\nu} = \frac{8}{\nu^2\pi^2} \frac{\cos \nu \frac{\tau_p}{\tau} \frac{\pi}{2}}{\left(1-\nu \frac{\tau_p}{\tau}\right)}, \tag{6.10}$$

for nonsalient rotor poles.

It is evident that the harmonics content is lower for nonsalient poles than for salient poles.

To reduce the harmonics for the salient pole excitation, the airgap above rotor poles may be gradually increased from the pole center to the pole shoe ends:

$$g\left(x_{\text{r}}\right) = \frac{g}{\cos \frac{\pi}{\tau} x_{\text{r}}} \tag{6.11}$$

This measure will lead to a less harmonics content in the emf induced by motion by the d.c. excitation flux density rotor in the stator (armature) a.c. windings.

Note: The PM airgap flux density distribution is very close to that of salient poles, but, for surface PM poles, the airgap g_{m} includes the PM radial thickness and is constant along the rotor periphery; it is not so for interior PM poles. Trapezoidal, rather than sinusoidal, PM airgap flux density distribution might be intended in variable frequency–fed PMSMs (the so-called brushless d.c. PM machines) with $q = 1$ or $q < 1$ tooth-wound windings, for rectangular bipolar current control.

We will continue by retaining the fundamental of the d.c. exciter or the PM airgap flux density in rotor coordinates, x_{r}:

$$B_{gF_1}\left(x_r\right) = K_{F_1} B_{gF_m} \cos\frac{\pi}{\tau} x_r \tag{6.12}$$

If the speed ω_r is constant, the relationship between the rotor x_r and the stator coordinate x_s is

$$\frac{\pi}{\tau} x_r = \frac{\pi}{\tau} x_s - \omega_r t - \theta_0 \tag{6.13}$$

Substituting Equation 6.12 in Equation 6.13 we obtain ($\theta_0 = 0$):

$$B_{gF_1}\left(x_s\right) = B_{gF_{m1}} \cos\left(\frac{\pi}{\tau} x_s - \omega_r t\right) \tag{6.14}$$

So, seen from the point of view of the stator, the d.c. rotor–excited or PM-produced flux density looks like a traveling wave at rotor speed.

The emf induced by this field in a stator phase winding is

$$E_{A1}\left(t\right) = -\frac{\mathrm{d}}{\mathrm{d}t} W_s k_{w_1} \int_{-\tau/2}^{\tau/2} l_{\text{stack}} \cdot B_{gF_1}\left(x_s, t\right) \mathrm{d}x \tag{6.15}$$

With Equation 6.15 we obtain

$$E_{A1}\left(t\right) = E_1 \sqrt{2} \cos \omega_r t \tag{6.16}$$

$$E_1 = \pi\sqrt{2}\left(\frac{\omega_r}{2\pi}\right) B_{gF_{m1}} \cdot l_{\text{stack}} W_s k_{w_1} \cdot \frac{2\tau}{\pi} \tag{6.17}$$

For three symmetric phases:

$$E_{A,B,C,1}\left(t\right) = E_1 \sqrt{2} \cos\left(\omega_r t - \left(i - 1\right)\frac{2\pi}{3}\right); \quad i = 1,2,3 \tag{6.18}$$

6.3.1 PM Rotor Airgap Flux Density

The extreme PM rotor configurations are shown in Figure 6.15a and b. A few analytical methods to calculate the PM airgap flux density, including the stator slot openings, have been introduced [12], but ultimately the 2D or 3D finite element method (FEM) has to be used for new configurations, at least for the validation of results.

Results such as in Figure 6.17 are obtained for surface PM (nonsalient) poles, interior PM (salient) poles, and surface PM poles, distributed or tooth-wound coil windings, $q < 0.5$.

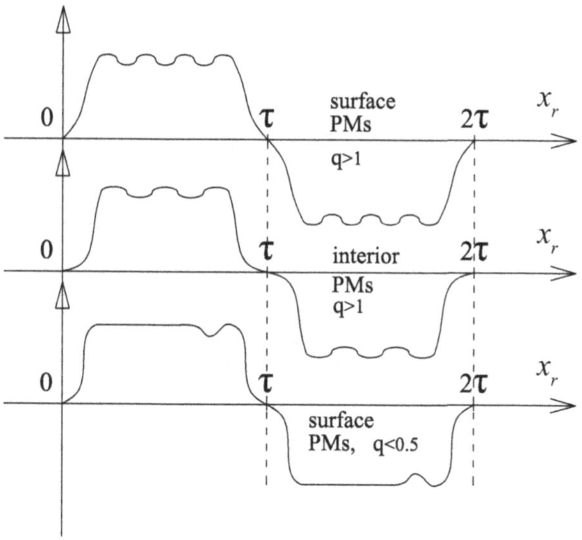

FIGURE 6.17 PM rotor–produced airgap flux density distribution.

6.4 TWO-REACTION PRINCIPLE VIA GENERATOR MODE

Let us start with a d.c. rotor–excited three-phase SG with no load that is driven at speed ω_r (Figure 6.18a). When a balanced a.c. load is connected to the stator, the emfs that occur in the stator have the angular frequency, ω_r. So balanced three-phase stator currents at this frequency are expected.

The phase shift angle between the three-phase emfs and currents, Ψ, depends on the nature of the load (power factor) and on machine parameters (Figure 6.18b).

We may now decompose each phase current into two components, one in phase with the emfs and one at 90° to it: I_{Ad}, I_{Bd}, I_{Cd}, I_{Aq}, I_{Bq}, and I_{Cq}. With sinusoidal emf and current-time variations, phasors may be used for steady-state in SGs. The *d*-axis phase components of phase currents I_{Ad}, I_{Bd}, and I_{Cd} produce a traveling mmf, F_{ad} (x_s, t), aligned to the rotor poles (or to the d.c. excitation flux density) but opposite in sign if defined as

$$F_{ad}\left(x_s,t\right) = -F_{adm}\cos\left(\omega_1 t - \frac{\pi}{\tau}x\right) \tag{6.19}$$

Similarly, for *q* axis, phase components whose mmf is aligned with the interpole rotor axis is obtained as

$$F_{aq}\left(x_s,t\right) = F_{aqm}\cos\left(\omega_1 t - \frac{\pi}{\tau}x - \frac{\pi}{2}\right) \tag{6.20}$$

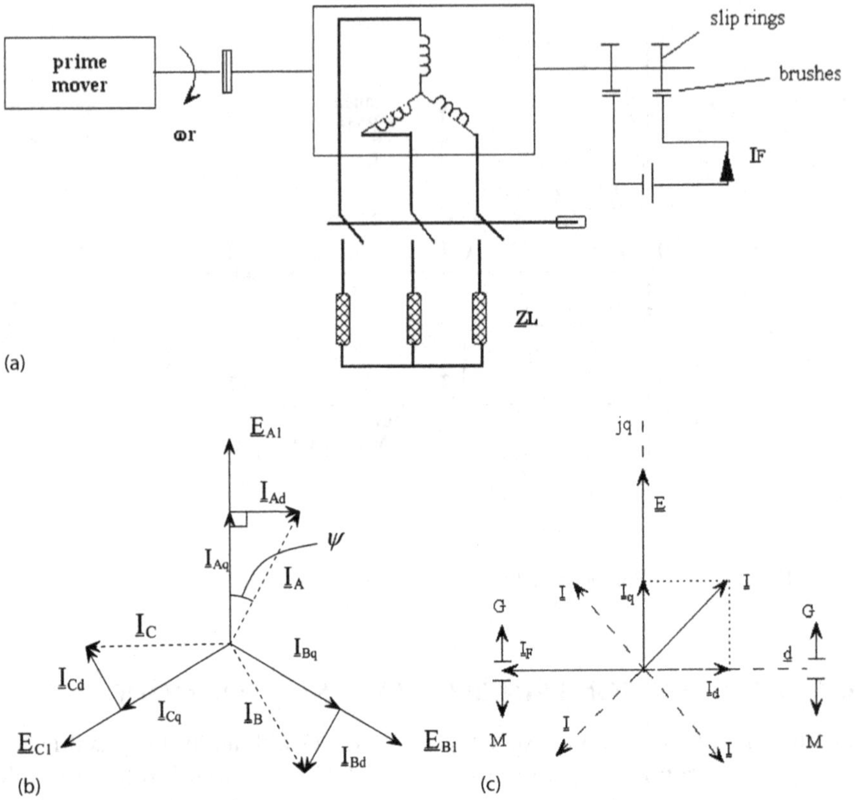

FIGURE 6.18 SG principle: (a) SG at load, (b) emfs and current phasors, and (c) Generator (G)–Motor (M) operation mode divide.

Comparing Equation 6.20 with Equation 6.4:

$$F_{\text{adm}} = \frac{3\sqrt{2}I_{\text{d}}W_{\text{s}}k_{W_{1s}}}{\pi p_1}; \quad I_d = I\cos\delta_i = \left|I_{\text{A,B,C},d}\right|; \quad \psi = \frac{\pi}{2} - \delta_{\text{i}} \tag{6.21}$$

$$F_{\text{aqm}} = \frac{3\sqrt{2}I_q W_{\text{s}}k_{W_{1s}}}{\pi p_1}; \quad I_q = I\sin\delta_{\text{i}} = \left|I_{\text{A,B,C},q}\right| \tag{6.22}$$

The emfs in Equations 6.16 through 6.18 may be expressed as phasors:

$$\underline{E}_{\text{A,B,C}} = -j\omega_{\text{r}}M_{F_a} \cdot \underline{I}_{\text{F A,B,C}} \tag{6.23}$$

Equation 6.23 shows that the emfs are produced by the d.c. rotor excitation through motion. In other words, they are produced by a fictitious three-phase a.c. stalled winding flowed by symmetric fictitious currents I_{FA}, I_{FB}, and I_{FC} of frequency ω_{r}.

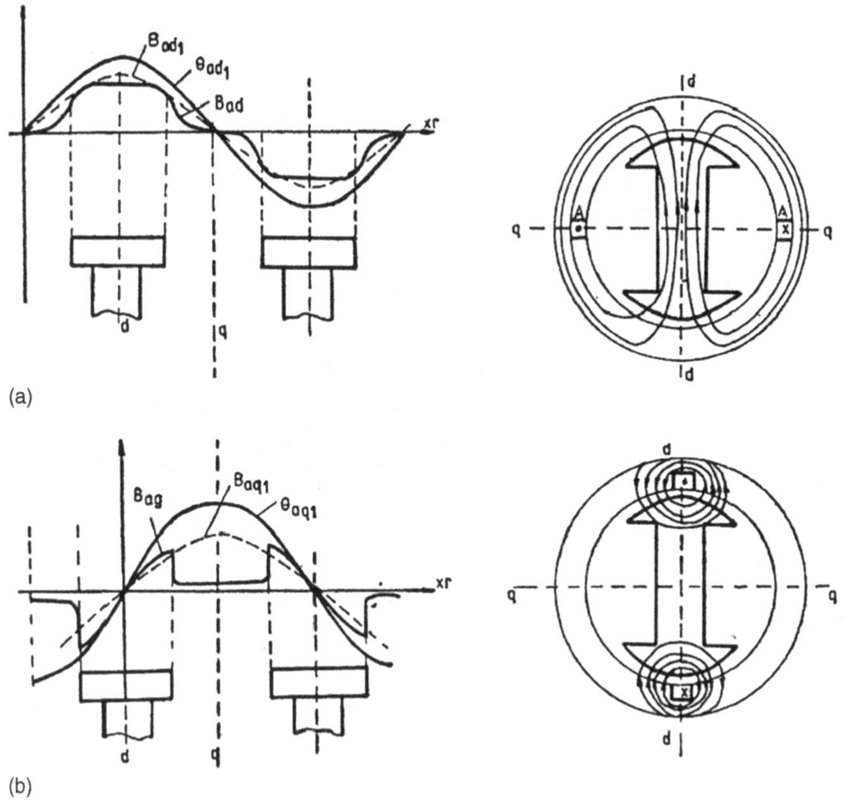

FIGURE 6.19 Phasor diagrams of SMs at unity power factor for (a) generator and (b) motor.

Comparing Equations 6.6 and 6.7 and having I_F as the RMS values of $I_{FA,B,C}$, with Equation 6.23, the mutual inductance, M_{Fa}, is

$$M_{F_a} = \mu_0 \frac{\sqrt{2}}{\pi} \frac{W_s W_F k_{w_{1s}} \cdot \tau l_{stack}}{g k_c (1 + k_s)} \cdot k_{F_1} \tag{6.24}$$

$$W_F = \frac{n_{cp}}{2} W_{CF}; \quad \text{for nonsalient rotor poles} \tag{6.25}$$

The phasor diagram in Figure 6.19a concerns one-phase phasors and the fictitious field current I_F, which is ahead of emf E (Equation 6.23) by 90°, along axis d. Now, in the case of nonsalient rotors the active powers P_{elm} and reactive powers Q_{elm} are produced by the interaction between $E_{A,B,C}$ and $I_{A,B,C}$:

$$\underline{S}_n = P_{elm} + jQ_{elm} = 3\text{Re}\left(\underline{E} \cdot \underline{I}^*\right) + 3\text{Imag}\left(\underline{E} \cdot \underline{I}^*\right) \tag{6.26}$$

Motoring and generating modes are defined solely with respect to the active power (here, positive for the generator and negative for the motor mode). The reactive power, Q_{elm}, may either be positive or negative depending on the excitation (field) current, I_F, level. For an underexcited machine $Q_{elm} < 0$ and for an overexcited machine $Q_{elm} > 0$. So the SG has an extraordinary property of switching from the leading to the lagging power factor just by changing the field current, I_F. This is the key property for achieving voltage control in electric power systems under load source variations.

6.5 ARMATURE REACTION AND MAGNETIZATION REACTANCES, X_{dm} AND X_{qm}

In the previous section, we have introduced the *d* and *q* rotor axes–aligned stator winding mmfs for steady state: $F_{ad}(x_s, t)$ and $F_{aq}(x_s, t)$. Consequently, their airgap flux density distributions, B_{ad} and B_{aq}, can be calculated simply if the equivalent airgap variation along the rotor periphery is defined mathematically. Figure 6.20 and b illustrates these configurations for axes *d* and *q* both, for the salient pole rotor configuration (the nonsalient pole configuration is a particular case of the latter and the PM rotors fall into the same category).

Extracting the fundamentals B_{ad1} and B_{aq1} from the nonsinusoidal airgap flux densities B_{ad} and B_{aq} leads to

$$B_{ad1} = \frac{2}{2\tau}\int_0^{\tau} B_{ad}(x_r)\sin\frac{\pi}{\tau}x_r\, dx_r \tag{6.27}$$

$$\begin{aligned} B_{ad}(x_r) &= 0; \quad 0 \le x_r \le \left(\frac{\tau - \tau_p}{2}\right) \text{ and } \left(\frac{\tau + \tau_p}{2}\right) \le x_r \le \tau \\ &= \frac{\mu_0 F_{adm}\sin\frac{\pi}{\tau}x_r}{k_c g(1 + k_{sd})}; \quad \left(\frac{\tau - \tau_p}{2}\right) < x_r < \left(\frac{\tau + \tau_p}{2}\right) \end{aligned} \tag{6.28}$$

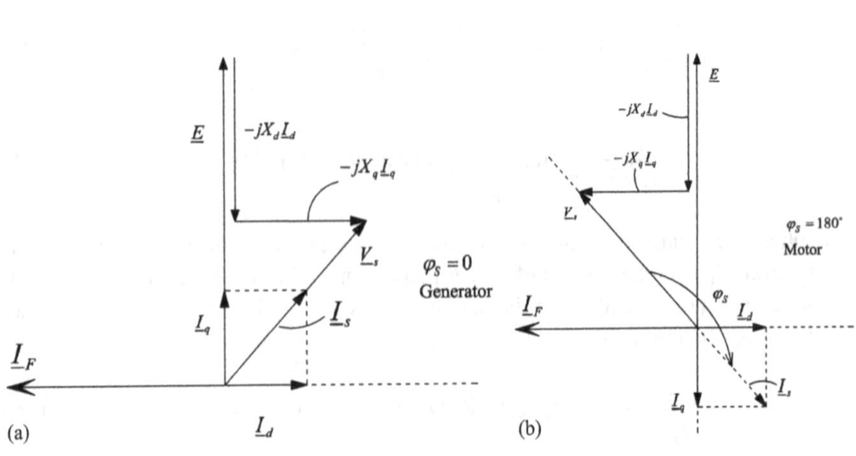

FIGURE 6.20 *d* and *q* axes armature-reaction flux densities and their flux lines.

$$B_{\text{aq1}} = \frac{2}{2\tau}\int_0^\tau B_{\text{aq}}(x_r)\sin\frac{\pi}{\tau}x_r\,dx_r \tag{6.29}$$

with

$$B_{\text{aq}}(x_r) = \begin{cases} \dfrac{\mu_0 F_{\text{aqm}}\sin\frac{\pi}{\tau}x_r}{k_c g(1+k_{sq})}; & 0 \le x_r < \dfrac{\tau_p}{2} \text{ and } \left(\tau - \dfrac{\tau_p}{2}\right) < x_r < \tau \\ \dfrac{\mu_0 F_{\text{aqm}}\sin\frac{\pi}{\tau}x_r}{k_c(6g)}; & \dfrac{\tau_p}{2} < x_r < \left(\tau - \dfrac{\tau_p}{2}\right) \end{cases} \tag{6.30}$$

The airgap between rotor poles for the q axis has been taken as $6g$ (other approximations are welcome).

Finally:

$$B_{\text{ad1}} = \frac{\mu_0 F_{\text{adm}}\cdot k_{\text{d1}}}{k_c g(1+k_{\text{sd}})}; \quad k_{\text{d1}} = \frac{\tau_p}{\tau} + \frac{1}{\pi}\sin\frac{\tau_p}{\tau}\pi < 1 \tag{6.31}$$

$$B_{\text{aq1}} = \frac{\mu_0 F_{\text{aqm}}\cdot k_{\text{q1}}}{k_c g(1+k_{\text{sq}})}; \quad k_{\text{q1}} = \frac{\tau_p}{\tau} - \frac{1}{\pi}\sin\frac{\tau_p}{\tau}\pi + \frac{2}{3\pi}\cos\frac{\tau_p}{\tau}\frac{\pi}{2} < 1 \tag{6.32}$$

For obtaining a uniform airgap, the condition is $k_{\text{d1}} = k_{\text{q1}} = 1$, as is for induction machines. In general, $k_{\text{d1}} = 0.8 - 0.92$ and $k_{\text{q1}} = 0.4 - 0.6$ for regular salient pole rotor d.c.-excited SMs.

Note: For interior PM rotors similar expressions may be derived, but with dedicated equivalent airgap variations and with $k_{\text{d1}} < k_{\text{q1}}$ if the PMs are placed along axis d.

The magnetization inductances along the two axes correspond to B_{ad1} and B_{aq1} and may be derived directly in relation to the already known magnetization inductance, L_{1m}, of IMs (with a uniform airgap), because only the coefficients k_{d1} and k_{q1} are new:

$$L_{\text{1m}} = \frac{6\mu_0\cdot(W_s\cdot k_{W_{1s}})^2\cdot\tau\cdot L_e}{\pi^2\cdot k_c\cdot g\cdot(1+k_s)\cdot p_1}; \quad B_{\text{g1}} = \frac{\mu_0\cdot F_{\text{1m}}}{k_c\cdot g\cdot(1+k_s)} \tag{6.33}$$

$$L_{\text{dm}} = L_{\text{1m}}\cdot k_{\text{d1}} = L_{\text{1m}}\cdot\frac{B_{\text{ad1}}}{B_{\text{g1}}}; \quad L_{\text{qm}} = L_{\text{1m}}\cdot k_{\text{q1}} = L_{\text{1m}}\cdot\frac{B_{\text{aq1}}}{B_{\text{g1}}} \tag{6.34}$$

Magnetic saturation is denoted by the coefficient k_s (k_{sd} and k_{sq} on axes d and q), but more involved treatments are used in industry. So now we state that the symmetric stator current components, I_d and I_q, self-induce a.c. emfs $\underline{E}_{\text{ad}}$, $\underline{E}_{\text{aq}}$, in the stator phase are:

$$\underline{E}_{\text{ad}} = -j\omega_r L_{\text{dm}}\underline{I}_d; \quad \underline{E}_{\text{aq}} = -j\omega_r L_{\text{qm}}\underline{I}_q \tag{6.35}$$

In other words, the above terms are valid strictly for balanced stator voltages and currents under steady state.

As for the IMs, stator cyclic inductances (with all phases flowed by balanced currents), called synchronous inductances L_d and L_q, are

$$L_d = L_{\text{dm}} + L_{\text{sl}}; \quad L_q = L_{\text{qm}} + L_{\text{sl}} \tag{6.36}$$

$$\underline{E}_d = -j\omega_{\text{r}} L_d \underline{I}_d; \quad \underline{E}_q = -j\omega_{\text{r}} L_q \underline{I}_q \tag{6.37}$$

L_{sl} is the stator phase leakage inductance (same as for IMs).

6.6 SYMMETRIC STEADY-STATE EQUATIONS AND PHASOR DIAGRAM

Based on the expression of phase emfs $E_{\text{A,B,C}}$ (Equation 6.23), produced by three fictitious a.c. stator currents, $\underline{I}_{F\text{A,B,C}}$, and also on Equations 6.35 through 6.37, where the armature reaction is sensed in the stator phases as two self-induced emfs, E_{d} and E_{q}, the phasor phase equation of SM—as a generator—under symmetric steady state—is

$$\begin{gathered}\underline{I}_s R_s + \underline{V}_s = \underline{E} + \underline{E}_d + \underline{E}_q; \quad \underline{E} = -j\omega_{\text{r}} M_{\text{Fa}} \underline{I}_{\text{F}} \\ \underline{E}_d = -jX_d \underline{I}_d; \quad \underline{E}_q = -jX_q \underline{I}_q; \quad X_d = \omega_{\text{r}} L_d; \quad X_q = \omega_{\text{r}} L_q \\ \underline{I}_s = \underline{I}_d + j\underline{I}_q; \quad \underline{I}_d = \frac{\underline{I}_{\text{F}}}{I_{\text{F}}} I_d; \quad \underline{I}_q = -j\frac{\underline{I}_{\text{F}}}{I_{\text{F}}} I_q\end{gathered} \tag{6.38}$$

The last equation in 6.38 clarifies the fact that I_{F} and I_d phasors are along the same direction ($I_d \gtrless 0$) and that I_q is lagging for the generator mode and leading for the motor mode by 90°. There are three such equations (one for each phase), with all phasors shifted by ± 120° (electrical). We may combine all emfs into one term:

$$\begin{gathered}\underline{I}_s R_s + \underline{V}_s = -j\omega_{\text{r}} \underline{\Psi}_{s0} = \underline{E}_{\text{res}} \\ \textit{where} \\ \underline{\Psi}_{s0} = M_{\text{Fa}} \underline{I}_{\text{F}} + L_d \underline{I}_d + L_q \underline{I}_q\end{gathered} \tag{6.39}$$

which is the resultant phase flux-linkage phasor.

Multiplying Equation 6.39 by $\underline{I}_s^*$ we may directly obtain the output active and reactive powers of the generator:

$$\begin{gathered}3\underline{V}_s \underline{I}_s^* = P_s + jQ_s \\ P_s = 3X_{\text{Fa}} \cdot I_{\text{E}} I_q - 3I_s^2 R_s + 3(X_d - X_q) I_d I_q \\ Q_s = -3X_{\text{Fa}} \cdot I_{\text{F}} I_d - 3(X_d I_d^2 + X_q I_q^2)\end{gathered} \tag{6.40}$$

As we neglected the core losses so far, the first and last terms added in Equation 6.41 represent the electromagnetic (active) power:

$$P_{elm} = T_e \cdot \frac{\omega_r}{p_1} = 3\omega_r M_{Fa} I_F I_q + 3\omega_r \left(L_d - L_q\right) I_d I_q \tag{6.41}$$

So the electromagnetic torque, T_e, is

$$T_e = 3p_1 \left[M_{Fa} I_F + \left(L_d - L_q\right) I_d\right] I_q; \quad M_{Fa} I_F = \Psi_{PM_d} \tag{6.42}$$

With PMs placed along axis d (instead of electric d.c. excitation), the PM flux linkage (as seen from the stator phases), Ψ_{PM_d}, comes along, when $L_d < L_q$ as well.

The electromagnetic torque has two components:

- The interaction torque (between the excitation [or PM] field and the stator mmf)
- The reluctance torque (for $L_d \neq L_q$) due to stator-produced magnetic energy variation with the rotor position via rotor magnetic saliency

The leading power factor in Equation 6.40 means the positive reactive power Q_s and may be obtained only for negative, demagnetizing, I_d. Changing the sign of torque T_e implies changing the sign of I_q in Equation 6.44, which in fact means a shift of 180° in its phase, as the stator phase current component along the direction of d.c.-excited or PM-produced emf. For large SMs, we may neglect the stator resistance when analyzing the phasor diagram. For the motoring and generating modes, the relations in Equation 6.38 lead to simplified phasor diagrams as in Figure 6.19, for unity power factor.

Unity power factor may be maintained when load (I_q) varies only if the field current can be varied. For the nonsalient pole rotor SMs, $X_d = X_q = X_s$, and, thus, the voltage equation in Equation 6.38 becomes

$$\underline{I}_s R_s + \underline{V}_s = \underline{E} - jX_s \underline{I}_s \tag{6.43}$$

which leads to the equivalent circuit in Figure 6.21.

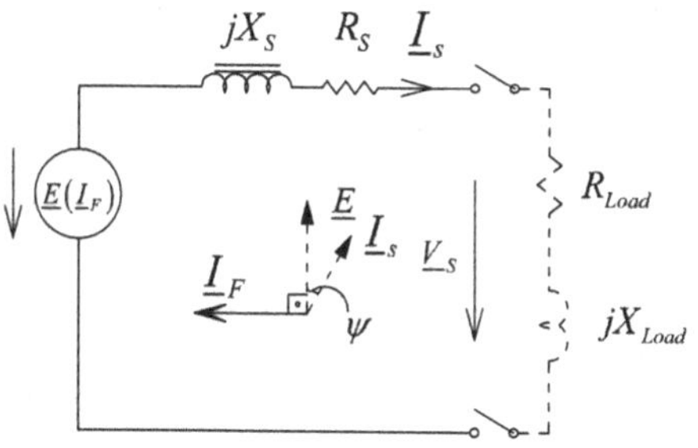

FIGURE 6.21 Equivalent circuit of SM with nonsalient pole rotor.

For nonsalient pole rotor ($X_d \neq X_q$) SMs, the equivalent circuit is more complicated to include the reluctance torque (power). For PMSMs, simply I_F = const.

6.7 AUTONOMOUS SYNCHRONOUS GENERATORS

ASGs have found many applications in automobiles, trucks, buses, diesel-electric locomotives, avionics (at 400 Hz), vessel, cogeneration, wind generators, standby sources for telecom, hospitals, and remote areas, etc. The power per unit varies from a few mega volt amperes to 1 kW or less. A few characteristics of ASGs that define their operation are

- No-load saturation curve ($E(I_F)$, $I_s = 0$, n_1 = const.)
- Short-circuit curve (I_{sc} (I_F), $V_s = 0$, n_1 = const.)
- External (load) curve (V_s (I_s) for n_1 = const., cos φ_s = const.)

6.7.1 No-Load Saturation Curve/Lab 6.1

The no-load saturation (magnetization) curve may be obtained through computation by the same analytical method as for the brush–commutator machine, point by point, for given pole flux Φ_p. The FEM computation of this curve is also common. This curve may also be obtained through a dedicated test (which is standardized).

$$E_1\left(I_F\right) = \frac{\omega_r}{\sqrt{2}}\frac{2}{\pi}\tau B_{gFm}\left(I_F\right)\cdot k_{1F} \times l_{stack}\cdot W_s k_{w_{1s}}\ \left(V_{(RMS)}\right) \tag{6.44}$$

The experimental arrangement (Figure 6.22a) consists a prime mover (a variable speed drive with refined speed control) that drives the ASG at $n_1 = \frac{f_{1n}}{p_1}$. A variable d.c.–d.c. source can provide a variable field I_F within very large limits (1–100). The measured variables are the stator emf E_1, the frequency f_1, and the field current I_F. For a monotonous variation of I_F from zero up and down, even the cycle of hysteresis of $E_1(I_F)$ is obtained (Figure 6.22b). The test should be done for up to E_{1max}/V_n = (1.2–1.5), depending on the lowest power factor of load envisaged for rated voltage and current. The average (dotted) curve in Figure 6.22b constitutes the no-load saturation curve. For maximum emf, E_{1max}, it is essential that the maximum field current be limited, as the cooling system of the machine should handle the situation adequately.

6.7.2 Short-Circuit Curve: ($I_{sc}(I_F)$)/Lab 6.2

The short-circuit curve is obtained experimentally with the SG already short-circuited at stator terminals and driven at rated speed, $n_1 = f_{1n}/p_1$, by a refined speed-controlled prime mover (Figure 6.23a). The curve is linear (Figure 6.23b) due to a low amount of magnetic saturation in the SG on account of the "purely" demagnetized (I_d only) content of the stator current, I_{sc}, if R_s is neglected (for no torque: $I_q = 0$) (Figure 6.23c and d). From Equation 6.43 with $I_{sc} = I_d$, $V_s = 0$:

$$\underline{E}_1\left(I_F\right) = jX_{dmsat}\underline{I}_{3sc} \tag{6.45}$$

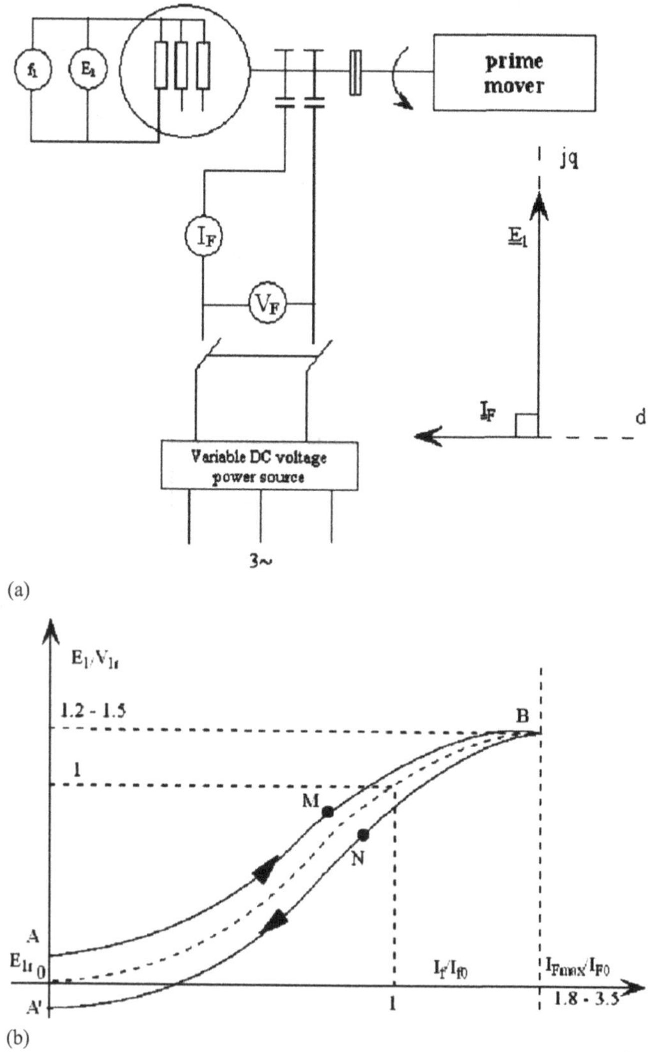

FIGURE 6.22 No-load-curve test: (a) test rig and (b) the characteristic curves.

The resultant emf produced by the main (airgap) field in the machine is given by

$$\underline{E}_{1\text{res}}\left(I_{\text{F}}\right) = \underline{E}_{1}\left(I_{\text{F}}\right) - jX_{\text{dmunsat}}\underline{I}_{3\text{sc}} = -jX_{\text{sl}}\underline{I}_{\text{sc}3} \tag{6.46}$$

$E_{1\text{res}}$ refers to a low value of E_1 in the no-load curve (Figure 6.23e), which justifies nonsaturation claims for the magnetic circuit. An equivalent I_{F0} field current may be defined for it:

$$I_{F_0} = I_{\text{F}} - I_{3\text{sc}} \cdot \frac{X_{\text{dm}}}{X_{\text{Fa}}} \tag{6.47}$$

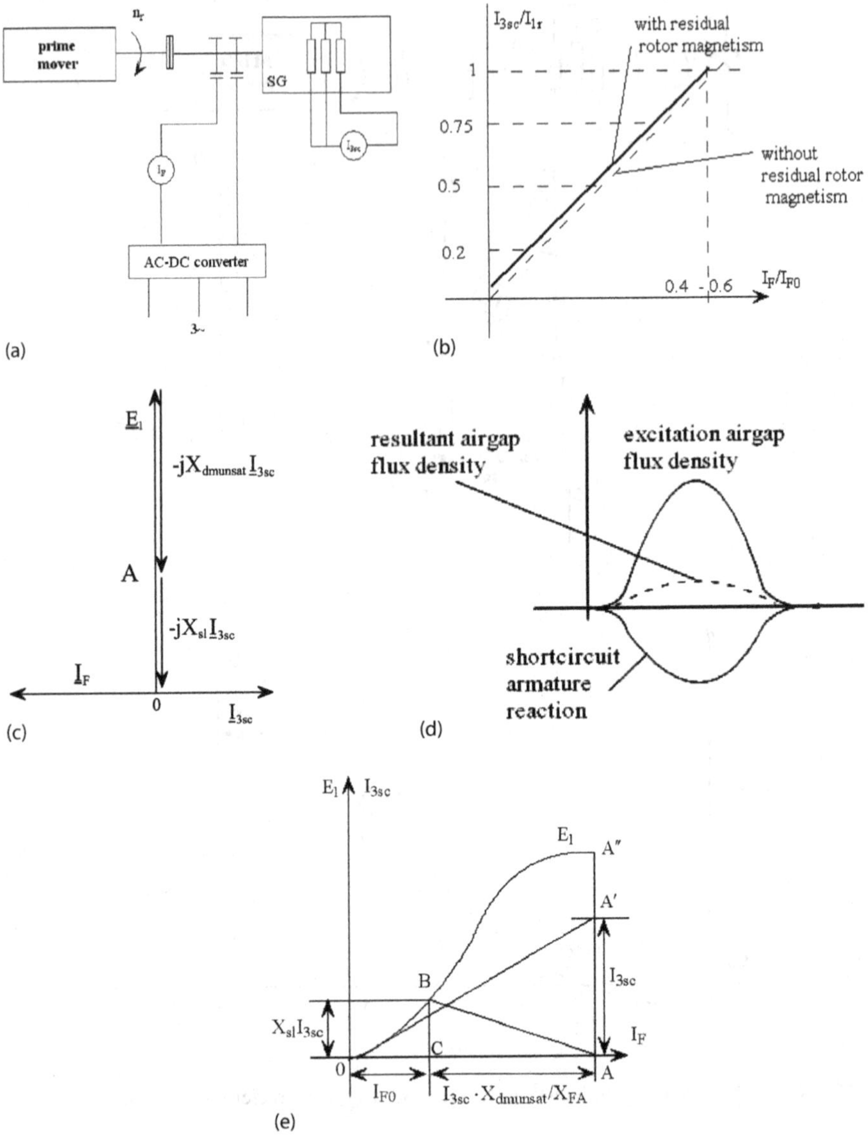

FIGURE 6.23 Short-circuit curve: (a) test rig, (b) the curve, (c) phasor diagram, (d) flux density, and (e) short-circuit triangle.

This is how the short-circuit triangle ABC (Figure 6.23e) was born.

While X_d in Equation 6.46 corresponds to I_{F_0}, the saturated value corresponds to I_F, and is obtained from

$$X_{dsat} = \frac{E_1(I_F)}{I_{3sc}(I_F)} = \frac{\overline{AA''}}{\overline{AA'}} \tag{6.48}$$

6.7.3 Load Curve: $V_s(I_s)$/Lab 6.3

The load curve refers to terminal line (or phase) voltage, V_s, versus line current, I_s, for given speed, field current, and load power factor. For practical applications, the speed is kept constant through the speed controller (governor) of the prime mover, while the voltage is controlled to remain constant within given limits of load variation, through field current control. The load curves should only make sure that the design (tested) ASG will be capable to, say, provide the rated output voltage for the rated load at the lowest-settled lagging power factor level. For small- and medium-power SGs, a direct $R_L L_L$ a.c. load may be applied (as in Figure 6.24). Then the voltage equation (in Equation 6.36) is completed with the load equation:

$$\underline{V}_s = R_L \underline{I}_s + j\omega_r L_L \underline{I}_s; \quad \cos\varphi_L = \frac{R_L}{\sqrt{R_L^2 + \omega_r^2 L_L^2}} = \text{const.} \tag{6.49}$$

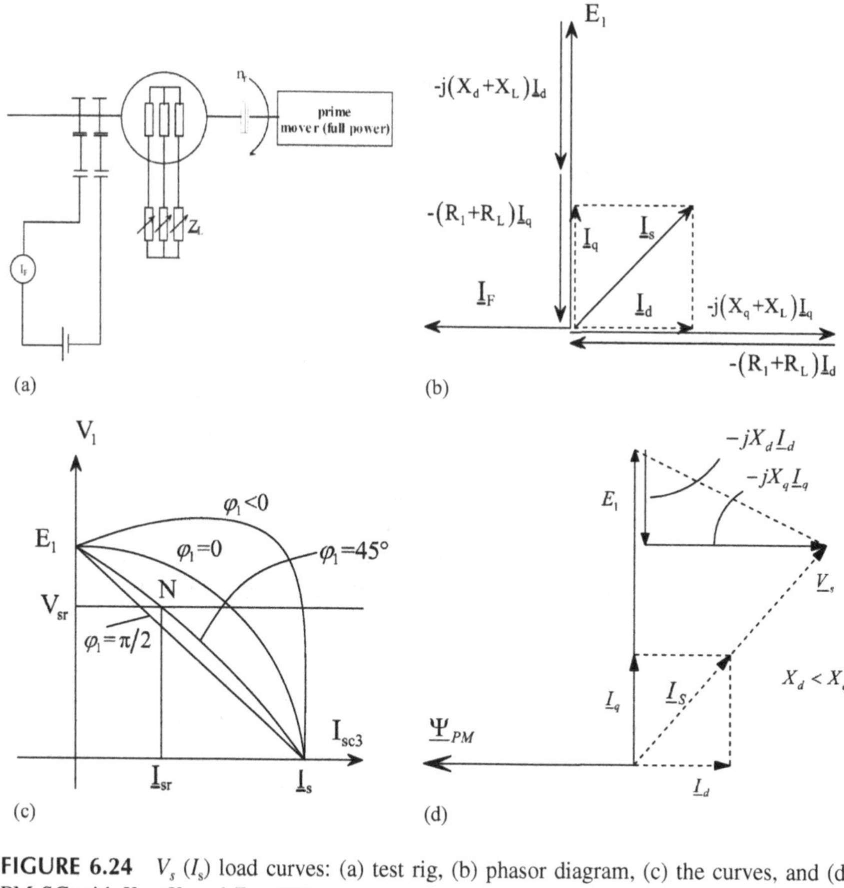

FIGURE 6.24 V_s (I_s) load curves: (a) test rig, (b) phasor diagram, (c) the curves, and (d) PM-SG with $X_d < X_q$ and $E_1 = (V_s)_{In}$.

to obtain

$$
\begin{aligned}
E_1(I_F) &= I_q(R_s + R_L) + (X_d + X_L)(-I_d); \quad I_d \gtrless 0 \\
0 &= (R_s + R_L)(-I_d) - (X_q + X_L)I_q; \quad I_q > 0
\end{aligned}
\tag{6.50}
$$

Assigning values to I_F, $\cos\varphi_L$, and then to $Z_L = R_L/\cos\varphi_L$, the values of I_d and I_q may be calculated from Equation 6.50. Then the active and reactive powers, P_s and Q_s, can be calculated, together with the terminal voltage:

$$
V_s = \frac{R_L}{\cos\varphi_L} I_s; \quad I_s = \sqrt{I_d^2 + I_q^2} \tag{6.51}
$$

The no-load curve, $E_1(I_F)$, is used when I_F varies. It is also true that X_d (and even X_q) varies with magnetic saturation but this aspect is beyond our aim here. Load curves such as in Figure 6.24c are obtained.

The voltage regulation ΔV_s pointed out for the transformer or autonomous induction generator, is substantial for ASGs, because X_d and X_q are large:

$$
\Delta V_s = \frac{E_1 - V_s}{E_1} = \frac{\text{no-load voltage} - \text{load voltage}}{\text{no-load voltage}} \tag{6.52}
$$

If, for $\cos\varphi_{sc} = 0.707$ lagging minimum power factor, the rated voltage is still desired at the rated load current; it follows clearly from Figure 6.24 that

$$
\frac{X_{d\,\text{sat}}}{Z_n} = x_{d\,\text{sat}} < 1; \quad Z_n = V_n/I_n \tag{6.53}
$$

where x_d is the d-axis reactance in pu values. For $x_{d\,\text{sat}} < 1$, large airgap values are required. But this means that a larger d.c. excitation mmf per pole would be required, and, thus, the excitation Joule losses would also be large. For PMSGs, the option of controlled field current is lost, but for interior PM pole rotors ($X_d < X_q$) with inversed saliency, it is possible, for the resistive load, to obtain zero voltage regulation at the rated load by design (Figure 6.24d). For some applications this is enough, provided speed is maintained constant by the prime mover. SGs on automobiles are likely to operate either with diode rectifiers or with forced controlled rectifier loads. For such situations, the above considerations are a basis for further developments [13].

Example 6.1: Autonomous SG

Let us consider an autonomous SG with S_n = 1MVA, $2p_1$ = 4, x_d = 0.6(pu), x_q = 0.4(pu), and V_{nl} = 6 kV (star connection). Calculate the emf E_1 (per phase) at the rated current to preserve the rated voltage, for resistive ($\cos\varphi_L = 1$), resistive–inductive ($\cos\varphi_L = 0.707$), and resistive–capacitive ($\cos\varphi_L = 0.707$) loads.

Solution:

We have to go back to the general phasor diagram for nonunity power factor conditions and zero losses (Figure 6.25).

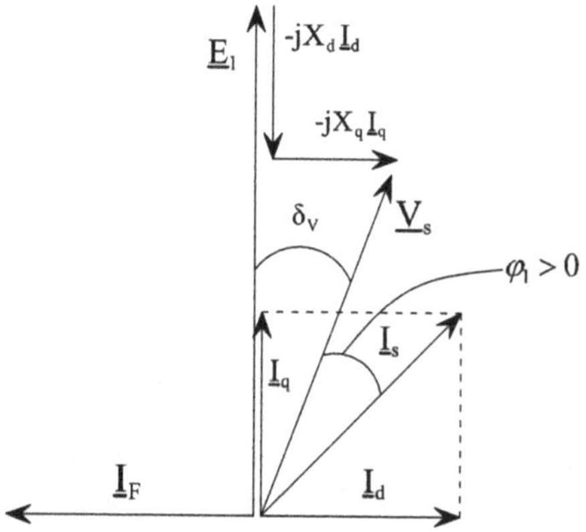

FIGURE 6.25 Phasor diagram of SG for zero losses.

The rated current I_n is defined here for unity power factor (and zero losses):

$$I_n = \frac{S_n}{\sqrt{3}V_{nl}} = \frac{1\times10^6}{\sqrt{3}\times6\times10^3} = 96.34\ [\text{A}] \tag{6.54}$$

From the rated current and cos $\varphi_L = 1$, the resistive load R_L is

$$R_{\text{load}} = \frac{V_{nl}\cdot\cos\varphi_L}{\sqrt{3}I_n} = \frac{6000\times1}{\sqrt{3}\times96.34} = 36\ [\Omega];\quad Z_n = \frac{V_{nl}/\sqrt{3}}{I_n} = 36\ [\Omega] \tag{6.55}$$

Then, from Equation 6.50, with $X_{\text{load}} = 0$, and Equation 6.54 we may calculate E_1, I_d, and I_q:

$$\frac{(-I_d)}{I_q} = \frac{X_q}{R_{\text{load}}} = \frac{x_q}{R_{\text{load}}}\frac{V_{nl}/\sqrt{3}}{I_n} \tag{6.56}$$

But, from $I_n = \sqrt{I_d^2 + I_q^2}$; $I_n = I_q\sqrt{\left(\frac{I_d}{I_q}\right)^2 + 1}$

$$I_q = I_n/\sqrt{\left(\frac{I_d}{I_q}\right)^2 + 1} = \frac{96.34}{\sqrt{1+0.4^2}} = 89.45\,[\text{A}]$$
$$I_d = -0.4\times I_q = -0.4\times89.45 = -35.78\,[\text{A}]$$

So, from Equation 6.50 the emf E_1 is

$$E_1 = I_q\times R_L + X_d(-I_d) = 89.45\times36 + 0.6\times36\times35.78$$
$$= 3993\,[\text{V}]$$

The rated phase voltage $V_{nph} = V_{nl}/\sqrt{3} = 6000/\sqrt{3} = 3468.2\,[\text{V}]$.

So the rated voltage regulation

$$(\Delta V)_{I_n,\cos\varphi_L=1} = \frac{E_1 - V_{nph}}{E_1} = \frac{3993-3468.2}{3993} = 0.131 = 13.1\%$$

For the other values of the load factor, the computation routine is the same as above and voltage regulation will be larger for the resistive–inductive load and negative for the resistive–capacitive load ($E_1 < V_{nph}$).

6.8 SYNCHRONOUS GENERATORS AT POWER GRID/LAB 6.4

Every SG, new or repaired, has to be connected to the power grid. The power grid is made of all SGs connected in parallel and the power lines, with step-up and step-down transformers, that deliver electric energy to various consumers.

Let us assume that the power grid has "infinite power" or zero internal (series) impedance, and thus its voltages, amplitudes and phases, do not change when an additional SG is connected. The connection of an SG to the power grid has to take place without large current (power) transients. To do so, the SG and power grid line voltages have to have the same sequences, amplitudes, and phases (frequencies also). This synchronization process is done today automatically through so-called digital synchronizers (Figure 6.26), which do coordinated voltage, frequency (speed), and phase control to minimize connection transients.

In a university lab, however, manual synchronization is performed, with two voltmeters and 3 × 2 light bulbs in series between the SG and the power grid. To adjust the SG voltages, the field current, I_F, is modified, while frequency (phase)-refined adjustment is done until all light bulbs are dark. Then the three-phase power switch is closed. There will be some transients and the machine rotor oscillates a little and then settles down at synchronous speed. To deliver active power by the generator, the

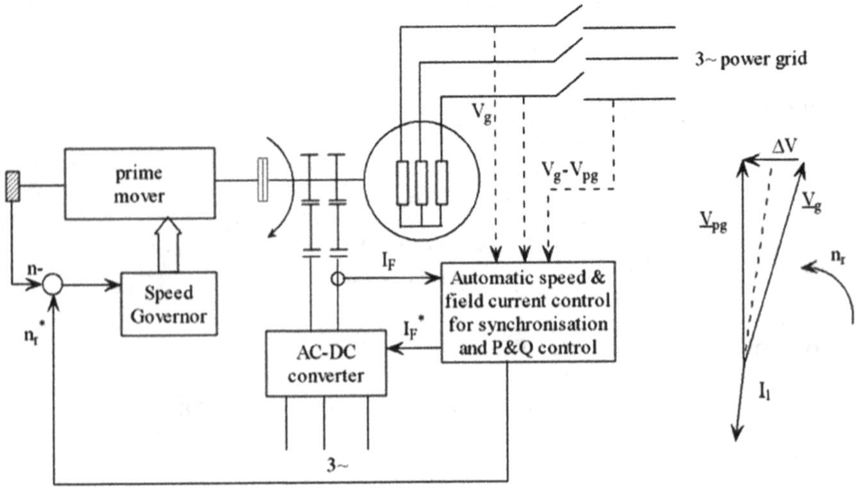

FIGURE 6.26 Connecting the SG to the power grid.

prime mover power is increased (hence water or fuel intake is increased in real turbines). To deliver reactive power, the field current is increased.

The SG operation at the power grid is characterized by three main curves:

- Active power/angle curve, $P_e\ (\delta_V)$
- V-shaped curves ($I_s\ (I_F)$, P_s = const., V_s = const., $n = n_1$ = const.)
- Reactive power capability curves, $Q_s\ (P_s)$

6.8.1 Active Power/Angle Curves: $P_e(\delta_V)$

The voltage power angle, δ_V, is the phase shift angle between the emf, E_1, and the phase voltage, V_s (see Figure 6.25). The δ_V concept (variable) can be used either for autonomous or for power grid operation, but the latter case is typical as it relates directly to the SG static and dynamic stabilities (still to be defined). The active and reactive powers have already been expressed in Equations 6.40 for a lossless SG. Here we add the I_d and I_q components expressions from Figure 6.25:

$$I_d = \frac{V_s \cos\delta_V - E_1}{X_d}; \quad I_q = \frac{V_s \sin\delta_V}{X_q} \tag{6.57}$$

From Equations 6.40, and Equation 6.57, the active (electromagnetic) and reactive, P_s and Q_s, powers are

$$P_s = \frac{3E_1V_s \sin\delta_V}{X_d} + \frac{3}{2}V_s^2\left(\frac{1}{X_q} - \frac{1}{X_d}\right)\sin 2\delta_V \tag{6.58}$$

$$Q_s = \frac{3E_1V_s \cos\delta_V}{X_d} - 3V_s^2\left(\frac{\cos^2\delta_V}{X_d} + \frac{\sin^2\delta_V}{X_q}\right) \tag{6.59}$$

Note again that for PMSMs $E_1 = \omega_r\Psi_{PM}$ while for d.c.-excited SGs $E_1 = \omega_r M_{Fd} I_F$; $E_1 =$ 0 for reluctance ($X_d > X_q$) SMs.

The $P_s\ (\delta_V)$ and $Q_s\ (\delta_V)$ curves from Equations 6.58 and 6.59 are shown in Figure 6.27.

As the voltage equation's association of signs was taken for the generator mode, negative active power means the motor mode. Positive voltage power angle (from zero to ⊕180°) means generator operation and negative voltage power angle (from zero to ⊖180°) refers to motor operation. However, not all these zones provide stable operation.

A few remarks are in order:

- δ_V increases up to δ_{VK}, which corresponds to the maximum active power available (for given E_1 (field current) value).
- The reactive power (for constant E_1) changes from leading to lagging when δ_V increases either in motor or in generator modes.

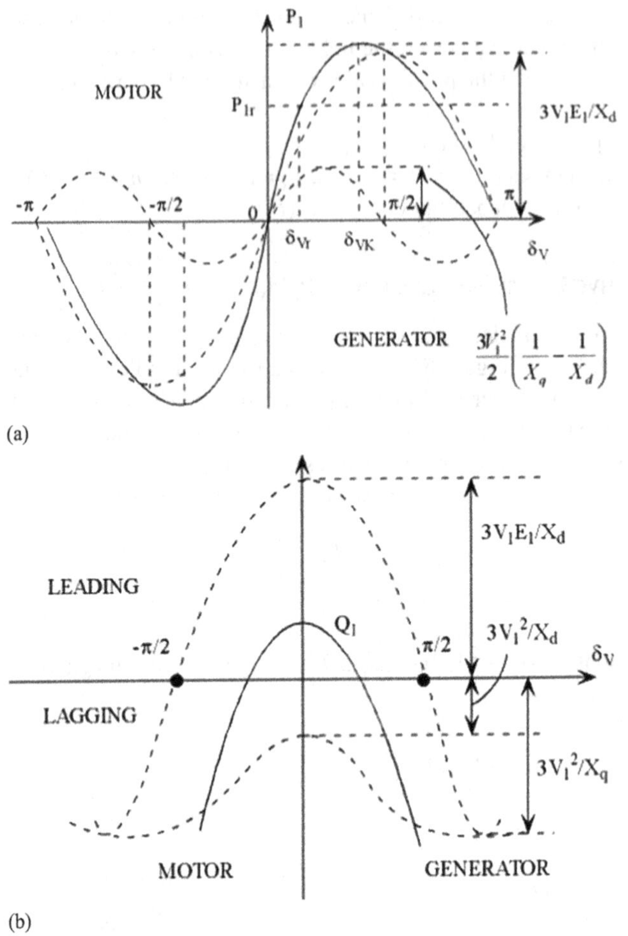

FIGURE 6.27 (a) Active and (b) reactive powers, P_s and Q_s, *vs.* voltage power angle, δ_V.

- For the salient pole rotor, $\delta_{VK} < 90°$.
- The rated (design) voltage power angle is chosen in the 22°–30° range for nonsalient pole rotor SGs and 30°–40° range for salient pole rotor SGs, because of the larger inertia in the latter.

6.8.2 V-Shaped Curves

The V-shaped curves represent a family of I_s (I_F) curves drawn for constant terminal voltage (V_S), and speed ($\omega_r = \omega_1$), with active power P_s as a parameter. The computation of the V-shaped curves makes use of the P_s expression 6.58 and the no-load curve, $E_1(I_F)$, with known values of synchronous reactances, X_d and X_q. As the terminal voltage is kept constant, the total flux linkage, $\Psi_s \approx V_s/\omega_1$, is constant, and, thus, X_d and X_q do not vary much when I_F varies in steps, to build the V-shape curve.

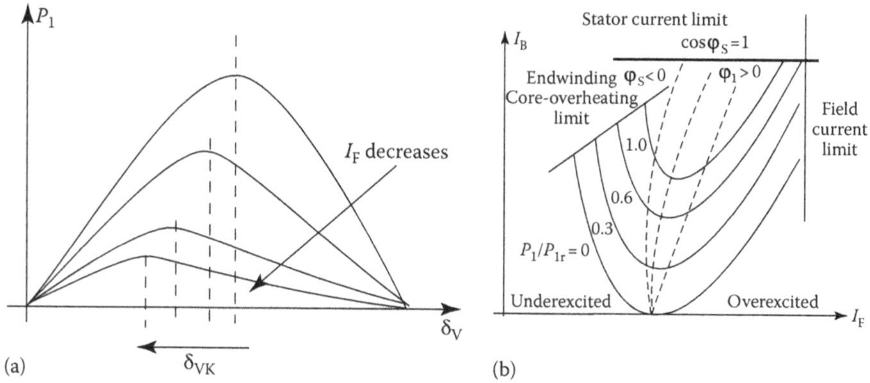

FIGURE 6.28 V-shaped curves: (a) P_s (δ_V) curves and (b) I_s (I_F) – V – curves.

The computation sequence is straightforward:

- Assign gradually decreasing values of I_F (from the maximum value allowed due to cooling reasons) and, for a given value of P_s, calculate the power angle, δ_V, from Equation 6.58 after "reading" E_1 from $E_1(I_F)$ curve, which is already known.
- Then, from Equation 6.57, with E_1 and δ_V known, I_d and I_q are calculated; finally the stator phase current is calculated using $I_s = \sqrt{I_d^2 + I_q^2}$.
- When decreasing I_F, at some point, $\delta_V = \delta_{VK}$; this is the last point on the V-shaped curves, to provide at least static stability (Figure 6.28).

The minimum stator current (for given power and voltage) corresponds to unity power factor:

$$P_s = 3V_s I_s \cos\varphi_s \tag{6.60}$$

With the machine underexcited (smaller I_F values) the power factor is lagging; it is leading for larger I_F values. Unity power factor corresponds to $Q_s = 0$ in Equation 6.59:

$$\left(E_1\right)_{Q_s=0} = V_s\left(\cos\delta_V + \frac{X_d}{X_q}\frac{\sin^2\delta_V}{\cos\delta_V}\right) \tag{6.61}$$

Maintaining unity power factor with increases in active power (and δ_V) corresponds to an increase in E_1 (or in I_F).

6.8.3 Reactive Power Capability Curves

The maximum limitation of I_F is due to thermal reasons. However, a machine's overall and hot spot temperatures depend on both currents: I_F and I_s. Also, I_s, I_F, and δ_V determine the stator flux, and thus the core losses. So, when the reactive power demand from an SG is increased, it implies an increase in the field current (I_F) and, thus, at some point, the stator current (I_s) and the active power have to be limited also.

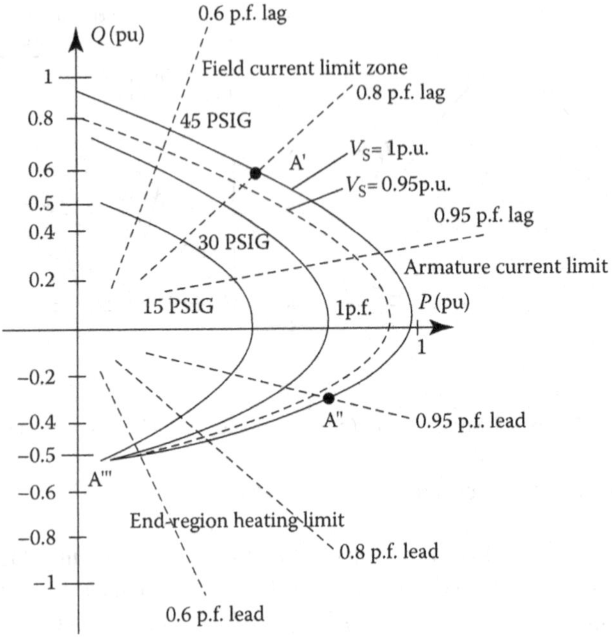

FIGURE 6.29 Reactive/active power limit curves for a hydrogen-cooled SG.

The computation process for the V-shaped curves may be taken just one step further to calculate the reactive power from Equation 6.59, and then represent the Q_s (P_s) curves (Figure 6.29).

For underexcited machines $Q_s < 0$, but for this case, the d-axis reaction field of the stator (of I_d) is added to the d-axis field of excitation. In the end turns and end core zones, a large total a.c. magnetic field is produced, which creates additional, end-region, stator core losses that limit the maximum absorbed reactive power by the machine (Figure 6.29).

6.9 BASIC STATIC- AND DYNAMIC-STABILITY CONCEPTS

We have inferred in the previous section that an SG can deliver active power up to the critical voltage power angle, $\delta_{VK} \le \pm 90°$, if its loading is going up or down very slowly.

Static stability is the property of an SG to remain in synchronism in the presence of very slow load variations. It can be noticed that as long as the electric power, P_s, delivered by an SG increases with an increase in the prime mover output (mechanical power), the machine remains statically stable.

In other words, as long as $\partial P_s/\partial \delta_V > 0$ the machine is statically stable (Equation 6.58):

$$P_{ss} = \frac{\partial P_s}{\partial \delta_V} = 3E_1 V_s \cos \delta_V + 3V_s^2 \left(\frac{1}{X_q} - \frac{1}{X_d} \right) \cos 2\delta_V \tag{6.62}$$

P_{ss} is called the synchronizing power; as long as P_{ss} is positive the SM is statically stable ($P_{ss} = 0$ for δ_{VK}!). When the field current decreases (E_1 decreases), P_{ss} decreases, and thus δ_{VK} increases and reduces the statically stable region.

Dynamic stability is the property of an SM to remain in synchronism (with the power grid) for quick variations of the shaft (mechanical) power or of the electric power (load). In general, when the shaft inertia (of a prime mover and an electric generator) is large, the power angle (δ_V), speed (ω_r) transients for load variations are much slower than electric transients. So, for basic calculations, during sudden mechanical loads, the SG may still be considered at steady state with torque T_e:

$$T_e = \frac{P_s \times p_1}{\omega_r} = \frac{3p_1}{\omega_r}\left[\frac{E_1 V_s \sin\delta_V}{X_d} + \frac{V_s^2}{2}\left(\frac{1}{X_q} - \frac{1}{X_d}\right)\sin 2\delta_V\right] \tag{6.63}$$

Let us now consider the case of the nonsalient pole rotor SG to simplify T_e ($X_d = X_q$) (Figure 6.30).

Neglecting also the rotor cage torque during transients (which would otherwise be beneficial), the shaft motion equations during mechanical (shaft) load variation from P_{sh1} to P_{sh2} (Figure 6.30) are

$$\frac{J}{p_1}\frac{d\omega_r}{dt} = T_{shaft} - T_e; \quad \omega_r - \omega_{r0} = \frac{d\delta_V}{dt} \tag{6.64}$$

Multiplying Equation 6.64 by $d\delta_V/dt$, one obtains

$$d\left(\frac{J}{2p_1}\left(\frac{d\delta_V}{dt}\right)^2\right) = \left(T_{shaft} - T_e\right)d\delta_V = \Delta T \cdot d\delta_V = dW \tag{6.65}$$

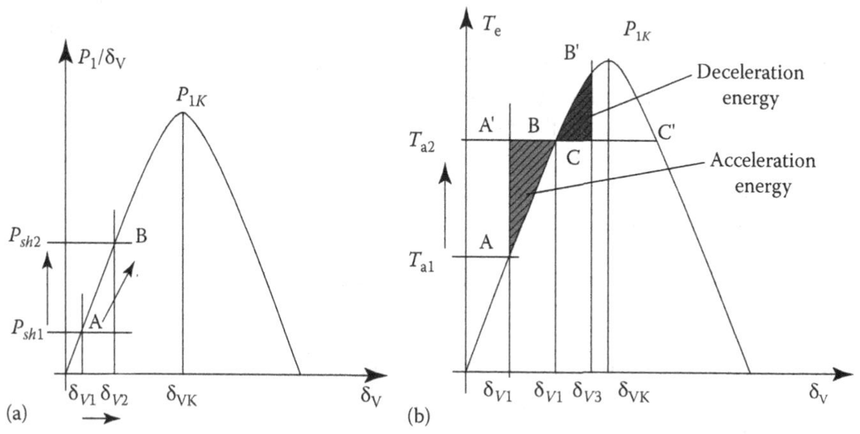

FIGURE 6.30 Dynamic stability: (a) P_s (δ_V) and (b) the criterion of equal areas.

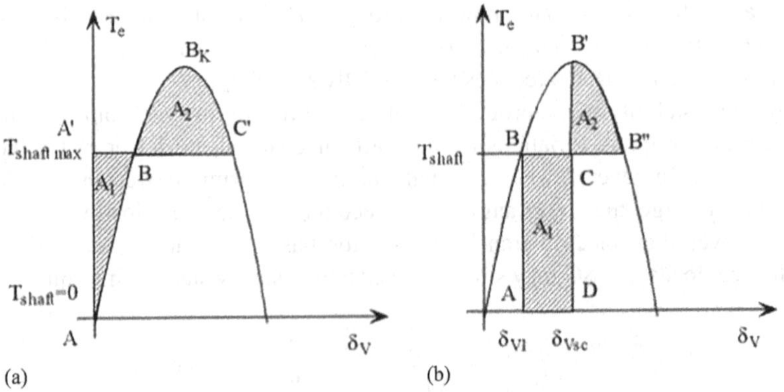

FIGURE 6.31 Dynamic-stability limits: (a) maximum allowable shaft torque step up from zero and (b) maximum short-circuit clearing time (angle $\delta_{Vsc} - \delta_{V1}$).

Equation 6.65 only illustrates that the kinetic energy variation of the mover (shaft) is translated into an acceleration area AA′B and a deceleration area BB′C:

$$W_{AB} = \int_{\delta_{V1}}^{\delta_{V2}} \left(T_{shaft} - T_e\right) d\delta_V > 0 \tag{6.66}$$

$$W_{BB'} = \int_{\delta_{V2}}^{\delta_{V3}} \left(T_{shaft} - T_e\right) d\delta_V < 0 \tag{6.67}$$

Only until the two areas are equal ($W_{AB} \le W_{BB'}$) to each other the SG will come back from point B′ to B, after a few attenuated oscillations. Attenuation comes from the rotor-cage asynchronous torque neglected here. This is the so-called criterion of equal areas.

The maximum shaft power step up from zero that can be accepted (to preserve the return to synchronism), corresponds to the case when point A is in the origin and point B′ is in C′ (Figure 6.31a). On the other hand, a practical situation is related to the short-circuit clearing time from the no-load operation (Figure 6.31b).

During the short circuit, the electromagnetic torque is considered zero, so the machine tends to accelerate (δ_V increases). At point C (in Figure 6.31b), the short circuit is cleared, and, thus, the electromagnetic torque reoccurs and the machine decelerates.

Synchronism recovery takes place if

$$\text{Area}\left(\text{ABCD}\right) \le \text{Area}\left(\text{CB'B''}\right) \tag{6.68}$$

Example 6.2: Short-Circuit Clearing Time, t_{sc}

Let us consider that a turbogenerator with $x_d = x_q = 1$ (pu), $S_n = 100$ MVA, $f_n = 50$ Hz, and $V_{nl} = 15$ kV operates at a power angle $\delta_V = 30°$ and unity power factor. The inertia in seconds is $H = \frac{J}{S_n}\left(\frac{\omega_1}{p_1}\right)^2 = 4$s. A sudden three-phase short circuit occurs and its electric transients are so fast that they are neglected here. Calculate the short-circuit clearing angle δ_{sc} and time t_{sc} to save the synchronous operation after fault clearing.

Solution:

As shown above, criterion (6.68) is to be met to preserve synchronism after fault clearing.

We simply calculate and make the two areas in criterion (6.68) equal to each other.

$$\frac{T_{ek}}{2}\left(\delta_{V_{sc}} - \pi/6\right) = \int_{\delta_{V_{sc}}}^{5\pi/6}\left(T_{ek}\cdot\sin\delta_V - \frac{T_{ek}}{2}\right)d\delta_V \tag{6.69}$$

$$T_e = \frac{3p_1}{\omega_1}\frac{E_1V_s\sin\delta_V}{X_d} = T_{ek}\sin\delta_V \tag{6.70}$$

for $\delta_V = \pi/6$; $T_{en} = T_{ek}/2$.

From Equation 6.69, $\delta_{Vsc} \approx 1.3955$ rad $\approx 80°$.

Now we go back to the motion Equation 6.64:

$$\frac{J}{p_1}\frac{d^2\delta_V}{dt^2} = T_{\text{shaft}\,n} = T_{en} = S_n\frac{p_1}{\omega_1} \tag{6.71}$$

or

$$H\cdot\frac{1}{\omega_1}\frac{d^2\delta_V}{dt^2} = 1 \tag{6.72}$$

and arrive at the solution (H = 2 s, $\omega_1 = 2\pi \cdot 50$ rad/s):

$$\delta_V(t) = \pi\times 25\times\frac{t^2}{2} + At + B \tag{6.73}$$

At $t = 0$, $\delta_V(0) = \pi/6$, and at $t = t_{sc}$, $\delta_V = \delta_{Vsc} \approx 1.3955$ rad; in addition, at $t = 0$, $\left(\frac{d\delta_V}{dt}\right)_{t=0} = 0$, because $(\omega_r)_{t=0} = \omega_{r0} = \omega_1$. Finally, $A = 0$ and $B = \pi/6$. So the time t_{sc} for δ_{Vsc} is

$$\delta_{V_{sc}} = \frac{\pi\times 25\times t_{sc}^2}{2} + \pi/6 = 1.3955 \text{ rad}$$

The short-circuit maximum clearing time to preserve synchronism is (from Equation 6.73)

$$t_{sc} = 0.14 \text{ s}$$

This subsecond value is not very far from industrial reality and shows how sensitive are SGs to transients because the rotor frequency is zero (d.c. or PM excitation in the rotor).

The presence of cage winding on the SG rotor produces some damping of rotor oscillations in transients and complicates the math at stability.

6.10 UNBALANCED LOAD STEADY STATE OF SGS/LAB 6.5

SGs connected to the grid and in autonomous operation operate on unbalanced voltage power grids or unbalanced three-phase current (impedance) loads, respectively. In the general case, the a.c. stator three-phase voltages and the currents are not balanced. To simplify the situation, let us consider the case of unbalanced currents:

$$I_{A,B,C}(t) = I_{A,B,C}\cos(\omega_1 t - \gamma_{A,B,C});\quad I_A \neq I_B \neq I_C;\quad \gamma_A \neq \gamma_B \neq \gamma_C \neq 120^\circ \quad (6.74)$$

For constant (or absent) magnetic saturation, we may use the method of symmetrical components (Figure 6.32):

$$\underline{I}_{A+} = \frac{1}{3}\left(\underline{I}_A + a\underline{I}_B + a^2\underline{I}_C\right);\quad a = e^{j\frac{2\pi}{3}} \quad (6.75)$$

$$\underline{I}_{A-} = \frac{1}{3}\left(\underline{I}_A + a^2\underline{I}_B + a\underline{I}_C\right);\quad \underline{I}_{A0} = \frac{1}{3}\left(\underline{I}_A + \underline{I}_B + \underline{I}_C\right) \quad (6.76)$$

$$\underline{I}_{B+} = a^2\underline{I}_{A+};\quad \underline{I}_{C+} = a\underline{I}_{A+};\quad \underline{I}_{B-} = a\underline{I}_{A-};\quad \underline{I}_{C-} = a^2\underline{I}_{A-} \quad (6.77)$$

For the direct component, the voltage Equation 6.36 is valid:

$$\underline{V}_{A+} = \underline{E}_{A+} - jX_d\underline{I}_{dA+} - jX_q\underline{I}_{qA+} \quad (6.78)$$

The d.c. excitation (or PM) rotor produces symmetric (balanced) emfs: E_{A+}, E_{B+}, and E_{C+}. The inverse components of stator currents I_{A-}, I_{B-}, and I_{C-} produce an mmf that travels at opposite rotor speed ($-\omega_r$). Thus, the relative angular speed of inverse mmf is $2\omega_r$, and so is the frequency of its induced currents in the rotor cage and in the field circuit (if its supply source accepts a.c. currents). This is a kind of induction (asynchronous) machine behavior at slip S_-:

$$S_- = \frac{-\omega_r - \omega_r}{-\omega_r} = 2 \quad (6.79)$$

Let us denote the equivalent negative sequence small impedance at $S = 2$, Z_-. As $E_{A-} = 0$ (symmetric emfs), the negative sequence equations are

$$\underline{I}_{A-}\underline{Z}_- + \underline{V}_{A-} = 0;\quad \underline{Z}_- = R_- + jX_- \quad (6.80)$$

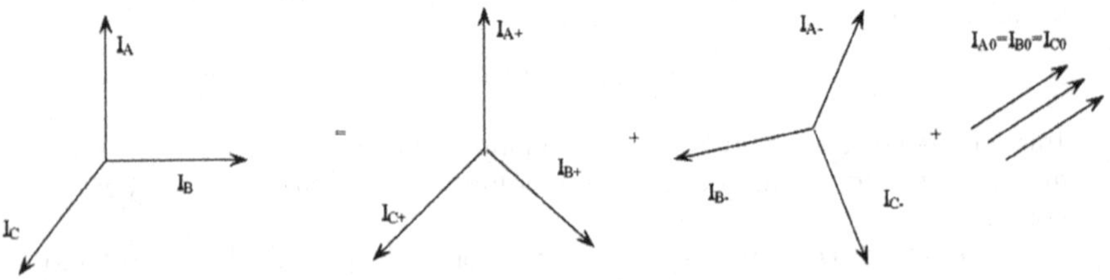

FIGURE 6.32 The symmetrical components of three-phase a.c. currents.

The homopolar components $I_{A0} = I_{B0} = I_{C0}$ produce a zero-traveling (fix) field because the three-phase windings are spaced at 120° with each other. So they do not interact with the positive and negative sequence components. Their corresponding impedance is $Z_0 = R_0 + jX_0$:

$$X_0 \le X_{sl}; \quad R_s < R_0 < R_s + R_{irons} \tag{6.81}$$

where
X_{sl} is the stator phase linkage reactance and
R_{irons} is the stator core series equivalent resistance.

In general, for a cage rotor d.c.-excited SM:

$$X_d > X_q > X_- > X_{sl} > X_0 \tag{6.82}$$

The stator phase equation for I_{A0} is similar to Equation 6.80:

$$jX_0\underline{I}_{A0} + \underline{V}_{A0} = 0 \tag{6.83}$$

The total voltage of phase A is

$$\underline{V}_A = \underline{V}_{A+} + \underline{V}_{A-} + \underline{V}_0 \tag{6.84}$$

or

$$\underline{V}_A = \underline{E}_{A+} - jX_dI_{dA+} - jX_qI_{qA+} - \underline{Z}_-\underline{I}_{A-} - jX_0\underline{I}_{A0} \tag{6.85}$$

Similar equations hold for phases B and C. Once the phase load impedances are given and the machine parameters are known, the current asymmetric components may be calculated. The above machine parameters may be calculated (in the design stage) or measured on the fabricated machine.

6.10.1 Measuring X_d, X_q, Z_-, and X_0/Lab 6.6

We will introduce here only some basic testing methods to determine X_d, X_q, Z_-, and X_0. To measure X_d and X_q (though not heavily saturated), the SG with open field circuit ($I_F = 0$), supplied with symmetric positive sequence voltages, is rotated close to synchronous speed: $\omega_r \ne \omega_1$ (pole-slipping method), but

$$\omega_r = (1.01 - 1.02)\omega_1 \tag{6.86}$$

So, the currents induced in the rotor cage at (0.01–0.02)ω_1 frequency are negligible. Recording phase A voltage and current, $V_A(t)$ and $I_A(t)$, (Figure 6.33), we notice that they pulsate with the slip (small) frequency of $(0.01 - 0.02)\omega_1$, due to the fact that $X_d \ne X_q$:

$$X_d \approx \frac{V_{A_{max}}}{I_{A_{min}}}; \quad X_q = \frac{V_{A_{min}}}{I_{A_{max}}} \tag{6.87}$$

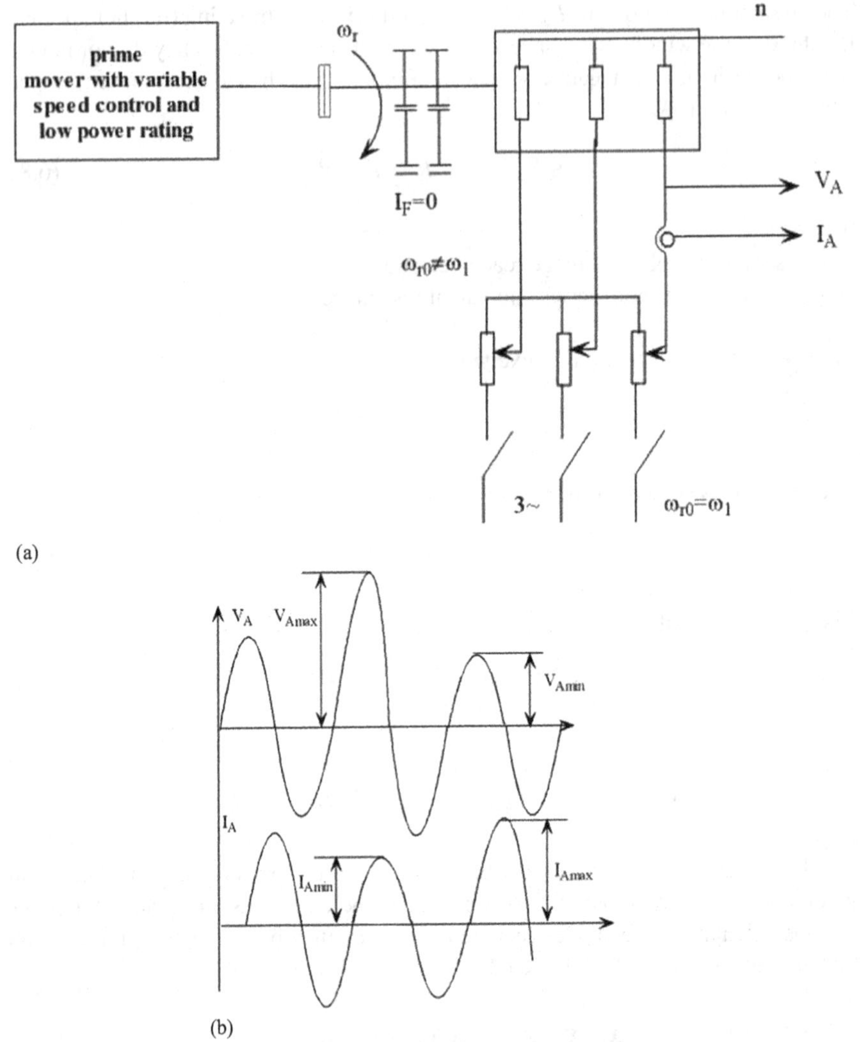

FIGURE 6.33 Pole-slipping method: (a) test arrangement and (b) voltage and current recordings.

The voltage amplitude pulsates because the supply transformer power is considered relatively small.

The negative sequence impedance, Z_-, may be measured by driving the SG at synchronism ($\omega_r = \omega_1$) but supplying the stator with negative sequence ($-\omega_1$) small voltages (as Figure 6.33), and with the field circuit short-circuited.

Measuring power, current, and voltage on phase A: P_{A-}, I_{A-}, and V_{A-}, we calculate

$$|\underline{Z}_-| = \frac{V_{A-}}{I_{A-}}; \quad R_- = \frac{P_{A-}}{I_{A-}^2}; \quad X_- = \sqrt{|\underline{Z}_-|^2 - (R_-)^2} \tag{6.88}$$

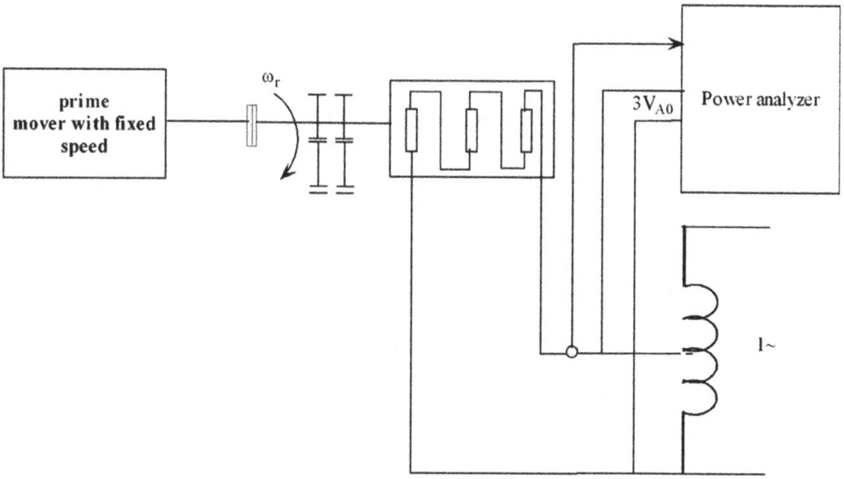

FIGURE 6.34 Homopolar impedance Z_0 measurements.

A similar arrangement with all phases in series and a.c. supply, at zero speed (or at synchronous speed) (Figure 6.34) can be used to measure Z_0 (the homopolar impedance):

$$\underline{Z}_0 = \frac{V_{A0}}{3I_{A0}}; \quad R_0 = \frac{P_0}{3I_{A0}^2}; \quad X_0 = \sqrt{|\underline{Z}_0|^2 - R_0^2} \tag{6.89}$$

The voltages V_{A-} and V_{A0} in the equations above should be made small to avoid large currents, and thus avoid machine overheating.

Note: For PM cageless rotors, the negative sequence impedance is almost equal to the positive one. The zero sequence (homopolar) impedance still has $X_0 \le X_{sl}$.

Example 6.3: The Phase-to-Phase Short Circuit

Let us consider a lossless two-pole SG with S_n = 100 kVA, V_{nl} = 440 V (star connection), f_1 = 50 Hz, $x_d = x_q$ = 0.6 pu, x_- = 0.20 pu, and x_0 = 0.12 pu that is connected in three-phase, two-phase, and one-phase short circuits. Calculate the phase current RMS values in the three cases.

Solution:

As shown earlier in this chapter, the three-phase short-circuit current, I_{3sc}, is

$$I_{3sc} = \frac{E_1}{X_d} \tag{6.90}$$

Rated short-circuit three-phase current is obtained for $E_1 = V_{nl}/\sqrt{3}$.

First the reactance norm $X_n = \dfrac{V_{nl}/\sqrt{3}}{I_n}$, with $I_n = S_n/\sqrt{3}V_{nl} = 100\times 10^3/(440\sqrt{3}) =$ 131.37 A, is $X_n = \dfrac{440}{\sqrt{3}\times 131.37} = 1.936\ \Omega$.

$$I_{3sc} = \frac{V_{nl}/\sqrt{3}}{x_d \times X_n} = \frac{440}{\sqrt{3} \times 0.6 \times 1.936} = 218.95\ \text{A} > I_n$$

Here, we may also introduce the short-circuit ratio:

$$\frac{I_{3sc}}{I_n} = \frac{1}{x_d} = \frac{1}{0.6} = 1.66 \tag{6.91}$$

For a single-phase short circuit:

$$\underline{I}_{A+} = \underline{I}_{A-} = \underline{I}_{A0} = \frac{\underline{I}_{1sc}}{3}$$

So, from Equation 6.85, with $V_A = 0$:

$$I_{1sc} = \frac{3E_{A+}}{X_s + X_- + X_0} = \frac{3 \times 440/\sqrt{3}}{(0.6 + 0.2 + 0.1) \times 1.936} = 436.86\ \text{A}$$

For the phase-to-phase short circuit, the computation is a bit more complicated, but it starts from Equation 6.75 to Equation 6.85 for $I_A = 0$ and $I_B = -I_C = I_{2sc}$ ($V_B = V_C$):

$$I_{A+} = \frac{j}{\sqrt{3}} \underline{I}_{2sc} = -I_{A-}; \quad \underline{I}_{A0} = 0$$

$$\underline{V}_A = \underline{E}_{A+} - jX_+ \underline{I}_{A+} - \underline{Z}_- \underline{I}_{A-} = \underline{E}_{A+} - \frac{j}{\sqrt{3}} I_{2sc} \left(jX_+ - \underline{Z}_- \right) \tag{6.92}$$

$$V_B = a^2 \underline{V}_{A+} + a \underline{V}_{A-} = V_C = a \underline{V}_{A+} + a^2 \underline{V}_{A-}$$

Finally,

$$\underline{I}_{2sc} = \frac{j\underline{E}_A \sqrt{3}}{\left(jX_+ + \underline{Z}_- \right)}; \quad \underline{V}_A = -2\underline{V}_B = \frac{2j}{\sqrt{3}} \underline{I}_{2sc} \underline{Z}_- \tag{6.93}$$

Apparently, Equation 6.93, with V_A and I_{2sc} measured, directly yields Z_-, while the first expression in Equation 6.93 would then lead to the calculation of X_+. This may be feasible only if the fundamentals of the measured variables are extracted first. However, with $X_+ = X_d$ from the three-phase short circuit, we may calculate Z_- from Equation 6.93.

In our example, $Z_- = jX_-$, so, from Equation 6.93:

$$I_{2sc} = \frac{E_A \sqrt{3}}{X_+ + X_-} = \frac{440}{(0.6 + 0.2) \times 1.936} = 284.09\ \text{A} \tag{6.94}$$

Comparing I_{3sc}, I_{2sc}, and I_{1sc}, we may write

$$I_{3sc} < I_{2sc} < I_{1sc} \tag{6.95}$$

6.11 LARGE SYNCHRONOUS MOTORS

As already alluded to, SMs may work also as motors, provided their voltage power angle becomes negative (E_1 lagging V_s) (Figure 6.35). The speed of the SM is constant with load (torque) if frequency $f_1 = p_1 n_1$ = const. So the grid-operated SM has to be started in the induction-motor mode with the field circuit connected to a resistor until stable asynchronous speed, $\omega_{ras} = (0.95 - 0.98)\omega_1$, is obtained. After that, the field circuit is disconnected from the resistor and connected to the d.c. source when, after a few oscillations, it eventually synchronizes. For smaller PM cage rotor or reluctance cage rotor SMs, direct connection to the grid is performed, and, at least under light loads, these SMs should synchronize, but (again) using the induction-motor mode during acceleration.

6.11.1 Power Balance

To ease the mathematical expressions, so far we considered lossless SMs. In reality, all-electric machines have losses and, even if their efficiencies are good, they have to be considered in the cooling system design. The difference between the SM electric input P_{1e} and its mechanical output P_{2m} is the summation of losses, $\sum p$:

$$\sum p = P_{1e} - P_{2m} = p_{\cos} + p_{\text{iron}} + p_s + p_{\text{mec}}; \quad p_{\cos} = 3R_s I_s^2 \tag{6.96}$$

The electric losses take place in stator windings ($p_{\cos}$) and in the stator iron core (p_{iron}), and there are also additional losses (p_s) due to stator space harmonics rotor-induced currents, etc. Iron losses depend approximately on the stator flux, Ψ_s:

$$\left|\underline{\Psi}_s\right| = \left|M_{Fa}\underline{I}_F + L_d\underline{I}_d + L_q\underline{I}_q\right| \tag{6.97}$$

and frequency ω_r that, for power grid operation, are constant, as

$$V_s \approx \omega_r \left|\underline{\Psi}_s\right| \tag{6.98}$$

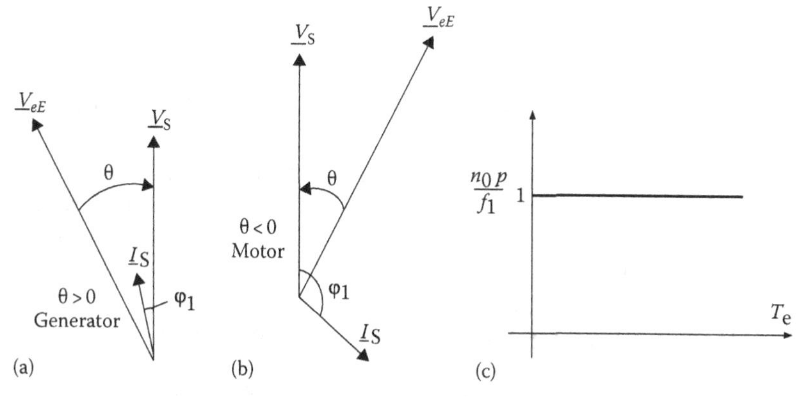

FIGURE 6.35 (a) SG, (b) motor, and (c) speed/torque curve.

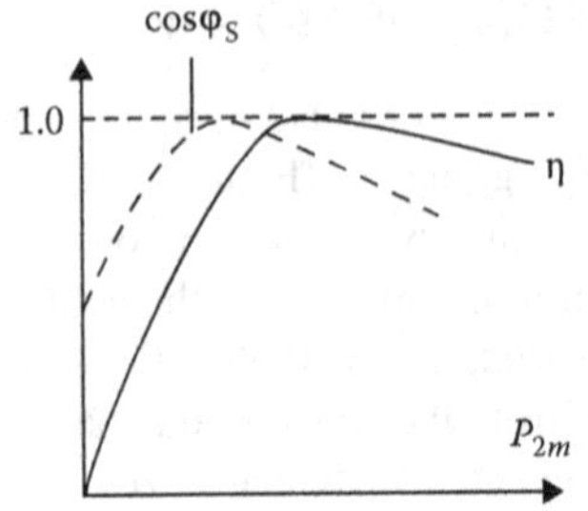

FIGURE 6.36 Efficiency, η, and power factor, cos φ_s, versus output power, P_{2m}, for I_F = const., ω_r = const., and V_s = const.

The field circuit losses, $p_{exc} = R_F I_F^2$, come frequently from a separate supply; this is not so for automotive alternators. For large motors, we still can neglect losses when calculating I_d and I_q currents:

$$I_d \approx \frac{V_s \cos\delta_V - E_1}{X_d}; \quad I_q \approx \frac{V_s \sin\delta_V}{X_q}; \quad I_s = \sqrt{I_d^2 + I_q^2} \tag{6.99}$$

Also, for the active and reactive powers, P_s and Q_s, Equations 6.58 and 6.59 still hold. So the power factor, cos φ_s, is

$$\cos\varphi_s = \sqrt{1 - \frac{Q_s^2}{(3V_s I_s)^2}} \tag{6.100}$$

As already explained, for given field current, i_F, and speed, E_1 is given, and, for given values of power angle (negative for motoring since Equations 6.36 are still used), we can calculate P_s, Q_s, cos φ_s, I_s, and then all losses.

So the efficiency is

$$\eta = \frac{|P_s| - \Sigma p}{|P_s|} = \frac{P_{2m}}{|P_s|} \tag{6.101}$$

A qualitative graphic representation of η and cos φ_s versus P_{2m} (output (shaft) power) is shown in Figure 6.36.

The power factor starts by being leading at low load and then becomes lagging. By close-loop controlling of the field current, it is feasible to maintain a desired leading power factor angle from 5(10)% to 100% of full load, in order to compensate the reactive power needed in the SM supply area.

6.12 PM SYNCHRONOUS MOTORS: STEADY STATE

The PMSMs may have a cage on their rotors and be operated directly at the power grid, or the former may lack any cage on their rotors, and thus, by necessity, they

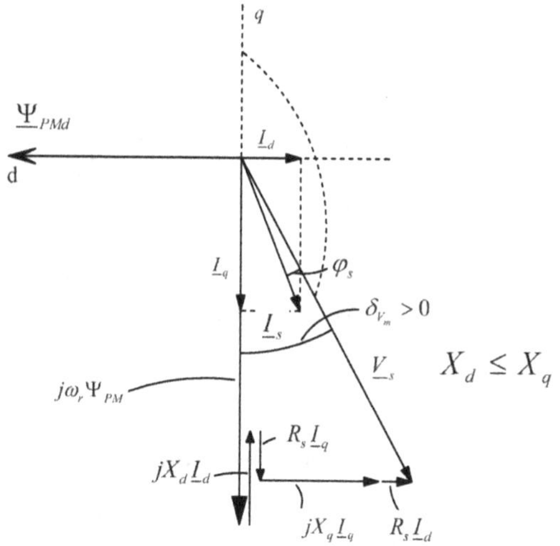

FIGURE 6.37 PMSM phasor diagram.

have to be supplied from PWM frequency changers, as $f_1 = p_1 n_1$. In the latter case, they start and operate in synchronizm as the stator currents frequency and phase angles are locked to the rotor position.

Here, we investigate only the steady state. The PMSMs are built for very small to large powers (torque) per unit, but, in general, both copper and iron losses, p_{co} and p_{iron}, are directly considered in the model. The efficiency is to be treated as in the previous section. We just take Equation 6.36, replace $M_{Fa}I_F$ by Ψ_{PMd}, and use the sink sign convention (IR-V):

$$\left(\underline{I}_d + j\underline{I}_q\right)R_s - \underline{V}_s = -j\Psi_{PM_d}\omega_r - jX_d\underline{I}_d - jX_q\underline{I}_q;\quad X_d < X_q \tag{6.102}$$

First we represent Equation 6.102 in the phasor diagram of Figure 6.37 when the power angle is positive for motoring and so is the electromagnetic power and torque.

From Figure 6.37:

$$\begin{aligned} V_s\cos\delta_{V_m} - \omega_r\Psi_{PM} &= -R_sI_q + X_dI_d;\quad X_d = \omega_rL_d;\quad I_q > 0 \\ V_s\sin\delta_{V_m} &= X_qI_q - R_sI_d;\quad X_q = \omega_rL_q;\quad I_d < 0 \end{aligned} \tag{6.103}$$

From Equation 6.103, for V_s constant and δ_{Vm} given (as a parameter), we may calculate I_d, I_q, and then the electromagnetic torque (from Equation 6.42):

$$T_e = P_{elm}\frac{p_1}{\omega_r} = 3p_1\left(\Psi_{PM} + \left(L_d - L_q\right)I_d\right)I_q \tag{6.104}$$

The core losses may be approximately considered proportional to V_s^2 irrespective of the frequency or the load, and, thus, they may be added when the efficiency is calculated.

TABLE 6.3
PMSM Performance

δ_{Vm}	I_d	I_q	$I_s = \sqrt{I_d^2 + I_q^2}$	T_e	$P_e = \frac{\omega_r}{p_1} T_e p_{exc} = R_F I_F^2 E_1 = V_{nl}/\sqrt{3}$	
[°]	[A]	[A]	[A]	[Nm]	[W]	$\eta \cos\varphi_s = \frac{P_e}{3V_s I_s}$
0	0	0	0	0	0	
30	−4.78	5.545	7.32	28.9	4537.0	0.941
45	−9.75	7.63	12.38	48.95	7685.7	0.943
60	−15.84	9.07	18.25	72.86	11139.0	0.952
90	−30.24	9.63	31.75	120.1	18857.0	0.902

Example 6.4: A PMSM with Interior PM Rotor

A PMSM with an interior PM rotor has the data: V_{nl} = 380 V (star connection), p_1 = 2, f_1 = 50 Hz, X_d = 7.72 Ω, X_q = 18.72 Ω > X_d, and R_s = 1.32 Ω. Let us calculate I_d, I_q, I_s, T_e, P_e, and $\eta \cos\varphi_s$ for δ_{Vm} = 0°, 30°, 45°, 60°, and 90°. Neglect all but copper losses in the stator.

Solution:

We directly apply Equations 6.103 and 6.104, the efficiency definition in section 6.11, and the power factor angle, φ_s, from the phasor diagram in Figure 6.37:

$$\varphi_s = \delta_V - \tan^{-1}\left(\frac{-I_d}{I_q}\right) \tag{6.105}$$

The computation routine is straightforward and the results are shown in Table 6.3.

The performance is unusually good ($\eta \cos\varphi_s$ product is high) despite high I_d (demagnetization) current levels. Also up to δ_{Vm} = 90°, the torque increases because the machine has reverse saliency ($X_d < X_q$), and the rated power grid operation at δ_{Vm} = 60°–80° should be acceptable.

Note: Recently, a few major companies replaced induction motors up to a few hundred kilowatt and speeds below 500 rpm ($2p_1$ = 10,12) in variable speed drives for conveyors, etc., by PMSMs in order to increase $\eta \cos\varphi$, and thus cut the PWM frequency changer (converter) kVA, and, hence, its costs and the energy bill.

Example 6.5: The Reluctance Synchronous Motor (RSM)

An RSM with a multiple flux barrier rotor has a rotor cage provided for grid-connected operation (Figure 6.38a). The phasor diagram (with Ψ_{PMd} = 0 and I_d > 0) is shown in Figure 6.38b. Let us consider V_{nphase} = 220 V, I_n = 5 A, $2p_1$ = 4 poles, f_1 = 50 Hz, x_d = 2.5 pu, $x_q = x_d/5$, and r_s = 0.08 (pu). Calculate I_d, I_q, I_s, T_e, P_e, $\eta \cos\varphi_s$, and $\cos\varphi_s$ for δ_V = 0°, 10°, 20°, and 30°.

Solution:

With

$$X_n = V_{nph}/I_n = 220/5 = 44\ \Omega$$

$$X_d = x_d \cdot X_n = 2.5 \times 44 = 110\ \Omega$$

$$X_q = X_d/5 = 22\ \Omega;\quad R_s = r_s \cdot X_n = 0.08 \times 44 = 3.52\ \Omega$$

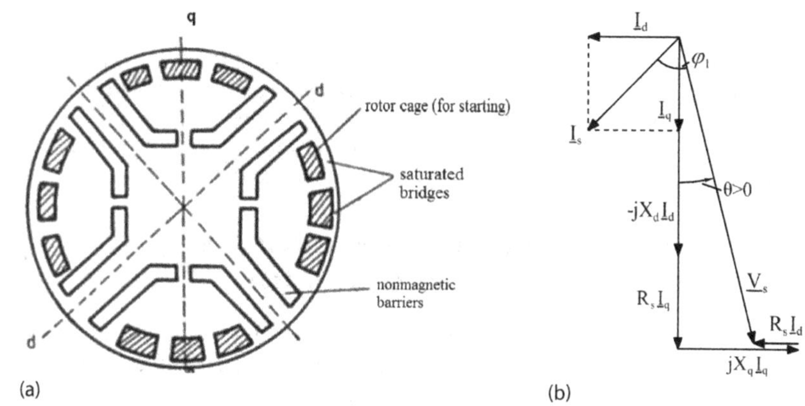

FIGURE 6.38 (a) An RSM with multiple flux barrier rotor and (b) phasor diagram.

from the phasor diagram:

$$X_dI_d + R_sI_q = V_s\cos\delta_{V_m};\quad \varphi_s = \delta_{V_m} + \tan^{-1}\left(\frac{I_d}{I_q}\right);\quad I_d,\ I_q > 0$$
$$X_qI_q - R_sI_d = V_s\sin\delta_{V_m};\quad I_s = \sqrt{I_d^{\,2} + I_q^{\,2}} \tag{6.106}$$

T_e is obtained from Equation 6.104, with $\Psi_{PMd} = 0$:

$$T_e = 3p_1\left(L_d - L_q\right)I_dI_q \tag{6.107}$$

$$\eta \approx \frac{T_e \cdot \omega_r/p_1}{T_e \cdot \omega_r/p_1 + 3R_sI_s^{\,2}} \tag{6.108}$$

The computing routine is straightforward and the results are as in Table 6.4.

As seen from Table 6.4, the power factor is below 0.7 and the $\eta\cos\varphi_s$ product is also notably smaller than for PMSMs. For RSMs, in power grid operation, the

TABLE 6.4
RSM Performance

δ_{V_m} [°]	I_d [A]	I_q [A]	I_s [A]	T_e [Nm]	P_e [W]	$\eta\cos\varphi_s$	$\cos\varphi_s$
0	1.987	0.318	2.01	1.06	166.7	0.1264	0.158
10	1.903	2.041	2.79	6.53	1025.2	0.5479	0.59
20	1.734	3.701	4.095	10.78	1692.1	0.628	0.693
30	1.563	5.25	5.696	13.80	2165.0	0.5866	0.67

Note: Please align the equations numbers to the right side of sheets by visiting your equations style

(*Continued*)

Example 6.5: (Continued)

rated value of δ_{Vm} is $\delta_{Vmax} = 20°–30°$. However, the cost of RSMs is lower, and, in some applications, RSMs may have a higher efficiency than IMs with the same stators. When part of a variable speed drive, and supplied at variable frequency (from zero), the rotor cage of RSM is eliminated and, in home appliance like applications, RSM might be preferred cost-wise to IM or PMSM.

6.13 LOAD TORQUE PULSATIONS HANDLING BY SYNCHRONOUS MOTORS/GENERATORS

Diesel engine SGs and compressor loads for SMs are typical industrial situations where the shaft torque varies with the rotor position. It is also important to see how the SMs handle sudden load torque changes or asynchronous starting. At least for large machines (with large inertia), electric transients are usually faster than mechanical transients, and, thus, electric equations for the steady-state case can still be used. This is what we call here mechanical transients. For a generator:

$$\frac{J}{p_1}\frac{d\omega_r}{dt} = T_{shaft} - T_e + T_{as}; \quad \frac{d\delta_V}{dt} = \omega_r - \omega_1 \tag{6.109}$$

$$T_e \approx \frac{3p_1}{\omega_1}\left[\frac{V_s E_1 \sin\delta_V}{X_d} + \frac{V_s^2}{2}\left(\frac{1}{X_q} - \frac{1}{X_d}\right)\sin 2\delta_V\right] \tag{6.110}$$

T_{as} is the asynchronous torque due to the rotor cage and the rotor field circuit a.c.-induced currents (when $\omega_r \neq \omega_1$). Provided the initial values of variables ω_r and δ_V are known and the input variables V_s, E_1, and T_{shaft} evolution (in time or with speed) is given, any mechanical (slow) transient process can be solved through Equations 6.109 and 6.110 by numerical methods. However, linearization of Equations 6.109 and 6.110 can shed light on phenomena and offer a feeling of magnitudes which is essential for engineering insight. The asynchronous torque, around synchronism, varies almost linearly with slip speed, $\omega_2 = \omega_r - \omega_1$:

$$T_{as} = -K_a(\omega_r - \omega_1) = -K_a\frac{d\delta_V}{dt} \tag{6.111}$$

Positive T_{as} means motoring ($\omega_r < \omega_1$) and negative T_{as} corresponds to generating ($\omega_r > \omega_1$).

The synchronous torque T_e (Equation 6.110) may be linearized around an initial value, δ_{V0}:

$$\begin{aligned} &\delta_V = \delta_{V0} + \Delta\delta_V; \quad T_{shaft} = T_{a0} + \Delta T_a \\ &T_e = T_{e0} + T_{es} \times \Delta\delta_V \\ &T_{es} = \left(\frac{\partial T_e}{\partial \delta_V}\right)_{\delta_{V0}}; \quad T_{e0} = (T_e)_{\delta_{V0}} = T_{a0} \end{aligned} \tag{6.112}$$

From Equations 6.109 through 6.112:

$$\frac{J}{p_1}\frac{d^2(\Delta\delta_V)}{dt^2} + K_a\frac{d(\Delta\delta_V)}{dt} + T_{es}\Delta\delta_V = \Delta T_a \tag{6.113}$$

For small power angle deviations, $\Delta\delta_V$, around δ_{V0}, a second-order model for transients has been obtained. From it we distinguish the eigenvalues $\underline{\gamma}_{1,2}$:

$$\gamma_{1,2} = -\frac{K_a p_1}{2J} \pm \sqrt{\left(\frac{K_a p_1}{2J}\right)^2 - \omega_0^2}; \quad \omega_0 = \sqrt{\frac{p_1 T_{es}}{J}} \tag{6.114}$$

where ω_0 is the so-called proper (eigen) angular frequency of the generator/motor, and it acts when there is no cage on the rotor, $((\gamma_{1,2})_{Ka=0} = \pm j\omega_0)$. It depends (through T_{es}) on the power angle (δ_{V0}), the level of excitation ($\bar{I}_F$), and machine inertia (J). It is in the range of a few (3 to 1) Hz or less for the largest-power SGs. When an SM is connected to the grid, and it has a rotor cage, all terms in Equation 6.113 are active (non-zero).

Now, if the shaft torque has pulsations (in diesel engines, ICEs, or compressor loads for SMs):

$$\Delta T_a = \Sigma T_{av}\sin(\Omega_v t - \Psi_v) \tag{6.115}$$

there is an amplification $K_{m\nu}$ of rotor angle–oscillation amplitudes with respect to the case of autonomous generators ($T_{es} = 0$) with no rotor cage ($K_a = 0$):

$$\frac{J}{p_1}\frac{d^2\Delta\delta_V}{dt^2} = \Sigma T_{av}\sin(\Omega_v t - \Psi_v) \tag{6.116}$$

with the solution:

$$\Delta\delta_{Vva}(t) = -\frac{p_1}{J}\frac{T_{av}}{\Omega_v^2}\sin(\Omega_v t - \Psi_v) \tag{6.117}$$

When solving Equation 6.113 we get

$$\Delta\delta_{Vv}(t) = \frac{T_{av}\sin(\Omega_v t - \Psi_v - \varphi_v)}{\sqrt{\left(\frac{J\Omega_v^2}{p_1} - T_{es}\right)^2 + (K_a\Omega_v)^2}}; \quad \varphi_v = \tan^{-1}\frac{K_a\Omega_v}{\frac{J}{p_1}\Omega_v^2 - T_{es}} \tag{6.118}$$

And $K_{m\nu}$ (the mechanical resonance module) is

$$K_{mv} = \frac{(\Delta\delta_{Vv})_{max}}{(\Delta\delta_{Vva})_{max}} = \frac{1}{\sqrt{\left(1 - \frac{\omega_0^2}{\Omega_v^2}\right)^2 + K_{dv}^2}}; \quad K_{dv} = \frac{K_a p_1}{J\Omega_v} \tag{6.119}$$

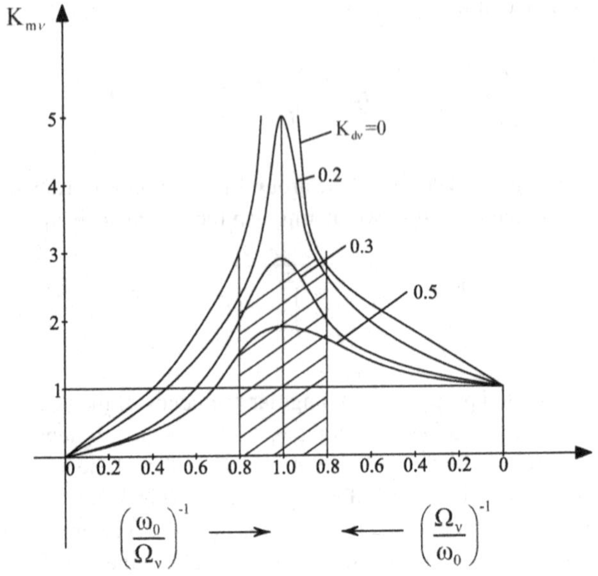

FIGURE 6.39 $K_{m\nu}$ (mechanical resonance module).

where $K_{d\nu}$ is the known damping coefficient in second-order systems.

A graphical representation of Equation 6.119 is shown in Figure 6.39. It is visible that for

$$1.25 > \frac{\omega_0}{\Omega_v} > 0.8 \tag{6.120}$$

the amplification of oscillations is large. A strong damper cage in the rotor (K_a – large) leads to a reduction in the amplitude of rotor-angle oscillations that is in accordance with diesel-engine or compressor drives. An inertial disk on the shaft is also beneficial. The power angle and speed oscillations lead also to stator current oscillations and input electric power oscillations, which have to be limited. Grid-connected SMs are in general sensitive to shaft torque pulsations due to the danger of loosing synchronizm.

6.14 ASYNCHRONOUS STARTING OF SMS AND THEIR SELF-SYNCHRONIZATION TO POWER GRID

Asynchronous starting of grid-connected SMs is illustrated in Figure 6.40.

Large SMs are started as asynchronous motors by connecting the field circuit terminals to a 10 R_F resistance (R_F is the field circuit resistance) to limit the Georges' effect (asymmetrical circuit rotor effect [see Chapter 5]), and, thus, provide for smooth passing of the machine through 50% rated speed. Then, when speed stabilizes (at slip: $S = 0.03 - 0.02$), the field circuit is switched to the d.c. source. The field current reaches reasonable values in tens of milliseconds at best, but let us neglect this process and provide full I_F. When I_F occurs, the emf, E_1, has a certain phase shift, δ_{V0}, with respect to the terminal (power grid) voltage. It may be any value. The best situation, $\delta_{V0} = 0$, and turning into motoring is illustrated in Figure 6.40b.

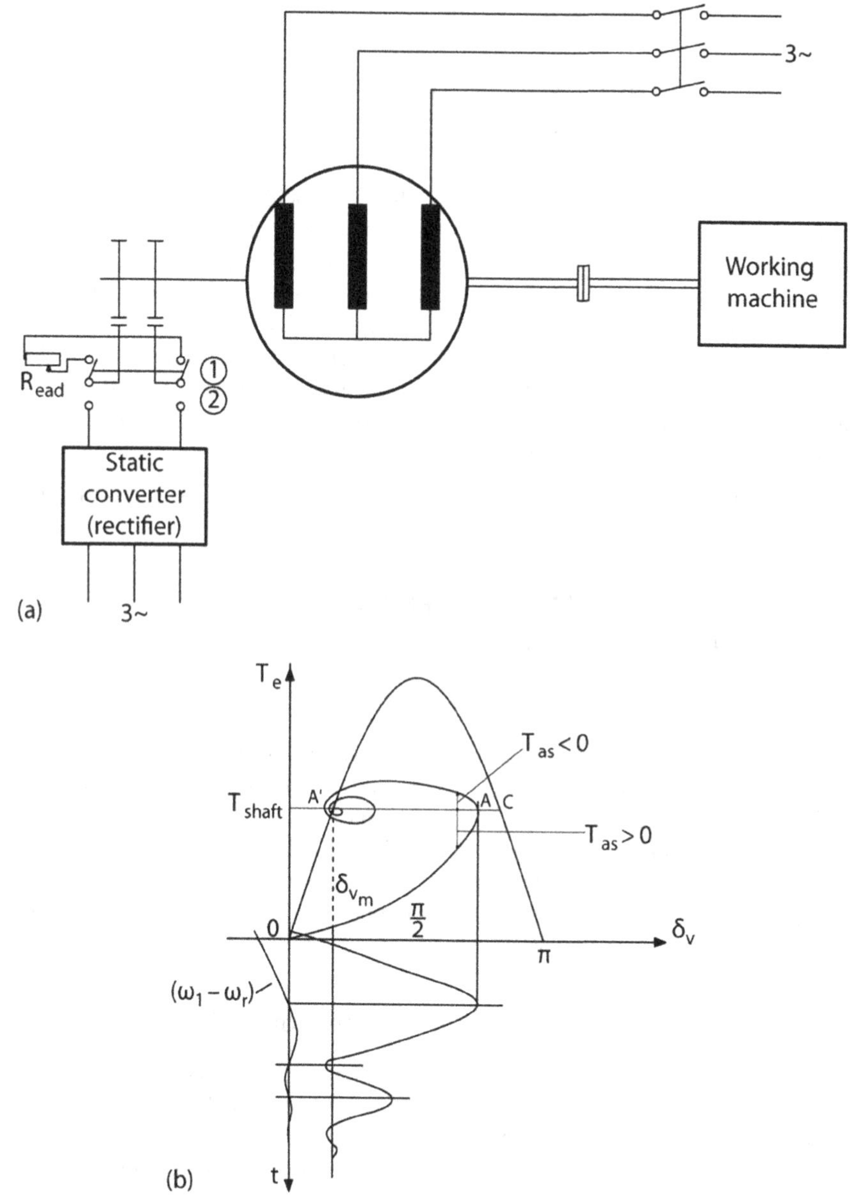

FIGURE 6.40 Asynchronous starting: (a) basic arrangement and (b) transients of self-synchronization from zero initial power angle.

Equations below, rewritten for motoring, illustrate Figure 6.40b completely:

$$\frac{d\delta_{V_m}}{dt} = \omega_1 - \omega_r; \quad T_{as} = K_a\left(\omega_1 - \omega_r\right)$$
$$\frac{J}{p_1}\frac{d\omega_r}{dt} = -\frac{J}{p_1}\frac{d^2\delta_{V_m}}{dt^2} = T_e + T_{as} - T_{shaft} \tag{6.121}$$

$$T_e = \frac{3p_1V_s}{\omega_1}\left[\frac{E_1}{X_d}\sin\delta_{V_m} + \frac{V_s}{2}\left(\frac{1}{X_q} - \frac{1}{X_d}\right)\sin 2\delta_{V_m}\right] \tag{6.122}$$

with δ_{V0} and ω_{r0} as initial assigned values. The self-synchronization process may be solved by Equations 6.121 through standard numerical procedures with shaft torque (T_{shaft}) given as a function of speed (or even on rotor position). This way, for heavy-starting applications (large coolant circulation pumps in nuclear electric power plants), the self-synchronization process may be simulated simply. It is also feasible to measure the a.c. field current during the asynchronous operation and start synchronization when the latter passes through zero, which corresponds to $\delta_{V0} = 0$, because, at low slip frequencies, the field circuit looks strongly resistive. This way safe (flawless) starting under load is ensured.

Note: Speed control by frequency control is mandatory with SMs. This aspect will be treated in the companion book.

6.15 SINGLE-PHASE AND SPLIT-PHASE CAPACITOR PM SYNCHRONOUS MOTORS

Single-phase PMSMs without a cage on the rotor and with a parking magnet for self-starting are supplied with a PWM inverter from zero frequency upwards (Figure 6.41), and are typical for low-power, light-start, variable-speed applications from sub watt to hundreds of watts and high speeds (30,000 rpm and more). On the other hand, if the stator is provided with two distributed windings at a space angle shift of 90° (like for the induction motor) while the rotor has PMs and a cage, the machine may be connected directly to the power grid. It starts as an induction machine motor and then eventually self-synchronizes on its own, up to a certain load torque and inertia level. This is the split-phase PMSM.

6.15.1 Steady State of Single-Phase Cageless-Rotor PMSMs

The single-phase (PWM inverter fed) PMSM (Figure 6.41) under steady state is rather easy to model as there are no windings on the rotor, and the parking magnet influence on steady state may be neglected. For a sinusoidal emf in the stator phase (produced through motion by the PMs on the rotor) and sinusoidal terminal voltage, the machine voltage equation is straightforward:

$$\underline{V}_s = R_s\underline{I}_s + \underline{E}_s + j\omega_1 L_s\underline{I}_s; \quad \underline{E}_s = j\omega_r\Psi_{PM}; \quad \omega_r = \omega_1 \tag{6.123}$$

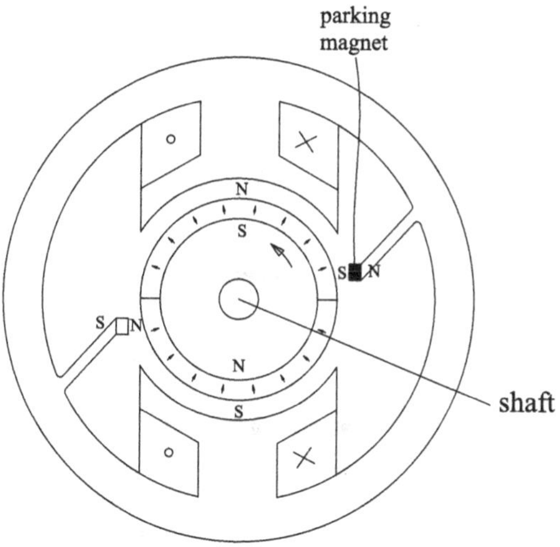

FIGURE 6.41 Single-phase PMSM with parking magnet and PWM inverter frequency control (no cage on the rotor), $2p_1 = 2$.

Now with the sinusoidal emf, E_s:

$$E_s(t) = E_{s1}\cos\omega_r t \tag{6.124}$$

With sinusoidal voltage $\underline{V}_s$, the current $\underline{I}_s$ is sinusoidal:

$$I_s(t) = I_{s1}\cos(\omega_r t - \gamma) \tag{6.125}$$

The electromagnetic torque T_e—in the absence of rotor magnetic saliency—is

$$T_e = \frac{p_1 E_s(t) I_s(t)}{\omega_r} = \frac{p_1}{\omega_r}\frac{E_{s1}I_{s1}}{2}\left(\cos\gamma + \cos(2\omega_r t - \gamma)\right) \tag{6.126}$$

So the torque pulsation is 100%.

Now at zero stator current, there is a cogging torque due to slot openings in the stator. The stator has in fact two slots and $2p_1 = 2$ poles (Figure 6.42). So the number of cogging torque (T_{cogg}) periods is the $LCM(2,2) = 2$:

$$T_{cogg} = T_{cogg\,max} \cdot \cos(2\omega_r t - \gamma_{cogg}) \tag{6.127}$$

If, by proper design, $\gamma_{cogg} = \gamma$ and

$$T_{cogg\,max} = -\frac{p_1}{\omega_r}\frac{E_{s1}I_{s1}}{2} \tag{6.128}$$

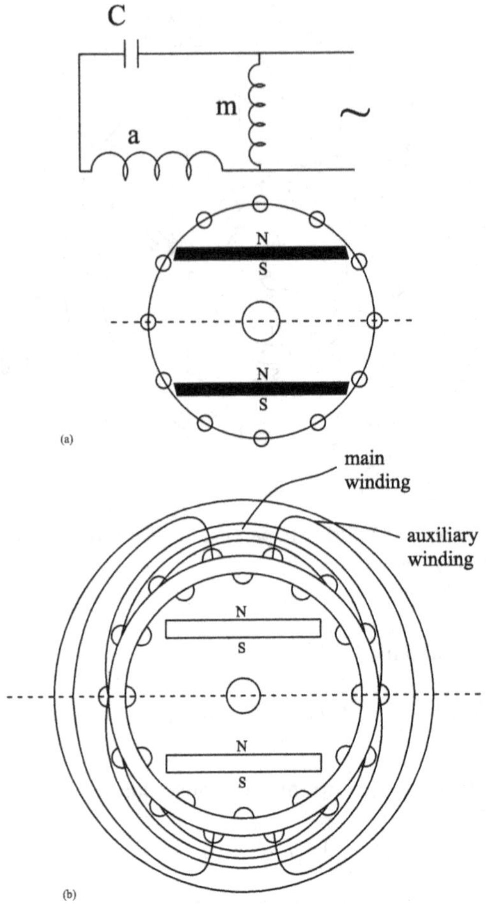

FIGURE 6.42 Grid-operated PMSM with two-pole split-phase capacitor (with PMs and a cage in the rotor).

the resultant torque becomes constant:

$$T_e + T_{cogg} = \frac{p_1}{\omega_r}\frac{E_{s1}I_{s1}}{2} \tag{6.129}$$

So the torque has lost its pulsations but only at, say, rated current.

The phasor diagram is straightforward (Figure 6.43a). The core losses, p_{iron}, are given by

$$p_{iron} \approx \frac{\omega_r \Psi_s^2}{R_{iron}}; \quad \Psi_s = \sqrt{\Psi_{PM}^2 + (L_s I_s)^2 - 2\Psi_{PM} L_s I_s \cos(\delta_V - \varphi)} \tag{6.130}$$

R_{iron} may be obtained from experiments or in the design stage.

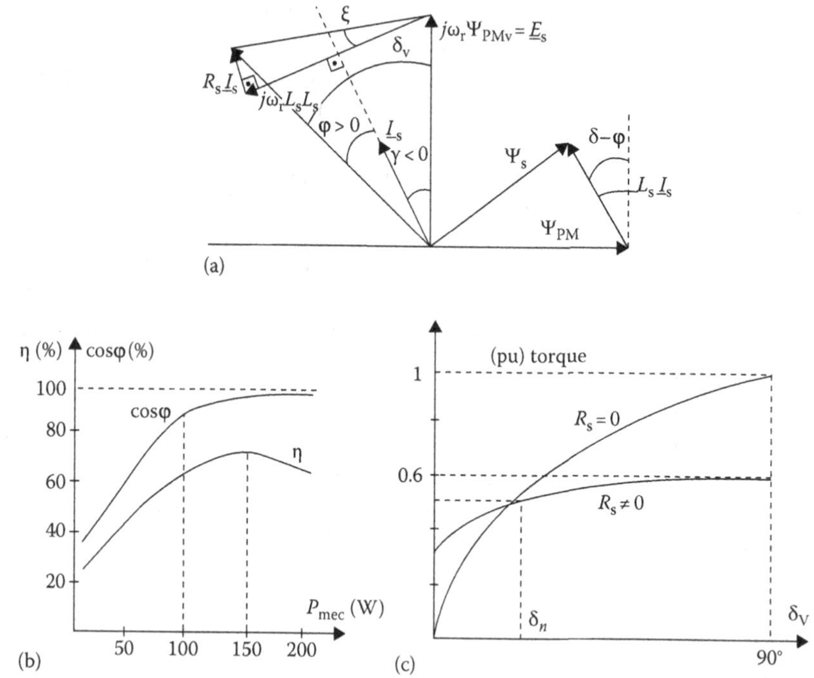

FIGURE 6.43 (a) Phasor diagram of single-phase PMSM, (b) efficiency and power factor for constant voltage and frequency, and (c) pu torque versus power angle, δ_V.

The voltage power angle, δ_V, is defined as for three-phase machines, and I_s is (from Figure 6.43a)

$$I_s = \frac{\sqrt{V_s^{\,2} + E_s^{\,2} - 2V_s E_s \cos\delta_V}}{Z} \tag{6.131}$$

$$Z = \sqrt{R_s^2 + \omega_1^2 L_s^2}; \quad \tan\xi = \frac{R_s}{\omega_1 L_s}; \quad \cos(\varphi + \xi) = \frac{E}{IZ}\sin\delta_V \tag{6.132}$$

The mechanical power, P_{2m}, is

$$P_{2m} = V_s I\cos\varphi - I_s^2 R_s - \frac{\omega_r^2 \Psi_s^2}{R_{iron}} - p_{mec} \tag{6.133}$$

where p_{mec} denotes the mechanical losses (in watt) $\left(V_s = \frac{V_{s1}}{\sqrt{2}}, I_s = \frac{I_{s1}}{\sqrt{2}}\right)$.

Efficiency η is

$$\eta = \frac{P_{2m}}{V_s I_s \cos\varphi} \tag{6.134}$$

The above model, with δ_V as a parameter, allows for calculating the stator current (I_s), cos φ, T_e (average), efficiency, and the power factor.

Figure 6.43c shows the average torque T_e (in pu), η, and cos φ (Figure 6.43b) versus power angle (δ_V) for a 150 W machine with and without considering the stator winding losses (R_s).

The stator resistance limits notably the peak torque (Figure 6.43c). Higher efficiency (lower resistance, R_s) for this power range is feasible at the price of higher copper weight.

For a split-phase capacitor PMSM (or RSM) with cage rotor modeling for steady state, the symmetrical component method may be applied (as for the split-phase capacitor IM), but for $S = 0$ and by adding the emf E_s for the positive sequence component (see [14] for details).

Note: Test procedures for SMs.

The testing methods for synchronous (also for d.c. or induction) motors/generators may be classified as standard and research tests.

Tests of a more general nature are included under standards that are renewed periodically to reflect the progress in the art. The IEEE standards 115-1995 represent a comprehensive set of tests for SMs. They may be classified into

- Acceptance tests
- Steady-state performance tests
- Parameter estimation (for transients and control)

See [13] in Chapter 8 for a synthesis of IEEE standards 115-1995. Quite a few test procedures have been introduced earlier in this chapter under the logo "Lab." Advanced testing for parameter identification will be treated in the companion book (on transients, design, optimal design, FEA).

6.16 PRELIMINARY DESIGN METHODOLOGY OF A THREE-PHASE SMALL AUTOMOTIVE PMSM BY EXAMPLE

General specifications:

- Base power $P_b = 100$ W
- Base speed $n_b = 1800$ rpm
- Maximum speed $n_{max} = 3000$ rpm
- Power at maximum speed: P_b
- DC voltage $V_{dc} = 14$ V (automobile battery)
- Supply: PWM inverter; maximum line voltages as in Figure 6.44
- Star connection of stator phases

Additional specifications related to duty cycle, motor-cooling system, constraints related to volume, efficiency at base power, and cost of active materials in the motor may be added. Here a preliminary design methodology by example is introduced.

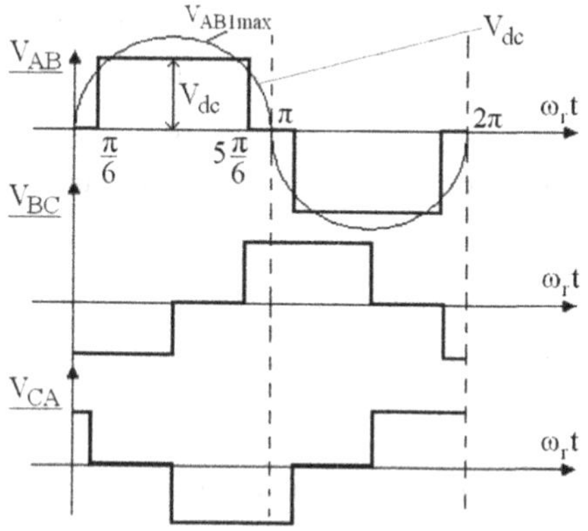

FIGURE 6.44 Ideal line voltages for six-pulse PWM inverter operation.

Maximum phase voltage:
According to Figure 6.44, the fundamental V_{phmax} for the maximum phase voltage is

$$\left(V_{\text{ph max}}\right)_{\text{RMS}} = \frac{V_{\text{line max}}}{\sqrt{6}} = \frac{1}{\sqrt{6}}\cdot\frac{4}{\pi}\left(\sin\frac{2\pi}{3}\right)\cdot V_{\text{dc}}$$
$$= \frac{\sqrt{2}}{\pi}V_{\text{dc}} = \frac{\sqrt{2}}{\pi}\times 14 = 6.28\ \text{V}\left(\text{RMS}\right) \tag{6.135}$$

Interior stator diameter, D_{is}, and stator stack length, l_{stack}:
Here, the sizing of the machine is based on the tangential specific force f_t = (0.2 – 1.2) N/cm² (for small torque [sub 1 Nm] levels).
The base torque T_{eb}, provided from n_b = 0 rpm to n_b = 1800 rpm, is

$$T_{eb} \approx \frac{P_b}{2\pi n_b} = \frac{100}{2\pi\times 1800/60} = 0.5308\ \text{Nm} \tag{6.136}$$

The ratio of the stack length, l_{stack}, to the stator interior diameter, D_{is}, is $\lambda = l_{stack}/D_{is}$ = 0.3 – 3. For l_{stack}/D_{is} = 1.0:

$$T_{eb} = \frac{D_{is}}{2} f_t \cdot \pi D_{is} \cdot \frac{l_{stack}}{D_{is}} \cdot D_{is};\quad f_t = 1\ \text{N/cm}^2 \tag{6.137}$$

$$D_{is} = \sqrt[3]{\frac{2T_{eb}}{\lambda\pi f_t}} = \sqrt[3]{\frac{2\times 0.5308}{1\times\pi\times 1\times 10^4}} = 3.24\times 10^{-2}\text{m}$$
$$= l_{stack} \tag{6.138}$$

We have to choose now the number of poles. For maximum speed and $2p_1 = 4$, $f_{max} = p_1 n_{max} = 2\times\frac{3000}{60} = 100$ Hz. For this frequency, 0.5 mm thick laminated silicon steel can still be used for the magnetic cores of the stator (and the rotor).

To reduce the copper weight (and losses) and the fabrication costs (which are proportional to the number of coils in the stator winding), a six-slot stator and four-pole rotor configuration is used (Figure 6.45).

The PM span on the rotor, b_{PM}, is chosen almost equal to the stator slot pitch, τ_s, to reduce the cogging torque.

PM sizing and coil mmf $n_c I_b$ computation:
Let us consider that the PM flux linkage in the stator, 2 coils per phase, varies sinusoidally with the maximum of Ψ_{PM}:

$$\Psi_{PM_{max}} = B_{gPM} \times b_{PM} \times l_{stack} \times 2n_c \tag{6.139}$$

where B_{gPM} is the PM airgap flux density.

$$\Psi_{PM}(\theta_{er}) = \Psi_{PM_{max}} \cdot \sin\theta_{er}; \quad \theta_{er} = p_1 \cdot \theta_r \tag{6.140}$$

where

θ_r is the geometric angle
θ_{er} is the electric angle

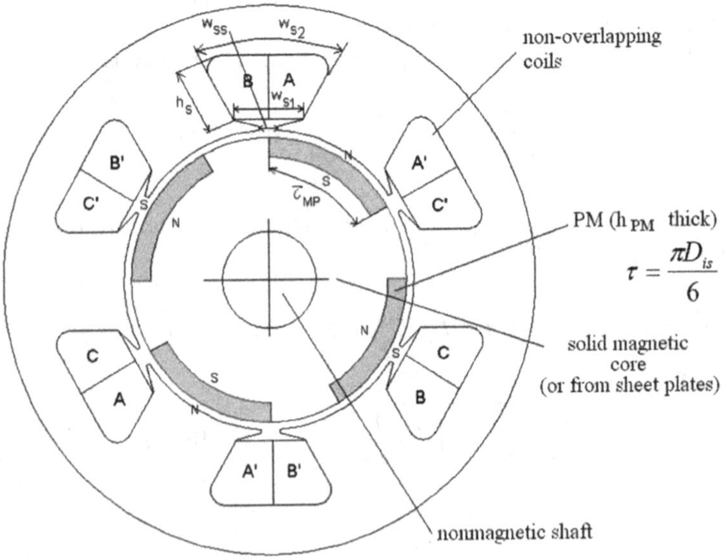

FIGURE 6.45 PMSM with six slots and four poles.

As for the d.c. brush PM machine, B_{gPM} is

$$B_{gPM} \approx \frac{B_r}{1+k_{fringe}} \times \frac{h_{PM}}{h_{PM}+g} \tag{6.141}$$

where the fringing factor $k_{fringe} = 0.1–0.2$.

For NdFeB PMs ($B_r = 1.2\ T$ and $H_c = 900$ kA/m) with an airgap $g = 0.5 \times 10^{-3}$ m and $h_{PM} = 4g = 2 \times 10^{-3}$ m, Equation 6.141 yields

$$B_{gPM} = \frac{1.2}{(1+0.1)} \times \frac{2}{2+0.5} = 0.827\ \text{T}$$

The slot opening $b_{os} = 2 \times 10^{-3}$ m.

The maximum PM flux in one phase is given by Equation 6.139. Substituting the above values in Equation 6.139, we obtain

$$\Psi_{PM_{max}} = 0.872 \times \frac{\pi \times 3.24 \times 10^{-2}}{6} \times \frac{2}{3} \times 3.24 \times 10^{-2} \times 2n_c = 6.3874 \times 10^{-4} \times n_c \tag{6.142}$$

The torque for $I_d = 0$ and $I_q = I_b$ (RMS) is

$$T_{eb} = 3p_1 \frac{\Psi_{PM_{max}}}{\sqrt{2}} \times I_b \tag{6.143}$$

So, from Equations 6.142 and 6.143:

$$n_c I_b = \frac{0.5308 \times \sqrt{2}}{3 \times 2 \times 6.3874 \times 10^{-4}} = 195.28 \text{ A turns(RMS) per coil}$$

Stator slot sizing:

There are two coils in each slot, and thus the active area of stator slot, A_{co}, is

$$A_{co} = \frac{2n_c I_b}{k_{fill} \times j_{cob}} = \frac{2 \times 195.28}{0.4 \times 6.5 \times 10^{+6}} = 136.22 \times 10^{-6}\text{m}^2 \tag{6.144}$$

where current density $j_{cob} = 6.5$ A/mm² and slot filling factor $k_{fill} = 0.4$. The stator tooth and top-slot width:

$$b_{t1} = b_{s1} \approx \tau_s/2 = \pi \times 3.24 \times 10^{-2}/12 = 8.478 \times 10^{-3}\ \text{m} \tag{6.145}$$

Choosing the active slot height $h_{su} = 12 \times 10^{-3}$ m, the bottom width of slot b_{s2} is

$$b_{s2} = \frac{\pi(D_{is}+2h_{su})}{6} - b_{s1} = \frac{\pi(32.4+2\times 12)10^{-3}}{6} - 8.478 \times 10^{-3}$$
$$= 21.04 \times 10^{-3}\ \text{m}$$

We may now check the active slot area A_{cof}:

$$\begin{aligned} A_{cof} &= \frac{(b_{s1}+b_{s2})}{2}\times h_{su} = \frac{(8.478+21.038)10^{-3}}{2}\times 12\times 10^{-3} \\ &= 147.58\times 10^{-6}\ \text{m}^2 \\ &> A_{co} \end{aligned} \tag{6.146}$$

So, the slot sizing holds.

Stator yoke height, h_{ys}, and outer diameter, D_{out}:
Let us accept $B_{ys} = 1.4$ T in the stator yoke. Then the yoke height, h_{ys}, is

$$h_{ys} = \frac{B_{gPM}\times\tau_{PM}}{2B_{ys}} = \frac{0.872\times\pi\times 3.24\times 10^{-2}}{2\times 1.4\times 4} = 7.92\times 10^{-3}\ \text{m} \tag{6.147}$$

It is $\tau_{PM} > b_{PM}$ to avoid magnetic saturation under load.

So the outer stator diameter, D_{out}, is

$$\begin{aligned} D_{out} &= D_{is} + 2h_{su} + 2h_{sw} + 2h_{ys} = 3.24\times 10^{-2} + 2\times 1.2\times 10^{-2} \\ &+ 2\times 1.5\times 10^{-3} + 2\cdot 7.92\times 10^{-3} = 75.24\times 10^{-3}\ \text{m} \end{aligned} \tag{6.148}$$

$D_{is}/D_{out} = (32.4\times 10^{-3})/(75.24\times 10^{-3}) = 0.43$, which is not close to 0.6-0.65 that is considered close to the maximum efficiency design (see IM design in Chapter 5). So a redesign with a smaller f_t may be needed.

Machine parameters:
The phase resistance, R_s, is

$$R_s = \rho_{co}\frac{l_{turn}\times 2\times n_c^2}{(n_c I_b / j_{co})} \tag{6.149}$$

The turn length, l_{turn}, is

$$\begin{aligned} l_{turn} &\approx 2(l_{stack} + 1.25\tau_s) = 2(3.24 + 1.25\times 1.695)\times 10^{-2} \\ &\simeq 0.107\ \text{m} \end{aligned}$$

$$R_s = \frac{2.3\times 10^{-8}\times 0.107\times n_c^2\times 2}{195.28\times 10}\times(6.5\times 10^6) = 1.641\times 10^{-4} n_c^2 \tag{6.150}$$

The phase inductance comprises the main inductance L_m, the leakage inductance L_{sl}, and the coupling inductance L_{12}:

$$\begin{aligned} L_m &\approx 2n_c^2\mu_0\frac{(\tau_s - b_{os})}{h_{PM}+g}l_{stack} = \frac{2\times 1.256\times 10^{-6}(16.95-2)10^{-3}\cdot 3.24\times 10^{-2}}{(2+0.5)\times 10^{-3}} \\ &\times n_c^2 = 4.86\times 10^{-7} n_c^2 \end{aligned} \tag{6.151}$$

The phase mutual inductance $L_{12} \approx -L_m/3$ and the leakage inductance is approximated here by $L_{sl} = 0.3L_m$. So the synchronous inductance L_s is

$$L_s = L_m - L_{12} + L_{sl} = 4.86\times10^{-7} n_c^2 \left(\frac{4}{3}+0.3\right) = 7.92\times10^{-7} n_c^2 \tag{6.152}$$

We may now calculate the copper losses for base torque, T_{eb}:

$$P_{cob} = 3R_s I_b^2 = 3\times1.641\times10^{-4}\left(n_c I_b\right)^2 \approx 19\text{W} \tag{6.153}$$

Neglecting iron and mechanical losses, the efficiency at base power and speed would be

$$\eta_b = \frac{P_b}{P_b + P_{cob}} = \frac{100}{100+19} = 0.84 \tag{6.154}$$

Number of turns per coil, n_c:

Let us first draw the phasor diagram for base speed and torque (with pure I_q control $[I_d = 0]$) (Figure 6.46a). The frequency at base speed is $f_b = n_b \times p_1 = 1800 \times 2/60 = 60$ Hz. The emf E_1 (RMS) is

$$E_1 = \frac{\omega_{1b}\Psi_{PM_{max}}}{\sqrt{2}} = 2\pi60\times\frac{6.3874\times10^{-4}\times n_c}{\sqrt{2}} = 0.1707\times n_c \tag{6.155}$$

$$\begin{aligned} V_{ph\,max} &= \sqrt{\left(E_1 + R_s I_b\right)^2 + \left(\omega_{1b} L_s I_b\right)^2} = n_c\sqrt{0.2027^2 + 0.0583^2} \\ &= 6.28\text{ V} \end{aligned} \tag{6.156}$$

So the number of turns, n_c, is

$$n_c \approx 30 \text{ turns/coil} \tag{6.157}$$

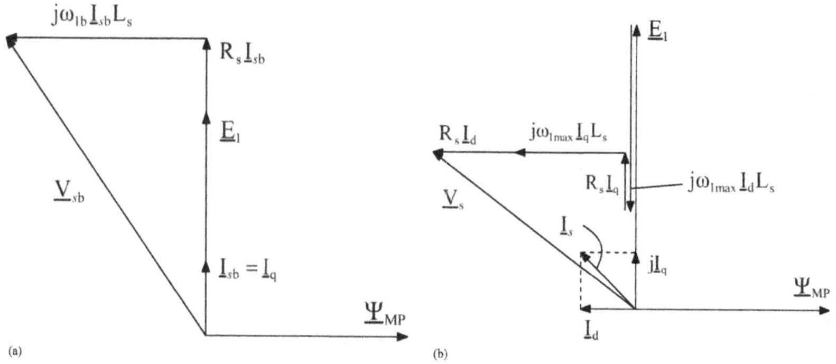

FIGURE 6.46 The phasor diagram: (a) at base speed (pure I_q control, $I_d = 0$) and (b) at maximum speed ($i_d < 0$).

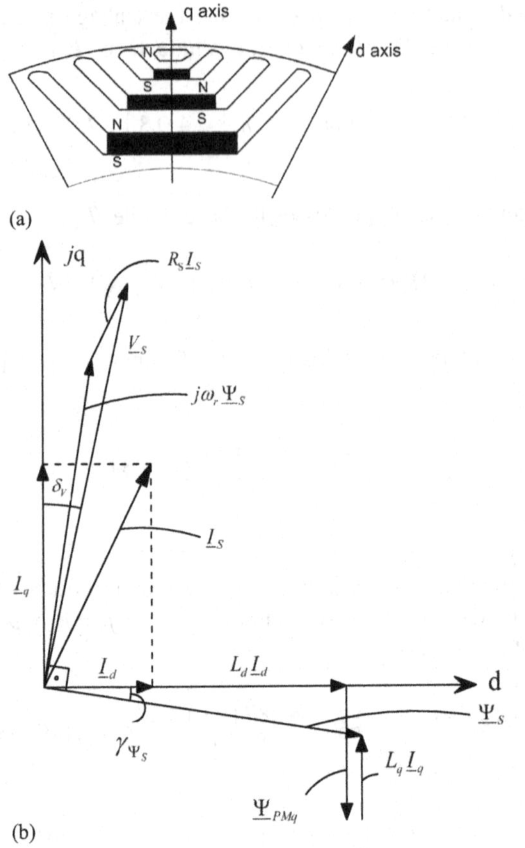

FIGURE 6.47 (a) PM-assisted RSM and (b) its phasor diagram.

The RMS phase base current is

$$I_b = (n_c I_b)/n_c = 6.51\ \text{A} \tag{6.158}$$

The apparent input power, S_n, is

$$S_n = 3V_{ph\,max} \times I_b = 3 \times 6.28 \times 6.509 = 122.629\ \text{VA} \tag{6.159}$$

So the base speed power factor, $\cos \varphi_b$, is

$$\cos \varphi_b = \frac{P_b}{\eta_b S_b} = \frac{100}{0.84 \times 122.629} = 0.97 \tag{6.160}$$

Maximum speed torque (P_b) capability verification:
To maintain the base power at 3000 rpm for which the frequency $f_{max} = 3000 \times 2/60 = 100$ Hz, we first calculate the required torque:

$$\left(T_e\right)_{n_{max}} = \frac{P_b}{2\pi n_{max}} = \frac{100}{2\pi \times 3000/60} = 0.318 \text{ Nm} \tag{6.161}$$

So the I_q current needed is

$$\frac{\left(I_q\right)_{n_{max}}}{I_b} = \frac{\left(T_e\right)_{n_{max}}}{T_{eb}} \tag{6.162}$$

So $\left(I_q\right)_{n_{max}} = 6.509 \times \frac{0.318}{0.5308} = 3.9$ A.

We may now check what torque can be produced at rated current $I_b = 6.509\ A$. So I_d is

$$I_d = \sqrt{I_b^2 - \left(I_q\right)_{n_{max}}^2} = \sqrt{6.509^2 - 3.9^2} = 5.21 \text{ A} \tag{6.163}$$

So, from the phasor diagram in Figure 6.46b:

$$V_s = \sqrt{\left(E_1 + R_s I_q - \omega_{1\max} L_s I_d\right)^2 + \left(R_s I_d + \omega_{1\max} \times L_s I_q\right)^2}$$

$$= \sqrt{\begin{aligned}&\left(0.1707 \times \frac{3000}{1800} \times 30 + 1.641 \times 10^{-4} \times 30^2 \times 3.9 - 2\pi 100 \times 7.92 \times 10^{-7} \times 30^2 \times 5.21\right)^2 \\ &+ \left(1.641 \times 10^{-4} \times 30^2 \times 5.21 + 2\pi 100 \times 7.92 \times 10^{-7} \times 30^2 \times 3.9\right)^2\end{aligned}}$$

$$= \sqrt{6.825^2 + \left(0.755 + 0.4364\right)^2} = 6.928\ V > V_{ph\max} = 6.28 \text{V}$$

So, with even a strong negative (demagnetization) I_d, the machine cannot provide constant power, P_b, from 1800 to 3000 rpm at base current. Notice that at base speed (1800 rpm) $I_d = 0$. Larger stator current, for larger negative I_d may do it.

6.17 SINGLE-PHASE PM AUTONOMOUS A.C. GENERATOR WITH STEP-CAPACITOR VOLTAGE CONTROL: A CASE STUDY

Motivation:
Auxiliary services such as on tracks, buses, trolleybuses, street cars, subway, trains need single-phase a.c. generators at constant speed in the new range with coarse voltage control at high efficiency.

Here such a system with a single-phase a.c. PM generator with step capacitor voltage control is discussed to get a flavor of what a potentially practical solution implies.

6.17.1 Introduction

Among potential solutions we enumerate here:

- D.c. Excited single-phase synchronous generator with a cage on the rotor (Figure 6.48), which implies brushes to feed the d.c. excitation and has a

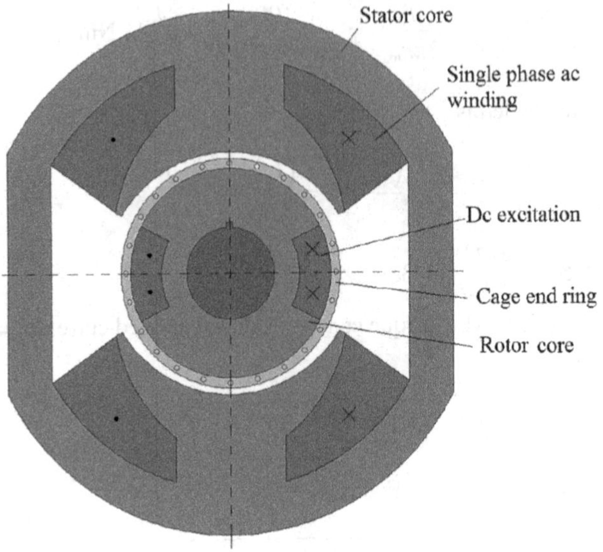

FIGURE 6.48 D.c. excited single-phase synchronous generator with a cage on rotor.

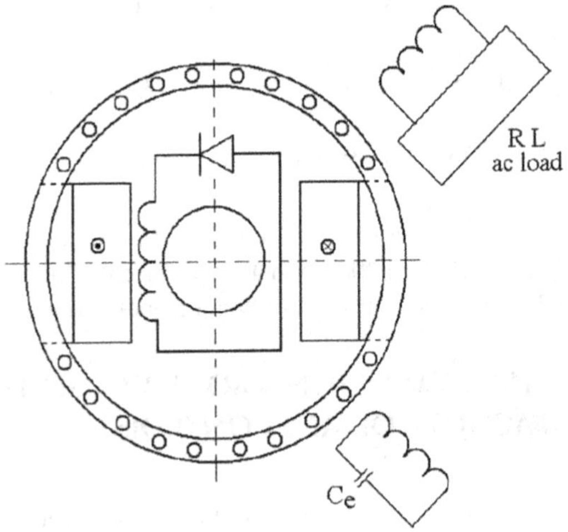

FIGURE 6.49 Single-phase induction generator with capacitor vector control.

diminished efficiency while the cost is reduced by simply using two coils for two poles, to simplify fabrication.

- Single-phase induction generator with capacitor connected auxiliary winding and variable capacitor in the main winding for voltage regulation (Figure 6.49).

  In this case the system is brushless but efficiency and size are not premium.
- Double-saliency stator-d.c. excited generator with 2 armature (a.c.) coils on the stator, too (Figure 6.50).

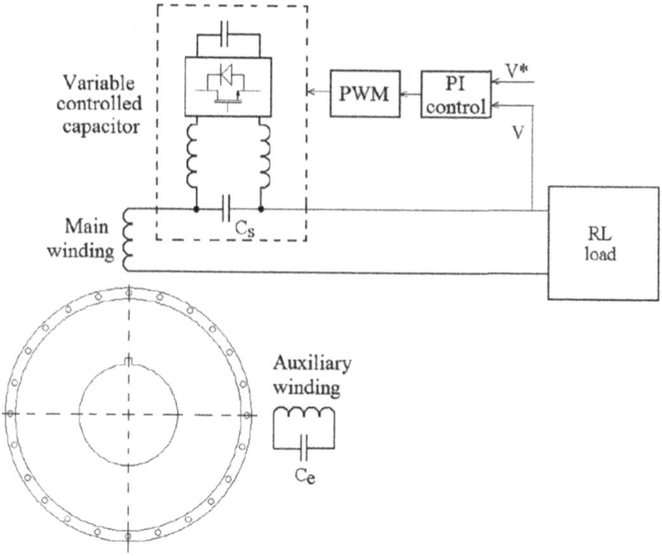

FIGURE 6.50 d.c. + a.c. single stator doubly-salient synchronous generator.

This time only half of the homopolar d.c. excitation flux produces/induces emf (voltage) in the stator a.c. coils, and thus the machine is larger though brushless; also the machine reactance (in p.u.) is large (small airgap) and thus voltage regulation (reduction with load) is large and thus the d.c. excitation circuit has to be designed larger (for minimum envisaged lagging power factor load).

6.17.2 The Proposed Configuration Characterization

In view of the above Ref [] introduced single-phase PM generator (Figure 6.51) with two poles (for 50 Hz at 3000 rpm) where output voltage is controlled rather constant with load by capacitor step control. It has again only two a.c. coils on stator.

To reduce the reactive power rating of the step-controlled capacitor inversed saliency ($L_d < L_q$)—as in PMSM—is produced by magnetic saturation on load.

Also in general, with surface PMs: L_d, $L_q < 1$ (p.u.) for smaller voltage regulation effort.

The phase equation in steady state (no rotor cage as the PM edduy currents are neglected since we may segment the PM poles tangentially and axially) is simply:

$$\underline{V}_s + R_s\underline{I}_s = E_1 - jX_dI_d - jX_qI_q; \quad X_d < X_q \tag{6.165}$$

The corresponding phasor diagram is shown in Figure 6.52 in the "happy" situation when, due to $x_d \ll x_q$ the full resistive load voltage $V_s = E_1$! Even if so, voltage regulation is required for lagging (leading) power factor loads.

From the phasor diagram we extract the conditions for $E_1 = V_{sn}$ for I_{sn} for resistive load:

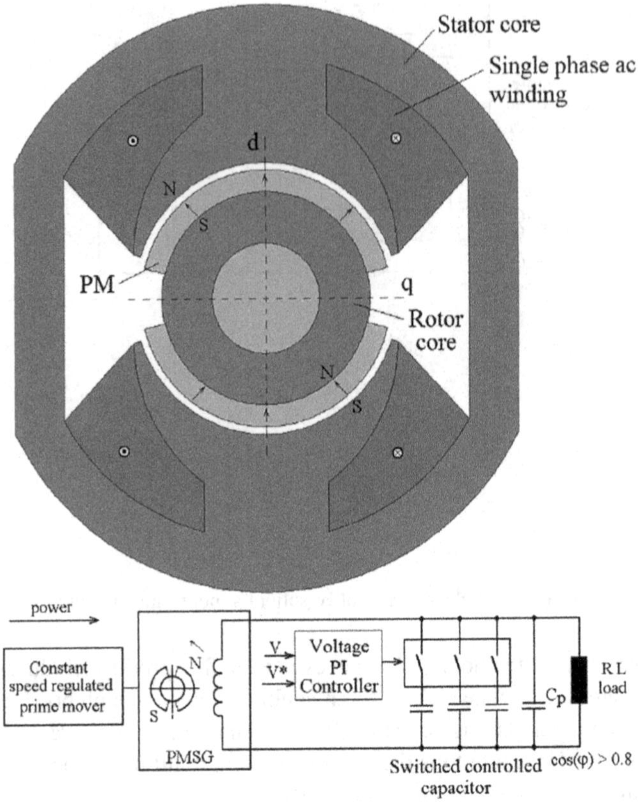

FIGURE 6.51 Single-phase PM generator.

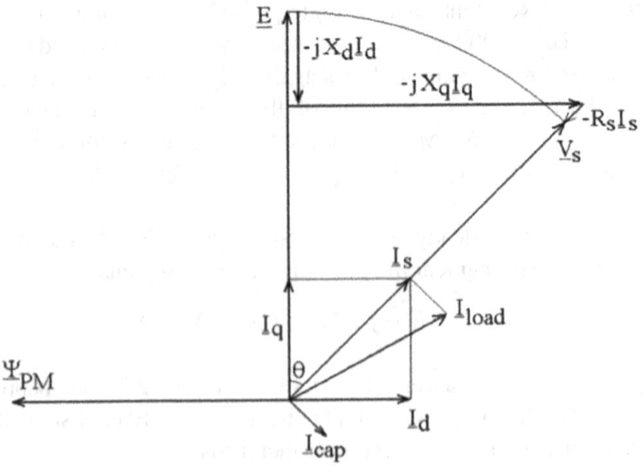

FIGURE 6.52 Single-phase PM generator phasor diagram for resistive load when $V_s = E_1$!

$$E_1 - X_d I_{sn} \sin \theta_n = V_{sn} \cos \theta_n$$
$$V_{sn} \sin \theta_n = X_q I_{sn} \cos \theta_n \tag{6.166}$$

For $x_d = 0.48$ p.u. and $x_q = 1.0$ p.u. ($x_q I_{sn} = V_{sn}$, $x_d I_{sn} = V_{sn}$).

The conditions in (6.166) are met with $E_1 = 1.05\ V_{sn}$. Thus the balance for active and reactive power is

$$V_{sn} I_{sn} + R_s I_{sn}^2 = E_1 I_{qn} - \left(X_d - X_q\right) I_{dn} I_{qn}$$
$$Q_{1n} = E_1 I_{dn} + X_d I_{dn}^2 + X_q I_{qn}^2 = 0 \tag{6.167}$$

For a less demanding applications (in terms of voltage regulation for various loads) no capacitor is needed for voltage regulation in the ±10% down to 0.85 lagging power factor.

A preliminary design for 3 kW at 3000 rpm; 50 Hz led to tentative d.c. output data as in Table 6.…

The key design issues is to produce the required E_1, X_d, X_q and for that FEA is required.

Figure 6.53 shows the FEM results for no-load (Figure 6.53a) at full resistive load (Figure 6.53b), the phase inductance (Figure 6.53c) from FEM and approximated with 1,2,3 terms:

$$L_s\left(\theta_r\right) = L_0 + L_1 \cos\left(2\theta_r - \varphi_1\right) + L_2\left(4\theta_r - \varphi_2\right) + L_3\left(6\theta_r - \varphi_3\right) \tag{6.168}$$

Finally, we obtained: $R_s = 0.92\Omega$, $\Psi_{PM} = 0.948$ Wb, L_0, L_1, φ_1, L_2, φ_2, L_3, φ_3, at 16A rms = 21.9 mH, 12.6/141.6 mH/deg; 4.6/165 mH/deg; 2.2/–92.88 mH/deg.

At least two L_0, L_1 have to be retained in equation (6.168) which will be used in the circuit modeling of generator for control which starts from the equation:

$$E_1\left(\theta_r\right) = V_s + R_s I_s + \frac{d}{d\theta_r}\left(L_s\left(\theta_r, I_s\right) \cdot I_s\right); \quad \frac{d\theta_r}{dt} = \omega_r \tag{6.169}$$

But before turning to controlled performance let offer FEM data on cogging torque and full load torque for rated pure resistive load of 2.5 kW (Figure 6.54).

6.17.3 Sample Step Capacitor Results

For control the step capacitor control (Figure 6.55) produced results as in Figure 6.56.

The control results look satisfactory with only 3 capacitor in parallel (in different combinations).

6.17.4 Experimental Effort

The full scale prototype test platform is shown in Figure 6.57.

As expected experimental results showed secondary effects neglected in the simulations even on no load (Figure 6.58).

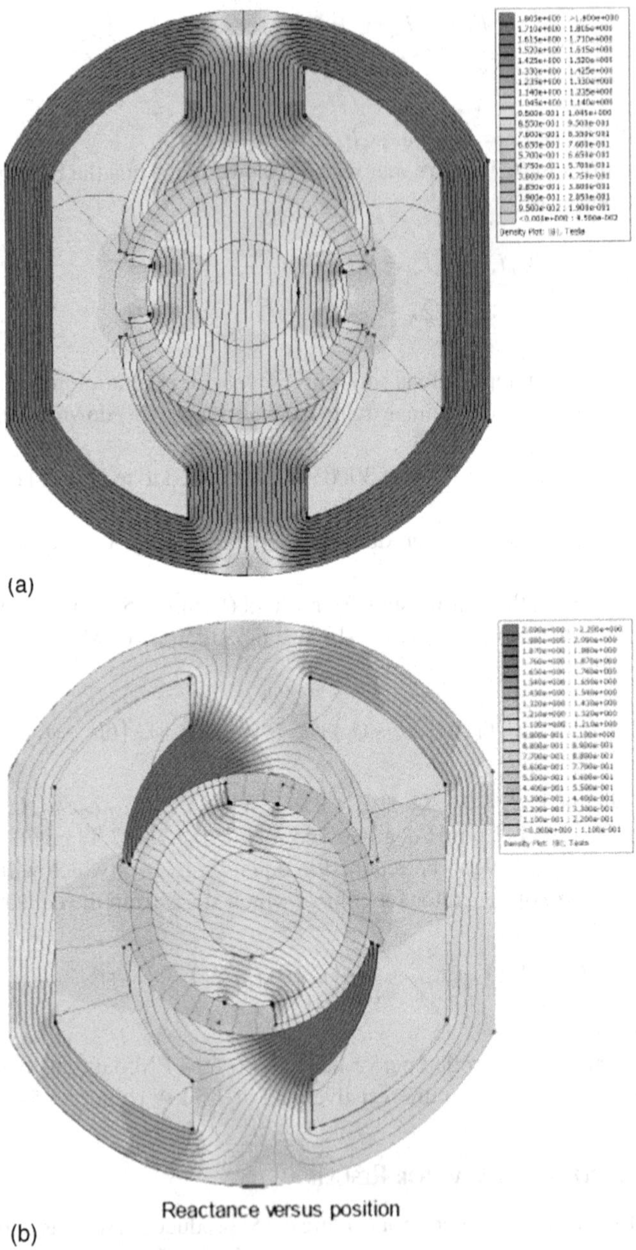

FIGURE 6.53 2D FEM field and inductance results: (a) PM field in axis d, (b) PM field with rated current in q axis

(Continued)

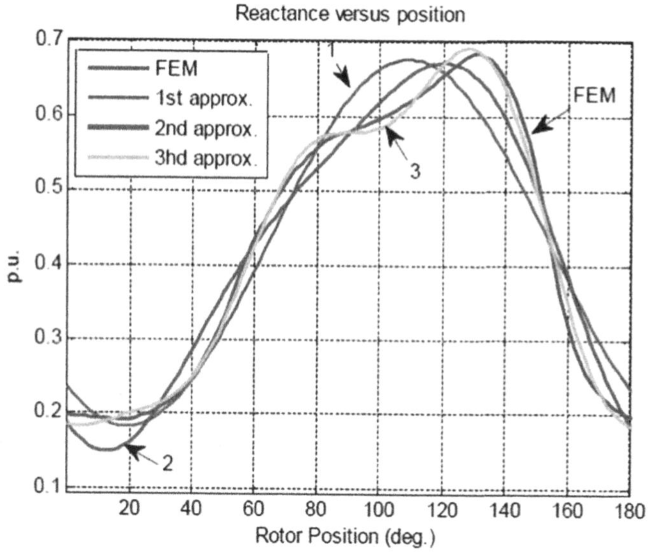

(c)

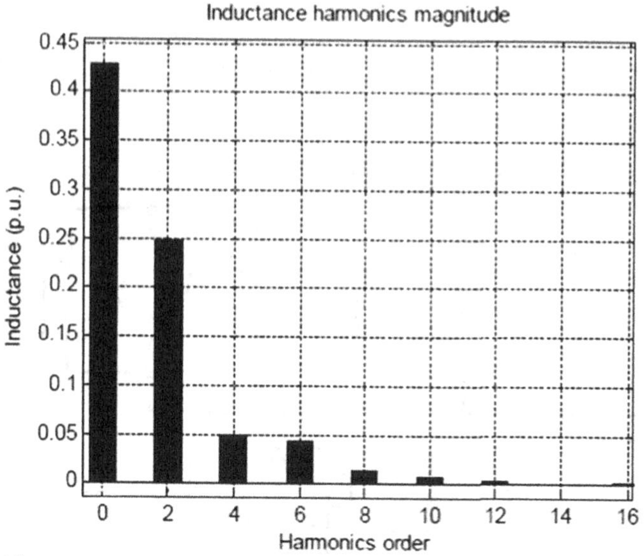

(d)

FIGURE 6.53 (Continued) (c) machine reactance in p.u. versus rotor position, (d) machine inductance harmonics.

On load (Figure 6.59) the load current is rather sinusoidal but the generator current is not.

Experiments show the output voltage versus current characteristics under severely (0.7 lagging) low power factor, at 50 Hz and 60 Hz (Figure 6.60).

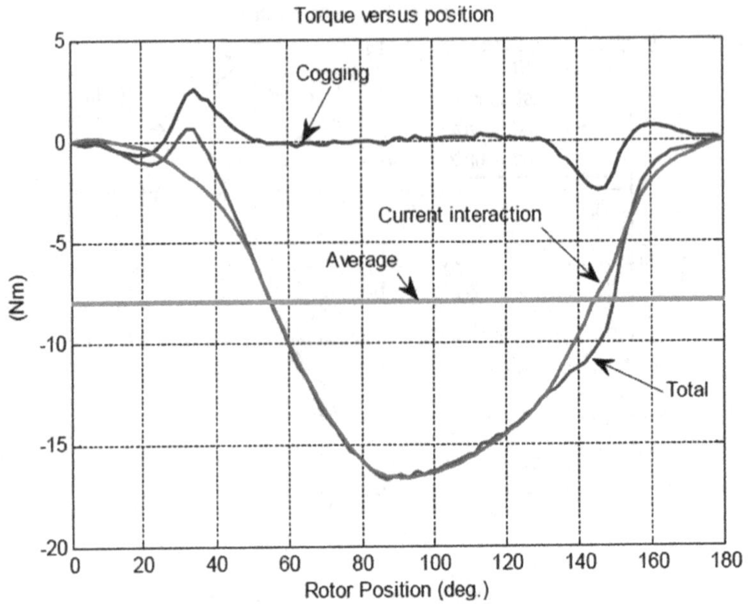

FIGURE 6.54 Cogging and torque ripple at 2.5 kW 50 Hz/3000 rpm with sinusoidal (I_q) current.

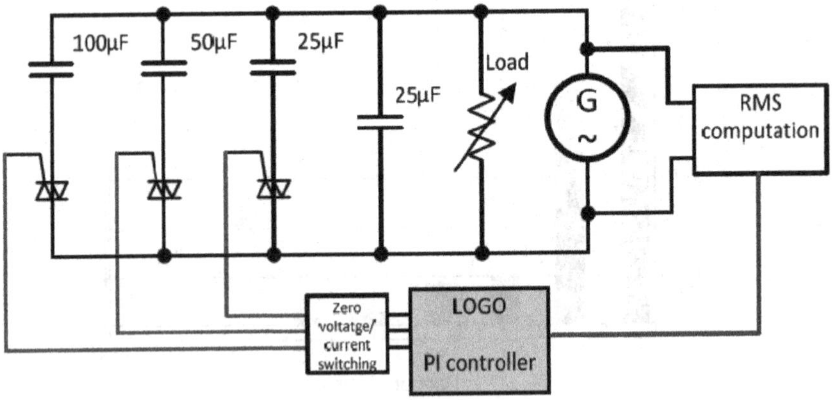

FIGURE 6.55 The step capacitor voltage control scheme.

It is evident that in these extreme power factor conditions only at 60 Hz (3600 rpm) the generator can keep the voltage regulation lower than (5–6)% up to rated power.

Finally the experiments lead to measurements of losses and at 60 Hz an efficiency of 90% was obtained at 2.5 kW power with resistive load.

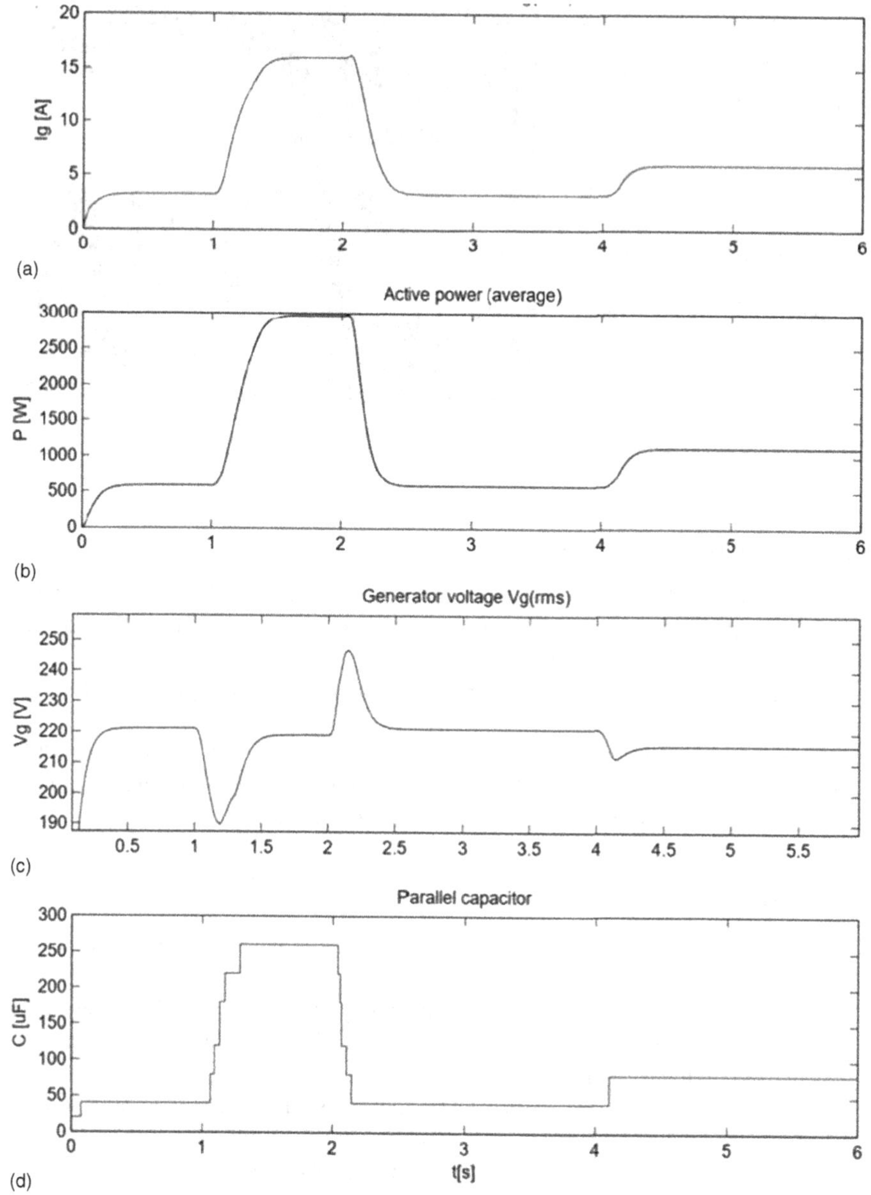

FIGURE 6.56 Step capacitor voltage control sample simulation results: (a) generator current amplitude, (b) active power, (c) generator voltage, (d) parallel capacitor versus time for large step resistive load variation.

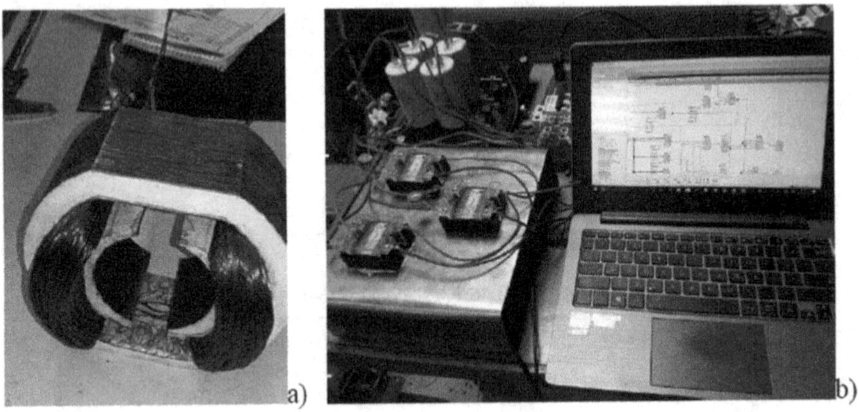

FIGURE 6.57 The prototype 1 phase PM generator test platform: (a) the generator stator, (b) control system.

6.17.5 Experimental Effort

The prototype produced less PM emf than in simulations which lead to lower than desired/calculated performance. As FEM was used most probably the PM quality was not the one claimed by the provider.

It means that further refinment efforts are needed until a fully practical prototype is made.

However, the rather convenient active cost of 66 USD, for 2.5 kW at 50 Hz (3000 rpm) generator with a ±7% (so far) voltage regulation at 0.8 lagging power factor and around 90% efficiency should be considered a solid start.

6.18 SUMMARY

- SMs have a 3 (2)-phase a.c. winding on the stator and a d.c. (or PM) or a variable reluctance rotor.
- As the rotor is d.c. and the stator is a.c. at frequency f_1, the speed at steady state, $n_1 = f_1/p_1$, is constant with the load ($2p_1$ denotes the number of poles on the stator and on the rotor, in general).
- SMs are used as generators in electric power systems or as autonomous sources for wind energy conversion, cogeneration, and stand-by power, or on automobiles, trucks, diesel–electric locomotives, vessels, and aircraft.
- SMs are used as motors from sub watt power (as single-phase machines) to 50 MW, 60 kV variable speed drives (for gas compressors).
- The PMSMs/PMSGs with variable frequency (speed) control via power electronics constitute the most dynamic chapter in industrial, transportation, home appliance energy and motion control for energy savings and better productivity.
- Linear versions of SMs are applied for MAGLEV transport, industrial carriers, and, with oscillatory motion, for electromagnetic ICE valves, active suspension damping, and compressor drives of refrigerators.
- AC-distributed windings are typical (with q [slots/pole/phase] > 1), but non-overlapping coil (concentrated) windings are used for PMSMs when $q \leq 1/2$

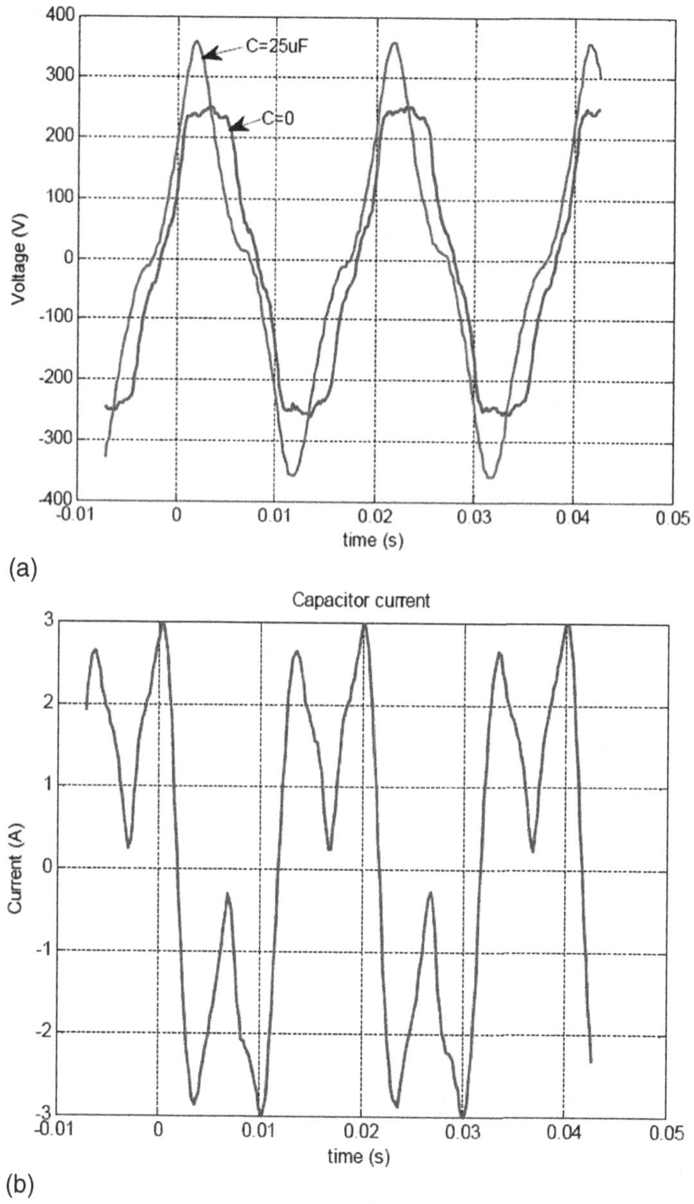

FIGURE 6.58 No-load tests at 3000 rpm: (a) output voltage waveform with and without the C = 25μF filter capacitor, (b) compensation capacitor C current.

and the stator has N_s slots (teeth) and the rotor $2p_1$ poles ($N_s = 2p_1 \pm 2K$; $K = 1, 2\ldots$).

These kinds of PMSMs are characterized by lower copper losses, machine size, and cogging (zero current) torque pulsations.

- The number of cogging-torque periods per revolution is given by the LCM of N_s and $2p_1$: the larger the LCM, the smaller the cogging torque.

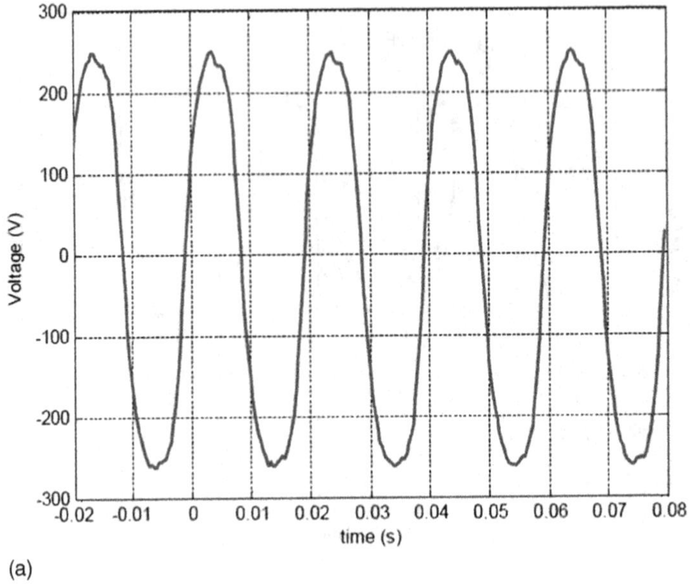

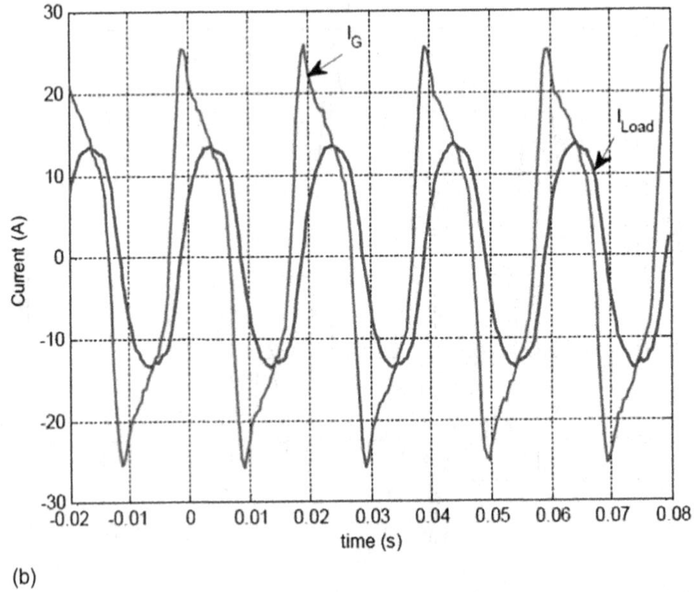

FIGURE 6.59 Measured output voltage, (a), and output and generator currents waveforms, (b).

- The stator emf, E_s, of d.c. excitation (or PMs) in the rotor is sinusoidal or trapezoidal (PM rotor with $q = 1$ stator windings or $q < 1/2$). A trapezoidal emf recommends rectangular (120° wide) a.c. current control with two phases active at any time (except for phase commutation intervals), and provides reasonable torque pulsations and simplified rotor-position-triggered control.

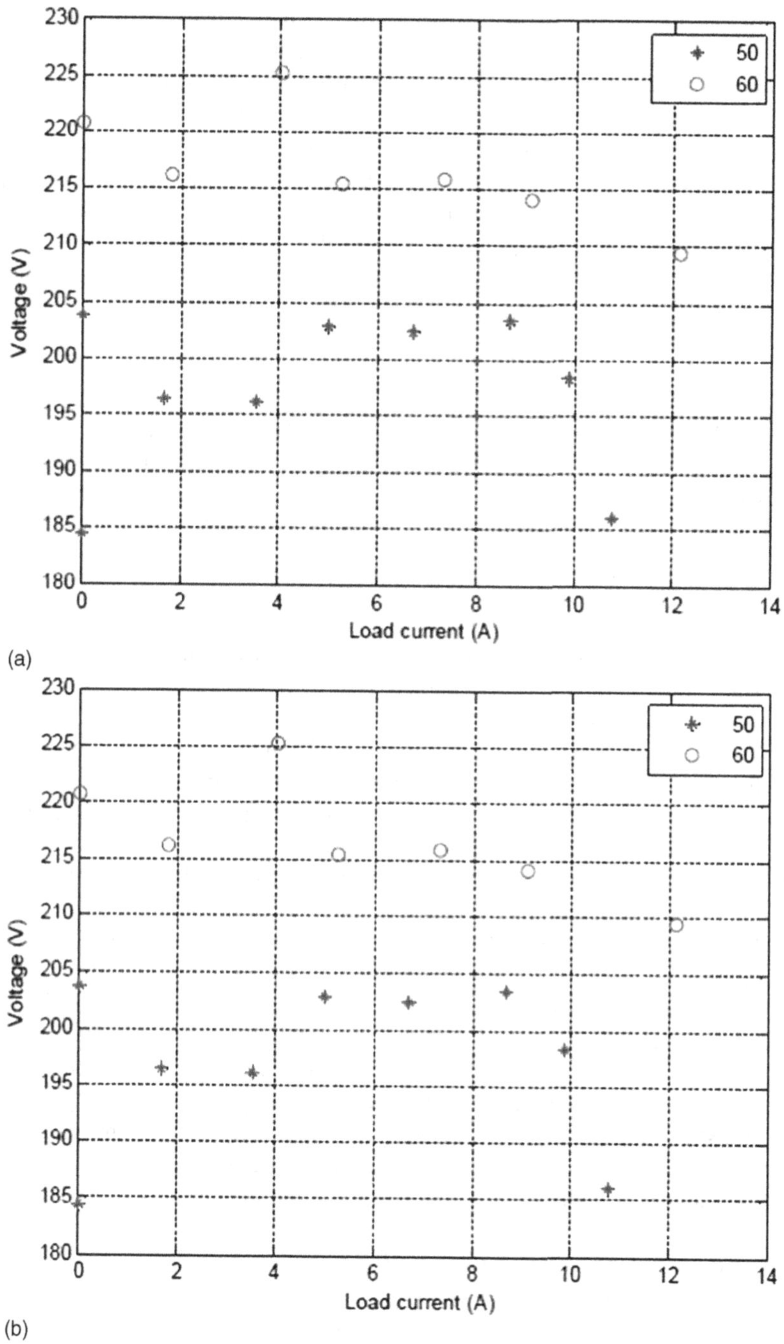

FIGURE 6.60 Measure voltage (current) load curves at 0.7 (ag. power factor load for 50 Hz and 60 Hz).

- During steady state, the stator currents do not induce any voltage (and currents) in the rotor (which may have d.c. [or PM] excitation and a cage in grid-operated applications) because the stator mmf travels at rotor speed.

  Consequently, the stator (armature) winding current may be decomposed in two components: I_d and I_q in each phase. These components produce their air-gap flux density aligned to the rotor pole (axis *d*) or to the interpole (axis *q*) to manifest themselves as two magnetization synchronous reactances, $X_{dm} > X_{qm}$; if we add the leakage reactances, the so-called synchronous reactances X_d and X_q are obtained. These are valid for steady-state symmetric stator currents.

  So the voltage circuit (per phase) becomes straightforward:

$$\underline{I}_s R_s + \underline{V}_s = \underline{E}_s - jX_d \underline{I}_d - jX_q \underline{I}_q$$

  For $X_d = X_q = X_s$ the nonsalient pole rotor configuration is obtained.
- The "internal" impedance of the SG is in fact related to X_d and X_q, which in PU terms is $x_s = X_s{*}I_n/V_n = 0.5 - 1.5$ (pu) for d.c.-excited machines and smaller for PM pole rotor machines

$$\left(x_s < 0.5\ \text{pu}\right).$$

  So voltage regulation is substantial in SGs for autonomous applications.
- Autonomous generators are characterized by no-load saturation, short-circuit, and external (load) curves.
- The d.c. excitation circuit has been designed to maintain rated terminal voltage for rated current and the lowest-assigned power factor (above 0.707, lagging).
- Most SGs are connected to the power grid through a synchronization routine and under least-possible transients.
- Active power delivery by an SG is obtained by increasing the shaft power of the prime mover (turbine).
- Reactive power delivery/absorption is obtained through field current, I_F, control.
- The voltage power angle δ_V (E_s, V_s) is the crucial variable, and statically stable operation is provided up to $|\delta_V| = \delta_{VK} \leq 90°$, when the SG is tied to the power grid.
- Static stability is the property of a SM/SG to remain in synchronism for slow shaft-power or electric load variations.
- PMSMs, with interior (salient) PM poles, are characterized by inverse saliency, $X_d < X_q$, and, thus, they may operate safely even close to $|\delta_V| = 90°$.

  The power angle $\delta_V < 0$ for motoring for the source (generator) association of signs.
- Grid operation is characterized by P_s (δ_V), Q_s (δ_V), I_s (I_F) (V curves), and Q_s (P_s) (reactive power envelopes).
- The electromagnetic torque, T_e, of SMs has an interaction term (between the d.c. excitation [or PM] field and the stator mmf) and a reluctance term (when $X_d \neq X_q$).
- SGs may operate with unbalanced loads but they behave differently with respect to positive, negative, and zero (homopolar) stator currents sequences by $X_+ \to X_s > X_- > X_0$ reactances.

- A limited degree of the negative current, $I_-/I_+ < 0.02 - 0.03$, is acceptable for SGs at power grids, to avoid overheating of rotor cages by induced currents at $2f_1$ frequency.
- Steady-state, three-phase, two-phase, and one-phase short-circuit operations are not always dangerous for SMs but

$$I_{3sc} < I_{2sc} < I_{1sc}.$$

- The ratio $1/x_d = \frac{I_{3sc}}{I_n}$ is a specification parameter as it defines the maxim power capability of SMs.
- The SM has to be started as an induction motor, which then self-synchronizes at power grid [14]. SMs may be operated at leading power factor to compensate reactive power requirements in local power grids.
- SG connection to grid by its controllers time (300 seconds) may be reduced 10 times by using a 1% rating PWM converter at SG terminals [15].
- SMs at power grid are characterized by efficiency and power factor versus output power, and by the V-shaped (I_s (I_F)) curves.
- PMSMs can operate at power grid or supplied by a PWM inverter at variable frequency (speed) [16].
- A very good product of the power factor and efficiency may be obtained with PMSMs even with a large number of pole pairs at (low speed) medium-power range.
- Large anisotropy ($L_d/L_q > 3$-5)—multiple flux barrier —rotor motors may perform well at a low initial cost either at power grid (with a cage on the rotor) or fed from variable frequency PWM inverters, from 100 W to hundreds of kW.
- Single-phase PMSMs with $N_s/2p_1 = 1$, and tooth-wound windings and no cage on the rotor, but with a rotor parking magnet or tapered airgap (for starting) are used from subwatt to kilowatt power range in association with PWM inverters, for variable speeds.
- Alternatively, distributed a.c. stator two-phase windings, cage rotor PMSMs in power grid direct connection may be used as split-phase capacitor motors for a strictly constant speed (with variable load), and better efficiency or a smaller motor volume.
- A few lab tests for SMs are introduced in this chapter and reference is made to the IEEE standard 115-1995 where a plethora of test procedures for acceptability, performance, and parameter estimation are described.
- A preliminary electromagnetic design methodology for a six-slot/four-pole, three-phase PMSM for automotive applications is presented by a numerical example.
- Reference is made to other, special, configurations of SMs, rotary or linear, with pertinent literature suggested for further reading.

6.19 PROPOSED PROBLEMS

6.1 A salient pole rotor synchronous hydrogenerator that has $S_n = 72$ MVA, $V_{nline} =$ 13 kV (star connection), $2p_1 = 90$ poles, $f_1 = 60$ Hz, and $q_1 = 3$ slots/pole/phase is considered lossless. The stator interior diameter $D_{is} = 13$ m, the stator stack length $l_{stack} = 1.4$ m, the airgap under pole shoes $g = 20 \times 10^{-3}$ m, Carter coefficient $k_c = 1.15$, $\tau_p/\tau = 0.72$ (rotor pole shoe/pole pitch), and the saturation coefficient $k_s = 0.2$; it is a single-layer bar stator winding (diametrical span coils: $Y/\tau = 1$). It operates at the power grid.

Calculate

a. The stator winding factor, k_{w1}
b. The d and q magnetization reactances X_{dm} and X_{qm} if the number of turns per phase $W_s = p_1 \times q_1 \times n_c = 45 \times 3 \times 1 = 115$ turns/phase
c. x_d and x_q in pu
d. E_s, I_d, I_q, P_s, and Q_s for $\cos\varphi_1 = 1$ and $\delta_V = 30°$
e. The no-load airgap flux density, B_{gF}, and the corresponding excitation mmf per $W_F I_F$ required for it

Hints:
Use $k_{w1} = \sin \pi/6(q \sin(\pi/6q))$, Equations 6.31 and 6.32,
$x_d = X_d \cdot V_{nph}/I_n$; $I_n = S_n/(\sqrt{3}V_{nl})$, Figure 6.19a, Equations 6.57 through 6.59, 6.15, and 6.6 and 6.7.

6.2 An aircraft lossless nonsalient pole rotor SG with the parameters $x_d = x_q = 0.6$ (pu), $S_n = 200$ KVA, $V_{nl} = 380$ V, $f_1 = 400$ Hz, and $2p_1 = 4$ operates at rated voltage on a balanced resistive load at rated current.

Calculate

a. The rated phase current for star stator connection,
b. X_d and X_q in Ω,
c. The load resistance, and
d. E_1, I_d, I_q, and δ_V (power angle).
e. For rated current and 0.707 lagging power factor calculate I_d, I_q, I_s, and V_s (terminal voltage) for the same E_1 as above.
f. Calculate the voltage regulation, ΔV_s, for (d) and (e).

Hints: See Equations 6.49 through 6.53 and Example 6.1.

6.3 For the SG in Problem 6.2, calculate the V-shaped curves (I_s (I_F)) if

$$E_1 = aI_F - bI_F^2 \tag{6.164}$$

and if E_1 is as in problem 6.2 for $I_F = 50$ A
and if E_1 is $E_1/2$ for $I_F = 15$ A.
Use $I_F = 10, 20, 30, 40, 50, 60$, and 70 A
Hints: Use (6.164) for the no-load saturation curve to find a and b and then, for $P_s/S_n = 0.0, 0.3, 0.6$, and 1.0, calculate δ_V from Equations 6.57 and 6.58 for I_F given (E_1 from (6.164)) in 5 A steps from 10 to 70 A.

6.4 A nonsalient pole rotor synchronous autonomous generator with $S_n = 50$ KVA, $V_{nl} = 440$ V (star connection), $f_1 = 60$ Hz, $2p_1 = 2$, and $x_s = x_d = x_q = 0.6$ PU, $x_- =$

0.2 PU, and $x_0 = 0.15$ PU operates at rated current with only phase A connected to a resistive load ($I_B = I_C = 0$) for no-load phase voltage $E_1 = 300$ V (RMS).

Calculate

a. The rated phase current I_n, X_s, X_+, X-, and X_0 in Ω
b. I_{A+}, I_{A-}, and I_{A0}
c. V_A (phase voltage for phase A)
d. V_B and V_C (phase voltages on open phases B and C)

Hints: See Equations 6.75 through 6.80, Equation 6.85, and Equation 6.77.

6.5 A PMSM with surface 4 PM poles is fed from a PWM inverter at variable frequency.

For $V_{nl} = 380$ V (star connection), $f_n = 50$ Hz, $X_d = X_q = 6\ \Omega$, and $R_s = 1\ \Omega$, calculate I_d, I_q, I_s, P_e, T_e, η, and cos φ for $\delta_V = 0°, 15°, 30°, 45°, 60°, 75°$, and 90°. Perform the same calculations for $V_{nl} = 100\ V$ and $f = 14$ Hz.

Hint: See Example 6.1.

6.6 A multiple flux barrier four-pole rotor PMSM has weak (ferrite) PMs in the flux barriers (Figure 6.47) and the parameters $L_d = 200$ mH, $L_q = 60$ mH, $\Psi_{PMq} = 0.215$ Wb, $2p_1 = 4$ poles, $(V_n)_{phase} = 220$ V, and zero losses. Calculate I_d, I_q, I_s, T_e, and cos φ versus $\delta_V = 0°, 10°, 30°, 45°, 60°, 70°, 80°, 90°$, and 100°, for $n = 1500$ rpm.

Hints: Notice that the PMs are placed in axis q, and thus the phasor diagram is different from that of Figure 6.37 (see Figure 6.47b).

So,

$$V_s \sin \delta_V = \omega_r L_q I_q - \Psi_{PM_q} \omega_r ; R_s \approx 0$$
$$V_s \cos \delta_V = \omega_r L_d I_d$$

6.7 A three-phase lossless PMSG with the data $E_1 = 250$ V (phase, RMS), $V_{nl} = 220\sqrt{3}$ V (star connection), $X_d = 10\ \Omega$, $X_q = 20\ \Omega$, $2p_1 = 4$, and $f_1 = 60$ Hz is driven at constant speed and balanced with a resistive load. Calculate the variation of voltage with power angle, δ_V, until the voltage regulation becomes zero again because of inverse saliency ($V_s = E_1$), and the corresponding delivered power.

Hints: Use Equation 6.50 and Figure 6.24b.

6.8 A large lossless SM has the data $P_n = 5$ MW, $f_1 = 60$ Hz, $(V_n)_{phase} = 2.2$ kV, $2p_1 = 2$ poles, $x_d = x_q = x_s = 0.6$ (pu), an inertia $H = \frac{J}{2}\left(\frac{\omega_1}{p_1}\right)^2 \times \frac{1}{P_n} = 10\,\text{s}$, and $E_1 = 2.5$ kV (phase).

Calculate

a. The rated current, I_n, and X_s (in Ω)
b. I_d, I_q, I_s, P_s, and Q_s at $\delta_V = 30°$
c. The electromagnetic torque (T_e) at $\delta_V = 30°$
d. The asynchronous torque coefficient K_a if at $S = 0.01$, the asynchronous torque $T_{as} = 0.3 \times (T_e)_{\delta V = 30°}$
e. The eigen frequency of the SG, ω_{0n}, and its variation when E_1 changes from 2.5 kV to 1.25 kV at $\delta_V = 30°$

f. The module of mechanical resonance, $K_{m\nu}$, for $\Omega_\nu = 2\pi \times n_1/2$ and for ω_{0n}; n_1 is the machine speed

Hints: See Section 6.13.

6.9A single-phase $2p_1 = 2$ pole PMSM (driven by a PWM inverter and having a parking PM for safe starting) has the data $V_{sn} = 120$ V, $f_1 = 60$ Hz, the PM emf $E_s = 0.95\ V_{sn}$, $R_s = 3\ \Omega$, and $L_s = 0.05$ H.

Calculate

a. The synchronous speed at $f_1 = 60$ Hz
b. The stator current at $\delta_V = 0°, 15°, 30°, 45°, 60°$, and $90°$
c. The power factor and efficiency (only copper losses count for δ_V as under (b))
d. The average torque versus δ_V if cogging torque is neglected

Hints: See in Section 6.15, Equations 6.123 through 6.134 and the computation routine described there.

6.10 Redo the preliminary three-phase PMSM design in Section 6.16 for the specifications $P_b = 2000$ W, $n_b = 1800$ rpm, $n_{max} = 3000$ rpm (at P_b), and $V_{dc} = 42$ V.

Hints: Follow Section 6.16.

REFERENCES

1. Ch. Gross, *Electric Machines*, Chapter 7, CRC Press, Taylor & Francis Group, New York, 2006.
2. M.A. Toliyat and G.B. Kliman (eds.), *Handbook of Electric Motors*, 2nd edn., Chapter 5, Marcel Dekker, New York, 2004.
3. T. Kenjo and S. Nagamori, *PM and Brushless DC Motor*, Clarendon Press, Oxford, U.K., 1985.
4. T.J.E. Miller, *Brushless PM and Reluctance Motor Drives*, Clarendon Press, Oxford, U.K., 1989.
5. S.A. Nasar, I. Boldea, and L.E. Unneweher, *PM, Reluctance and Selfsynchronous Motors*, CRC Press, Boca Raton, FL, 1993.
6. D.E. Hanselman, *Brushless PM Motor Design*, McGraw-Hill, Inc., New York, 1994.
7. E.S. Hamdi, *Design of Small Electric Machines*, John Wiley & Sons, New York, 1994.
8. J.F. Gieras and M. Wing, *PM Motor Technology*, Marcel Dekker, New York, 2002.
9. T.J.E. Miller, *Switched Reluctance Motors and their Control*, Clarendon Press, Oxford, U.K., 1993.
10. I. Boldea and S.A. Nasar, *Linear Electric Actuators and Generators*, Cambridge University Press, Cambridge, U.K., 1997.
11. I. Boldea and S.A. Nasar, *Linear Motion Electromagnetic Devices*, CRC Press Taylor & Francis Group, New York, 2001.
12. I. Boldea, *Variable Speed Generators*, Chapter 10, CRC Press, Taylor & Francis Group, New York, 2005.
13. I. Boldea, *Synchronous Generators*, Chapter 4, CRC Press, Taylor & Francis Group, New York, 2005.
14. I. Boldea, T. Dumitrescu, and S.A. Nasar, Steady state unified treatment of capacitor A.C. motors, *IEEE Trans.*, EC-14(3), 1999, 577–582.
15. R. Son, M. Jeon, "A study for starting characteristic analysis method of salient pole synchronous motors", *IEEE Trans.*, Vol. IA-53, No. 2, 2017, pp. 1627–1634.

16. S. Shah, H. Sun, D. Nikovski, J. Zhang, "VSC-based synchronizer for generators", *IEEE Trans.*, Vol. EC-33, No. 1, 2018, pp. 116–125.
17. P. Ponomarev, P. Lindh, J. Pyrhonen, "Effect of slot-and-pole combination on the leakage inductance and the performance of tooth-coil permanent-magnet synchronous machines", *IEEE Trans.*, Vol. IE-60, No. 10, 2013, pp. 4310–4317.

Index

A

AC distributed windings, 197–198, 370
cage rotor windings, 212
airgap flux density, 16, 121, 133, 149–165, 186–215, 237, 269–294, 310–318, 374–376
bar and ring resistances, 214
chording and distribution factor, 207
end ring and cage-bar currents, 214
factor components, 216
rotor cage geometry, 214
skewing factor, 214
total winding factor, 215
MMF space harmonics, integer q
chording factor, 212
linear current density, 122, 204
mmf distribution, 201, 203, 222, 312
peak value, 15, 33, 90, 204
winding distribution factor, 204
pole count changing ac 3-phase, 211
winding, 215–222, 244–251, 260–279, 287–375
practical one-layer ac 3-phase winding
bar (single turn) coils, 208
concentric coil phase belt, 208
identical (chain) coil phase, 208
self-induced emf angle shift, 207, 210
single-layer 24-slot, 4-pole winding, 207
slot-emf phasor star, 210
traveling magnetomotive force (MMF)
airgap flux density, 211–215, 237, 269–276, 294, 310–318, 356, 374–376
brush–commutator windings, 200
ideal multiphase MMF, 200
integer, 200–209, 279, 290
number of slots, 143, 146, 156, 193, 197, 200, 204, 290
two-phase ac windings, 211
active power, 11, 13, 16, 23, 44, 67–72, 76, 99, 109, 225–227, 235–236, 266, 271–272, 290–293, 317–321, 326, 328–332, 342, 363, 365, 374–375
active rotor electric machines
dc rotor and ac stator currents, 117
magnetic conergy, 117
matrix form, 87
motion equations, 333
primitive active rotor single-phase electric machine, 117
rotor cage, 333–337, 347, 363, 375
airgap flux density
armature reaction, 132–134, 156, 163, 192–193, 318–320
leakage factor, 156
leakage flux, 32, 53–55, 90, 153–158, 216–218
linear curve, 159
magnetic field, 5–7, 19, 27–44, 51–57, 102–114, 120–126, 136, 139, 149–153, 158, 165, 190, 197, 216, 219–220, 244, 289–291, 332
mmf, 37, 53–55, 77, 81–82, 96, 118, 121–123, 132–135, 149–165, 198–222, 244–266, 283, 306–336, 374
electric machine design, 186
electromagnetic force (EMF), 30, 141
stator diameter, 192, 276, 355, 358
stator excitation MMF, 149
Ampere's law
flux linkages and inductances, 53
magnetic field, 5–7, 19, 27–39, 44, 51–57, 112–123, 136, 158, 190, 216, 244, 289, 332
no-load magnetization curve, 62, 150–155, 217, 237
autonomous induction generator, 236–238, 326
autonomous synchronous generators, 322
characteristic curves, 177, 323
load curve, 322–330, 373–374
voltage equation, 127, 160, 174–177, 180–184, 325, 329, 350
voltage regulation, 51–52, 58, 68, 97, 107, 175, 291, 309, 326–328, 362–374, 377
$V_s(I_s)$ load curves, 325–326, 373–374
no-load saturation curve, 322, 376
phasor diagram, 31, 66–83, 182, 227, 263, 317, 324–327, 343, 359–363, 377
rated current, 24, 33, 54, 63–67, 101–109, 171, 183, 205, 240, 253, 274, 285, 326, 352–366, 374
short-circuit curve, 322–324
test rig, 183–184, 227, 323–325
autotransformers, 4, 28, 96, 106

B

brush-commutator machines, 6, 20, 120, 139–149, 199
airgap flux density, 16, 121, 133, 149–159, 165, 192, 211, 237, 269, 294, 318, 356, 374
armature reaction, 132, 156, 163, 192, 318–320
stator excitation MMF, 149
armature windings
coil span, 146, 202, 209, 278

double winding, 142
phase shift angle, 144, 315, 329
simple lap windings, $N_s = 16$, $2p_1 = 4$, 144–156
simple wave windings, $N_s = 9$, $2p_1 = 2$, 146–156
commutation process, 158–160
brush-segment contact resistance, 160
definition, 44, 63, 182, 199, 234, 256, 344
efficiency, 11–24, 67–70, 98–108, 126, 159, 171–174, 182, 240–293, 342–378
electromagnetic power, 17, 97, 166–194, 224, 293, 343
electromagnetic torque, 12–17, 112, 166–194, 224, 271, 291–294, 321, 334, 343, 351, 377
excitation circuit equation, 167
excitation losses, 159, 167
rotor coil current reversal, 160
separate excitation, 171, 178
transient inductance, 167
dc brush PM motor, 183
artificial loading, 184, 272
characteristic curves, 177, 323
closed-loop control, 183
cogging torque, 113, 156, 183–191, 351
countercurrent braking, 171–174
electromagnetic power, 97, 166, 191–194, 224, 343
equivalent circuit, 32, 60–70, 93–107, 165, 221, 288, 321
iron losses, 59, 105, 181, 285, 343
pole flux, 141, 151, 166, 177, 215
self-excitation regenerative braking, 177
speed control methods, 169, 251
starting and speed control, 171
steady-state, 7, 31, 60, 89, 127, 180, 240, 288, 315, 354, 374
twin motors, 183
electromagnetic force (EMF)
armature reaction, 134, 156, 192, 318
brush emf, 165
compensation winding, 163–165
motion emf, 112, 163, 180
resultant nonuniform flux density, 164
electromagnetic torque, 224–225, 237, 291, 321, 334, 351, 374
electromagnetic transients, 87–88
sudden short circuit, 27, 51, 87, 91
equivalent circuit and excitation connection, 165
handheld universal contemporary motor, 182
instantaneous steady-state torque, 180
machine voltage equation, 180, 350
no-load magnetization curve, 62, 150, 217, 237
Ampere's law, 54, 151, 199, 219
current density, 16, 33, 61, 122, 155, 204, 268, 357
magnetic flux line path, 151
pole body mmf, 154
rotor tooth mmf, 152
rotor yoke, 152, 187, 274, 282, 300
stator back iron, 149, 154, 276
torque, 6–24, 112–200, 235–306, 348–378
orthogonal (dq) model, 290
parameter estimation, 354, 375
preliminary design, 106, 186, 274, 354, 365
coil wire diameter, 189
copper losses, 24, 41, 62, 97, 189, 227, 299, 309, 359, 378
current density, 16, 38, 68, 101, 122, 155, 186, 192, 268, 357
PM stator geometry, 186
rotor diameter, 152, 186, 274
slot active area, 189
slot height left, 188
specifications, 274, 354
stack length, 24, 151, 186, 276, 294, 376
turns per rotor periphery, 188
silver-copper segments, 149
speed control, 181–182, 250, 291, 322, 350, 370
stator and rotor construction elements, 139
automobile engine-starter motors, 140
universal motor, 145–146
testing, 182, 271, 354
artificial loading, 184, 272
cogging torque, 113, 156, 183, 306, 351, 371
iron losses, 59, 105, 181, 227, 285, 294, 343
pole flux, 141, 150, 184, 215, 322
torque speed curve, 181, 243, 255

C

cage rotor windings, 212
airgap flux density, 121, 133, 149–186, 211–237, 269–318, 356
applications and topologies, 195, 297, 356
bar and ring resistances, 214
chording and distribution factor, 207
end ring and cage-bar currents, 214
factor components, 216
rotor cage geometry, 214
skewing factor, 214
total winding factor, 215
circuit models, 19, 22
cyclic inductance, 79, 320

D

dc brush PM motor, 183
artificial loading, 184, 272
characteristic curves, 177, 323

closed-loop control, 183
cogging torque, 113, 156, 183, 191, 306, 351, 365, 371
countercurrent braking, 171, 183
electromagnetic power, 97, 166, 171–194, 224, 256, 293, 343
equivalent circuit, 60, 93, 165, 221–295, 321
iron losses, 59, 105, 167, 181, 222, 227, 285, 341
pole flux, 184, 215, 322
self-excitation regenerative braking, 177
speed control methods, 169–171, 251
starting and speed control, 171–178
steady-state, 31, 60, 72–89, 127, 180, 226, 288, 320, 346, 374
current/speed curve, 169
ideal no-load speed, 169–193, 226
short-circuit torque, 169
speed/torque curve, 169
transients, 107, 169, 346, 354
twin motors, 183
distributed power systems, 22
structural diagram, 88
axis, 121, 157, 160–190, 315, 366, 377

E

electric machines
automotive electric motor applications, 21
electric energy, 111, 190, 328
coupled electric and magnetic circuits, 19
electric motor drive applications, 3
Faraday's law, 22, 30, 39, 58, 106–112
generator/motor operation mode, 3
electromagnetic torque, 5, 12, 24, 112–116, 132, 194, 224, 271, 343, 374
fixed magnetic field machines, 5
brush-commutator machines, 20, 120, 135
magnetic flux density, 152, 190, 217
speed and power limitations, 6
high-frequency electric transformers, 22, 99
IEEE and IEC standards, 20
losses and efficiency, 101, 186
magnetic composite soft materials, 22
methods of analysis, 19
induction machine (IM), 6
nameplate ratings, 17
physical limitations and ratings, 14
electromagnetic rotor shear stress, 15
global cost function, 16
insulation materials, 23
power grid, 18, 46, 77, 86, 198, 228, 235, 262, 309, 336, 374
types of transformers, 3
electric transformers
autotransformers, 28, 96, 106
cores, 15, 34, 43, 113, 195, 234, 290, 356
Ampere's and Faraday's laws, 49
copper conductor, 45
critical conductor height, 51
dc current conductors, 45
laminated magnetic core vicinity, 48
magnetic field, 49–50
Poynting vector, 50
Roebel bar, 51–52
single-phase distribution transformer, 49
single-phase power transformer, 47
slot leakage and ac resistance, 50
stacking pattern, 49
three-phase power transformer, 47
instrument transformers, 102–103
magnetic core ac
coil airgap, 32
emf ratio, 35
equivalent magnetic circuit, 33–34
Faraday's law, 34–35
flux (Gauss) law, 33
inductance, 34
magnetic circuit (Ampere's) law, 32–33
magnetic core, 33
magnetic permeability, 35
soft magnetic materials, 33
magnetization curve and hysteresis cycle
differential permeability, 40–41
frequency and graphical representation, 39
incremental permeability, 40–41
magnetic permeability, 38, 40
normal permeability, 40–41
no-load steady state ($I_2 = 0$)
core loss, 68
equivalent circuit, 67
magnetic saturation, 70–71
no-load magnetization curve, 69
no-load power, 68
phasor diagram and power breakdown, 67
permanent magnets, 41–42
power electronics
boost dc-dc converter, 106–107
contact-less battery charging system, 107–108
galvanic separation, 106
series-connected transformer, 107
static-power semiconductor-controlled switches, 106
two stage ac-ac power electronics converter, 107
preliminary transformer design
active material weight, 113
deliverables, 109
equivalent circuit, 113
losses and efficiency, 112
magnetic circuit sizing, 101–102
no-load current, 112

specifications, 108–109
windings sizing, 102–105
single-phase transformer, 31–32
circuit equations, core losses, 64–65
flux linkages and leakage inductances, 59–63
steady-state operation on load, 74–77
soft magnetic material losses
B-H curve, 45–46
core loss, 45, 47
eddy current losses, 43–44
hysteresis losses, 42–43
steady state and equivalent circuit
complex variables, 65
core loss series resistance, 66
input voltage, 65
load equation, 66
primary and secondary winding losses, 67
steady-state short-circuit mode, 71–73
three-phase transformers
general equations, 84–86
no-load current asymmetry, 82–83
paralleling, 91–92
phase connections, 78–81
unbalanced load steady state, 87–89
Y primary connection, 3-limb core, 83–84
transients
anti-overvoltage electrostatic transients, 102
electromagnetic (R,L) transients, 94–95
electrostatic (C,R) ultrafast transients, 99–102
forces, peak short-circuit current, 97–98
inrush current transients, 95–96
no load sudden short circuit, 96–97
windings
alternate windings, 57
concentric and biconcentric cylindrical windings, 57
conductor strand, insulation paper lapping, 54–55
continuously transposed cable, 54–55
corrugated tank, 58
disc windings, 55–56
foil windings, 56
layer and helical windings, 55–56
oil circulation, 57
transformer tank and oil conservator, 57–58
turn-ratio limited range regulation, 56
types, 55
electromagnetic (active) power, 321
boundary conditions, 44, 94, 271
electromagnetic (R,L) transients, 87
electro-magneto-mechanical energy conversion, 112
electromechanical time constant, 248
electrostatic (C,R) ultrafast transients, 92
distributed capacitor model, 93–94
microsecond front voltage propagation, 93
single-layer winding, 93
wave equation, 94
end-connection leakage inductance, 218
energy conversion, 251, 298, 370
active rotor electric machines, 117
dc rotor and ac stator currents, 117
magnetic conergy, 117
primitive active rotor single-phase electric machine, 117
coupling magnetic field, 111
cogging torque, 183
Faraday's law, 31, 43, 58, 106–112
mechanical energy conversion, 112–113
stored magnetic energy, 112, 135, 190
electromagnetomotive force (emf), 112
linear electric machines, 123, 129
linear compressor coil-mover PM linear motor, 128
linear oscillatory motor/generator, 125
loudspeaker/microphone, 126
three-phase linear flat synchronous machine, 124
passive rotor electric machines, 114
coenergy, 113–116, 190
constant instantaneous torque, 123
magnetic saliency, 114
reluctance synchronous machine, 116
stepper reluctance motor, 116
switched reluctance machines, 297
traveling field electric machines, 122
equivalent bar resistance, 214, 280

F

Faraday's law, 31, 43, 58, 106–112
electromagnetic induction, 2, 22, 97, 112
field circuit losses, 342
finite element method (FEM), 314
analysis, 19, 57–101, 251
electric machines, 314
magnetic flux lines, 55
fixed magnetic field machines, 5
brush-commutator machines, 120, 139–199, 322
magnetic flux density, 5, 62, 152, 190, 217
speed and power limitations, 6

G

Gauss law, 29, 267

H

hybrid electric vehicle (HEV), 22, 99
actuators, 22, 149, 298, 303

I

induction generator, 235, 253, 287, 326, 362
induction machines (IMs), 195
AC distributed windings, 197–198, 370
cage rotor windings, 212, 370
applications and topologies, 195, 297
autonomous generator mode, 236
capacitor split-phase induction motors, 262
dual capacitor IM, 263
equivalent impedances, 264
source voltage, 263
symmetrical components theory, 262
construction elements, 300
ball bearings, 197
single-phase supply capacitor, 196
slots, 196–221, 230, 268, 293, 304, 351, 370
wound rotor, 3-phase, 195
deep-bar and dual-cage rotors, 243–244
electromagnetic torque and motor characteristics, 237
electromagnetic power, 224, 240, 293, 343
rated current, 24, 54, 63–71, 86, 171, 183, 205, 240, 274, 285, 327, 361, 376
rated slip, 234, 275, 293
rated speed, 171–175, 211, 262, 348
rotor cage losses, 240
stator resistance, 240, 279, 321, 354
steady-state characteristics, 241
torque *vs.* slip S, 239
electromagnetic transients, 87
ideal no-load operation, 226, 236
active power, 23, 44, 68, 225, 235, 266, 290, 321, 363
iron losses, 59, 105, 167, 181, 227–228
phasor diagram, 263, 317, 343
supersynchronous operation, 226
leakage inductance, 228
airgap flux density, 270
differential leakage stator inductance, 218
double-layer, full-pitch windings, 268
dynamic end effect, 269–271
end-connection leakage inductance, 218
end effect thrust, 271, 294
equivalent goodness factor, 270
flat geometry, 267
primary length, 270
rectangular slot leakage field, 219
secondary current density path, 269
single layer windings, 200, 220
slot geometrical permeance coefficient, 219
static end effect, 267
transverse edge effect, 268
on load operation, 27, 234
main inductance, 59–61, 216, 358
Carter coefficient, 149, 217, 276, 376
cyclic main (magnetization) inductance, 217
magnetic saturation factor, 199, 217
magnetization characteristics, 218
no-load motor operation, 232
one rotor phase, 260
one stator phase, 257, 292
copper losses, 260, 299, 344, 359, 371, 378
parasitic (space harmonics) torques, 244
asynchronous parasitic torques, 245
complex number variables, 224
electromagnetic power, 97, 293, 343
electromagnetic torque, 112, 115, 135, 166, 224
flux linkages, 53–54, 113, 222, 247
mmf space harmonics, 200
reduced rotor windings, 222
slot skewing, 245, 291
symmetric steady state, 224, 259, 320
synchronous parasitic torques, 245–247, 290
preliminary electromagnetic design, 28, 58, 375
efficiency and power factor, 241, 271, 279, 353, 375
electric circuit, 19, 46, 60, 106, 190, 216, 287
magnetic circuit, 27, 41, 76, 101, 120, 274, 292, 323
main specifications, 274
parameters, 287–290
starting current and torque, 194
realistic analytical model, 232, 274
regenerative and virtual load testing, 271
frequency mixing method, 272
speed control methods, 251
breakdown torque, 220, 230, 240, 276
critical rotor frequency, 254
short-circuit reactance, 293
soft starter, 181, 249
torque–speed curves, 255
variable frequency and voltage, 293
wound rotor, 253
split-phase capacitor, 263, 350
starting methods, 247
additional rotor resistance, 237, 250
direct starting (cage rotor), 248
two-phase ac windings, 211
unbalanced supply voltages, 255, 291
zero speed operation, 228
rotor (slip) frequency, 228, 285
short circuit impedance, 66, 81–85
single-phase zero-speed testing, 229
starting current, 230, 249, 282, 291
starting torque, 140, 230

interaction torque, 117, 237, 321
iron loss resistance, 285

L

lagging power factor, 318, 325, 363–370
asynchronous starting, 292, 346
linear compressor coil-mover PM linear motor, 128
linear electric machines, 129, 136
airgap flux density, 133
double-layer, full-pitch windings, 268
dynamic end effect, 269
end effect thrust, 271
equivalent goodness factor, 270
linear compressor coil-mover PM linear motor, 128
loudspeaker/microphone, 126
BIL force, 126
mechanical resonance conditions, 127
steady-state harmonic motion, 127
tubular configuration, 123
primary length, 270
single layer windings, 200
load torque pulsations, 346
asynchronous torque, 346
damping coefficient, 348
mechanical resonance module, 347
proper (eigen) angular frequency, 347

M

magnetic fields
airgap, 2–34, 62, 99, 120, 132–140, 186–196, 215, 269–326, 356, 363, 374–376
magnetic circuit (Ampere's) law, 29
magnetic core, 29–55, 96, 101–104, 167, 195, 300, 356
magnetic permeability, 29, 54
magnetic field distribution, 19, 149, 219
magnetic materials, 15, 29, 34–40
magnetization curve and hysteresis cycle, 34
differential permeability, 36
graphical representation, 35
incremental permeability, 36
magnetic permeability, 40, 54
normal permeability, 35
permanent magnets, 37, 114, 121, 132
soft magnetic material losses, 38
B-H curve, 41
core loss, 285
eddy current losses, 38, 77, 107–111, 285, 310
hysteresis losses, 35–38
motion equation, 333, 335
induction machine (IM)
anisotropic rotor, 9–10
asynchronous machine, 6
heteropolar magnetic flux density, 5–8
self-synchronization, 8, 348–350, 375
wound rotor, 6–9, 195–197, 227, 244–260, 287
mutual inductance, 117, 317

O

optimal design, 155, 190, 354
dc brush PM motor ($\omega_b = 0$), 183
active and reactive power, 44, 68, 326, 342, 365
instantaneous torque, 118, 123, 182
magnetic saturation, 251, 256, 276–278, 283–286, 291, 319, 336, 358, 363
motion-induced voltage, 161, 222
motion emf, 112, 163, 180
motion equation, 333
stator equation, 224

P

Passive rotor electric machines, 114
coenergy, 113–116
constant instantaneous torque, 123
magnetic saliency, 321
switched reluctance machines, 10, 116, 297
permanent magnet synchronous generators (PMSGs), 298
phase coordinate model, 19
phase leakage inductance, 218, 320
phase voltage unbalance index, 256
phasor diagram, 35–36
PM synchronous and switched reluctance motors, 22
PM synchronous motors, 342, 350
electromagnetic torque, 343
interior PM rotor, 310, 319, 344
phasor diagram, 31, 66, 83, 182, 263, 317–327, 343, 359–364
pole changing winding, 211, 251, 291
pole pitch, 122, 145, 152, 163, 190, 198, 244, 276, 376
preliminary electromagnetic design, 28, 58, 116, 375
efficiency and power factor, 241, 271, 279, 353, 375
electric circuit, 19, 46, 60, 106–112, 190, 216, 274, 287
magnetic circuit, 19, 27, 76, 120, 274, 292, 323
airgap flux density, 275
Carter coefficient, 276
magnetic saturation, 276

magnetization reactance, 283
rated torque, 283
rotor diameter, 121, 152, 186, 274
rotor yoke, 284
stator back iron (yoke) height, 276
stator stack length, 276, 355, 376
stator wire diameter, 284
stator yoke weight, 284
tangential specific force, 15, 155, 186, 274, 355
tooth/slot pitch ratio, 276
top and bottom widths, 277
main specifications, 274
parameters
leakage reactances, 107, 281, 288, 374
reduction factor, 280
rotor aluminum bar resistance, 279
stator resistance, 236, 242, 279, 286, 321, 354
starting current and torque, 194, 282
preliminary transformer design
active material weight, 16, 105
equivalent circuit, 60, 87, 99, 106, 165, 177, 221, 229, 235–244, 288, 321–322
losses and efficiency, 11, 101, 186
magnetic circuit sizing, 101
no-load current, 62, 89, 96, 101, 217, 285
windings sizing, 102
filling factor, 102, 152, 192, 357
leakage reactance, 374
number of turns, 376
primary and secondary resistances, 104
primary emf, 102
radial width, 103
short-circuit rated voltage, 67, 97, 104
primitive active rotor single-phase electric machine, 117

R

reactive power, 363, 374
reluctance synchronous machine (RSM), 9

S

secondary winding, 59–61, 74, 92, 112, 195
shaft power, 12–14, 234, 286, 334, 342, 374
single-phase and split-phase capacitor PM, 350
single-phase transformer, 27–28, 53–54, 58, 68, 87, 95
alternate windings, 52, 55, 57–58
Ampere's law, 54, 151, 199, 219
cylindrical windings, 50, 55–58
definition, 44, 63, 182, 199, 234, 256, 344
primary and secondary mmfs, 53–55
resultant magnetic reluctance, 54
steady-state operation on load, 68
active and reactive power break down, 68
dual voltage single-phase residential transformer, 71
skewing leakage inductance, 218, 220, 282
skin effect parameter, 282
soft magnetic material losses, 38
B-H curve, 41
core loss, 48, 52, 58–69, 76, 95–107, 171, 194, 225, 257, 283–285, 331, 352
eddy current losses, 38, 77, 107, 310
hysteresis losses, 35–38
soft starter, 179, 247–252
electromagnetic torque, 12–17, 112–116, 135, 166–180, 224, 235, 237, 271, 321, 351, 374, 377
phase coordinate model, 19
star–delta switch, 249
stator copper losses, 24, 227, 248
stator teeth weight, 285
stator winding factor, 278
stator (armature) windings, 300
four-pole PMSM windings, 308, 310
LCM, N_s and, 310, 312
single-phase tooth-wound machines, 312–313
switched reluctance motor (SRM), 310–312
TF-PMSMs, 312–313
types, 306–307
super high-frequency models, 375
airgap flux density, 16, 121, 133, 149–165, 186–199, 211–215, 269–276, 294, 310–318, 356, 374–376
parameters, 78–79, 98–107, 175–191, 228, 236, 244–245, 261–291, 315, 337, 358, 376–377
synchronous inductances, 306, 320
applications and topologies, 195
automobile alternators, 297
hard disks, 299
linear SMs, 299, 303
open cross section, 305
permanent magnet synchronous generators (PMSGs), 298
power systems, 298, 318
special configuration PMSMs, 302
stepper machines, 297
switched reluctance machines, 297
armature reaction and magnetization reactances, 318
d and *q* axes, 318
magnetization inductances, 319
nonsinusoidal airgap flux densities, 318
synchronous inductances, 320
autonomous synchronous generators, 322
load curve, 322–330, 373
no-load saturation curve, 322, 376
short-circuit curve, 322

basic static-and dynamic-stability concepts, 332
  attenuation, 334
  short-circuit clearing time, 334
  synchronizing power, 333
  torque, 334
large synchronous motors, 341
  power balance, 68, 166–168, 232–234
  speed/torque curve, 169, 175
load torque pulsations, 346
  asynchronous torque, 245, 334, 346, 377
  damping coefficient, 348
  mechanical resonance module, 347–348
  proper (eigen) angular frequency, 347
PM synchronous motors, 342
  electromagnetic torque, 343
  interior PM rotor, 319, 344
  phasor diagram, 343
  reluctance synchronous motor (RSM), 344
preliminary design methodology, 354
  base current, 360
  general specifications, 354
  machine parameters, 358
  maximum phase voltage, 355
  maximum speed torque, 360
  number of turns, 359
  phasor diagram, 360
  PM flux, 356–357
  stator phase winding, 314
single-phase and split-phase capacitor PM, 350
  classification, 120, 218
  efficiency, 353
  electromagnetic torque, 352
  phasor diagram, 352–353
  two-pole split-phase capacitor, 352
  voltage equation, 350
stator (armature) windings, 300
  digital synchronizers, 328
  field current, 329
  four-pole PMSM windings, 306, 308
  reactive power capability curves, 329
  single-phase tooth-wound machines, 309
  SRM, 309
  TF-PMSMs, 309
  types, 304–305
  V-shaped curves, 329–332
synchronous reactances, 330, 374
  emfs and current phasors, 316
  load, 316
  mutual inductance, 19, 114–119, 216, 221, 291, 317, 359
  phasor diagram, 317, 320–327, 343, 352–360, 377
unbalanced 3-phase current loads
  impedance, 339
  phase-to-phase short circuit, 339
  pole-slipping method, 337
  slip, 338
  symmetrical components, 264, 336
  total voltage, 337

T

three-phase linear flat synchronous machine, 124
three-phase transformers, 72
  airgap flux density, 16, 121, 157, 199, 270, 294, 314, 376
  brush–commutator windings, 200
  general equations, 78
  ideal multiphase MMF, 200
  no-load current asymmetry, 75
  number of slots, 143, 193, 290
  paralleling, 66, 84
  phase connections, 72, 234
    connection schemes, 74
    primary line voltage, 74
    stator connection, 376
    zig-zag (Z) connections, 72
  unbalanced load steady state
    direct and inverse components, 81
    inverse transformation, 80
    secondary neutral point potential, 82
    symmetrical components superposition, 80
    zero sequence current, 81
    zero sequence impedance, 81–82

U

unbalanced load steady state
  impedance, 339
  phase-to-phase short circuit, 339
  pole-slipping method, 337
  slip, 196, 232, 251–262, 337, 346
  symmetrical components, 80, 262, 336
  total voltage, 337

V

voltage and current measurement transformers, 4
V-shaped curves, 329–332

W

wound rotor, 221

Z

zig-zag leakage inductance, 218

Printed in the United States
by Baker & Taylor Publisher Services